ScanFan Southern California

AF697624

SAN DIEGO COUNTY
LOS ANGELES COUNTY
ORANGE COUNTY
IMPERIAL COUNTY
RIVERSIDE COUNTY
VENTURA COUNTY
SAN BERNARDINO COUNTY
BAJA NORTE

29.0 to 54.0
118.0 to 174.0
406 to 512.0
806 to 956.0

ScanFan Publications™

ScanFan

San Marcos, California

ScanFan Southern California

This is an easy to use frequency directory for radio scanners. If you would like additional copies or if we can be of further assistance to you, please call or write:

ScanFan Publications
P.O.Box 544
San Marcos, CA 92079
Phone: 619-744-2303

This directory is designed to provide accurate information about radio scanning. Every effort has been made to provide accuracy, but no warranty of suitability, purpose, or fitness is implied. Neither the author nor ScanFan Publications assume any liability with respect to the use of the information contained in this directory.

COPYRIGHT © April 1994 by ScanFan Publications
ISBN 0-9620798-6-3 (5th Ed. 1994-95)

All rights are reserved. No part of this directory may be reproduced or transmitted in any form; electronically, mechanically, photocopying, or by any other means without prior permission from the publisher and author.

IMPORTANT NOTE TO SCANNER LISTENERS

California Penal Code 636.5 states that any unauthorized person who intercepts any police radio service communication, and who divulges such communication to anyone he knows to be a suspect in the commission of a crime with the intent to help the suspect escape arrest, trial, conviction or punishment, is guilty of a misdemeanor. This crime is punishable by a fine not exceeding one thousand dollars, or imprisonment in the county jail not exceeding six months, or both.

This book is dedicated to the memory of
Jim Rafferty

SAVE YOURSELF!

SAVE YOUR NATION!

SAVE THE WORLD!

WHAT TO DO ABOUT THE RISING NATIONAL DEBT
WHY CRIME IS RISING AND OUR STANDARD OF LIVING IS DECREASING
A NEW BOOK FROM - LUXOR PRESS!
SEND $10.00 TO LUXOR PRESS - P.O.BOX 544 - SAN MARCOS, CA 92079
!INCLUDES THE FEDERAL RESERVE ACT OF 1913

CREDITS

Thank you mom and dad: for your support; without which "ScanFan" would have never been.

CHIEF EDITOR	- Fred K.C. White
EDITOR	- C.B. White
PUBLISHER	- ScanFan Publications
COPYRIGHT SUPPORT	- Freling Baker ATTY.
COVER ART	- Dave Deal
PHOTOS	- Courtesy of Randy Lee
GENERAL DATA	- H. Abiff, James Bassett, Dave Bohmke, Brian Bussy, Donald Cones, Bill Cooper, Steve Gravits, Jim "Lucky" Cook, Randy Linn, Jerry Pickard, Erick Tillery, Brian Tryon, Mike Werner.

We would like to extend a special thank you to the following individuals and others who provided so much valuable assistance for this edition.

Patrick Allen
Paul Andregg
Travis Deal
Jack Green, Jr.
Ralph H. Fellows II
Bud Jamison
Brian Johnson
Randy Lee
David Mejia
Lara Quarantiello

AN INTRODUCTION

Welcome to the world of radio scanning! Much excitement is headed your way listening to services such as police, fire, emergency medical, news media, aircraft, maritime, and many others.

The data in ScanFan Southern California is the result of many hours of monitoring in the field and direct correspondence with radio system managers, electronic technicians, and other scanner enthusiasts.

HOW TO USE THIS DIRECTORY

Public safety data comes first containing county sheriff, county fire, emergency medical, and local police, fire, and government systems. Second are the State and U.S. Government sections. All other sections (Aircraft, Amateur Radio, Maritime, News Media, etc.) follow in alphabetical order. Please refer to the table of contents to aid you in using ScanFan Southern California.

We have added a frequency cross-reference as well which is the very last section of the directory. We hope that ScanFan Southern California is the best frequency directory you will ever buy.

WHAT IS A SCANNER?

A scanner by definition is a radio receiver capable of scanning the transmissions of two or more radio frequencies. These frequencies are used for various radio services such as local police and fire departments, paramedics, federal government, aircraft, maritime, railroads, taxi cabs, buses, business, and amateur radio. You might overhear a police officer responding to a disturbance call, a firefighter on a traffic accident scene, a paramedic relaying medical information on a critical patient, federal agents surrounding a residence for a drug bust, a pilot talking with an airport control tower, or just the dogcatcher down the block. ScanFan is a directory which provides frequency data on these radio systems.

CHOOSING A SCANNER

There are numerous brands of scanners on the market. Popular brands include Uniden (Bearcat), Realistic, Regency, Cobra, Icom, J.I.L, Fox, Ace, and Yaesu. Scanner prices may range from $100 for a basic model to well over $5000 for the most sophisticated models. The average price is about $250 for a brand new, full featured, high performance scanner. Several factors should be taken into consideration before you purchase a scanner.

FREQUENCY RANGES

An important consideration in buying a scanner is determining what frequency ranges you want to be able to monitor. The primary radio bands that almost every scanner covers are 30-50, 144-174, and 440-512 MHz. It is in these bands that you will hear most police, fire, emergency, business, maritime and federal communications.

Some scanners include the civilian aircraft band (108-136 MHz), federal government UHF band (406-420 MHz), and the 806-950 MHz band as well. A few scanners and receivers on the market have continuous coverage from 25-1300 MHz allowing one to hear the F.M. broadcast band (88-108 MHz), 1.25 meter ham band (222-225 MHz), and the military aircraft band (225-400 MHz). It is highly recommended that you buy a scanner with 806-950 MHz coverage since many systems are moving to this frequency range. Remember, the more frequency bands you have, the more stations you will be able to monitor. Here is a general chart of the VHF and UHF frequency spectrum. All frequencies are in megahertz.

25-50	- VHF Low Band
50-54	- Six Meter Ham Band
54-72	- Television Channels 2-4
72-76	- VHF-Mid Band
76-88	- Television Channels 5 and 6
88-108	- F.M. Broadcast Band
108-137	- VHF-Air Band
136-174	- VHF High Band
136-138	- Satellites
138-144	- Military Systems
144-148	- Two Meter Ham Band
148-150	- Military Systems
150-162	- Land Mobile Systems - Local & Private
162-174	- Land Mobile Systems - Federal
174-218	- Television Channels 7-13
218-220	- Maritime Systems (midwestern U.S.)
220-222	- Land Mobile Systems ACSSB
222-225	- 1.25 Meter Ham Band
225-400	- Military Aircraft / Satellites
400-420	- Land Mobile Systems - Federal
420-450	- 75 Centimeter Ham Band
450-470	- UHF Band - Land Mobile
470-512	- UHF-TV Band - Land Mobile & TV 14-20
512-806	- UHF TV audio Channels 21-69
806-821	- Land Mobile Systems - Repeater Inputs
824-846	- Cellular Telephones - Mobiles
850-869	- Land Mobile Systems - Repeater Outputs
870-879.990	- Cellular Telephones - Channels 1 thru 333
880-889.980	- Cellular Telephones - Channels 334 thru 666
890-891..480	- Cellular Telephones - Data Channels
891.5-893.970	- Cellular Telephones - Wireline
892-894	- Common Carrier
894-896	- Reserved
896-901	- Land Mobile Systems - Repeater Inputs
902-928	- 33 Centimeter Ham Band
928-935	- Paging Systems
935-940	- Land Mobile Systems - Repeater Outputs
945-950	- Studio Transmitter Links (STL's)
1260-1300	- 23 Centimeter Ham Band

AN INTRODUCTION

CRYSTAL OR PROGRAMMABLE?

A programmable scanner is the most versatile and will offer the highest amount of performance and features for the price. Immediate access to thousands of frequencies is available with a programmable model. Crystal-type scanners require an individual crystal for each frequency you want to monitor and can't be changed without replacing crystals. Consider a crystal scanner only if you will monitor a few set frequencies.

BASE, MOBILE, OR HANDHELD?

Scanners are available in a variety of models to meet the individual needs of the listener. If you will be doing most of your monitoring at home or work, a base scanner will probably be the best. Mobile scanners are useful if you want to keep abreast of highway conditions while driving; you can listen to traffic helicopters, highway patrol units, and Caltrans maintenance crews. You can also listen to fire calls to hear about traffic accidents and then avoid them by taking an alternate route.

If you are constantly in and out of your vehicle and away from a permanent location or think you'd like to listen to communications at various sporting events or air shows, a handheld model may be the best. A handheld scanner is also helpful if you are a member of the newsmedia and need to keep up to date on the progress of a crime or fire incident.

SCANNER FEATURES

Today's scanners have all "the bells and whistles". Popular features include search mode, scan delay, memory banks, and priority channel. Here is a brief description of each feature.

Search Mode is a method whereby a scanner is programmed to find unknown frequencies by searching between a high and low frequency limit. Many scanners have selectable search increments of 5, 12.5, or 30 kHz.

Scan Delay function holds a channel for about 2 seconds after a transmission ends. Therefore, you won't miss the reply to a radio conversation.

Memory Banks store several programmed frequencies. You may scan just one or all the memory banks desired. For example, you could program one bank of channels with San Diego Police frequencies and another bank with frequencies for Orange County Fire Department. Memory banks are a very handy feature.

Priority Function checks your favorite channel every few seconds so you don't ever miss a transmission.

Sound Squelch enables the scanner to resume scanning if it tunes a frequency that has a carrier but no sound or modulated signal.

S-Meters display the relative strength of received signals.

Emission Modes for popular receivers include FM-narrow, FM-wide, AM, and SSB. Most land mobile systems above 30 MHz use the FM-narrow mode. The FM Broadcast Band (88-108 MHz) and television stations use FM-wide. The AM Broadcast Band (540-1710 kHz) and the 108-136 and 225-400 MHz aircraft bands use the AM mode. Single-sideband (SSB) is used chiefly below 30 MHz and in the amateur radio bands. The more modes that your radio can receive, the more stations you will be able to hear.

Tone Decoders are used to guard radio systems from interference and other users. The unit decodes subaudible tones, allowing you to listen to one business or agency on a frequency being shared with other licensees. Since the correct tone is required to open the squelch, interference from intermod, cross-modulation, and images is reduced. ScanFan includes many tones used by various agencies.

Preamplifiers are one of the most misunderstood accessories for a scanner. They are designed to boost weak signals coming to the receiver. Unfortunately, a preamplifier will many times just increase the receiver noise level and the amount of intermodulation. Preamplifiers work best when they have a low noise figure and are used with scanners and circuits not highly susceptible to interference.

WHICH IS THE BEST SCANNER?

No one scanner is the best for all applications. Some models have many features, but leave much to be desired in performance. Others have very high performance, but few features. Generally the most popular scanners have the most desired features and good to excellent performance.

The Realistic PRO-2006 has good performance, many features, and a reliable track record. Radio Shack is discontinuing this unit for the new PRO-2007 base scanner. Ace has the high performance AOR-2500 and AOR-3000 scanners with just as many features, and the price range is from $500-$900. The Uniden BC580XLT, BC760XLT, BC855XLT and the BC950XLT have superb sensitivity, but are more subject to intermodulation; they are best for mobile applications.

Icom's R-7000, R-7100 and R-9000 are dream radios. The units have superior receivers, excellent reliability, and many desirable features. The price of these radios is about $1000 for the Icom R-7100 and about $5000 for the R-9000.

The Realistic PRO-43 is hot! It is one of the few high performance handheld scanners that has the 225-400 MHz band. It lists for about $400 and listeners seem to be very pleased with this handy unit. The Uniden 2500XLT is another wide-coverage handheld scanner that is popular with listeners.

If you take time to do a little research, read some magazine reviews, and ask for advice from fellow listeners, you will find the best scanner for you.

AN INTRODUCTION

TECHNICAL SPECIFICATIONS

Certain terms are used to describe the performance of a scanner or receiver. One should carefully look over a scanner's stated specifications to determine how it will perform.

Sensitivity refers to the weakest readable signal the receiver will respond to. The lower the measurement in microvolts, the better. Good sensitivity is about 0.5 microvolts or less for 12-dB Sinad on the VHF and UHF bands.

Selectivity refers to the receiver's ability to reject strong signals on adjacent frequencies. For example, if you have your scanner tuned to 154.370 MHz, you wouldn't want to hear a signal on the adjacent frequency 154.385 MHz. Average selectivity for a scanner is about -50 dB at plus or minus 30 kHz. The Realistic Pro-2006 has a selectivity on NFM of -50 dB at plus or minus 15 kHz; very good selectivity for a scanner.

Dynamic range is the ability of a receiver to reject intermodulation but still respond to weak signals. Intermodulation or "intermod" for short is interference generated in the receiver when strong signals mix together, producing spurious signals. Most scanners are very "open" to nearly all strong signals and intermod is a common problem. This is why you may hear strong paging signals on some public safety frequencies. Most scanners have such poor intermod immunity that dynamic range is usually not mentioned on specification sheets. The Icom R-7000 & R-9000, and Realistic Pro-2004, Pro-2005 and PRO-2006 have a good degree of dynamic range. Also consider using a notch filter to remedy the problem, available from Grove Enterprises, if you are receiving interference from one nearby transmitter.

Images are generated in every receiver. They are twice the intermediate frequency. Typical offsets are 21.4, 21.6, or 21.7 MHz from the actual frequency, depending on the scanner. One may use images to hear signals outside a scanner's range by adding or subtracting the image from the frequency you wish to hear. The following are the image offsets of popular scanners:

Ace Scanners:	21.400
Icom Receivers:	21.400 (2nd)
Realistic Scanners:	21.400
Regency Scanners:	21.400 or 21.600
Uniden (Bearcat) Scanners:	21.600 or 21.700

In metropolitan areas such as Los Angeles, images can be an annoying problem. For example, you may hear Beverly Hills FD on 482.050 MHz while using a Realistic scanner. A quick check will show that it is an image and the actual frequency is 460.650 MHz (21.4 MHz lower). Hence, the ability of a scanner to reject images is very desirable.

In conclusion, if you suspect a station is outside its normal frequency range, you might want to check, it might be an image.

SCANNER ANTENNAS

Antennas come in many shapes and sizes depending on the application. Handheld scanners have telescoping or flexible rubber whips which are fine for local reception of strong signals. If you want to increase your listening range, connect a mobile or base antenna to your handheld scanner. Most handheld scanners are equipped with a BNC male antenna connector for this purpose.

In mobile installations, a scanner antenna should be located on either the roof or trunk lid for omni-directional coverage. Mounted elsewhere (fender, bumper, etc.) the antenna will favor a particular direction. Highly recommended mobile antennas include the Diamond D-505, the Grove ANT-4, Radio Shack's magnetic mount (Model 20-219), and all scanner antennas made by the Antenna Specialists Co. Other mobile antennas to consider are Sky Scan from Britain and the Spectra made by Austin Antennas.

If an outside antenna is undesirable, the vehicle's existing AM/FM antenna may be used with satisfactory results. However, a multicoupler such as the Grove ANT-63 or Paradynamics PDC-63 will have to be installed.

Base antenna systems are unlimited in potential size and design. If you want to have only one antenna that covers an extremely wide frequency range, consider using a discone antenna such as the Icom AH-7000. If peak performance is desired in a certain band/s, try a dual-band omni antenna or a directional beam antenna. Some commercial and amateur radio antennas work great for reception on the public safety bands. AEA, Comet, Diamond, Cushcraft, and High Gain are popular brands of quality amateur antennas. Highly recommended commercial antennas are made by Antenna Specialists, Celwave, Cushcraft, and Decibal Products.

If you want a low-profile and low cost antenna for general reception, the Antenna Specialists MON-58, Grove ANT-5, and Radio Shack model 20-176 are proven performers. The Antennacraft SA-1-175KT, the Antenna Specialist AV-801, and the Butternut SC-3000 antennas have excellent track records for the 30-50, 144-174, and 440-512 MHz bands.

For VHF-low band reception, an antenna cut for the specific frequency range is required to yield satisfactory results. Most scanner antennas are way too small to have good reception on the 30-50 MHz frequencies. Therefore, many listeners have difficulty picking up CHP dispatch and mobiles in the 42 MHz range. A quarter-wave whip cut for 42 MHz (about 6 ft long) is an excellent antenna. The Butternut SC-3000 is probably the best-performing scanner antenna on the market for reception in the VHF-low band.

In conclusion, your scanner reception is dependent on the time and money you wish to spend on your system. First, try to determine what services you want to monitor, then research different antennas that cover the desired frequency ranges.

AN INTRODUCTION

COAXIAL CABLE

The type of cable you use is important since there is always a loss of signal strength between the antenna and the radio. The amount of loss is dependent on the frequency and length, and the type of cable. A fifty-foot run of RG-58/U, suitable for a CB or shortwave installation, will simply not be as effective as the same length of RG-213 in the VHF and UHF bands. At 800 MHz and higher, the coaxial cable type may mean the difference between good reception and no reception at all. Premium cable such as Belden 9913, Intl. 9086, or equivalent is recommended for all cable lengths over fifty feet when monitoring 800 MHz frequencies and above.

RADIO SYSTEMS

Radios operate on frequencies. A **frequency** is a specific point in the electromagnetic spectrum. Radio frequencies begin slightly above the level of hearing and extend to almost the frequency of light. A radio system consists of one or more stations which are base, mobile, or stationary.

The simplest type of radio system is a **one frequency simplex** system. All radios in the system operate on the same frequency and each unit has to wait a turn to talk. An example is a citizens band radio. The range of two base or mobile units is limited by the power and frequency used because no relay or repeater is used. Most one frequency simplex systems are located in the VHF-Low band where satisfactory range is possible by simplex communications. When additional range is necessary, a one frequency system may implement one to several remote receivers and transmitters on high mountain tops or tall buildings. Wires or highly directional radio links usually connect the dispatch center to the remote site. The dispatch center then chooses the remote site best suited for communicating with the mobile unit.

One example of this system is the San Diego County Sheriff's Office. If a mobile unit is in Escondido or San Marcos, dispatch may use Mount Palomar, KNCJ-401. If the mobile unit is in the east county, dispatch may use the Mount Laguna site, KNCJ-406. In this way, radio coverage is provided to all areas of San Diego County.

Most systems that demand very reliable coverage over a wide area use several radio sites. For example, many 800 MHz businesses require multiple radio sites (e.g., Mt. Otay, Mt. Palomar, Mt. Woodson for San Diego County).

The next type of system is a **two frequency simplex** or **semi-duplex** radio system. One frequency is used for talking and another for receiving. An operator may talk or listen, but not both at the same time. Mobile units are unable to talk directly to one another as well. Many taxi cabs use two frequency simplex systems. For example, Oceanside Cab Co. uses 152.405 for dispatch and 157.710 for receiving mobiles. C.H.P. mobiles use a semi-duplex system: mobiles transmit on one frequency while dispatch operates on the base frequency. Mobiles may also go to a car-to-car channel by talking simplex on the dispatch frequency. By this method, mobile units may talk with one another while still listening for calls from dispatch.

A **full-duplex** radio system is just like a telephone. One frequency is used for receiving and another for transmitting and both at the same time. Cellular telephones use a full-duplex system. For example, each car phone transmits in the 824-847 MHz range while the base side transmits in the 869-892 MHz range. The full-duplex system is all automatic, no switching between receive and transmit frequencies is required. A full-duplex system may employ one or more remote receive/transmit sites as well.

The most common type of radio systems use **repeaters**. A repeater is a sophisticated device located on a tall mountain top or high building that receives the radio transmissions from one frequency (input) and rebroadcasts them on another frequency (output). The normal range of a mobile or portable unit is generally limited to a maximum line-of-sight distance. However, if a repeater on a mountain is used, the useful range is greatly extended. Los Angeles P.D. uses repeaters on all their dispatch channels. For example, the L.A.P.D. Central Area #1 has an **input** frequency of 509.7375 MHz and an **output** of 506.7375 MHz. The **output frequency** is the only one you need to monitor to hear all the radio traffic on the channel. However, if a mobile or portable unit is nearby, then you will hear transmissions on the input frequency. Land mobile systems above 440 MHz usually employ repeaters with some exceptions.

When frequencies are known to be repeated, an "**R**" is denoted after the listing. Standard offsets for input frequencies are as follows:

30-50 MHz - The input frequency usually is about .5 to 2 megahertz separated from output frequency (e.g., 45.940 base / 45.420 mobile for SDSO).

50-54 MHz - The input frequency is usually .5 megahertz lower than the output frequency (e.g., 51.900 base / 51.400 mobile).

144-148 MHz - The input frequency is usually .6 megahertz from the output frequency (e.g., 146.730 base /146.130 mobile).

148-174 MHz - The input frequency is usually separated within 3 megahertz from the output frequency (e.g., 158.730 base / 155.790 mobile for Monrovia PD).

AN INTRODUCTION

222-225 MHz - The input frequency is exactly 1.6 megahertz from the output frequency (e.g., 224.140 base / 222.540 mobile).

440-470 MHz - The input frequency is exactly 5 megahertz from the output frequency (e.g., 461.225 base / 466.225 mobile).

470-512 MHz - The input frequency is exactly 3 megahertz higher than the output frequency (e.g., 471.3375 base / 474.3375 mobile).

851-896 MHz - The input frequency is exactly 45 megahertz lower than the output frequency (e.g., 856.675 base / 811.675 mobile).

935-940 MHz - The input frequency is exactly 39 megahertz lower than the output frequency (e.g., 939.500 base / 900.500 mobile).

ScanFan lists the input frequencies for many agencies or you can find them fairly easily. However, the **output frequencies** usually have a stronger signal and are the ones you want to program into your scanner.

SKIP PROPAGATION

If you listen between 30-50 MHz, you may hear some strange signals at times. Perhaps you'll hear unfamiliar names and places, strange accents, and even foreign languages. You are hearing skip; radio signals that reflect off the ionisphere and are heard thousands of miles away from origin. With skip, it is easy to hear the highway patrols in Georgia and South Carolina, fire departments on the east coast, and police departments in the southern states. Most skip propagation is attributed to events that occur in the sun. Sunspot activity increases and decreases in cycles. When a sunspot cycle is at its peak, ionization is at its maximum and radio waves in the VHF-low band easily reflect high off the ionisphere instead of continuing into space. We are currently in sunspot cycle 22 which is past its peak and skip signals are heard occasionally.

CONVENTIONAL SYSTEMS

Conventional systems operate on one or more frequencies which are each designated for specific uses. For example, one frequency may be used for dispatch, one for car-to-car communications, and another for a repeater input. If one channel is busy and shared with other users, there will be times when the user will have to wait a turn to talk. If several stations try to use the frequency at the same time, the result is usually chaos. A conventional system may be either one or two frequency simplex, semi or full duplex, base repeated or mobile relay.

TRUNKING SYSTEMS

A trunking system generally consists of 5 or more repeated frequencies that may be assigned to business or government entities in the 851-868.9875 and 935-940 MHz ranges. All stations and mobile units share a group of frequencies. However, certain frequencies may be restricted to data, telephone interconnect or area use.

When a microphone is keyed, the site controller (computer) automatically selects an open channel. One or more frequencies are used by the site controller to instruct the participating units. The control channel routinely changes to save wear on a repeater and is easily identified; it usually sounds like a loud buzzer or gun.

Trunking has several advantages over a conventional system. For example, in a city where the police, fire, and public works departments each have their own channels, the use of frequencies is generally inefficient. The police dispatch channels may be so busy that calls to officers are delayed while other city channels are barely used. In a conventional system, units may have to search for an open channel. Also, units and agencies sharing a system in the conventional mode inadvertently cause interference and overhear one another.

Trunking solves these problems since each agency and its units are automatically switched by the site controller to a clear frequency after each transmission. Therefore, several users may share one system without the interference and lack of privacy found in a conventional system. Trunking systems ensure the efficient use of frequencies and allow many more units to utilize the radio.

Here are some illustrations that should help you understand trunking and its merits:

Suppose two or more fire chiefs who contact each other on a dispatch channel decide to talk on a fire tactical channel. They rotate the dials of their radios to the chosen channel. The site controller processes the request and automatically interconnects the units when they transmit. The units on the particular fire tactical channel will not receive interference from other channels.

Two police detective units desire to talk privately about a case. Both units go to a private channel and enter in each other's radio I.D. number. The site controller again processes the request and interconnects only the two units. The site controller provides a clear frequency for each transmission. The two detectives will not be overheard nor receive interference from any other unit that is utilizing the system.

A towing company has a radio system on the UHF business band that operates on a repeater shared with other licensees. Tow trucks out in the field are sometimes "stepped on" by other licensees on the same frequency making communication difficult. Later the company switches to a trunking system and all of its units never receive interference from shared users.

AN INTRODUCTION

TYPES OF TRUNKING SYSTEMS

Trunking systems may be Special Mobile Radio (SMR) or licensed to a sole user. A SMR is owned and operated by an entrepreneur for profit and will have multiple users known as fleets. Each fleet of a SMR pays a monthly fee that is based upon its number of units, air time, and coverage area. A user such as a construction company is not required to have an FCC license, but operates under the authority of the SMR. There are only about twenty five SMR's in San Diego County due to restrictions placed on 800 MHz stations within 68.4 miles of Mexico. However, hundreds of SMR's exist in other areas of Southern California.

The typical SMR in the **856-866 MHz** range consists of five frequencies that are separated each **1 MHz** apart. In the text of ScanFan, a notation such as **"856-860.675 T"** indicates that the five frequencies are **856.675, 857.675, 858.675, 859.675,** and **860.675.**

SMR's in the **935-940 MHz** range are designated ten frequencies that are each spaced **12.5 kHz** apart. For example, one system has the frequencies: **935.0125, 935.025, 935.0375, 935.050, 935.0625, 935.075, 935.0875, 935.100, 935.1125, and 935.125 MHz.**

The FCC requires a trunking system to have a minimum of 100 mobile units for each frequency or 500 mobiles for a typical five-channel system. Therefore, only large companies such as San Diego Gas & Electric or Orange County Fire Department are eligible to be sole users of a trunking system.

PUBLIC SAFETY TRUNKING

The systems used by public safety agencies are usually more complex than the average SMR. The radios are programmed to provide for maximum flexibility, use, and convenience. The police, fire, and public works departments within a community often co-exist on a five frequency system. The trunking technology also allows each agency to have as many channels as required.

Therefore, the police do not need to be concerned about having enough channels for uses such as dispatch, tactical, interagency, and surveillance. The fire department also has ample room for all their communication needs (e.g., fireground, mutual aid, and telephone-interconnect). Each department within a city may also be given a channel (e.g., parks, lifeguards, building, engineering, streets, sewage, and water). Talk-around channels (simplex) are always available to all users for short-range communications.

When a trunking system is shared by several departments or agencies, each fleet is assigned a priority level. For example, emergency traffic has the top priority in San Diego's trunking system, followed by police patrol units and fire dispatch down the line to school buses which have the lowest priority level. The use of priority levels ensures that units involved in emergency or life-threatening situations have immediate access to the radio even if the system is very busy.

Another unique feature of trunking is that each radio in the system has an identification number. If a radio has been lost or stolen, its transmitter and receiver can be disabled and it becomes essentially useless. The radio may even be reprogrammed from the communications center by remote control! After seeing its many advantages, it is no wonder why so many agencies desire to have a trunking system.

Trunking systems are not without their disadvantages. They are very expensive to install and maintain. Many repeater sites are usually required to provide adequate coverage for mobiles and portables. Sometimes it is not prudent to trunk dispatch frequencies that are busy perhaps 90% of the time. Theoretically, ten trunked frequencies would be required to gain just one free channel. The best compromise is to trunk the frequencies that receive lesser use while keeping busy dispatch channels in a less expensive conventional format. If the site controller should malfunction, then the most vital communications are entirely unaffected.

MONITORING A TRUNKING SYSTEM

How well one may monitor a trunking system varies. The fewer number of trunked frequencies and system users the better. The first step in monitoring a trunking system is to program in all the frequencies used and lockout any control and data channels.

Remember to unlock any control channels from the previous day that are now being used for voice communications. They tend to change once every twelve or twenty-four hours. Many control channels change at twelve midnight.

Don't use the delay feature of your scanner on any memories since the frequency may change between transmissions. Listen for the agency you want to monitor. When pauses occur between transmissions, keep scanning to track the active frequency. It may be necessary to manually hit the scan button to receive the same conversation, especially when monitoring a SMR. Fortunately, most public safety systems are easier to monitor due to fewer users sharing the system. Writing down notes is also helpful; you may establish a pattern for the control channels and units on certain frequencies.

We at ScanFan encourage you to try monitoring trunking systems. A good time to listen to public safety and SMR systems is at night when many security companies are transmitting and daytime services and businesses are off the air. Give trunking a try. You may be surprised at what interesting things you hear.

A GUIDE TO THIS DIRECTORY

MONITORING AND THE LAW

The following information is provided as a guide. Neither ScanFan Publications or its author assumes any liability for its interpretation or use.

The Electronic Communications Privacy Act of 1986 forbids the monitoring of certain services receivable on scanners. Violations are punishable by fine and/or imprisonment. The law forbids monitoring (1) Encrypted or scrambled signals, (2) Mobile telephones and voice paging, (3) Studio/transmitter links, and (4) Paid subsidiary carrier authorization (SCA).

You may monitor any radio signal that is "readily accessible to the public" including public safety, federal and state, aircraft and maritime, amateur radio and CB, business and industrial, and AM/FM broadcasters.

Uniform Code 705 forbids: (1) divulging what you hear to anyone who is not a party to the broadcast, (2) making use of information you hear for your own personal gain, and (3) making use of information overheard to commit a crime. Severe punishment or fines may result for violations.

California law also prohibits the monitoring of cordless telephones in addition to cellular telephones. There may be other local or state laws restricting scanning or use of scanners depending on where you live.

A new federal law may restrict the 800 MHz frequency coverage of some scanners. More details regarding monitoring laws can usually be found in monthly scannner publications such as Monitoring Times.

SCRAMBLERS

Scramblers are used to prevent eavesdropping on communications by unauthorized individuals. Very few private and local government agencies use them. The cost effectiveness is just not warranted. It is far easier to operate using code words, communicate on discrete frequencies, or just keep radio silence.

However, all U.S. government agencies of sensitive nature (FBI, DEA, Secret Service, U.S. Marshal, etc.) must use the Digital Encryption Standard (DES). DES converts a signal to a digital code, which cannot be recognized as a voice when heard on a scanner or unequipped radio. The digital code sounds like a rush of static over the receiver. DES has 720,000,000,000,000,000 possible different codes, therefore; decifering it is very difficult. Federal agencies with DES essentially have complete privacy from eavesdropping.

Fortunately, scrambling is used only about 20% of the time in federal communications. Most communications are in the clear. After a crime has been committed, such as a bank robbery, it is generally better for the public to know about the incident. Also very few listeners actually use a scanner to break the law. Most scanner listeners are just hobbyists who want to be more aware of the happenings around them.

PUBLIC SAFETY LISTINGS

ScanFan Southern California is easy to use. Public Safety data for San Diego, Los Angeles, Orange, Imperial, and Riverside County is first. Next are state and federal government listings. The most active frequencies are always **bolded.**

In each county, you will find frequencies for police, fire and emergency services. First in order is the sheriff's office and county fire department followed by emergency medical services (American Red Cross, ambulances, paramedics, search & rescue, and station "X"). Next are county government agencies. Finally, all cities are listed alphabetically.

Public Safety is where you will find radio data for a city's police department, fire department, local government, and any other municipal system. Local government may include street department, sanitation department, building department, lifeguards, animal control, water district, etc. Sometimes a local government frequency will be used for police surveillance operations or serve as a police tactical channel. Often a large city will have dozens of licensed frequencies. For example, the city of San Diego has frequencies for its police, fire, local government, convention center, lifeguards, stadium, and zoo to name a few.

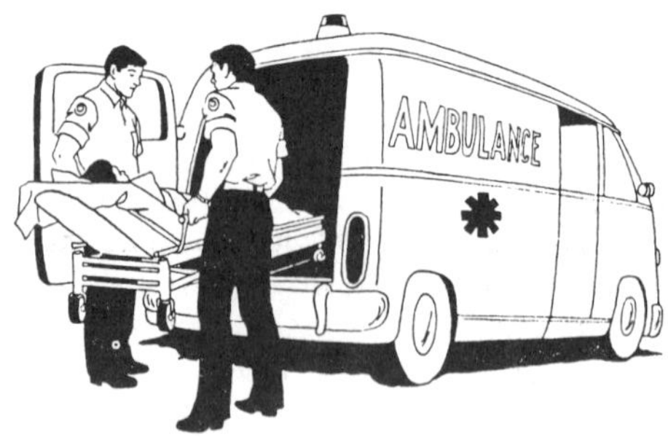

Police departments are of primary interest among scanner listeners. In addition to **dispatch** channels, specific frequencies may be designated for special types of communications. **Inquiry** channels are used for checking on wanted persons, stolen vehicles, stolen property, and criminal and driving records. **Car-to-car** and **tactical** channels are used for officers to talk directly with one another. The distance of hearing non-repeated car-to-car and tactical channels is usually 5-10 miles with a standard scanner and indoor antenna. Greater distances will be obtained when a gain or an external antenna is used.

For SWAT and surveillance operations, a police department may have several pools of low-power frequencies. FCC regulations allow a law enforcement agency to operate on almost any frequency it chooses provided transmissions are low-power and do not cause harmful interference. For example, many police detectives operate handie-talkies on frequencies throughout the 851-896 MHz range for surveillance operations.

A GUIDE TO THIS DIRECTORY

PUBLIC SAFETY

Fire Departments are also very popular to listen to. You will hear calls for accidents, fires, and medical aids on **dispatch** channels. More than 80% of fire calls are for medical aids. **Fireground** and **tactical** channels are used for communications between fire personnel on the incident scene. **Mutual Aid channels** are shared between all departments either countywide or statewide for calls involving more than one department. In the State of California, channels **White 1, White 2**, and **White 3** have been set aside as fire mutual aid channels. **Command** channels provide a communications link between the incident commander and the dispatch center. **Remote radio links** are often used between the radio site and the dispatch center. By listening to a remote link, you will hear portables and mobiles that are normally difficult to hear.

STATE OF CALIFORNIA

State operated agencies such as highway patrol, forestry, fish & game, parks & recreation, and state police may be found in this section. The listings have been greatly expanded. Frequencies may be found for the entire Southern California area for all agencies such as forestry and highway patrol.

U.S. GOVERNMENT

Federal government information is the most difficult of all to obtain and verify because data regarding these radio systems is not available for public view. Despite the limited sources, we have tried to have a very accurate list of U.S. Government frequencies. All branches from the Agriculture Department to the Treasury Department are included.

U.S. MILITARY

Air Force, Army, Navy, and Marine Corps. are all found in this one section. The verification of frequency data for military radio systems is far more complex than any other radio service. Changes occur routinely and some frequencies remain dormant for years before suddenly coming to life. Land mobile services for military installations are located primarily in the 138-144, 148-150.8, and 406-414 MHz ranges.

The UHF military aircraft band (225-400 MHz) is one area few radio hobbyists have explored. Most communications in this band are related to military aircraft navigation and aircraft tactics. The AM mode is the primary emission type, however, narrow and wideband FM may be found at times. These 7000 military channels are allocated on even 25 kHz increments (e.g.. 243.000, 305.675 etc.). ScanFan greatly appreciates all input on military systems.

AIRCRAFT

The aircraft section lists frequencies for the civilian aircraft band which occupies the spectrum from 108-136 MHz. All emissions are AM mode. **Please note: the aircraft section serves as a general reference and is not intended for in-flight navigational purposes.**

Aeronautical Radio (ARINC) is used by airline companies for flight operations, cargo movement, maintenance, passenger services, phone patches, and administrative matters.
Airline Ground Communications are low power radios used by airline companies for baggage, cargo handling, maintenance, ramp control, and airport security.
Approach / Departure Control are specific frequencies manned by an air traffic controller who directs the movement of approaching and departing aircraft outside the VFR.
Automatic Terminal Information Service (ATIS) is a continuous recorded broadcast of weather and runway conditions that is updated hourly.
The **Control Tower** is a facility, located at major airports that directs and monitors the movement of aircraft within its controlled air space.
The **Ground Control** is a traffic controller who directs taxiing aircraft and utility vehicles (fuel trucks, cargo vehicles, and crash/fire vehicles) at major airports.
Unicom is a radio service located at both small and large airports that permits pilots to contact ground facilities. At airports without a control tower, the unicom operator gives field advisories and reports of other air traffic.
Vortac is a communications facility that provides electronic course guidance for aircraft.

The following abbreviations are used in the aircraft listings:

A = ATIS
AC = Approach Control
A/DC = Approach / Departure Control
CD = Clearance Delivery
CT = Control Tower
DC = Departure Control
FSS = Flight Service Station
GC = Ground Control
ILS = Instrument Land System
M = Multicom (122.9)
S = Security (airports)
U = Unicom
V = Vortac
W = Weather Advisories

A GUIDE TO THIS DIRECTORY

AMATEUR RADIO SERVICE

Amateur Radio is a non-commercial radio service which provides personal and emergency communications for more than 1 million operators worldwide. To obtain a ham license, you must pass either a 55 question exam or a 25 question exam with Morse Code. Many people have become licensed hams with only a few weeks of study. Included in the amateur radio section are all popular repeater and simplex frequencies for Southern California.

OPEN REPEATER This repeater is available to all licensed amateur radio operators. A tone is required for access if it is shown.

CLOSED REPEATER To access this repeater requires membership in a sponsored club or private group. A tone is usually required.

PRIVATE REPEATER To access this repeater requires membership in a private club. Membership is at the discretion of the private group. Tones are usually used.

TONE A subaudible tone transmitted on the input frequency that opens the repeater when recognized. Also called PL for "Private Line." It is used to reduce interference when two nearby repeaters share the same frequency and guard against unwanted signals opening a repeater.

MINUS: indicates a standard input that is downshifted.
PLUS: indicates a standard input that is upshifted.

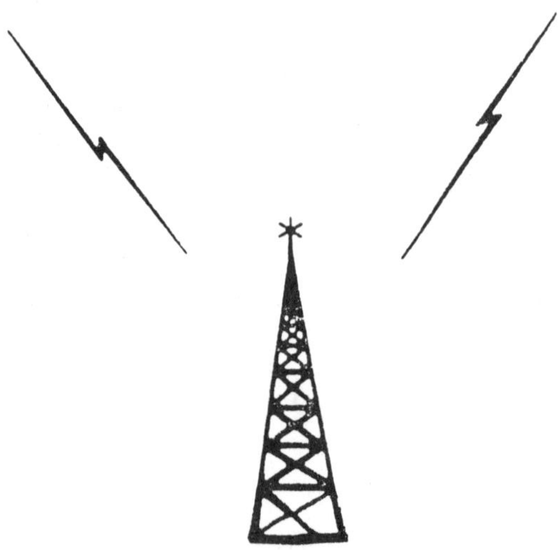

BROADCASTERS

Here you will find A.M. & F.M. broadcast stations, television stations, and traveler's information services (TIS). Many changes have taken place in the last few years. Stations are constantly being bought and sold, switching frequencies and call signs, and going off the air. We have tried to include some non-local AM stations usually heard at night (e.g., KOA Denver; 850 kHz).

BUSINESS

There are more business radio licensees than any other radio service. Keeping track of the thousands of business frequencies in Southern California is a full-time job that is only getting harder. A business may change frequency/ies as often as it desires, FCC records do not always reflect the correct data, and many businesses are now using SMR trunking systems. Despite the difficulties, we at ScanFan have tried to produce an accurate list for business radio systems. Included in the business section are frequencies used by alarm companies, amusement parks, hotels, industrial plants, and shopping centers.

GENERAL MOBILE RADIO SERVICE

The General Mobile Radio Service (G.M.R.S.) consists of eight repeater frequency pairs and seven low power simplex frequencies. The repeater output frequencies are 462.550, 462.575, 462.600, 462.625, 462.650, 462.675, 462.700, and 462.725 MHz. The seven low power frequencies include 462.5625, 462.5875, 462.6125, 462.6375, 462.6625, 462.6875, and 462.7125 MHz.

The frequency 462.675 is used nationally by Radio Emergency Associated Communications Teams (R.E.A.C.T.). These volunteer teams regularly report road and weather conditions and summon assistance for disabled motorists.

Readers who are more interested in REACT may write or call: Southern Calif. REACT Council, P.O. Box 2638, Garden Grove, CA 92643. Phone (310) 519-4477.

LAND TRANSPORTATION

Included in this section are armored transport services, bus lines, taxi cabs, towing companies, and trucking & delivery services. Towing companies make for informative listening since they respond to many traffic accidents. One note on taxi cabs; mobiles will be usually found five megahertz above the base frequency.

A GUIDE TO THIS DIRECTORY

MARITIME

Beyond the excitement of the Coast Guard and rescue operations, there are ship to shore calls, harbor patrols, tug boats, port operations and recreational boats you can monitor. The following is a guide to the VHF maritime band.

Channel 16 is the most important maritime frequency to put in your scanner. All vessels are required to monitor this channel when not engaged in communications on a working channel. Distress calls and contacts between vessels and shore stations are first initiated on channel 16. The idea is to establish contact on channel 16, then quickly move to a working channel.

The U.S. Coast Guard may be regularly monitored on channels 21, 22, 23, 81, 82, and 83. Additional U.S. Coast Guard frequencies are found in the U.S. Government section.

Channel 6 is established as a safety channel. Vessels contact one another about potential navigational and weather problems here. It's also used during search & rescue operations for inter-vessel communications and aircraft.

Communications related to the movement and safety of vessels in or near ports are commenced on channels 9, 68, 69, 70, 71, 72, and 78; the busiest channels are 68 and 70.

When looking at the chart of VHF marine channels, you will notice that many channels are allocated for use only in international ports and are not authorized for use in the U.S. Many of the international channels conflict with U.S. business and public safety frequencies. However, it is common to hear radio operators using these unauthorized frequencies for their private communications. In fact, the entire VHF maritime band is subject to illegal use by drug trafficers, survivalists, and some truckers.

NEWS MEDIA

Newsmedia stations are interesting to monitor as reporters, photographers, and TV news crews race to the scene of newsworthy incidents. Often they provide valuable details from sketchy police and fire calls. You will gain a valuable insight into the production of newstories and coordination of remote broadcasts (live shots).

Newsmedia stations are found in the following frequency ranges: 25.870-26.470, 152.870-153.350, 161.640-161.760, 174-216, 450-450.990 and 455.010-455.990 MHz. Some stations are using 800 MHz trunking and cellular telephones.

Activity in the 26 MHz band is small compared to the VHF and UHF bands. This band suffers from noise, skip interference, and freebanders who use it. All communications are simplex in this band, therefore; handheld units are generally not used. The 26 MHz band is most applicable for long-distance base to mobile communications during low sunspot years. In major metropolitan areas, it is used for backup systems and remote broadcast pickup.

Frequencies in the 174-216 MHz range are authorized to broadcast stations for low power applications, such as wireless microphones and cueing devices. The range on these transmissions is generally limited to 1/4 of a mile, but if you are located near a live shot or a broadcast studio, it is possible to listen to these communications.

The 450 MHz band is the hub of all news media communications. Here is where you will find most news traffic reports, and coordination of remote broadcasts. Unlike the 26 MHz and VHF-High band, repeaters dominate. Portable units (handhelds, walkie-talkies) are very popular. Typical range is 50-75 miles for mobiles and 30-50 miles for portables depending on the local terrain.

If you own a scanner with the 940-950 MHz frequency range, you may want to try listening to Studio Transmitter Links (STL's). STL's are used to relay audio and video from remote transmitters (live shots) to the main studio. All transmissions of STL's are in the wide FM mode so your receiver must have that mode. Remote broadcasters make reference to these frequencies and additional ones in the 2, 7, 13, and 22 GHz bands when setting up live shots. A yagi or parabolic-type antenna directly aimed at the transmitter antenna may pick up these communications.

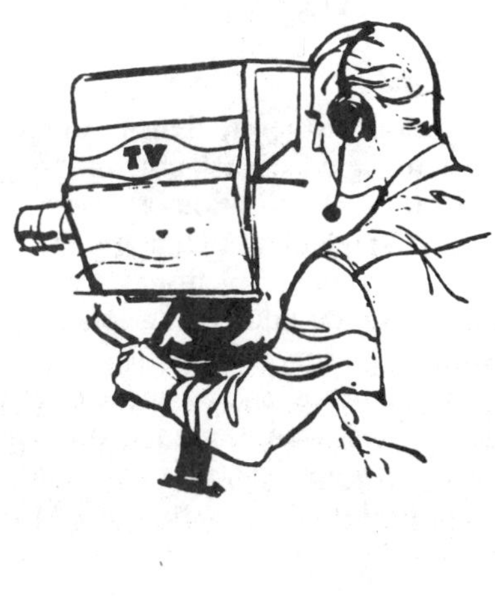

A GUIDE TO THIS DIRECTORY

SCHOOLS

Community and private colleges, and selected school districts may be found in this section. You will hear high school security guards, campus police and fire, student escort services, parking attendants, and school bus drivers transporting students. Many schools have plans to move their systems to the 800 MHz range.

UTILITIES

This section includes operation and security frequencies used by Imperial Irrigation District, S.D.G.& E., Southern Calif. Edison, telephone companies, and water districts. During storms and severe weather conditions, these frequencies come to life as work crews repair downed electrical wires and help clear flooded areas.

NATIONWIDE

Many frequencies are used nationwide for the same purposes. Included in this section are frequencies for the citizen's band, cordless telephones, sports events and many others.

RADIO CODES

All known radio codes used by local law enforcement have been listed. Abbreviations and radio lingo have been included to aid the listener in understanding what is being said over the air. Many agencies use a set of 10-codes plus additional unique codes. Input from listeners is greatly appreciated.

FREQUENCY CROSS-REFERENCE

In this section, you will find many frequencies listed in numerical order. This type of listing is very helpful if you are searching for unknown signals or maybe just trying to identify one station. Due to lack of space, some nongovernment stations are not listed.

WHAT'S NEW?

We believe ScanFan Southern California to be accurate at time of publication. However, changes in radio systems are always occurring and a certain amount of errors are always unavoidable in a publication of this type.

This year ScanFan includes **much more** data regarding input frequencies and subaudible tones. If you have a scanner or receiver that is equipped with a tone decoder, this information will be very helpful.

During the last year, most county agencies in Los Angeles made a smooth transition to a 40 channel trunked system on 800 MHz. Municipalities in Los Angeles also continue to move to the UHF-T band (470-512 MHz). For example, Burbank completed a quick move from VHF to 470 MHz in late 1993. The Los Angeles PD expansion is also on schedule and activity is occurring on the 484 MHz frequencies. The new Astro radios from Motorola offer both analog and digital compatability. The plan is to eventually have all L.A.P.D. communications utilize digital modulation. Los Angeles City Services (not public safety) also will switch to a 40 channel 800 MHz system.

Orange County was recently granted 63 frequencies on 800 MHz for the sheriff, marshal, and other law enforcement agencies. Motorola received the contract for constructon of the trunked and digital system. Only limited testing has occurred so far on the new frequencies. The other substantial development in Orange County is the new Sheriff South Dispatch for the City of San Clemente.

The countywide trunking system in San Bernardino is basically complete after five years of construction and it is functioning beautifully. A radio system is rarely constructed that covers such a large geographical area and is specifically tailored to the needs of the individual radio user. Public officials from across the world including the outback of Australia are using the San Bernardino system as a model.

A number of San Bernardino 800 MHz frequencies are still reserved for future expansion or emergency backup systems. ScanFan has focused on those frequencies where activity is presently heard. Unfortunately the Valley and Westend Systems are almost impossible to monitor during the day due to the large number of trunked frequencies.

Riverside County continues to make a slow transition to its new 800 MHz system and communications are being simulcast on the old VHF frequencies. When listening to the 800 MHz system it is vital that your scanner is programmed to the correct frequencies; there are areas of overlapped coverage by more than one cell site and you may need to do some experimenting.

San Diego's trunking system is in the last stages of being installed and provides communications for its police, fire, medics, lifeguards, and school buses. Signals are strong for listeners and the MDT system has reduced a lot of routine radio traffic that before took valuable airtime. Overall the administration is very pleased with the new system.

There are plans for a San Diego and Imperial regional radio system for the sheriff, county fire, and numerous cities. Some testing is already underway for the state of the art system which will be use more than 65 radio sites and will cost about $85 million. El Cajon PD is another department that is currently changing to 800 MHz. Lastly, the military bases in San Diego County are all moving to a new 406 MHz trunking system.

Officials in Ventura county are also pushing for some type of regional radio system on 800 MHz. The recent earthquakes revealed the inadaquacies of the VHF radio system for communications during a major disaster. The desire is to have a state-of-the-art radio system wherebye the Ventura Sheriff, local police and fire, and all county services may freely intercommunicate.

A GUIDE TO THIS DIRECTORY

WHAT'S NEW?

The FCC's new Part 88 "Spectrum Refarming Rules" are also going to change radio systems as we know them. Every radio in 150-174 and 450-512 MHz bands that is under part 90 will eventually become obsolete as tighter channel spacing and frequency stability standards are imposed. This change will require new types of scanners to deal with the reduced bandwidth and the closer spaced frequencies. Ultimately more stations and services will be on the air to monitor than ever before.

Digital radios whereby voice is converted into a digital or binary format before transmission are rapidly appearing on the radio scene. The demand for narrower bandwidths, increased speed for data transfer, and greater efficiency of the spectrum guarantees that this technology is here to stay. This will be the biggest challenge ever for listeners.

OTHER HELPFUL PUBLICATIONS

Monitoring Times
Grove Enterprises, Inc.
P.O. Box 98,
Brasstown, NC 28902
Telephone: 1-704-837-9200

Monitoring Times is a monthly publication devoted to scanning, shortwave, utility stations, and other topics.

Popular Communications
76 North Broadway,
Hicksville, NY 11801
Telephone: 1-516-681-2922

Popular Communications is a monthly magazine that focuses on scanning, shortwave, utility stations, CB, satellites, amateur radio, broadcasting, and other topics.

RCMA Scanner Journal
P.O. Box 542
Silverado, CA 92676

A monthly journal packed with frequency information for all parts of the country, FCC topics, reviews, and much more.

Scanner Modification Handbook I & II
By Bill Cheek
C.R.B. Research Books
P.O. Box 56
Commack, NY 11725

Bill Cheek covers the topic of scanner modifications very thoroughly. His instructions are so detailed that a person without any prior electronic experience can perform most of the modifications found in his book.

Military Radio Systems
By Robert Kelty
Mobile Radio Resources
1224 Madrona Ave.
San Jose, CA 95125

The most comprehensive and accurate book available detailing systems used by the Air Force, Army, Navy, and Marine Corps. Send $37.50 post paid to receive this directory by priority mail.

WE NEED YOUR HELP

ScanFan plans to print a new edition every one to two years. We will strive to make each new edition better than the preceding one. To properly accomplish this goal; however, we need your assistance. The best way to maintain accurate data is for communications professionals and other monitoring enthusiasts to contribute information. If you have any comments or suggestions, we would appreciate a telephone call or a letter. Any ideas on how to make ScanFan easier to use will also be welcome. You may also want to add your name to our mailing list to be informed of new products.

To comment on ScanFan San Diego, to send or receive updates and corrections or add your name to our mailing list, please write or call us at:

ScanFan Publications
P.O. Box 544
San Marcos, Ca.
92079-0544
Telephone: (619) 744-2303 Fax: (619) 471-2415

OTHER PRODUCTS

ScanFan also carries "Scannermates" which are handy, easy to use, laminated frequency guides. They cover police, fire, paramedic, emergency, and local government frequencies. Selected frequencies are also included for the State of California, Federal Government, business, land transportation, and others. There is even a section on radio codes. The guides are extremely accurate, being updated several times a year. They are flexible, waterproof (you can drop a cup of coffee on one), and they won't fade if you leave one out in the sun. Every scanner listener who wants instant access to the most active and popular frequencies, or every listener who is on the go should have one for his or her area! They are arranged by county for different areas in Southern California. Call for price and availability.

TABLE OF CONTENTS

SAN DIEGO COUNTY

DESCRIPTION **PAGE NUMBER**

TABLE OF CONTENTS

LOS ANGELES COUNTY

DESCRIPTION | **PAGE NUMBER**

TABLE OF CONTENTS

ORANGE COUNTY

IMPERIAL COUNTY

TABLE OF CONTENTS

RIVERSIDE, SAN BERNARDINO, VENTURA

DESCRIPTION **PAGE NUMBER**

OTHER AREAS

DESCRIPTION **PAGE NUMBER**

STATE OF CALIFORNIA

DESCRIPTION **PAGE NUMBER**

TABLE OF CONTENTS

U.S. GOVERNMENT

DESCRIPTION **PAGE NUMBER**

TABLE OF CONTENTS

NONGOVERNMENT

DESCRIPTION **PAGE NUMBER**

TABLE OF CONTENTS

NONGOVERNMENT

DESCRIPTION **PAGE NUMBER**

MISCELLANEOUS

DESCRIPTION **PAGE NUMBER**

FREQUENCY CROSS-REFERENCE

DESCRIPTION **PAGE NUMBER**

SAMPLE LISTING

THE FOLLOWING EXAMPLE IS A GUIDE TO THE TERMS AND ABBREVATIONS USED THROUGHOUT THE DIRECTORY. MUCH ADDITIONAL INFORMATION MAY ALSO BE FOUND IN THE INTRODUCTION SECTION.

ARCADIA

CHANNEL (a)	INPUT (b)	OUTPUT (c)	TONE (d)	CALL SIGN (e)
Police Primary (F-1)	510.0625	**507.0625R**	110.9	KXN-419
Police Direct (F-2)		507.0625	110.9	KXN-419
Police Secondary (F-3)	509.5625	**506.5625R**	110.9	KCX-723
Police Direct (F-4)		506.5625	110.9	KCX-723

(a) CHANNEL DESIGNATION AND USE OF CHANNEL.

1. PRIMARY INDICATES MAIN DISPATCH AND OPERATIONS
2. DIRECT DENOTES NO USE OF REPEATER
3. "FREQUENCY 1 THRU ..." ARE ABBREVIATED AS (F-1), (F-2),..., ETC.

(b) INPUT FREQUENCY (IF ANY)

(c) OUTPUT FREQUENCY

THE MOST ACTIVE OUTPUT FREQUENCIES ARE BOLDED.

(d) CTCSS OR DIGITAL TONE USED ON THE CHANNEL

TONES ARE USED TO GUARD RADIO RECEIVERS FROM ON-CHANNEL AND ADJACENT CHANNEL INTERFERENCE. TONES LISTED HAVE BEEN VERIFIED IN THE FIELD USING STANDARD TEST EQUIPMENT.

(e) FCC LICENSED CALL SIGN

STANDARD ABBREVIATIONS

AVL = VEHICLE LOCATION SYSTEM
MDT = MOBILE DATA TERMINAL
MULTI = MULTIPLE CTCSS TONES
R = REPEATED
S = SIMPLEX
T = TRUNKED

THE PUBLIC SAFETY INFORMATION HAS BEEN PERSONALLY VERIFIED WHENEVER FEASIBLE OR HAS BEEN PROVIDED BY A RELIABLE SOURCE THAT IS DIRECTLY INVOLVED WITH INSTALLATION, MAINTENANCE OR FREQUENCY COORDINATION OF THE RADIO SYSTEM. SOME NONGOVERNMENT AND FEDERAL LISTINGS ARE SUBJECT TO VERY FREQUENT CHANGE AND VERIFICATION AND RESEARCH IS A CONTINUOUS PROCESS.

TONE SQUELCH DATA

FREQUENCY (Hertz)	MOTOROLA "Private Line"	JOHNSON "Call Guard"	DIGITAL TONES (DTS)		
67.0	XZ	001	023	306	703
71.9	XA	002	025	311	712
74.4	WA	003	031	315	723
77.0	XB	004	032	331	731
79.9	WB	005	043	343	732
82.5	YZ	006	047	346	734
88.5	YA	007	054	351	743
91.5	ZZ	008	065	364	754
94.8	ZA	009	071	365	
97.4	ZB	010	072	371	
100.0	1Z	012	073		
103.5	1A	013	074	411	
107.2	1B	014		412	
110.9	2Z	015	114	413	
114.8	2A	016	115	423	
118.8	2B	017	116	431	
123.0	3Z	018	125	432	
127.3	3A	019	131	445	
131.8	3B	020	132	464	
136.5	4Z	021	134	465	
141.3	4A	022	152	466	
146.2	4B	023	155		
151.4	5Z	024	156	503	
156.7	5A	025	162	506	
162.2	5B	026	165	516	
167.9	6Z	027	172	532	
173.8	6A	028	174	546	
179.9	6B	029		565	
186.2	7Z	030	205		
192.8	7A	031	223	606	
203.5	M1	032	226	612	
206.5	8Z	033	243	624	
210.7	M2	034	244	627	
218.1	M3	035	245	631	
225.7	M4	036	251	632	
229.1	9Z	037	261	654	
233.6	M5	038	263	662	
241.8	M6	039	265	664	
250.3	M7	040	266		
254.1	6Z	041	271		

THE ABOVE TONES HAVE BEEN STANDARDIZED BY THE ELECTRONIC INDUSTRIES ASSOCIATION (EIA). THE MOTOROLA TRADENAME "PL" OR "PRIVATE LINE" IS THE MOST WELL-KNOWN TERM IN THE RADIO INDUSTRY.

SAN DIEGO COUNTY

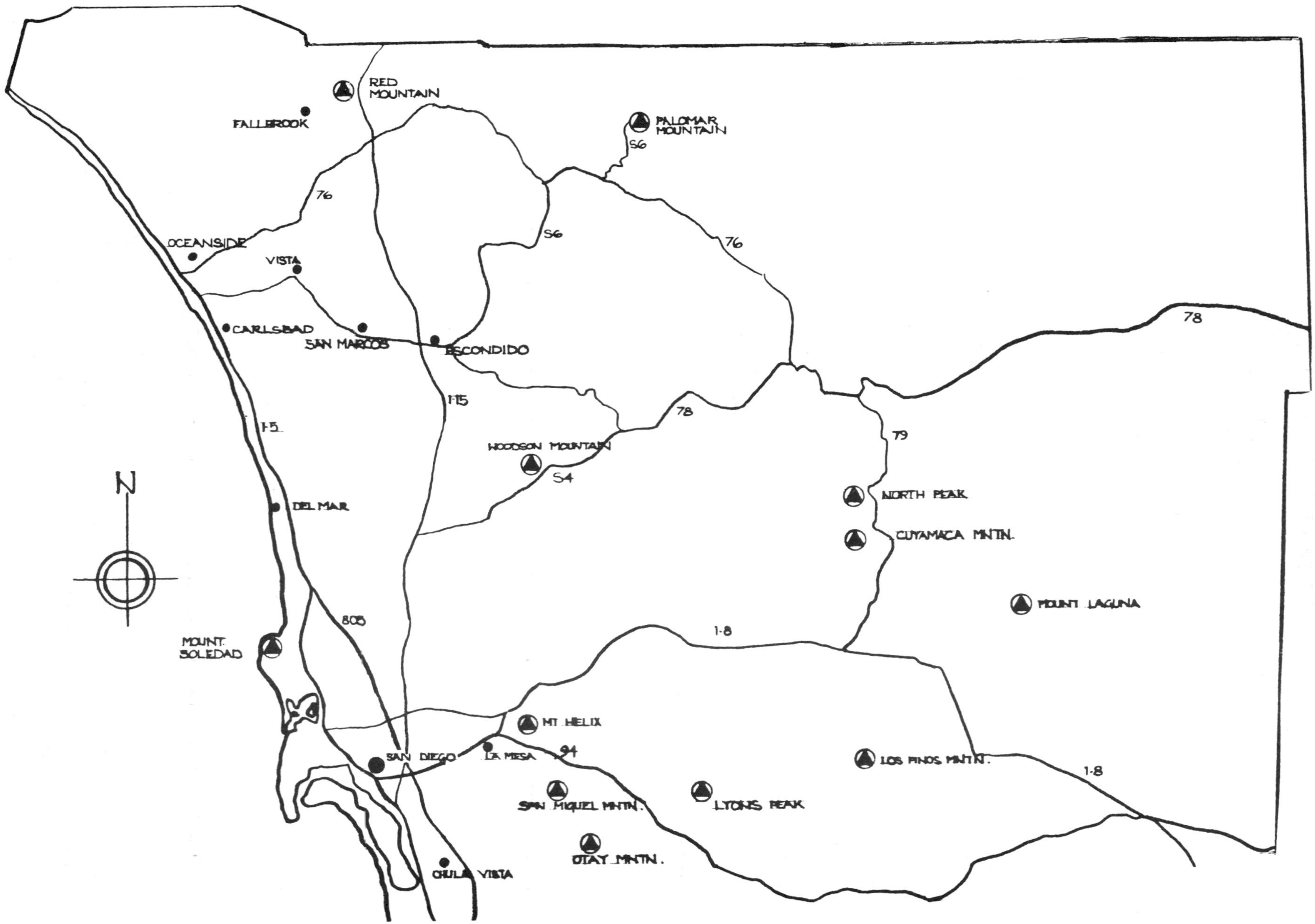
RED MOUNTAIN
FALLBROOK
PALOMAR MOUNTAIN
S6
76
OCEANSIDE
VISTA
S6
76
CARLSBAD
SAN MARCOS
ESCONDIDO
78
I-15
78
I-5
WOODSON MOUNTAIN
79
S4
N
DEL MAR
NORTH PEAK
CUYAMACA MNTN.
MOUNT LAGUNA
805
MOUNT SOLEDAD
I-8
MT HELIX
SAN DIEGO
LA MESA
94
LOS PINOS MNTN.
I-8
SAN MIGUEL MNTN.
LYONS PEAK
OTAY MNTN.
CHULA VISTA

SAN DIEGO COUNTY

SHERIFF'S OFFICE

PRIMARY CHANNELS

CHANNEL	USE	FREQUENCY
(F-1)	Car-to-Car	**453.725**
(F-2)	AREA 2 - North Coastal	**453.750 R**
(F-3)	AREA 3 - North Inland	**453.925 R**
(F-4)	AREA 4 - East / Central	**453.500 R**
(F-5)	AREA 5 - South Bay	**453.950 R**
(F-6)	Administration	**453.400 R**
(F-7)	Records Inquiry	**453.575 R**
(F-8)	AREA 6 - Central	**453.100 R**
(F-9)	Tactical North	**453.425 R**
(F-10)	Tactical South	**453.825 R**
(F-11)	Tactical North Direct	453.425
(F-12)	Tactical South Direct	453.825
(F-24)	CLEMARS UHF	460.025
none	Countywide Tactical	453.975
none	Cuyamaca Peak	**45.940 R**
none	Los Pinos Peak	**45.780 R**
none	Tactical-Alpine	453.100

Notes for frequencies:

Dispatching for the Central Area is on F-8 during the day and F-4 or the Low Band repeaters at night. Dispatching sometimes occurs on F-10 during peak hours to relieve radio congestion from F-4 and F-5. Also F-2 and F-3 are often simulcast (dispatched together). These variances are often due to a lack of Sheriff dispatchers.

Administration includes Central Field Operations, Office of Special Investigations, Narcotics Task Force, detentions transportation, and countywide radio pagers.

Frequencies will vary for channels 13-23 depending on the type of unit (patrol, detective, detention, etc.) and the service area (North Inland, South Bay, etc.).

Channel 24 (460.025) is rarely used, but is available to all units throughout the county for car-to-car, tactical, and interagency communications.

The VHF- low band repeaters on 45.780 and 45.940 MHz are located on Los Pinos and Cuyamaca Peaks respectively. These lower frequencies greatly extend the radio range of Sheriff units in the rugged back-country areas of the county where the UHF radio system lacks coverage.

The PL tone for all Sheriff frequencies is 146.2 hertz. Inputs are a standard 5 MHz above the output frequencies except for the low band repeaters.

SAN DIEGO COUNTY

SHERIFF'S OFFICE

REGIONAL RADIO SYSTEM 1996

The Sheriff will be one of the many users of the new San Diego County regional radio system for all county services and most municipalities. Radio traffic is presently being simulcasted on a few of the new frequencies.

861.150		**867.4125**	
862.150		**867.4375**	Mexican Border Area
863.150		**867.450**	
866.0625	Mexican Border Area	**867.5875**	Mexican Border Area
866.1375		**867.6375**	
866.4125		**867.7125**	
866.4375	Mexican Border Area	**867.950**	
866.6375		**868.075**	
866.9125		**868.1375**	
866.950		**868.4125**	
867.075		**868.4375**	Mexican Border Area
867.1375		**868.950**	Mexican Border Area

INTERAGENCY CHANNELS

CHANNEL	USE	FREQUENCY
BLUE - County Mutual Aid	Backup Dispatch	153.995
CALCORD - Statewide	Interagency	156.075
CLEMARS - Statewide	Interagency	154.920
CLEMARS - Statewide	Interagency	460.025
CLERS - Regional	Interagency	453.675 R
NLEMARS - Nationwide	Interagency	155.475

SPECIAL OPERATIONS CHANNELS

NAME	FREQUENCY
Detectives - Surveillance	453.875
Detectives - Surveillance	453.975
Detectives - Surveillance	458.725
Narcotics Enforcement Detail	
Task Force Cache - Confirmed	**460.025**
Task Force Cache - Confirmed	**460.125**
Task Force Cache - Confirmed	**460.2375**
Task Force Cache - Confirmed	**460.4875**
Search & Rescue - Primary VHF	155.160
Search & Rescue - Secondary VHF	155.235
Search & Rescue - Primary UHF	453.100
Special Enforcement Detail - Primary	**453.250**
Special Enforcement Detail - Alternate	453.125
Special Enforcement Detail - Alternate	460.125

SAN DIEGO COUNTY

SHERIFF'S OFFICE

DETENTION FACILITIES

NAME	FREQUENCY
Barrett Honor Camp - Security F1	453.325
Barrett Honor Camp - Security F2	453.825
Barrett Honor Camp - Security TAC	153.995
Central Detention Facility - Security	453.625
Central Detention Facility - Paging	37.040
El Cajon Detention Facility - Security F1	453.700
El Cajon Detention Facility - Security F2	453.825
Imperial Beach Detention Center - Security	453.600
Las Colinas - Security	460.125
Morena Honor Camp - Security	153.995
Otay Mesa - Security F1	460.125
San Jose Honor Camp - Security	153.995
South Bay Detention Facility - Security F1	453.275
South Bay Detention Facility - Security F2	453.850
Vista Detention Facility - Security F1	453.325
Vista Detention Facility - Security F2	453.725
Vista Detention Facility - Paging	37.040
Prisoner Escape Tactical - Mutual Aid BLUE	153.995

AERIAL SUPPORT TO REGIONAL ENFORCEMENT AGENCIES

A.S.T.R.E.A.

AIRCRAFT - 4 McDonald Douglas 500D's, 2 Bell 47G's, 2 twin-engine airplanes and a single-engine airplane. The Bell 47's are used primarily for training. The Sheriff's Office also has recently obtained four twin-engine fixed-wing aircraft, two Hughes UH-1E helicopters, and a UH-1H helicopter.

CAPABILITIES - 500D's have top speed of 150 miles per hour, carrying capacity of one thousand pounds, and operational altitude of ten thousand feet.

CREW - A crew consists of one pilot and one observer. The pilot controls the helicopter and operates the VHF-Air Band radio. The observer has tactical control (operates police radio, maneuvers lights, and handles weaponry). Currently there are four pilot-observer crews that work seven days a week. One A.S.T.R.E.A. is usually assigned during each daylight shift.

SAN DIEGO COUNTY

SHERIFF'S OFFICE

A.S.T.R.E.A.

EQUIPMENT - Multiband radio with VHF, UHF, and 800 MHz coverage. A powerful public address system and a lighting system capable of illuminating an area the size of a football field from an altitude of five-hundred feet. A.S.T.R.E.A.'s helicopters each carry four lights mounted together, totaling 3 1/2 million candlepower. Note: A.S.T.R.E.A. does not operate on its own radio frequency.

HELIPORTS - Gillespie Field in El Cajon is the primary heliport for A.S.T.R.E.A.. Another heliport is available in Descanso. All Sheriff stations have makeshift heliports that are available when requested.

PURPOSE - A.S.T.R.E.A.'s primary purpose is to assist law enforcement agencies in the apprehension of criminal suspects. Other types of calls include medical transport, search & rescue, recovery of stolen vehicles, locating lost persons, photographing crime scenes, surveillance, making fire drops, traffic control, and disaster scene evaluation. During fires, A.S.T.R.E.A. will act as air control for tankers doing water drops.

REQUIREMENTS - Pilots: Deputy Sheriff with commercial pilot rating, two hundred helicopter flight hours and two years of patrol experience.
Observers: Deputy Sheriff with two years of patrol experience.

AREAS SERVED

AREA 2 - Cardiff, Del Mar, Encinitas, Leucadia, Olivenhain, Rancho Santa Fe, San Elijo, and Solana Beach.

AREA 3 - Bonsall, De Luz, Escondido rural, Fallbrook, Rainbow, San Marcos, and Vista.

AREA 4 - Anza Borrego, Campo, Cuyamaca Rancho, Dulzura, Jacumba, Julian, Lake Henshaw, Mount Laguna, Mount Palomar, Ocotillo Wells, Pala, Pauma Valley, Potrero, Poway, Ramona, Rincon, Santa Ysabel, Tecate, and Valley Center.

SAN DIEGO COUNTY

SHERIFF'S OFFICE

AREAS SERVED

AREA 5 - Bonita, Calavo Gardens, Casa De Oro, Castle Rock, El Cajon rural, Grossmont, Imperial Beach, Jamul, La Presa, Lemon Grove, Lincoln Acres, Otay Mesa, Spring Valley, Sunnyside, and Sweetwater Reservoir.

AREA 6 - Alpine, Crest, El Cajon, and Lakeside.

CALL SIGN IDENTIFICATION

LOCATION	CALL SIGN
Borrego Springs	KNCJ-404
Boulevard	KNCJ-416
Cuyamaca Peak	KNCJ-402
Encinitas	KNCJ-418
Fallbrook	KNCJ-414
Imperial Beach	KNCJ-408
La Mesa	KNGS-956
Lemon Grove	KNCJ-411
Los Pinos	KNCJ-405
Mobile Identification	KA-88385
Mt. Laguna	KNCJ-406
Mt. Palomar	KNCJ-401
Mt. Woodson	KNCJ-419
National City	KNGR-699
Oceanside	KNGR-698
Red Mountain (Fallbrook)	KNCJ-417
San Diego (Dispatch Center)	KNCK-664
San Marcos Peak	KNCJ-415
Santee	KNCJ-409
Spring Valley	KNCJ-412
Tecate Peak	WNAR-999
Vista	KNCJ-410
Vista	KNGX-598
Vista	KNGY-554

SAN DIEGO COUNTY

SHERIFF'S OFFICE

UNIT IDENTIFICATION SYSTEM

1. The first two digits define the service area:

10-19	Administrative
10	San Diego County
20-29	Coastal Area
20	North Coast unincorporated
21	City of Del Mar
22	Unassigned
23	Solana Beach
24	Unassigned
25	Encinitas
30-39	North Inland
30	Vista unincorporated
31	City of Vista
32	City of Fallbrook
33	Unassigned
34	Unassigned
35	City of San Marcos
36	Escondido unincorporated
40-49	Central Inland
40	Poway unincorporated
41	City of Poway
42	Unassigned
43	Unassigned
44	Ramona
45	Harbison Canyon
50-59	Central Area
50	El Cajon, Lakeside
51	City of Santee
52	Alpine & Crest
60-69	South Bay
60	Bonita, Lemon Grove rural
61	City of Lemon Grove
70-79	Border Area
70	Chula Vista, Otay
71	City of Imperial Beach
80-89	East Inland
80	Pauma, Valley Center
81	Unassigned
82	Borrego, Julian, Warner
83	Unassigned
84	Pine Valley
85	Unassigned
86	Campo, Tecate
90	Radio Training

2. The third digit represents the type of unit:

A = Academy
B = Personnel
C = Communications
D = Detentions Transportation
E = Ambulances, Special Equipment
F = Fraud
G = Internal Affairs
H = Homicide
I = Investigator
J = Juvenile
K = Canine / Public Affairs
L = Lab
M = Motorcycle
N = Narcotic
O = Special Investigations
P = Patrol
R = Resident Deputy
T = Traffic
V = Vice
W = Transportation
X = Arson, Bomb Squad
Z = Reserve

SAN DIEGO COUNTY

SHERIFF'S OFFICE

UNIT IDENTIFICATION SYSTEM

3. The fourth digit, which is seldom heard, defines the officer's rank:

   C = Captain
   L = Lieutenant
   S = Sergeant

4. The fifth and six digits represent the vehicle requested from 1-99.

5. The last digit indicates the shift:

   A = Day Shift
   B = Swing Shift
   C = Night Shift

6. Special Units

   Unit 1 = Sheriff Jim Roache

An example of this unit identification system is: **35 P S 1 A**

35	- Represents the service area:	**City of San Marcos**
P	- Represents the type of unit:	**Patrol**
S	- Represents the officer's rank:	**Sergeant**
1	- Represents the vehicle requested:	**Vehicle Number One**
A	- Represents the shift:	**Day Shift**

Another example of the identification system is: **20 T 2**

20	- Represents the service area:	**Coastal Cities**
T	- Represents the type of unit:	**Traffic**
2	- Represents the vehicle requested:	**Vehicle Number Two**

All letter digits are pronounced phonetically (e.g., PAUL for P). Most identifications consist of less than five digits. To keep identification time as short as possible, only the necessary digits are used.

SAN DIEGO COUNTY

FIRE DEPARTMENTS

DISPATCH ZONES

ZONE NUMBER	TITLE	HEADQUARTERS	FREQUENCY
1 & 2	Rancho	Rancho Santa Fe	**154.235**
3.	San Diego	San Diego	**154.310**
4.	Heartland	El Cajon	**800 MHz**
5.	South Bay	Chula Vista	**154.415**
6.	Monte Vista	El Cajon (CDF)	**151.190 R**
7.	Inland Volunteer	Kearny Mesa	**154.175**

AREAS SERVED

ZONE 1 & 2 - Bonsall, Cardiff, Deer Springs, Del Dios, Del Mar, De Luz, Elfin Forest, Encinitas, Escondido rural, Fallbrook, Leucadia, Rainbow, Rancho Santa Fe, San Marcos, Solana Beach, and Vista.

ZONE 3 - City of San Diego and Poway.

ZONE 4 - Alpine, Bostonia, Crest, El Cajon, Lakeside, La Mesa, Lemon Grove, Santee, and Spring Valley.

ZONE 5 - Bonita-Sunnyside, Chula Vista, Coronado, Imperial Beach, Lower Sweetwater, Montgomery, National City, and Otay.

ZONE 6 - District 1 = De Luz, Miller, Red Mountain.
District 2 = Flinn Springs, Grossmont/Mt. Helix, Monte Vista.
District 3 = Dulzura, Lyons Valley, Potrero, Sycuan.
District 4 = Campo, Whitestar.
District 5 = Cuyamaca, Julian.
District 6 = Ramona, Warner Springs, Witch Creek.
District 7 = Rincon, San Marcos, Valley Center 1 & 2.
Plus numerous reservations and remote areas.

ZONE 7 - Barona, Borrego Springs, Boulevard, Campo, Cuyamaca, Deerhorn Valley, Del Dios, Descanso, Dulzura-Barrett, Harbison Canyon, Jacumba, Jamul, Julian, Lake Morena, Mount Laguna, Mount Palomar, Pala, Potrero, Ramona, Ranchita, Rincon Pauma, San Pasqual, Shelter Valley, Sunshine Summit, Sycuan, Tecate, and Warner Springs.

Notes:

The PL tone 103.5 Htz is used for the above dispatches. National City FD will soon join the citywide 800 MHz system. San Diego FD simulcasts on 154.310 and 800 MHz.

SAN DIEGO COUNTY

FIRE DEPARTMENTS

ZONE 1 & 2 RANCHO

CHANNEL	FREQUENCY
Fire Dispatch	**154.235**
Fire Secondary	**154.385**
Fire Command RED	**155.085**
Fireground Tactical 1 GRAY	**154.355**
Fireground Tactical 2 YELLOW	153.890
Fireground Tactical 3 BROWN	154.250
Fireground - North County Fire	153.815
Fire Tactical "Low Band"	46.460
Remote Link - Carlsbad Fire	453.3375R
Remote Link - Mutual Aid RED	453.0375R
Remote Link - North County Fire	453.8625R
Remote Link - Rancho Fire	453.6375R

ZONE 3 SAN DIEGO

CHANNEL	FREQUENCY
San Diego Fire "Station F"	(See p.43)
Failsoft System	
Fire Dispatch	**860.025 R**
Fire & Medics	**859.025 R**
Fire Mutual Aid RED	155.085
Fireground Tactical GOLD	154.430
Fireground Tactical WHITE 1	154.280
Fireground Tactical WHITE 2	154.265

ZONE 4 HEARTLAND

CHANNEL	FREQUENCY
Fire Trunking System	**856.175 T**
Fire Trunking System	**856.200 T**
Fire Trunking System	**857.200 T**
Fire Trunking System	**858.175 T**
Fire Trunking System	**858.200 T**
Fire Trunking System	**859.200 T**
Fire Trunking System	**860.200 T**
Fire Conventional 1	**860.225 R**
Fire Conventional 2	**856.175 R**
Failsoft System	
Fire Dispatch	**858.175 R**
Fire Comm 1 & 2	**858.200 R**
Fire Tactical	**859.200 R**

SAN DIEGO COUNTY

FIRE DEPARTMENTS

ZONE 5 SOUTH BAY

CHANNEL	FREQUENCY
Fire Dispatch	**154.415**
Fireground Tactical 1 YELLOW	**153.890**
Fireground Tactical 2 GRAY	154.355
Fireground Tactical 3 BROWN	154.250
Fireground Tactical 4 GOLD	154.430

ZONE 6 MONTE VISTA

CHANNEL	FREQUENCY
CDF Fire Dispatch	**151.190 R**
RED NET - Air Tactical	**151.220**
Tactical 1 (reservations)	151.145
Tactical 5	151.250
Tactical 8	151.370
Tactical 11	151.460
Fire CDF Command 1	**151.355 R**
Fire CDF Command 2	**151.265 R**

For additional CDF frequencies, please see the State of California section.

ZONE 7 INLAND VOLUNTEER

CHANNEL	FREQUENCY
Fire Dispatch	**154.175**
Fireground Tactical 1 WHITE 3	**154.295**
Fireground Tactical 2 GRAY	154.355
Fireground Tactical 3 BROWN	154.250
Fireground Tactical 4 GOLD	154.430
Remote Link - Monument Peak	460.6375R

ALL ZONES

CHANNEL	FREQUENCY
Countywide Mutual Aid RED	**155.085**
State Fire Mutual Aid WHITE 1	**154.280**
State Fire Mutual Aid WHITE 2	**154.265**
State Fire Mutual Aid WHITE 3	**154.295**

SAN DIEGO COUNTY

FIRE DEPARTMENTS

COLOR CHANNELS

COLOR DESIGNATION	FREQUENCY
BROWN-Countywide	154.250
BROWN-North County FD	153.815
GOLD	154.430
GRAY	154.355
GREEN 1	154.235
GREEN 2	154.385
RED	155.085
WHITE 1	154.280
WHITE 2	154.265
WHITE 3	154.295
YELLOW	153.890

COLOR CHANNELS USE

BROWN - Tactical for San Diego County.

GOLD - Tactical primary for San Diego City Fire Department. (Replaced by 800 MHz)

GRAY - Tactical primary for North County Fire. Tone 114.8.

GREEN 1 - "Rancho Fire" primary dispatch. Tone 192.8

GREEN 2 - Old "Coastal Fire" dispatch, now Carlsbad FD dispatch. Secondary dispatch for "Rancho Fire."

RED - Countywide Mutual Aid channel, command for "Rancho Fire."

WHITE 1 - Statewide mutual aid channel, tactical for San Diego FD. Tactical primary for mutual aid calls.

WHITE 2 - Statewide mutual aid channel, tactical for San Diego FD. Tactical secondary for mutual aid calls.

WHITE 3 - Statewide Mutual Aid channel. Tactical primary for "Inland Volunteer Fire."

YELLOW - Tactical primary for "South Bay Fire" and "North County Fire." Also tactical for "Rancho Fire."

PL Tones: 192.8 (zone 1&2), 103.5 (zone 3), 141.3 (zone 4), 127.3 (zone 5), and 156.7 (zone 7/8/9). The tone 103.5 is also used countywide.

SAN DIEGO COUNTY

FIRE DEPARTMENTS

FREQUENCY ASSIGNMENTS

Zone 1 Rancho	Frequency	Tone
1000 = Deer Springs FPD	154.235	114.8
1100 = North County FPD	154.235	114.8
1200 = Vista FD	154.235	192.8
1300 = Escondido FD	154.325 / 850 MHz	103.5
1400 = San Marcos FPD	154.235	192.8
1500 = Rainbow VFD	154.235	114.8
1900 = Rincon Del Diablo FD	154.325 / 850 MHz	103.5

Zone 2 Rancho	Frequency	Tone
2100 = Oceanside FD	159.165	127.3
2200 = Carlsbad FD	154.385	127.3
2300 = Encinitas FD	154.235	192.8
2400 = Solana Beach FD	154.235	192.8
2500 = Del Mar FD	154.235	192.8
2600 = Rancho Santa Fe FPD	154.235	192.8
2700 = Camp Pendleton FD	138.600 R	167.9
2800 = Elfin Forest VFD	154.235	192.8

Zone 3 San Diego	Frequency*	Tone
3100 = San Diego FD	154.310	103.5
Poway FPD	154.310	103.5

*San Diego Fire Dispatch (trunked):
857.050 / 858.025 / 858.050 / 859.025 / 859.050 / 860.025 / 860.050

Zone 4 Heartland	Frequency*
4100 = Lakeside FPD	850 MHz
4200 = Santee FD	"
4300 = El Cajon FD	"
4400 = La Mesa FD	"
4500 = Lemon Grove FD	"
4600 = Spring Valley FPD	"
4700 = Alpine FPD	"
4800 = Crest FPD	"
4900 = Bostonia FPD	"

*Heartland Fire Dispatch (trunked):
856.175 / 856.200 / 857.200 / 858.175 / 858.200 / 859.200 / 860.200

SAN DIEGO COUNTY

FIRE DEPARTMENTS

FREQUENCY ASSIGNMENTS

Zone 5 South Bay	Frequency	Tone
5100 = Bonita-Sunnyside FPD	154.415	103.5
5200 = Chula Vista FD	154.415	103.5
5300 = Coronado FD	154.415	103.5
5400 = Imperial Beach FD	154.415	103.5
5500 = Lower Sweetwater FD	154.415	103.5
5600 = Montgomery FPD	154.415	103.5
5700 = National City FD	154.415	103.5

Zone 6 Monte Vista (151.190)

3300 = California Department of Forestry (C.D.F.)
6100 = CDF Engines, Ambulances, Rescue
6300 = Valley Center FD
6400 = CDF Duty Chiefs

Zone 7 Inland Volunteer

7000 = Julian District
7100 = Julian FPD (Local Calls)
7200 = Cuyamaca FPD (Local Calls)
7300 = Dulzura-Barrett VFD
7400 = Deerhorn Valley VFD
7500 = Lake Morena VFD
7600 = Campo VFD
7700 = Boulevard FD
7800 = Jacumba CSA
7900 = Potrero VFD

Zone 8 Inland Volunteer

8100 = Ocotillo Wells
8200 = Julian Cuyamaca FPD
8300 = Mount Laguna VFD
8400 = Pine Valley FPD
8500 = Descanso VFD
8600 = Harbison Canyon VFD
8700 = Unassigned
8800 = Cuyamaca FPD
8900 = Unassigned

Zone 9 Inland Volunteer

9100 = Unassigned
9200 = Del Dios VFD
9300 = San Pasqual VFD
9400 = Unassigned
9500 = Borrego Springs VFD
9600 = Rincon Pauma VFD
9700 = Palomar Mountain VFD
9800 = Pala VFD
9900 = Sunshine Summit VFD

If you live in the back country or a rural area in San Diego County, the best two fire frequencies to monitor are **151.190** (the dispatch for all State Forestry calls) and **154.175** (the dispatch for Inland Volunteer Fire).

SAN DIEGO COUNTY

FIRE DEPARTMENTS

MUTUAL AID IDENTIFICATION SYSTEM

1. The first two digits indicate the zone and the service area or the fire department. Refer to pages 12 and 13 for this numbering.

Zone 1 = Rancho (north inland)
Zone 2 = Rancho (north coastal)
Zone 3 = San Diego (San Diego FD)
Zone 4 = Heartland (east county)
Zone 5 = South Bay (south county)
Zone 6 = Monte Vista (C.D.F.)
Zone 7 = Inland Volunteer
Zone 8 = Inland Volunteer
Zone 9 = Inland Volunteer

2. The last two digits represent the type of equipment requested:

00-01 = Chief, Assistant & Division Chief
02-10 = Battalion Chiefs
11-29 = Engine Companies
50-59 = Water Tower without ladder
60-69 = Grass Rigs & Tankers
70-79 = Ladder Trucks, Water Trucks without ladder
80-89 = Utility Vehicles
90-99 = Ambulances, Paramedics, and Rescue

An example of the identification system is unit: **1411**

1	- Represents the zone area:	**Rancho**
4	- Represents the department:	**San Marcos**
11	- Represents the type of equipment:	**Engine Co.**

Another example of the identification system is unit: **2594**

2	- Represents the zone area:	**Rancho**
5	- Represents the department:	**Del Mar**
94	- Represents the type of equipment:	**Paramedic**

Note:

Many departments such as "Inland Fire" do not use the old four digit identification system for local calls and routine traffic. Exceptions include incidents requiring a mutual aid response such as a large brush or structure fire. "Rancho Fire" and "South Bay Fire" use the four digit system for both local and mutual aid calls.

SAN DIEGO COUNTY

Heartland Fire Stations & Equipment

Station	Department	Location	Equipment
1	Lakeside	9726 Riverview Rd	E1
2	Lakeside	11211 Valle Vista Rd	E2
3	Lakeside	14008 Business Rd	E3, E3s, M3, R3, B3
4	Santee (HQ)	8950 Cottonwood Ave.	E4, T4, B4, M4, LA4, WT4
5	Santee	9312 Carlton Hills Blvd.	E5, E5s, E205
6	El Cajon (HQ)	100 E Lexington Ave.	E6, T6, M6-1, M6-2
7	El Cajon	695 Tyrone Ave.	E7
8	El Cajon	842 N. Third	E8
9	El Cajon	1301 N. Marshal	E9, R9
10	Lemon Grove (HQ)	7853 Central Ave.	E10, E10s, E210
11	La Mesa (HQ)	8034 Allison Ave.	E11, T11, LS11
12	La Mesa	8834 Dallas St.	E12
13	La Mesa	9910 Grossmont Blvd.	E13, R13, OES192
14	San Miguel	8853 Troy St.	E14
15	San Miguel	2850 Via Orange Way	E15s
16	San Miguel	905 Gillespie Drive	E16, E16s
17	Alpine (HQ)	1834 Alpine Blvd.	E17, R17, E217
18	Unassigned		
19	Bostonia (HQ)	1273 Clarendon St.	E19, R19, A19
20	Santee	Not in Service	E20
21	San Miguel	10105 Vivera Dr.	E21
22	San Miguel	11501 Via Rancho SD	E22, B22
23	San Miguel	2140 Dehesa Rd.	E23
24	Alpine	Not In Service	E24
25	Alpine	Not in Service	E25
26	Lakeside	15245 Oak Creek Rd.	E26

Rancho Fire Stations & Equipment

Station	Location	Equipment
Deer Springs	Deer Springs Rd	E1111, E1112, BR1161
Del Mar	Fairgrounds	E2511, E2512, R2590, R2592
Elfin Forest (Volunteer)	Elfin Forest	E2811, E2812, E2813, BR2861
Encinitas #1 (HQ)	201 "C" Street	E2311, R2396, R2399
Encinitas #2	1867 MacKinnon	E2312, R2393
Encinitas #3	801 Orpheus	E2313, R2392
Encinitas #4	2011 Village Pkwy	E2314, E2315, BR2360
Rancho Santa Fe #1 (HQ)	19363 El Fuego Rd	E2611, BR2661, R2691
Rancho Santa Fe #2	Forest Ranch	E2612, BR2662, R2692
Rancho Santa Fe #3	Fairbanks Ranch	E2613, BR2663, R2693
San Marcos #1 (HQ)	333 Richmar Rd	E1411, E1421, BR1461, R1491
San Marcos #2	1250 Rancho Santa Fe	E1412, BR1462, R1492
San Marcos #3	404 Woodland Pkwy	E1413, R1490, OES 193
Vista #1 (HQ)	175 N. Melrose	E1211, E1212, E1215, BR1261, R1296
Vista #2	1050 Valley Dr.	E1216, BR1262, R1293
Vista #3	1966 E. Vista Way	E1213, R1295, OES195
Vista #4	2121 Thibodo Rd	E1214, BR195, R1294

B = Battalion
BR = Brush
E = Engine

OES = Office of Emergency Services
R = Rescue, Ambulance, Paramedics
T = Fire Truck

SAN DIEGO COUNTY

EMERGENCY MEDICAL SERVICES

AMBULANCE SERVICE

NAME	LOCATION	FREQUENCY	TONE	CALL SIGN
Air Evac Intl.				
Dispatch	San Diego	462.950 R	167.9	N/A
American	San Diego	850 MHz	none	SMR
Balboa	San Diego	856-860.725 T	none	SMR
Hartson's				
(F-1) Primary	San Diego	**155.205**	127.3	WNJK-829
(F-2) Secondary	San Diego	**155.280**	127.3	WNJK-829
(F-3) South Bay	Chula Vista	155.265	114.8	WNJK-829
(F-4) Supervisors	San Diego	155.220	85.4	WNJK-829
(F-5) East County	San Diego	155.385		WNJK-829
Tactical	San Diego	155.160	103.5	KM-3669
Future use	San Diego	150.775		WPAF-756
Paging Services	San Diego	453.075	none	N/A
Life Flight Helicopters*				
(F-1) Dispatch	San Diego	462.975 R	110.9	KNJB-878
(F-2) On Scene Tactical	San Diego	462.975	none	KNJB-878
UCSD Heliport Unicom	San Diego	123.075	none	WAD-9
UCSD Hospital Security	San Diego	154.515	88.5	KUL-276
Schaefer's	Southern Calif.	47.580	none	KQS-952
Schaefer's	Southern Calif.	43.000	none	WNGN-545
Scripps	San Diego	155.295		WYU-846
Sterling (Hartson owns)	San Diego	858.650 R		WNYQ-461

*LIFE FLIGHT HELICOPTERS

Life Flight is an emergency air ambulance that transfers critically ill and injured patients to the hospital. There are 2 full-time helicopters, based at U.C.S.D. Medical Center and Palomar Airport. A third helicopter is available on weekends at Gillespie Field in El Cajon.

The helicopter is equipped as a mobile emergency room, capable of transporting two patients. The helicopter is staffed with a pilot, flight nurse, and either a second flight nurse or a flight paramedic. The medical crew has extensive training in both prehospital and hospital care and in advanced life-saving procedures.

Life Flight is requested by first responders, public safety officers, or health professionals. Their dispatch contacts Life Flight Communication Center and the helicopter is enroute witin five minutes.

Enroute the crew is updated on patient condition and is given landing instruction from the first responders via the radio.

SAN DIEGO COUNTY

EMERGENCY MEDICAL SERVICES

*LIFE FLIGHT HELICOPTERS

While the first responders are waiting for the helicopter, they secure a landing site, extricate the patient, and initiate medical care. Patients are then prioritized (when multiple victims) and made ready for transport by the helicopter.

One first responder is appointed as the ground contact. They talk to the pilot over the radio giving landing site information, winds, obstructions, etc. The ground contact ensures safety at the landing site.

Once Life Flight arrives, the medical crew does a brief assessment, triages patients when needed and performs any necessary advanced procedures.

With the help of the first responders, the patient is loaded into the helicopter and transported to the appropriate hospital. Enroute the crew continues medical care and gives a report to the receiving hospital, updating them on patient status and vital signs.

Requesting Life Flight Service:

1. For serious medical emergencies where patients will have to wait long periods of time for routine ambulance or paramedic transport or when multiple serious injury patients are involved.

2. For serious medical emergencies where patients are located in areas inaccessable to routine ambulance or paramedic service.

3. When emergency medical technicians or paramedics at the scene feel that the helicopter medical team or faster transport will be of medical benefit to the patient.

4. When the patient requires the immediate services of a special or distant hospital or when nearby hospital emergency medical teams are on bypass.

Call Sign Identification:

1. Mt. Laguna KNCG-493
2. Mt. Palomar KNCG-495
3. Los Pinos KNCG-494
4. U.C.S.D. Medical Center KNCG-492
5. Communications Center (U.C.S.D.) KNJB-878

SAN DIEGO COUNTY

EMERGENCY MEDICAL SERVICES

LIFE FLIGHT PRESET CHANNELS*

CHANNEL	FREQUENCY
PRESET 1 - Station "X"	155.025
PRESET B1 - Fire Mutual Aid WHITE 3	**154.295**
PRESET 2 - Station "X"	155.325
PRESET 3 - Station "X"	155.175
PRESET 4 - San Diego Paramedics	153.785
PRESET 5 - Fire Mutual Aid RED	**155.085**
PRESET 6 - Police CLEMARS	154.920
PRESET 7 - Fire Tactical East BROWN	154.250
PRESET 8 - Inland Volunteer Fire	154.175
PRESET 9 - South Bay Fire	154.415
PRESET 10 - San Diego Police Tactical	155.685
PRESET 11 - U.C.S.D. Hospital Security	154.515
PRESET 12 - Fire Tactical North GRAY	**154.355**
PRESET 13 - Hartson's Ambulance	155.205
PRESET 14 - Carlsbad Fire Department	154.385
PRESET 15 - Fire Mutual Aid WHITE 1	**154.280**
PRESET 16 - Hospital Paramedics MED-1 Base	463.000
PRESET 17 - Hospital Paramedics MED-2 Base	463.025
PRESET 18 - Hospital Paramedics MED-3 Base	463.050
PRESET 19 - Hospital Paramedics MED-4 Base	463.075
PRESET 20 - Hospital Paramedics MED-5 Base	463.100
PRESET 21 - Hospital Paramedics MED-6 Base	463.125
PRESET 22 - Hospital Paramedics MED-7 Base	463.150
PRESET 23 - Hospital Paramedics MED-8 Base	463.175
PRESET 24 - L.A. & Orange County Dispatch 1	462.950
PRESET 25 - L.A. & Orange County Dispatch 2	462.975
PRESET 26 - Hospital Paramedics MED-1 Mobile	468.000
PRESET 27 - Hospital Paramedics MED-2 Mobile	468.025
PRESET 28 - Hospital Paramedics MED-3 Mobile	468.050
PRESET 29 - Hospital Paramedics MED-4 Mobile	468.075
PRESET 30 - Hospital Paramedics MED-5 Mobile	468.100
PRESET 31 - Hospital Paramedics MED-6 Mobile	468.125
PRESET 32 - Hospital Paramedics MED-7 Mobile	468.150
PRESET 33 - Hospital Paramedics MED-8 Mobile	468.175
PRESET 34 - L.A. & Orange County Mobile	467.950
PRESET 35 - L.A. & Orange County Mobile	467.975

Calcord (156.075) is used for air-to-ground contacts along the International border and in Mexico.

SAN DIEGO COUNTY

EMERGENCY MEDICAL SERVICES

AMERICAN RED CROSS

CHANNEL	FREQUENCY
CH-1 National Disaster Net	47.420
CH-2 Southern California	47.500
CH-3 Southern California	47.460
CH-4 Southern California	47.580
Liaison with REACT	462.675 R
Portable Repeaters	463.0125R
Portable Repeaters	463.1125R
Portable Radios	154.570
Portable Radios	154.600
Repeaters	**463.475 R**
Repeaters	**463.500 R**
Southern California	462.9375
Southern California	462.9625
Southern California	462.9875
Southern California	463.0125
Southern California	463.0375
Southern California	463.0625
Southern California	463.0875

San Diego Headquarters

South Bay Base (Chula Vista)
East County Base (El Cajon)
North Coastal Base (Oceanside)
North Inland Base (Escondido)

Unit Identification:

9479	MSS Vehicle - Camp Pendleton
9480	MSS Vehicle - Camp Pendleton
9481	MSS Vehicle - San Diego
9482	Truck - Dodge Ram
9483	Truck - Ford Courier
9484	Truck - Ford Custom F250
9485	Wagon - Motor Pool
9486	Wagon - Motor Pool
9487	Wagon - Disaster Services
9488	Wagon - Assistance Executive Director
9489	Wagon - Coronado
9490	DHS Vehicle
9491	Disaster Van - Imperial Valley
9492	Disaster Van - North Coastal
9493	Disaster Van - North Inland
9495	Disaster Van - South Bay
9496	Disaster Van - San Diego
9497	POV Mass Care
9498	POV Disaster Response
9499	POV Disaster Response

SAN DIEGO COUNTY

EMERGENCY MEDICAL SERVICES

HOSPITALS

NAME	LOCATION	FREQUENCY	CALL SIGN
Alvarado Community - Ops	San Diego	151.685	KNCQ-805
Alvarado Community - Paging	San Diego	453.125	WNJI-921
Balboa Naval - Medics	San Diego	143.525	USN
Bay General - Paging	Chula Vista	35.640	N/A
Bay General - Security	Chula Vista	464.675 R	WSS-267
Childrens - Paging	San Diego	31.040	N/A
Childrens - Paging	San Diego	154.625	KNGB-805
Chula Vista Community	Chula Vista	35.640	KNFR-796
Community Hospital	San Diego	35.640	N/A
Coronado	Coronado	463.425 R	WZK-832
El Cajon Valley	El Cajon	464.525 R	WSF-816
Grossmont	La Mesa	461.675 R	KMD-906
Grossmont	La Mesa	464.375 R	KMD-906
Hillside - Ops	San Diego	154.515	N/A
Kaiser - Paging	La Jolla	72.260	KNIN-887
Kaiser - Paging	San Diego	163.250	KZE-775
Kaiser - Security	San Diego	464.875	KB-74411
Mercy - Paging	San Diego	163.250	KOL-864
Mercy - Security	San Diego	464.525 R	KCH-700
Mesa Vista - Security	San Diego	154.600	KB-67411
Mesa Vista - Paging	San Diego	157.450	WYX-535
Mesa Vista - Ops	San Diego	463.550	N/A
Mesa Vista - Ops	San Diego	464.575	N/A
Palomar Memorial - Ops	Escondido	462.575 R	KAD-2887
Palomar Memorial - Ops	Escondido	462.725	KAD-2887
Palomar Memorial - Security	Escondido	457.525-457.600	KD-5964
Paradise Valley - Paging	National City	151.745	N/A
Paradise Valley - Ops	National City	155.355	N/A
Paradise Valley	National City	463.5625	WNDT-568
Paradise Valley	National City	464.4125	WNDT-568
Paradise Valley	National City	464.425	WNDT-568
Pomerado Memorial	Poway	464.375	WSX-567
San Miguel - Paging	San Diego	154.515	KNBJ-366
Scripps Clinic - Security	Borrego Springs	154.540	KNGX-493
Scripps Clinic - Operations	La Jolla	155.295	WYU-846
Scripps Clinic - Paging	San Diego	157.450	KRM-802
Scripps Memorial - Paging	Chula Vista	35.640	KNBS-893
Scripps Memorial - Paging	La Jolla	453.075	KNAK-758
Scripps Memorial	La Jolla	464.425	KSW-333
Scripps Memorial	La Jolla	464.525	KSW-333
Sharp Memorial - Ops	San Diego	151.655	N/A
Sharp Memorial - Ops	San Diego	151.955	N/A

SAN DIEGO COUNTY

EMERGENCY MEDICAL SERVICES

HOSPITALS

NAME	LOCATION	FREQUENCY	CALL SIGN
Sharp Memorial - Paging	San Diego	453.175	KNDZ-306
Sharp Memorial - Ops	San Diego	463.550 R	KNBX-397
Sharp Memorial - Ops	San Diego	464.325 R	KNBX-397
Tri-City - Paging	Oceanside	151.745	KQV-206
Tri-City - Operations	Oceanside	464.425 R	WNBD-746
U.C.S.D. - Paging	San Diego	152.0075	KOM-314
U.C.S.D. - Security	San Diego	154.515	KUL-276
U.S. Naval - Security	San Diego	148.725 R	USN
U.S. Naval - Paging	San Diego	163.000	USN
Veterans - Security (F-1)	La Jolla	168.525	USN
Veterans - Security (F-2)	La Jolla	168.625	USN
Veterans - Paging	San Diego	173.3875	USN

PARAMEDICS

CHANNEL	BASE FREQUENCY	MOBILE FREQUENCY
MED-1	463.000	468.000
MED-2	463.025	468.025
MED-3	463.050	468.050
MED-4	463.075	468.075
MED-5	463.100	468.100
MED-6	463.125	468.125
MED-7	463.150	468.150
MED-8	463.175	468.175

CHANNEL ASSIGNMENTS

HOSPITAL	LOCATION	CHANNEL	TONE
Bay General	Chula Vista	MED-8	162.2
El Cajon Valley	El Cajon	MED-5	162.2
Grossmont	La Mesa	MED-3	162.2
Mercy	San Diego	MED-2	162.2
Palomar Memorial	Escondido	MED-8	162.2
Paradise Valley	National City	MED-4	162.2
Scripps Encinitas	Encinitas	MED-5	162.2
Scripps Memorial	San Diego	MED-1	162.2
Sharp Memorial	San Diego	MED-6	162.2
Tri City Medical Center	Oceanside	MED-4	162.2
U.C.S.D. Medical Center	San Diego	MED-7	162.2
U.S. Naval	San Diego	MED-8	162.2

SAN DIEGO COUNTY

EMERGENCY MEDICAL SERVICES

SEARCH & RESCUE

NAME	FREQUENCY	CALL SIGN
County of San Diego (F-1)	**155.160**	KN-2871
County of San Diego (F-2)	155.235	KN-2871
East County Rescue	155.205	KZB-469
R.E.A.C.T. Teams	**462.675**	varies
Sheriff S.& R. (F-1)	**155.160**	KN-2871
Sheriff S.& R. (F-2)	153.995	KM-3669
Sheriff S.& R.	453.100	KA-88385

Station "R" = Sheriff Search & Rescue
PL Tones: 103.5, 110.9, 146.2 Htz

STATION "X"

CHANNEL	USE	FREQUENCY
1	EMS Coordination	155.025
2	Hospital - Paramedics	**155.175**
3	Hospital - Paramedics	**155.325**
4	Hospital - Paramedics	155.385
5	EMS Tactical	155.160
6	H.E.A.R.	155.340
none	EMS Tactical	155.235

HEAR = Hospital Emergency Administrative Radio

Hospitals participating in HEAR are located in Los Angeles, Orange, and Riverside Counties.

PL Tones: 103.5, 107.2 Htz

SAN DIEGO COUNTY

UTILITY SYSTEMS

AIR POLLUTION CONTROL DISTRICT

CHANNEL	FREQUENCY
Operations A - Handie-talkies	453.1125
Operations B - Handie-talkies	453.2375
Operations C - Handie-talkies	453.2875
Operations D - Handie-talkies	453.5375
Operations E - Handie-talkies	453.6875

PL Tone: 107.2 Htz

ANIMAL REGULATIONS OFFICE

CHANNEL	FREQUENCY
(F-1) North & South Shelters	**45.700**
(F-2) Central units	**45.900**
(F-3) Car-to-Car	45.380

Call Sign Identification:

Cuyamaca Peak KCW-737; Los Pinos KSU-377; Lyons Peak KMJ-521; Mount Laguna; Mount Palomar KSV-775; Mt. Woodson WNDB-644; San Diego KQT-977/KYQ-366.

PL Tone: 107.2 Htz

CIVIL DEFENSE

CHANNEL	FREQUENCY
(F-1) Operations Repeaters (input 37.900)	**37.980 R**
(F-2) Operations Direct	37.980

Call Sign Identification:

Coronado WNHR-421; El Cajon KNDH-888/WZX-777; Lakeside WZX-778; Los Pinos KWL-410; Lyons Peak KMK-862; North Peak KWL-409; Mount Laguna KMK-863; Mount Palomar KMK-861; Mount Woodson KNBC-930; National City KNJL-490; San Diego KNCK-638/KNCN-221/KNEE-240 KNEN-702/WNGX-493/WRA-403/WZJ-521

PL Tones: 114.8, 146.2 Htz

SAN DIEGO COUNTY

UTILITY SYSTEMS

DISTRICT ATTORNEY'S OFFICE

CHANNEL	FREQUENCY
(F-1) Operations Repeaters (input 159.135)	**155.715 R**
(F-2) Operations Direct	155.715
(F-3) District Attorney Tactical	159.030
(F-4) Marshal Tactical	158.865
(F-5) County Mutual Aid BLUE	153.995
(F-6) Mutual Aid CLEMARS	154.920
(F-7) Mutual Aid NLEMARS	155.475
(F-8) Statewide Calcord	156.075
San Diego Courthouse Security (input 156.210)	158.865 R

Call Sign Identification:

Alpine WHF-97; Cardiff KWE-632/WGL-424/WGW-64; Del Mar WGW-73; Lomas Santa Fe KWE-633; Los Pinos KNFX-734; Lyons Peak KQS-632; Mount Laguna; Mount Palomar; North Peak; Rancho Santa Fe WGR-412 WGW-74; San Diego KNAO-453/KQU-290/KTO-348/WBZ-312; San Marcos Peak KRA-340; Solana Beach WGW-75; Spring Valley WGL-423/WGW-76 WGW-77; Vista WGW-78.

PL Tone: 103.5 Htz

EMS COMMUNICATIONS SYSTEM

CHANNEL	FREQUENCY
(F-1) California State OES Region 4	**46.320 R**
(F-2) San Diego County Primary	**46.340 R**
Countywide EMS Communications	46.060
Countywide EMS Communications	46.140
Countywide EMS Communications	46.260
Countywide EMS Communications	46.280
Countywide EMS Communications	46.420
Countywide EMS Communications	46.440
Countywide EMS Communications	46.480

PL Tones: 103.5, 146.2 Htz

Repeater Pairs: 44.920 input for 46.320 output
44.640 input for 46.340 output

SAN DIEGO COUNTY

UTILITY SYSTEMS

GENERAL SERVICES DEPARTMENT

CHANNEL	FREQUENCY
(F-1) Primary Operations	**155.820**
(F-2) Pagers / Secondary (input 154.965)	**155.940 R**
(F-3) County Mutual Aid BLUE	153.995
(F-4) Fire Mutual Aid RED	155.085
Borrego Springs-Monument Peak Link	72.840 R
Mobile Data Terminals & Radio Technicians	861.150 R
Mobile Data Terminals	862.150 R
Mobile Data Terminals	863.150 R

Call Sign Identification:

Alpine KDJ-504; Campo WNGX-494/WNHR-829; Carlsbad KSO-646; Cuyamaca Peak KDJ-501/KMJ-736; Dulzura KMJ-945; Lake Henshaw KFM-381/KNBL-339; Los Pinos KNGZ-314; Moreno Village KNFL-979; Mount Laguna KMK-274; Mount Palomar KMJ-735; San Diego KIB-444 KMJ-734/KNAO-453/KTO-348; Solana Beach WNDB-632

Code Identification:

ADAM - Security Agents
ALPHA - Juvenile Court
BAKER - Adult Supervision
DELTA - General Services
GOLF - Parks and Recreation
NOVEMBER - Agriculture
OMEGA - County Med. Examiner
PROBE - Probation
REBEL - Flood & Sanitation
RED CROSS - Red Cross
TANGO - Humane Society
YANKEE - Office of Disaster Preparedness

PL Tones: 103.5, 107.2 Htz

LIFEGUARDS

CHANNEL	FREQUENCY
(F-1) Lifeguards 1 - North Area	**155.820**
(F-2) Lifeguards 2 - South South	**155.940**
(F-3) County Mutual Aid BLUE	153.995
(F-4) Fire Mutual Aid RED	155.085

Call Sign Identification:

Solana Beach WNDB-632

PL Tone: 107.2 Htz

SAN DIEGO COUNTY

UTILITY SYSTEMS

MARSHAL'S OFFICE

CHANNEL	FREQUENCY
(F-1) Operations Repeaters (input 159.135)	**155.715 R**
(F-2) Operations Direct	155.715
(F-3) Tactical	158.865
(F-4) Tactical	159.030
(F-5) Mutual Aid BLUE	153.995
(F-6) Mutual Aid CLEMARS	154.920
(F-7) Mutual Aid NLEMARS	155.475
(F-8) Statewide Calcord	156.075

PL Tone: 103.5 Htz

PUBLIC WORKS DEPARTMENT

CHANNEL	FREQUENCY
(F-1) Operations Primary	**173.390**
(F-2) Operations Secondary	173.210

Call Sign Identification:

Alpine WNHI-272; Cardiff WNHI-252/WNHI-266/WNHI-268/WNHI-270;
Del Dios WNHI-266/WNHI-250; Del Mar WNHI-259; Encinitas WNHI-269;
Lomas Santa Fe WNHI-267; Rancho Santa Fe WNHI-246/WNHI-248/WNHI-249;
WNHI-251/WNHI-253/WNHI-254/WNHI-261/WNHI-262/WNHI-263/WNHI-264;
Solana Beach WNHI-258, Spring Valley WNHI-245/WNHI-255-257, Vista WNHI-260

PL Tone: 103.5 Htz

ROAD DEPARTMENT

CHANNEL	FREQUENCY
(F-1) Highway Net 1	156.120
(F-2) Highway Net 2	156.225
(F-3) Highway Net 3	156.165

Call Sign Identification:

Cuyamaca Peak KMC-235; Monument Peak KMJ-737; Mount Israel KVV-892;
Mount Laguna KMJ-738; Mount Palomar KMC-234/KWH-633; Mount Woodson
KMC-240; San Diego KMC-242; San Marcos Peak KMJ-739/KRX-391;
Spring Valley KMC-239; Valley Center KCJ-941.

PL Tone: 107.2 Htz

SAN DIEGO COUNTY

CARLSBAD

CHANNEL	INPUT	OUTPUT	TONE	CALL SIGN
Police Dispatch (F-1)	155.250	**156.030 R**	103.5	KNHU-653
Police Car-to-Car (F-2)		155.835	103.5	KTT-848
Police Oceanside PD (F-3)		155.055	103.5	KNHU-653
Police Direct (F-4)		156.030	103.5	KNHU-653
Police Mutual Aid BLUE (F-5)		153.995	103.5	KTT-848
Police Tactical (F-6)		155.250	none	KNHU-653
Police CLEMARS (F-7)		154.920	103.5	KNJD-987
Police Fire RED (F-8)		155.085	103.5	KTC-877
Police Fire Liaison (F-9)		154.385	127.3	KTC-821
Police NLEMARS (F-10)		**155.475**	103.5	KNJD-987
Police Calcord (F-11)		156.075	none	KTT-848
Fire & Medics (2200)		**154.385**	127.3	KTC-821
Fire Tactical GRAY		154.355	114.8	KM-8396
Fire Command RED		155.085	103.5	KTC-877
Local Government (Building & Engineering) (Street Department)		45.600	103.5	KFR-737
Agua Hedionda Boat Patrol		156.850	none	N/A
School District - Buses	813.500	858.500 R	D311	WNGK-907
School District - Security		154.600	141.3	KB-71643
Water District		37.760	136.5	WSU-605

POLICE UNIT IDENTIFICATION

100-199 = Patrol units (1st Shift)
200-299 = Patrol units (2nd Shift)
300-399 = Patrol units (3rd Shift)
400 = Patrol Captain
401-499 = Patrol units (all shifts)
500 = Lieutenant
501-599 = Investigators
600 = Chief Investigator
601-699 = Administration
700 = Reserve Commander
701-799 = Reserve units
800 = Traffic Sergeant
801-830 = Traffic units
840-850 = Off Road Vehicles
912 = K-nine unit
914 = K-nine unit

Some units identify with a letter:

L = Lifeguards M = Motorcycle V = Parking

CARLSBAD POLICE PATROL AREAS

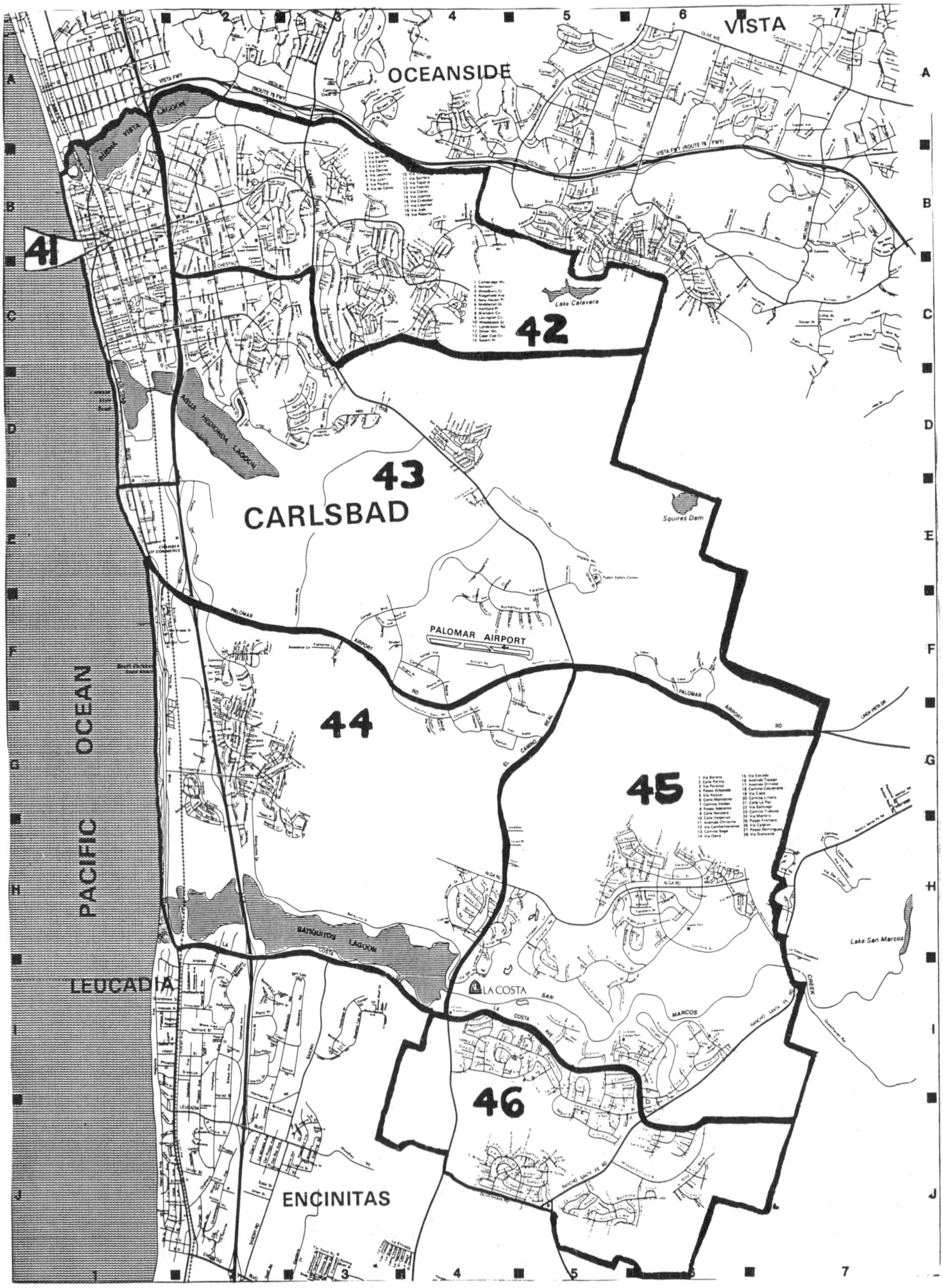

SAN DIEGO COUNTY

CHULA VISTA

CHANNEL	INPUT	OUTPUT	TONE	CALL SIGN
Police Primary (F-1)	158.835	**154.815 R**	103.5	KMG-284
Police CLEMARS (F-2)		154.920	103.5	KMG-284
Police Tactical (F-3)		158.925	103.5	KBU-410
Police Inquiry & Tactical (F-4)	155.640	155.595 R	103.5	KMG-284
Police Tactical (F-5)		155.640	103.5	KMG-284
Police Mutual Aid BLUE (F-6)		153.995	103.5	KMG-284
Police Direct (F-7)		154.815	103.5	KMG-284
Police Fire (F-8)		154.415	103.5	KRY-974
Police Fire YELLOW (F-9)		153.890	127.3	KRY-974
Police Calcord (F-10)		156.075	none	KMG-284
Police Surveillance Reported		153.965	none	none
Fire & Medics (5200)		154.415	103.5	KRY-974
Fire Mutual Aid WHITE 1		154.280	103.5	KRY-974
Fire Tactical YELLOW		153.890	127.3	KRY-974
Local Government (Engineering, Streets)	458.225	453.225 R	103.5	WNZE-511
Local Government (Animal Control)		155.865	103.5	N/A
School District Buses		155.265		KMA-344
School District Police		155.625	114.8	KMA-344

CORONADO

CHANNEL	INPUT	OUTPUT	TONE	CALL SIGN
Police Primary (F-1)	154.025	**154.845 R**	103.5	KMB-596
Police Direct (F-2)		154.845	103.5	KMB-596
Police Local Govt. (F-3)		154.115	103.5	KMB-596
Police Mutual Aid BLUE (F-4)		153.995	103.5	KVN-701
Police CLEMARS (F-5)		154.920	103.5	KTG-608
Police Tactical (F-6)		158.745	103.5	KMB-596
Police Tactical RED		155.250	103.5	KMB-596
Fire & Medics (5300)		**154.415**	103.5	KSR-889
Fire Mutual Aid WHITE 1		154.280	103.5	KSR-889
Fire Tactical YELLOW		153.890	127.3	KB-85097
Fireground		154.115	103.5	KVD-773
Local Government (Engineering) (Street Department)		39.900	74.4	N/A
Local Government (Enforcement Detail) (Lifeguards)		154.115	103.5	KVD-773
Port Operations VHF 12		156.600	none	KYR-567
School District - Buses		155.295	118.8	WNLG-556

SAN DIEGO COUNTY

DEL MAR

CHANNEL	INPUT	OUTPUT	TONE	CALL SIGN
Sheriff (25)	458.750	**453.750 R**	146.2	KNCJ-415
Fire & Medics (2500)		**154.235**	192.8	WNDX-572
Fire San Diego (F)		850 MHz	none	KMA-357
Fire Command RED		155.085	103.5	KRK-962
Fire Tactical GRAY		154.355	114.8	KM-8396
Local Government		45.120	156.7	WQR-663
(Lifeguards & Parks)				
(Public Works)				

EL CAJON

CHANNEL	INPUT	OUTPUT	TONE	CALL SIGN
Police Dispatch (F-1)	154.890	**155.625 R**	131.8	KDA-652
Police Inquiry & Tactical (F-2)	156.090	155.310 R	131.8	KDA-652
Police Tactical (F-3)		155.625	131.8	KDA-652
Police Tactical (F-4)		155.310	131.8	KDA-652
Police CLEMARS (F-5)		154.920	103.5	KMJ-866
Police Mutual Aid BLUE (F-6)		153.995	103.5	WNHG-583
Police Tactical (F-7)		155.790	131.8	KDA-652
Police NLEMARS (F-8)		155.475	103.5	KDA-652
Police Local Government (F-9)		155.985	131.8	WQN-933
Heartland Fire System	811.175	**856.175 T**	none	WNQA-931
Heartland Fire System	811.200	**856.200 T**	none	WNQA-931
Heartland Fire System	812.200	**857.200 T**	none	WNQA-931
Heartland Fire System	813.175	**858.175 T**	none	WNQA-931
Heartland Fire System	813.200	**858.200 T**	none	WNQA-931
Heartland Fire System	814.200	**859.200 T**	none	WNQA-931
Heartland Fire System	815.200	**860.200 T**	none	WNQA-931
Heartland Fire "Conventional 1"	815.225	860.225 R	none	WNQA-931
Heartland Fire "Conventional 2"	811.175	856.175 R	none	WNQA-931
Local Government		155.985	131.8	WQN-933
(Building, Engineering & Streets)				
(Code Enforcement)				
(Parks & Recreation)				

POLICE UNIT IDENTIFICATION

100-199 = Graveyard Patrol Shift
200-299 = Day Patrol Shift
300-399 = Afternoon Patrol Shift
400-499 = 5:30-12:00 PM Patrol Shift
700-799 = Administrative units
800-899 = Detectives & Narcotics
900-999 = Traffic units
1000's = Special Enforcement

El Cajon PD is in the process of joining the Heartland Fire Radio System.

SAN DIEGO COUNTY

ENCINITAS

CHANNEL	INPUT	OUTPUT	TONE	CALL SIGN
Sheriff (23)	458.750	**453.750 R**	146.2	KNCJ-418
Fire & Medics (2300)		**154.235**	192.8	KRE-368
Fire Command RED		155.085	103.5	KRF-399
Fire Tactical GRAY		154.355	114.8	KM-8396
Lifeguards North	154.965	155.940 R	107.2	WNDB-632
Union High School		151.925	123.0	KZS-283

ESCONDIDO

CHANNEL	INPUT	OUTPUT	TONE	CALL SIGN
POLICE / FIRE / PUBLIC WORKS				
Municipal Trunked System	816.250	**861.250 T**	none	WNNO-8837
Municipal Trunked System	817.250	**862.250 T**	none	WNNO-837
Municipal Trunked System	818.250	**863.250 T**	none	WNNO-837
Municipal Trunked System	819.250	**864.250 T**	none	WNNO-837
Municipal Trunked System	820.000	**865.000 T**	none	WNNO-837
Municipal Trunked System	820.250	**865.250 T**	none	WNNO-837
Talk-around & Secondary	816.000	**861.000 R**	none	WNQI-406
Mobile Data Terminals	819.150	864.150 R	none	WNQI-406
POLICE INTERAGENCY				
Police Dispatch - Failsoft	818.250	863.250 R	none	WNNO-837
Police Interagency Comm 1		866.5125	156.7	N/A
Police Interagency Comm 2		867.0125	156.7	N/A
Police Interagency Tac 1		867.5125	156.7	N/A
Police Interagency Tac 2		868.0125	156.7	N/A
Police Mutual Aid BLUE		153.995	103.5	WYV-340
Police Mutual Aid CLEMARS		154.920	103.5	KSK-436
Police Mutual Aid CLEMARS	823.5125	868.5125R	156.7	N/A
Police Mutual Aid CLERS	458.675	453.675 R	103.5	WDN-285
FIRE MUTUAL AID				
Escondido Dispatch - Failsoft	817.250	862.250 R	none	WNNO-837
Escondido Fire - VHF GREEN		**154.325**	103.5	KRM-691
Statewide Mutual Aid WHITE 1		154.280	103.5	KRM-691
Statewide Mutual Aid WHITE 2		154.265	103.5	KMR-691
Statewide Mutual Aid WHITE 3		154.295	103.5	KRM-691
County Fire Tactical GRAY		154.355	114.8	KRM-691
County Fire Command RED		155.085	103.5	KRL-425
UNIFIED SCHOOL DISTRICT				
Paging Services		35.960	none	KNEB-849
High Schools - Security		151.835	167.9	KTH-839
High Schools - Buses	811.475	856.475 R	none	WNUT-702
Transportation		151.865		KTH-839

SAN DIEGO COUNTY

FALLBROOK

CHANNEL	INPUT	OUTPUT	TONE	CALL SIGN
Sheriff (32)	458.925	**453.925 R**	146.2	KNCJ-417
Fire & Medics (1000)		**154.235**	114.8	KRE-373
Fire Secondary YELLOW		153.890	114.8	KMA-229
Fire Tactical BROWN		153.815	114.8	KRE-373
Local Government		153.815	103.5	KML-205
(Public Works)				
(Street Department)				
Union High School		155.265	110.9	WQU-361
Water & Utilities		173.250	192.8	WNLY-901

IMPERIAL BEACH

CHANNEL	INPUT	OUTPUT	TONE	CALL SIGN
Sheriff (71)	458.950	**453.950 R**	146.2	KNCJ-408
Coronado Police (F-1)	155.250	154.845 R	103.5	KMB-596
Coronado Police (F-2)		154.845	103.5	KMB-596
Fire & Medics (5400)		**154.415**	103.5	KRY-974
Fire Tactical YELLOW		153.890	114.8	KM-8396
Local Government		155.745	103.5	WQC-632
(Enforcement Detail)				
(Lifeguards)				
(Public Works)				
School District - Buses		155.265		WXP-422

LA MESA

CHANNEL	INPUT	OUTPUT	TONE	CALL SIGN
Police Primary (F-1)	156.510	**153.845 R**	103.5	KDX-347
Police CLEMARS (F-2)		154.920	103.5	KUN-548
Police Mutual Aid BLUE (F-3)		153.995	103.5	KVT-565
Police Direct (F-4)		153.845	103.5	KDX-347
Police Car-to-Car (F-5)		**155.145**	103.5	KDX-347
Police CLERS	458.675	453.675 R	103.5	WFK-767
Fire & Medics - Heartland Fire System		850 MHz	none	WNQA-931

POLICE UNIT IDENTIFICATION

100-199 = Patrol units (1st Watch)
200-299 = Patrol units (2nd Watch)
300-399 = Patrol units (3rd Watch)
401-403 = Traffic units
451-452 = Animal Control Officers
500-599 = Reserve units
711-715 = Investigators
751-759 = School Liaison Officers

SAN DIEGO COUNTY

LA MESA

CHANNEL	INPUT	OUTPUT	TONE	CALL SIGN
Local Government (Parks & Recreation) (Street Department)		155.145	103.5	KDX-347
Unified School Security		154.515	146.2	KNDV-858
Unified School Transportation	819.000	864.000 R	none	WNHS-767

LEMON GROVE

CHANNEL	INPUT	OUTPUT	TONE	CALL SIGN
Sheriff (61)	458.950	**453.950 R**	146.2	KNCJ-411
Fire & Medics - Heartland Fire System		850 MHz	none	WNQA-931
Local Government (Building & Engineering) (Street Department)		45.280	103.5	WQW-919

NATIONAL CITY

CHANNEL	INPUT	OUTPUT	TONE	CALL SIGN
Police Fire Trunking System	816.250	**861.250 T**	none	WPBD-652
Police Fire Trunking System	817.250	**862.250 T**	none	WPBD-652
Police Fire Trunking System	818.250	**863.250 T**	none	WPBD-652
Police Fire Trunking System	819.250	**864.250 T**	none	WPBD-652
Police & Mall Security	820.650	865.250 R	none	WPBD-652
Police CLEMARS		154.920	103.5	KUB-816
Police Mutual Aid BLUE		153.995	103.5	KVT-538
Police VHF		155.790	103.5	KMA-652
Fire VHF Dispatch (5700)		**154.415**	103.5	KRY-974
Fire Tactical YELLOW		153.890	127.3	KM-8396
Local Government (Building Department) (Parks & Recreation) (Street Department)		158.805	103.5	KLU-302
Bookmobile	465.3375	460.3375R	D051	WNXB-807
Engineering Department		37.180	141.3	WNJM-713
Friendship Manor Police Dept.	465.525	460.525 R		WNHU-995
Sweetwater Union High School	812.425	857.425 R	79.9	WNBY-264

SAN DIEGO COUNTY

OCEANSIDE

CHANNEL	INPUT	OUTPUT	TONE	CALL SIGN
Police Dispatch (F-1)	156.045	**155.055 R**	103.5	KUI-649
Police Secondary (F-2)	155.655	154.755 R	103.5	KNGL-579
Police Car-to-Car (F-3)		**154.865**	103.5	KB-24500
Police CLEMARS (F-4)		154.920	none	KNIV-534
Police Mutual Aid BLUE (F-5)		153.995	103.5	KO-8575
Police Detectives (F-6)		**155.565**	103.5	KNGL-579
Police Tactical (F-7)		155.415	103.5	KIU-649
Police Tactical (F-8)		155.490	103.5	KUI-649
Police Harbor Patrol (F-9)		156.015	103.5	KJP-496
Police NLEMARS (F-10)		155.475	none	KUI-649
Police Future Repeater	158.745	154.025 R	103.5	KIU-649
Police MDT's	819.150	864.150 R	none	WNSF-262
Fire & Medics (2100)		**159.165**	127.3	KTS-539
Fire Command RED		155.085	103.5	KTV-622
Fire Tactical GRAY		154.355	114.8	KTS-539
Lifeguards		155.880	103.5	KA-40955
Local Government (Humane Society) (Parks & Recreation) (Street Department) (Water Department)		**155.760**	103.5	WQN-944
Mira Costa Community College	469.575	464.575 R	127.3	KNBR-497
Public Library Data	458.1375	453.1375R	none	KD-41122
Small Craft Patrol		156.015	103.5	KJP-496
School District Buses		155.295	D032	WRD-858
School District Security		151.805		KA-56992
Utilities		45.120	103.5	KWV-739
Water Dept.- Point to Point	928.4125	952.4125R	none	WNKE-784

OCEANSIDE POLICE IDENTIFICATION

1. The first number digits generally correspond with the patrol areas and are followed by letters which define the shift or type of unit:

A = Day Shift Patrol
B = Evening Shift Patrol
C = Unassigned
D = Morning Shift Patrol
K = Canine
L = Administration
M = Motorcycle
N = Narcotic
O = Beach Patrol
U = 2-Man Unit
V = Parking

2. Some units have special identification:

200-299 = Investigators
300-399 = Harbor Patrol
700-799 = Beach Patrol

OCEANSIDE POLICE PATROL AREAS

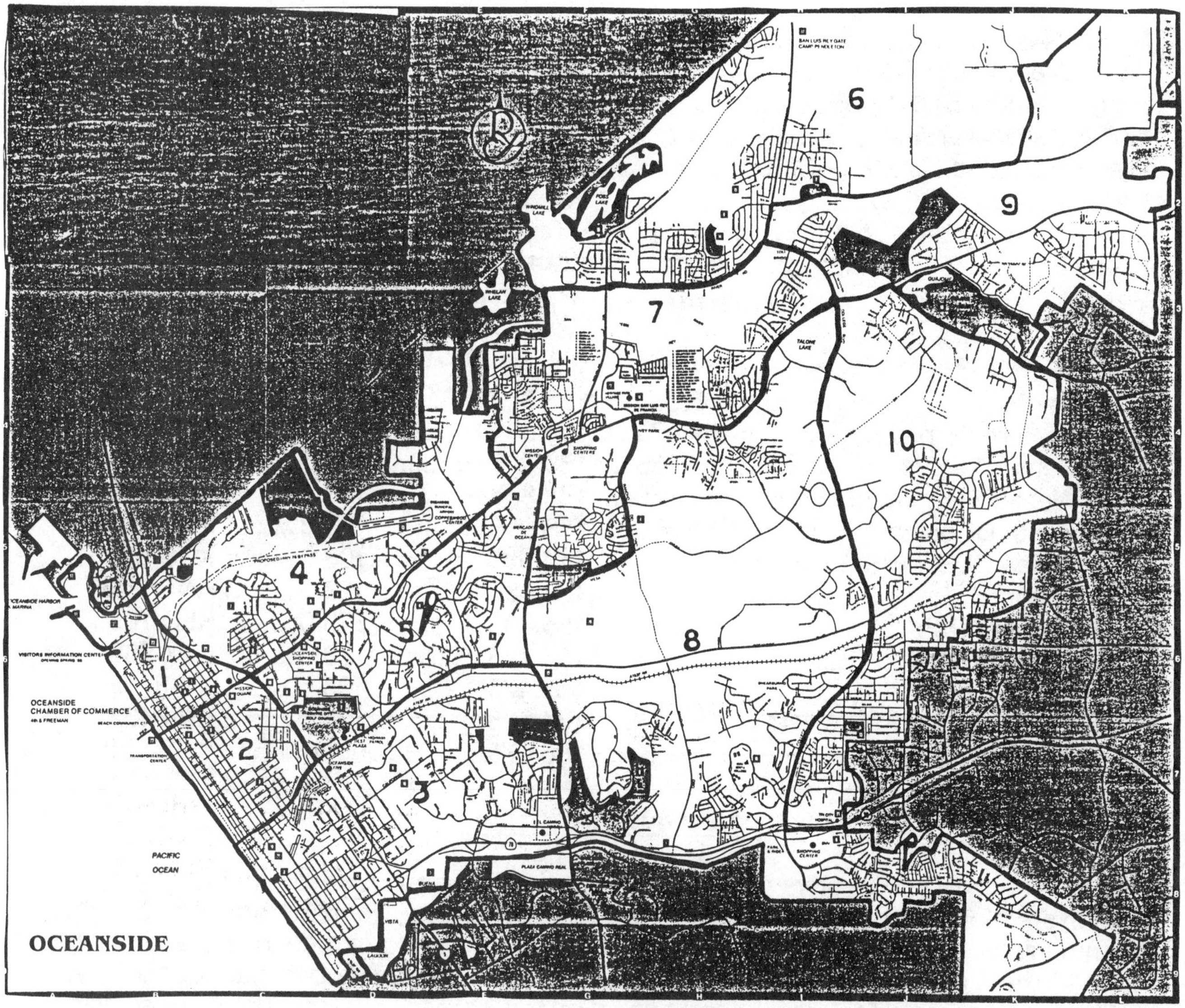

SAN DIEGO COUNTY

POWAY

CHANNEL	INPUT	OUTPUT	TONE	CALL SIGN
Sheriff (41)	458.500	**453.500 R**	146.2	KNCJ-419
Fire & Medics SDFD (57)		800 MHz	none	KNJJ-402
Unified School District - Buses	820.150	865.150 R	none	N/A
Unified School District - Security		151.895	D051	KNAV-340
Water & Utilities		153.635		KVD-799

SAN DIEGO POLICE DEPARTMENT

CHANNEL	INPUT	OUTPUT	TONE	CALL SIGN
(F-1) Southeastern Dispatch	154.950	158.730 R	123.0	KDJ-935
(F-2) Northeastern Dispatch	155.550	158.970 R	123.0	KMD-727
(F-3) Eastern Dispatch	154.875	159.090 R	123.0	KLI-385
(F-4) Central Dispatch	155.535	159.045 R	123.0	KXK-366
(F-5) Western Dispatch	155.370	158.895 R	123.0	KFE-583
(F-6) South Dispatch	154.650	158.910 R	123.0	KNBF-348
(F-7) Police CLEMARS		**154.920**	103.5	KUS-510
(F-8) Inquiry	154.785	154.055 R	123.0	KNFH-736
(F-9) Tactical "Tac"		155.685	123.0	KLD-705
(F-10) Detectives		154.725	123.0	KTX-765
(F-11) Police NLEMARS		155.475	103.5	KB-8432
(F-12) Police CLEMARS & SWAT		154.935	123.0	KB-8432
(F-13) Statewide CALCORD		156.075	none	KUS-510

New 800 MHz System

In the spring of 1990, voters approved proposition E, a bond measure to fund a new $35 million state-of-the-art radio system for San Diego police, fire, lifeguard, and school district communications. The shift to higher frequencies was done to reduce interference (some from Mexico) and utilize the many advantages of trunking and mobile data technology.

San Diego paramedics are fully operational on the 800 MHz system. The new radios are scheduled to be installed in all police and fire vehicles in 1994. Until all units have been equipped with the new radios and all system problems are resolved, transmissions will be simulcast on the old VHF channels and the new 800 MHz frequencies.

The Motorola system currently has two trunking groups on seventeen voice frequencies. Group 1 is set aside for general dispatching (police, fire, school buses). The control channel, some police, and all fire dispatches are found in Group 2. Otherwise there doesn't appear to be any specific pattern to the frequencies that will aid the listener but monitoring the current system is not incredibly difficult, but does require some effort. It is hoped that the dispatches will be placed in a conventional mode for the general public sometime in the future. Software and interfacing equipment for monitoring a trunked system may also be available on the market.

Specific members of the newsmedia may buy a Motorola radio to monitor the new system, but the cost is over twelve hundred dollars and the city may remotely lock out any channel to thwart listening.

SAN DIEGO COUNTY

SAN DIEGO POLICE FIRE SYSTEM

CHANNEL	INPUT	OUTPUT	TONE	SITES
GROUP 1				
- Dispatch	811.025	**856.025 T**	none	1,2,3,4,5
- Dispatch	811.050	**856.050 T**	none	1,2,3,4,5
- Dispatch	812.000	**857.000 T**	none	1,2,3,4,5
- Dispatch	812.025	**857.025 T**	none	1,2,3,4,5
- Dispatch	813.000	**858.000 T**	none	1,2,3,4,5
- Dispatch	814.000	**859.000 T**	none	1,2,3,4,5
- Dispatch	815.000	**860.000 T**	none	1,2,3,4,5
- Dispatch	817.050	**862.050 T**	none	1,2,3,4,5
- Dispatch	817.100	**862.100 T**	none	1,2,3,4,5
- Dispatch	818.050	**863.050 T**	none	1,2,3,4,5
GROUP 2				
- Control, FD & PD Tactical	814.025	**859.025 T**	none	1,2,3,4,5,6,7,8
- Control, FD & PD Tactical	814.050	**859.050 T**	none	1,2,3,4,5,6,7,8
- Control, FD & PD Tactical	815.025	**860.025 T**	none	1,2,3,4,5,6,7,8
- Control, FD & PD Tactical	815.050	**860.050 T**	none	1,2,3,4,5,6,7,8
- FD & PD Tactical	812.050	**857.050 T**	none	1,2,3,4,5,6,7,8
- FD & PD Tactical	813.025	**858.025 T**	none	1,2,3,4,5,6,7,8
- FD & PD Tactical	813.050	**858.050 T**	none	1,2,3,4,5,6,7,8
Conventional & Mutual Aid				
- International Calling		**866.0125**	156.7	10
- International Comm 1		**866.5125**	156.7	10
- International Comm 2		**867.0125**	156.7	10
- International Tac 1		**867.5125**	156.7	10
- International Tac 2		**868.0125**	156.7	10
- Permanent Repeaters	814.225	**859.225 R**	none	09
- Portable Repeaters	818.100	**863.100 R**	none	10
- Portable Repeaters	819.050	**864.050 R**	none	10
- Portable Repeaters	819.100	**864.100 R**	none	10
- Portable Repeaters	820.050	**865.050 R**	none	10
- Portable Repeaters	820.100	**865.100 R**	none	10
- Statewide CLEMARS		**154.920**	103.5	1,10
- Statewide CLEMARS	823.5125	**868.5125 R**	156.7	1, 10
- Statewide Calcord		**156.075**	none	1,10
Mobile Data Terminals System				
- System 1 - Central	811.075	**856.075 R**	none	1
- System 2 - Southeastern	812.075	**857.075 R**	none	2
- System 3 - Western	813.075	**858.075 R**	none	3
- System 4 - Northeastern	814.075	**859.075 R**	none	4
- System 5 - Mid City	815.075	**860.075 R**	none	5
- System 6 - Northern	816.050	**861.050 R**	none	6
- System 7 - Southern	816.100	**861.100 R**	none	7

SAN DIEGO COUNTY

SAN DIEGO POLICE SYSTEM

Below are the channels, alpha-numeric designators, and corresponding uses for the new San Diego PD Trunking System. The primary radio sites are also shown.

1.	NOR DIS 1	Northern District 1
2.	NOR DIS 2	Northern District 2
3.	SOU DIS 1	Southern District 1
4.	SOU DIS 2	Southern District 2
5.	EAS DIS 1	Eastern District 1
6.	EAS DIS 2	Eastern District 2
7.	WES DIS 1	Western District 1
8.	WES DIS 2	Western District 2
9.	CEN DIS 1	Central District 1
10.	CEN DIS 2	Central District 2
11.	N/E DIS 1	Northeastern District 1
12.	N/E DIS 2	Northeastern District 2
13.	S/E DIS 1	Southeastern District 1
14.	S/E DIS 2	Southeastern District 2
15.	MID DIS 1	Mid City District 1
16.	MID DIS 2	Mid City District 2
17.	NCW DIS 1	North City West District 1
18.	NCW DIS 2	North City West District 2
19.	SWAT 1	Special Weapons & Tactics
20.	SED	Special Enforcement Detail
21	INV 1	Investigations 1
22.	INV 2	Investigations 2
23.	INV 3	Investigations 3
24.	SD/MA	San Diego Mutual Aid
25.	PATCH 1	Phone Patches 1
26.	PATCH 2	Phone Patches 2
27.	SUPR	Supervisors
28.	SD/DR	Dynamic Regroup
29.	INQUIRY	Inquiry
30.	TAC 1	SDPD Tactical 1
31.	TAC 2	SDPD Tactical 2
32.	TAC 3	SDPD Tactical 3
33.	TAC 4	SDPD Tactical 4
34.	TAC 5	SDPD Tactical 5
35.	TAC 6	SDPD Tactical 6
36.	TAC 7	SDPD Tactical 7
37.	TAC 8	SDPD Tactical 8
38.	TAC 9	SDPD Tactical 9
39.	HP DIST 1	Harbor Police, Patrol Units
40.	HP DIST 2	Harbor Police, Airport
41.	HP TAC 1	Harbor Police, Tactical 1
42.	HP/DR	Harbor Police Dynamic Regroup
43.	NCP DIS 1	National City PD
44.	NCP DIS 2	National City PD
45.	NCP TAC 1	National City PD
46.	NCP D/R	National City PD
47.	ESC PD	Escondido PD
48.	ESC TAC 1	Escondido PD
49.	ESC TAC 2	Escondido PD
50.	ESC D/R	Escondido PD
51.	I CALL	Intl. Call Channel
52.	I COMM 1	Intl. Comm. 1
53.	I COMM 2	Intl. Comm. 2
54.	I TAC 1	Intl. Tac 1
55.	I TAC 2	Intl. Tac 2
56.	INT SERV	Intl. Service
57.	PORT RPT	Portable Repeater
58.	PT LOMA	Point Loma
59.	PALOMAR	Mt. Palomar Repeater
60.	CLEMARS	800 MHz CLEMARS

	Radio Sites	**Service Area**
1.	202 C Street Building	Central
2.	Encanto Sandpipe	Southeastern
3.	Mt. Soledad	Western
4.	Black Mountain	Northern
5.	Cowles Mountain	Eastern
6.	Mary Lane	Northern
7.	San Ysidro Peak	Southern
8.	Brown Field	Southern
9.	Point Loma	Western
10.	Portable	

(Also see P.37 for how the above sites are assigned for the new frequencies.)

SAN DIEGO COUNTY

SAN DIEGO POLICE DEPARTMENT

AREAS SERVED

(F-1) Southeastern - Areas south of Highway 94 and east of I-15. Includes Encanto, Logan Heights (east), and Paradise Hills.

(F-2) Northeastern - Bay Park, Del Mar Heights, Fiesta Island, La Jolla, Mission Beach, Miramar (north), Mira Mesa, Pacific Beach, Rancho Bernardo, Rancho Penasquitos, San Pasqual Park, Sea World, Sorrento Valley, Torrey Pines, and University City.

(F-3) Eastern - Areas north of Highway 94 and east of Highway 805. Includes City Heights, Clairemont, College Grove, Del Cerro, East San Diego, Grantville, Kearny Mesa, Kensington, Miramar, Mission Gorge, Mission Village, Montgomery Field, Normal Heights, S.D.S.U. area, San Carlos, San Diego Stadium, Sierra Vista, and Tierrasanta.

(F-4) Central - Balboa Park, Golden Hills, Logan Heights (west), and San Diego (downtown).

(F-5) Western - Fleetridge, Hillcrest, Linda Vista, Lindbergh Field, Loma Portal, Mission Beach, Mission Valley (east of I-5), North Park (west of 805), Ocean Beach, Old Town, Point Loma, Roseville, and University Heights.

(F-6) South - Brown Field, Nestor, Otay Mesa, Palm City, and San Ysidro.

New Patrol Areas

The Eastern and Northeastern Divisions are being subdivided into new divisions which will result in two new patrol areas for a total of eight districts.

Mid City District - Normal Heights and East San Diego (Beats 311-318).
North City West - North of Balboa Drive (Beats 111-131).

UNIT IDENTIFICATION SYSTEM

A. The first digit defines the station or area:

1 = Northern	4 = Southeastern	7 = Southern
2 = Northeastern	5 = Central	
3 = Eastern	6 = Western	

B. The second and third digit represents the unit's function:

00-09 = Commander	80-99 = Special Units
10-80 = Patrol	

SAN DIEGO COUNTY

SAN DIEGO POLICE DEPARTMENT

UNIT IDENTIFICATION SYSTEM

C. The last digit represents the type of unit:

A - Agent
B - 1st, 3rd watch, 1 officer extra unit
C - 1st, 3rd watch, 2 officers extra unit
D - Detective
E - DUI Enforcement
F - Special Event / Traffic Control
G - 1st, 3rd watch with trainee
H - Officer with trainee, same beat
I - Impound / Abatement
J - 1st, 3rd watch, one officer patrol
K - 1st, 3rd watch, two officer patrol
L - Lieutenant
M - Motorcycle patrol
N - Canine / Evidence unit
O - Community Service Officer
P - Parking Enforcement
R - School Task Force
S - Supervisors (sergeants)
T - Unassigned
U - 2nd watch officer with trainee
V - 2nd watch officer extra unit
W - 2nd watch, two officers extra unit
X - Extra patrol
Y - 2nd watch, one officer patrol
Z - 2nd watch, two officer patrol

An example of the identification system is: **511J**

2 Represents the station, which is: **Central**

11 Represents the unit's function, which is: **Patrol**

J Represents the type of unit, which is: **1st, 3rd watch, one officer patrol**

Note that the patrol areas, which are undergoing modification, correspond with the identification. Also the letter digits are pronounced phonetically.

SPECIAL UNITS

800 = Traffic Units / Extra Units
900 = Parking / Motor Units
1200 = S.W.A.T.
1300 = Gang Detectives
1400 = Code Compliance
1500 = Robbery Detectives
1600 = Homicide Detectives
1700 = Auto Theft Strike Units (ATSU)
1800 = Vice Squad
2000 = NEW designation
2700 = Drug Enforcement Admin.
2800 = NEW designation
2900 = NEW designation
3100 = NEW designation

ABLE = Airborne Law Enforcement (San Diego PD Helicopter)

SAN DIEGO COUNTY

SAN DIEGO POLICE DEPARTMENT

AIRBORNE LAW ENFORCEMENT (ABLE)

AIRCRAFT: Two Bell Jet Rangers, model 206B's.

CAPABILITIES: Maximum flying speed of 150 miles per hour, carrying capacity of one thousand pounds, and an operational altitude of ten thousand feet.

PURPOSE: ABLE's main duty is to provide assistance in the apprehension of criminal suspects in the City of San Diego limits.

EQUIPMENT: A multiband radio with the VHF, UHF, and 800 MHz band coverage. At night, ABLE uses the Nightsun, a 5 million candlepower light that literally turns the "night into day."

CREW: Seven full-time pilots and one secondary pilot. All of ABLE's personnel are crossed-trained as pilots and observers. A typical crew consists of one pilot and one observer.

HELIPORTS: ABLE is stationed in Montgomery Field in San Diego. All police stations and sub-stations in San Diego have makeshift heliports.

REQUIREMENTS: Must be a police officer with at least five years of patrol experience and possess a commercial pilot's license. The department then spends about ten thousand dollars to train each pilot.

SAN DIEGO POLICE BEATS

■ Beat 110
Clairemont
University City
Bay Ho

■ Beat 120
Mission Beach
Pacific Beach

■ Beat 130
La Jolla
Torrey Pines
Sorrento Valley
Del Mar Heights
Carmel Valley

■ Beat 210
San Pasqual
Rancho Bernardo
Carmel Mountain
Rancho Peñasquitos
Sabre Springs

■ Beat 220
Mira Mesa
Scripps Ranch

■ Beat 310
Grantville
Allied Gardens
Del Cerro
Lake Murray
San Carlos
Serra Mesa
Tierrasanta
Kearny Mesa

■ Beat 410
Valencia Park
Encanto
Emerald Hills
Skyline
Lomita
Jamacha

■ Beat 420
Alta Vista
North and South Bay Terraces
Paradise Hills

■ Beat 640
North Park

■ Beat 710
Tijuana River Valley
San Ysidro

■ Beat 720
Egger Highlands
Nestor
Otay Mesa
Palm City

■ Beat 810
Normal Heights
Kensington
Talmadge
College Area

■ Beat 820
City Heights

■ Beat 830
Rolando
El Cerrito
Darnall
Oak Park
Webster
Gateway

■ Beat 430
Mountain View
Southcrest
Shelltown
Mount Hope
Chollas View
Lincoln Park

■ Beat 510
Barrio Logan
Logan Heights
Memorial
Sherman Heights
Grant Hill
Stockton

■ Beat 520
Downtown
Marina
Gaslamp Quarter
Core-Columbia

■ Beat 530
Balboa Park
Cortez
Harborview
Golden Hill

■ Beat 610
Midway
Loma Portal
Point Loma
Ocean Beach
Old Town

■ Beat 620
Mission Valley
Morena
Bay Park
Linda Vista

■ Beat 630
Hillcrest
University Heights
Mission Hills
Midtown

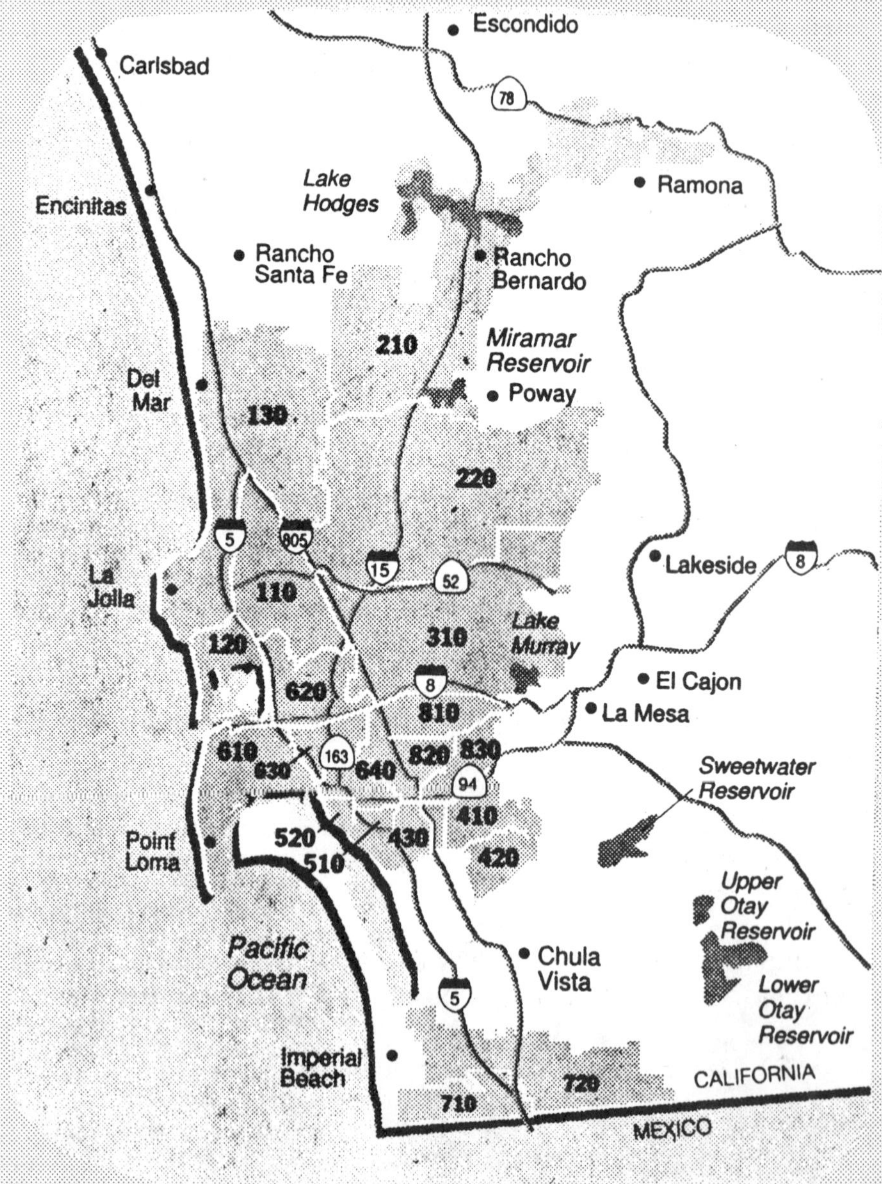

THE SAN DIEGO POLICE DEPARTMENT NOW HAS A TOTAL OF EIGHT DISTRICTS AND TWENTY-ONE PATROL BEATS WHICH ARE BASED ON THE NEEDS OF THE COMMUNITY. THE NEW DISTRICTS ARE NORTH CITY WEST (BEAT 130) AND MID CITY (BEAT 310).

SAN DIEGO COUNTY

SAN DIEGO FIRE DEPARTMENT

CHANNEL	INPUT	OUTPUT	TONE	CALL SIGN
Fire Dispatch	812.025	**857.025 T**	none	KNJJ-402
Fire Dispatch	813.025	**858.025 T**	none	KNJJ-402
Fire Dispatch	813.050	**858.050 T**	none	KNJJ-402
Fire Dispatch	814.025	**859.025 T**	none	KNJJ-402
Fire Dispatch	814.050	**859.050 T**	none	KNJJ-402
Fire Dispatch	815.025	**860.025 T**	none	KNJJ-402
Fire Dispatch	815.050	**860.050 T**	none	KNJJ-402
Fire Dispatch - Failsoft	815.025	**860.025 R**	none	KNJJ-402
Fire & Medics - Failsoft	814.025	**859.025 R**	none	KNJJ-402
VHF Dispatch		**154.310**	103.5	KMA-357
Mutual Aid FIREMARS	824.9875	**868.9875R**	156.7	KNJJ-402
Mutual Aid RED		**155.085**	103.5	KTX-766
Mutual Aid WHITE 1		154.280	103.5	KNIH-956
Mutual Aid WHITE 2		154.265	103.5	KA-4420
Mutual Aid WHITE 3		154.295	103.5	KA-4420

UNIT IDENTIFICATION SYSTEM

B	Battalion Chief
BR	Brush Rig
CR	Crash
E	Engine
F	Airport Foam Vehicle
HM	Hazmat
LT	Light (and air) truck
R	Rescue unit
T	Truck
U	Utility vehicle
W	Water tankers
X	Bomb disposal truck

Locations of Brush rigs, Light and Air, X-Ray 1, Foam, and Hazmat units will change based on conditions.

Engine, truck, and Battalion Chief locations will normally change only during multiple alarms or severe fire conditions, when they are moved to provide coverage for busy units.

All designators of "Fire" or any Battalion Chief above B6 are supervisory personnel who respond to multiple alarm and high rise fires, or other serious incidents.

Fire 61 through 69 are the Arson Investigators.

The Mobile Canteen responds to incidents where firefighters are expected to be on scene for more than 2 hours. Pioneer hook and ladder are a group of volunteers who operate the Mobile Canteen.

SAN DIEGO COUNTY

San Diego City Fire Stations

No.	Location	Thomas Bros.	Equipment, Comments
1	1st & B	65 C1	E1, T1, B1, E49, LT1, Arson 61-69
"	"	"	Chemical 1, X1, X2
3	State & Kalmia	60 B6	E3
4	8th & J Street	65 C2	E4, R4
5	9th & University	53 C3	E5, T5, B5
6	Twining & Palm	72 A4	E6, BR6
7	Crosby & National	65 D3	E7
8	Goldfinch & Washington	60 B3	E8
9	Ardath & Torrey Pines	44 B4	E9, HM1, HM2
10	62nd & Acorn	61 F2	E10, T10, B10
11	25th & Broadway	65 E1	E11, T11
12	Ozark & Imperial	66 C2	E12, T12, U12, B12, BR12
13	Fay & Nautilus	43A E3	E13
14	32nd & Lincoln	60 F3	E14, T14
15	Voltaire & Ebers	59 B3	E15,
16	L.J. Scenic & Via Casa Alta	44 B5	E16
17	Chamoune & Orange	61 C3	E17
18	Felton & Adams	60 F2	E18
19	I15 & Ocean View	66 A3	E19
20	Kemper & Midway	59 D3	E20, T20, U20
21	Mission Blvd & Grand	52 A4	E21, T21
22	Catalina & Savoy	59 B6	E22
23	Linda Vista & Comstock	53 B5	E23
24	Del Mar Hts & Mercado	34 B4	E24
25	Moreno & Napier	52 E6	E25, B25
26	54th & Krenning	61 D5	E26
27	Clairemont & Cole	44 E6	E27,
28	Aero & Kearny Villa	53 D3	E28, T28, W28, W1, W2, BR28, CR28
	(Repair & Training Facility)		LT9, F28,
29	San Ysidro & Cottonwood	74 A4	E29, T29, BR29
30	Coronado & Hollister	71 D5	E30
31	College & Camino Rico	54 E4	E31, LT31
32	Paradise Vly. & Briarwood	67 B4	E32
33	Rancho Bernardo Center	32 E1	E33, BR33
34	Cowles Mtn. & Navajo	55 C3	E34, BR34
35	Genesee & Eastgate Mall	44 E1	E35, T35, B35, U35
36	Mt. Abernathy & Chateau	53 B1	E36
37	Scripps Lake & Red Cedar	40 C2	E37
38	Camino Ruiz & New Salem	35 E2	E38, BR38
39	La Cuenta & Baroque	54 C2	E39, BR39
40	Paseo Montalban & Salmon R.	36 A5	E40, T40, BR1, BR2, Wildland 1
41	Carol Canyon & Miramar	35 D6	E41, T41
42	Lindbergh Field	60 A5	E42, R1, R2, R3, Oshsokh 1 (Foam truck)
Poway	Metate & Civic Center Dr.	37 B4	Squads
Poway	Old Coach & Espola	33 C2	Squads

SAN DIEGO COUNTY

SAN DIEGO CITY SERVICES

NAME	INPUT	OUTPUT	TONE	CALL SIGN
Airport Security (Brown & Montgomery Fields)		151.115	110.9	KMJ-703
Building Security		458.800	77.0	KA-87852
City Colleges	469.475	464.675 R	D732	N/A
Community Colleges PD (D)	458.450	453.450 R	85.4	KXD-597
Convention Center (F-1)	458.325	453.325 R	110.9	N/A
Convention Center (F-2)	458.900	453.900 R	110.9	N/A
Convention Center - Shows	466.175	461.175 R	D205	N/A
Electronic Department		151.115	110.9	KMJ-703
Electronic Department		151.205	110.9	KB-85467
Harbor Patrol Mission Bay	458.525	453.525 R	110.9	KEF-973
Harbor Patrol Paging		154.950	none	N/A
Lifeguards (F-1)	458.650	453.650 R	110.9	KIB-308
Lifeguards (F-2)		453.650	110.9	KIB-308
Lifeguards (F-3)		453.3875	110.9	KD-21784
Lifeguards (F-4)		453.4125	110.9	KD-21784
Lifeguards (F-5)		453.4375	110.9	KD-21784
Lifeguards Mission Bay	453.525	458.525 R	110.9	KEF-973
Lifeguards Mission Bay	458.775	453.775 R	110.9	WNLY-871
Lifeguards San Diego		154.995	110.9	KO-5713
Local Government 1 (Building, Engineering & Parks) (Solid Waste)		153.935	110.9	KQY-731
Local Government 2 (Engineering & Lifeguards)		154.995	110.9	KO-5713
Local Government 3 (Engineering)		158.865	110.9	WCL-777
Red Cross Dial-A-Ride		44.600	none	KUB-976
San Diego Jack Murphy Stadium				
Entrance A		461.6375	none	KB-32691
Entrance B		462.8875	none	KB-32691
Entrance C		464.5125	none	KB-32691
Entrance D		464.5625	none	KB-32691
Maintenance	458.200	453.200 R	110.9	WNCR-798
Non-Event 24 Hour Security		453.800	110.9	KA-87252
Non-Event 24 Hour Security		458.800	110.9	KA-87252
Paramedics		154.570		N/A
Parking Control		154.600		N/A
Security Primary		151.925		N/A
Security Tactical		151.955		N/A

SAN DIEGO COUNTY

SAN DIEGO CITY SERVICES

NAME	INPUT	OUTPUT	TONE	CALL SIGN
San Diego Zoo				
General Operations	469.425	464.425 R	N/A	KNCR-311
General Operations	469.425	464.575 R	N/A	WNLY-322
Maintenance		151.895	N/A	WNDL-585
Parking Control		154.570	N/A	KD-37270
Phone Interconnect	466.0125	461.0125R	114.8	KD-44530
Security Operations	458.200	453.200 R	114.8	KTV-721
Security Operations	458.800	453.800 R	114.8	KTV-721
Street Department (F-1)		151.115	110.9	KMJ-703
Street Department (F-2)		151.460	110.9	KNED-269
Transit District Trunked		856-860.325 T	none	KTL-687
Transit District Link	958.500	953.500 R	none	KTL-687
Unified Port District				
Harbor Police (F-1)		156.150	103.5	KNDL-450
Lindbergh Field Police (F-2)		155.895	103.5	KML-273
Police Tactical (F-3)		155.685	123.0	N/A
Police CLEMARS (F-4)		154.920	103.5	N/A
Police Fire WHITE 1 (F-5)		154.280	103.5	KA-4420
Police Tactical (F-6)		156.850	none	N/A
Future Trunking System	821.125	866.125 T	none	N/A
Future Trunking System	821.6125	866.6125 T	none	N/A
Future Trunking System	822.1125	867.1125 T	none	N/A
Future Trunking System	822.8875	867.8875 T	none	N/A
Unified School District				
Busing - See page 38 (Group 1)		800 MHz	none	WNHS-765
School Police (F-1)	458.300	453.300 R	84.5	WNFJ-501
School Police (F-2)	458.350	453.350 R	84.5	WNFJ-501
Security Portables		154.570		KB-73495
Water & Utilities		173.3375	110.9	KMK-721
Water & Utilities		173.350	110.9	KMK-721
Wild Animal Park				
Security (F-1)	458.200	453.200 R	114.8	KTV-721
Transportation (F-2)	458.800	453.800 R	114.8	KQR-954
Animal Services (F-3)	458.350	453.350 R	114.8	WNFJ-501
Special Events (F-4)	458.225	453.225 R	114.8	KFR-361
Phone Interconnect (F-5)	466.0125	461.0125R		KD-44530

SAN DIEGO COUNTY

SAN MARCOS

CHANNEL	INPUT	OUTPUT	TONE	CALL SIGN
Sheriff (35)	458.925	**453.925 R**	146.2	KNCJ-415
Fire & Medics (1400)		**154.235**	103.5	KSO-681
Fire Command RED		155.085	103.5	KMA-340
Fire Tactical GRAY		154.355	114.8	KB-83692
Local Government		158.805	192.8	KUI-757
(Building & Engineering)				
(Parks & Recreation)				
(Street Department)				
Lake San Marcos Security		151.715	151.4	KNCU-715
Palomar College Patrol (F-1)	469.475	464.475 R	D051	WNML-216
Palomar College Patrol (F-2)	464.425	469.425 R	D311	WNML-216
Palomar College Operations		151.655	88.5	KNEF-251
Unified School District - Buses		155.235	D351	WXK-701
Water District - Primary		153.650	127.3	KDW-858
Water District - Portables		158.175	127.3	KDW-858

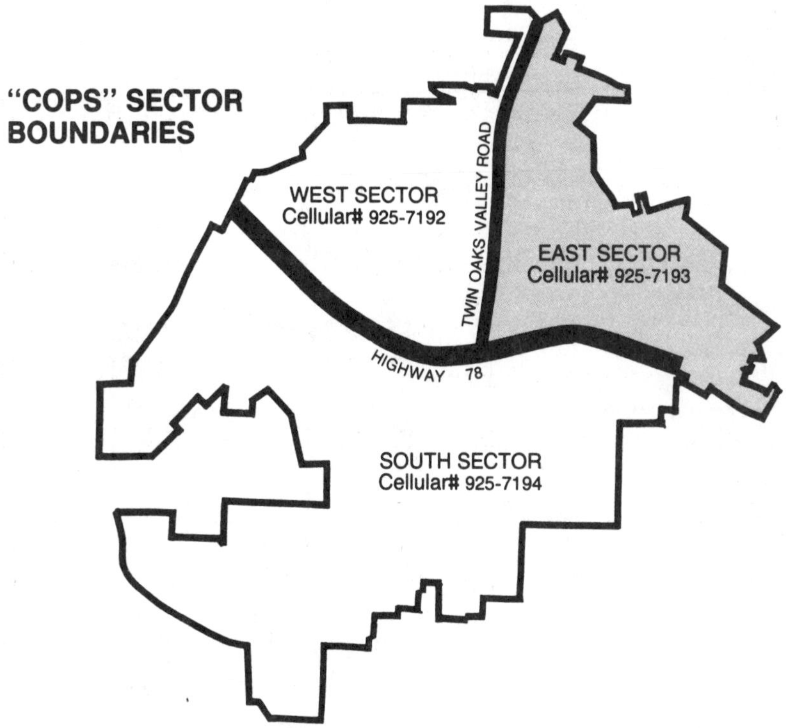

SAN DIEGO COUNTY

SANTEE

CHANNEL	INPUT	OUTPUT	TONE	CALL SIGN
Sheriff (51)	458.100	**453.100 R**	146.2	KNCJ-419
Fire & Medics - See Heartland Fire		800 MHz	none	WNQA-931

SOLANA BEACH

CHANNEL	INPUT	OUTPUT	TONE	CALL SIGN
Sheriff (23)	458.750	**453.750 R**	146.2	KNCJ-415
Fire & Medics (2400)		**154.235**	192.8	KNBP-832
Fire Command RED		155.085	103.5	KTI-545
Fire Tactical GRAY		154.355	114.8	KM-8396
Lifeguards 1		155.820	107.2	WNDB-632
Lifeguards 2		155.940	107.2	WNDB-632

SPRING VALLEY

CHANNEL	INPUT	OUTPUT	TONE	CALL SIGN
Sheriff (60)	458.950	**453.950 R**	146.2	KNCJ-412
Fire & Medics - See Heartland Fire		800 MHz	none	WNQA-931
Fire Mutual Aid WHITE 1		154.280	103.5	WNKM-738
Fire Remote Link	458.7875	453.7875	none	KRE-364

VISTA

CHANNEL	INPUT	OUTPUT	TONE	CALL SIGN
Sheriff (31)	458.925	**453.925 R**	146.2	KNGY-554
Fire & Medics (1200)		**154.235**	192.8	KNBP-832
Fire Command RED		155.085	103.5	KNAK-572
Fire Tactical GRAY		154.355	114.8	KM-8396
Local Government		158.775	110.9	KTK-761
(Building & Engineering)				
(Parks & Recreation)				
(Street Department)				
Local Government		856-860.800 T	none	SMR
Local Government - HT's		155.955	103.5	KB-66266
Unified Schools - Buses	822.000	862.000 R	none	WNHS-769
Unified Schools - Security		155.280	123.0	N/A
Vista Irrigation District (VID)		48.060	N/A	KMG-580

LOS ANGELES COUNTY

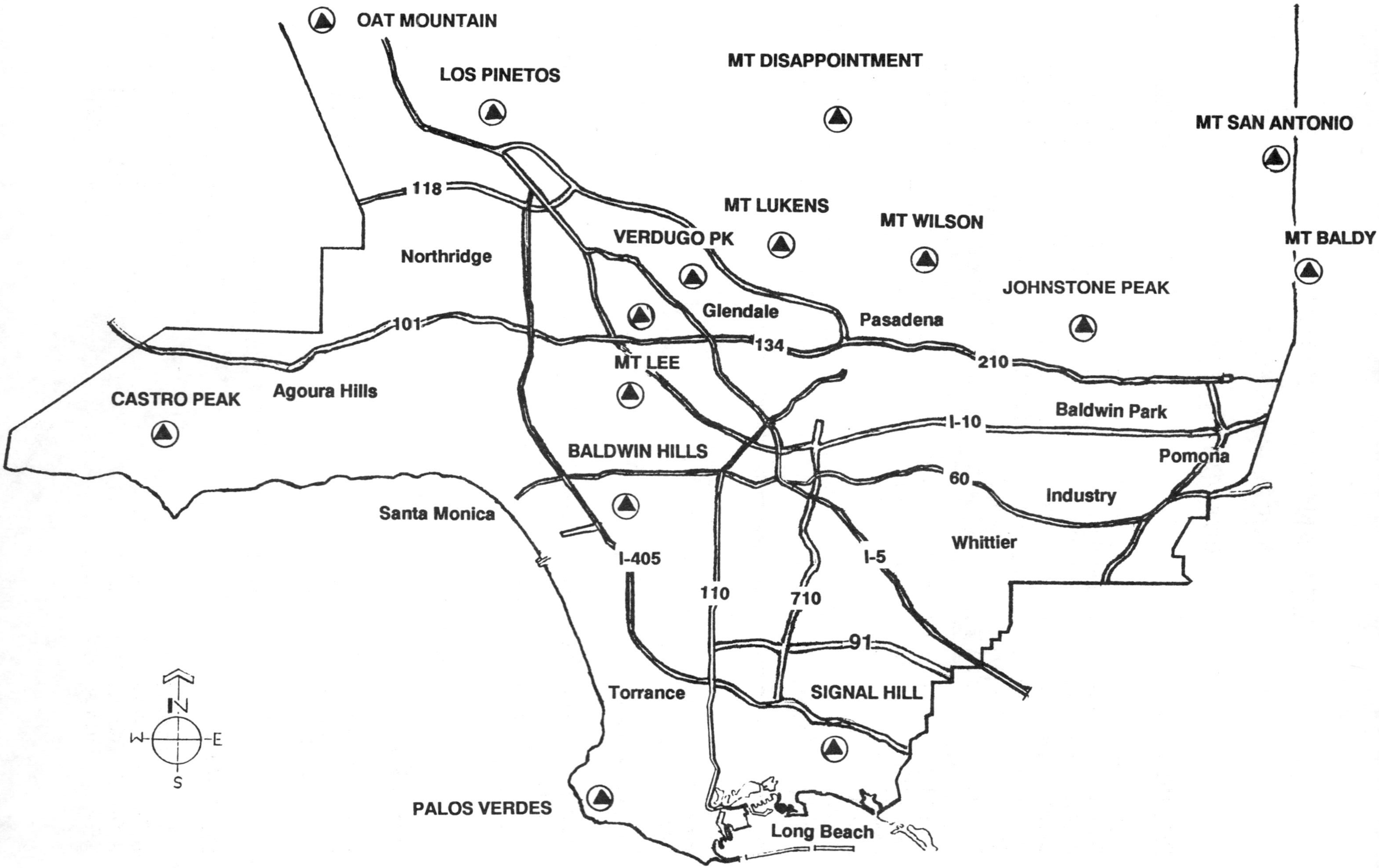
OAT MOUNTAIN
LOS PINETOS
MT DISAPPOINTMENT
MT SAN ANTONIO
118
MT LUKENS
MT WILSON
VERDUGO PK
MT BALDY
Northridge
JOHNSTONE PEAK
101
Glendale
Pasadena
134
MT LEE
210
CASTRO PEAK
Agoura Hills
Baldwin Park
I-10
BALDWIN HILLS
Pomona
60
Industry
Santa Monica
Whittier
I-405
I-5
110
710
91
Torrance
SIGNAL HILL
N
W
E
S
PALOS VERDES
Long Beach

LOS ANGELES COUNTY

SHERIFF'S OFFICE

DISPATCH CHANNELS

CHANNEL	BASE	MOBILE
DISPATCH 1 - Altadena (7), Crescenta Valley (12)	**483.9875**	486.9875
DISPATCH 2 - West Hollywood (9)	**483.3625**	486.3625
DISPATCH 3 - East Los Angeles (2)	**483.2125**	486.2125
DISPATCH 4 - Firestone (1), Lynwood (25)	**483.2625**	486.2625
DISPATCH 5 - Antelope Valley (11), Santa Clarita (6)	**482.8625**	485.8625
DISPATCH 6 - San Dimas (8), Walnut (29)	**482.9375**	485.9375
DISPATCH 7 - Avalon (18), Lomita (17)	**484.0375**	487.0375
DISPATCH 8 - City of Industry (14)	**483.6875**	486.6875
DISPATCH 9 - Carson (16), MetroRail Blue Line	**484.1625**	487.1625
DISPATCH 10 - Malibu (10), Marina Del Rey (27)	**482.9125**	485.9125
DISPATCH 11 - Temple City (5)	**482.9875**	485.9875
DISPATCH 12 - Lennox (3)	**483.4375**	486.4375
DISPATCH 13 - Norwalk (4), Pico Rivera (15)	**483.7625**	486.7625
DISPATCH 14 - Lakewood (13)	**483.1375**	486.1375

Numbers in parenthesis above refer to the sheriff station number.

Note: Mobile units will not normally be heard on the above base frequencies, except when the system is in a repeater mode. The dispatch frequencies are repeaterized in situations when mobile units are transmitting information that all other officers in the area need to hear. The familiar L.A.S.O. beeping tone is emitted on the base frequencies to notify other units that the mobile frequency is busy.

Input frequencies are the standard of exactly 3 MHz higher than the shown outputs (e.g., 485.8125 for 482.8125). The tone 186.2 Htz is used on most input and mobile frequencies. The radios for Sheriff Custody, Court Services, Jails, and Honor Camps have multiple tone assignments.

COUNTYWIDE TACTICAL CHANNELS "C TAC"

CHANNEL	FREQUENCY
C TAC 1 - Countywide Tactical	**482.8125R**
C TAC 2 - Countywide Tactical	**482.8375R**
C TAC 3 - Countywide Tactical	**483.0625R**

AREA TACTICAL CHANNELS "A TAC"

CHANNEL	FREQUENCY
A TAC 1 - Tactical North	**483.0875R**
A TAC 2 - Tactical West	**483.1625R**
A TAC 3 - Tactical East	**483.0125R**
A TAC 4 - Tactical South	**484.1125R**

LOS ANGELES COUNTY

SHERIFF'S OFFICE

LOCAL TACTICAL CHANNELS "L TAC"

CHANNEL	FREQUENCY
L TAC 1 - Lost Hills / Malibu / Marina Del Rey / W. Hollywood	483.9125R
L TAC 2 - Avalon / Lennox / Lomita	482.8875R
L TAC 3 - Lakewood	483.3875R
L TAC 4 - City of Industry	483.9625R
L TAC 5 - Carson, Transit Services Bureau*	483.4625R
L TAC 6 - East Los Angeles	483.4875R
L TAC 7 - San Dimas / Temple / Walnut	483.2375R
L TAC 8 - Antelope Valley / Crescenta Valley / Santa Clarita	483.1125R
L TAC 9 - Firestone / Lynwood	483.8625R
L TAC 10 - Norwalk / Pico Rivera	484.0125R

*Police for MetroRail Blue Line

INVESTIGATORS & SPECIAL OPERATIONS

CHANNEL	FREQUENCY
SV 1 - Secure Voice Operations (La Basin)	483.8875R
SV 2 - Secure Voice Operations (La Basin)	482.9625R
SV 3 - Secure Voice Operations North	483.8875R
INV 1 - Investigators Countywide	483.3125R
INV 2 - Investigators North / Central	483.3375R
INV 3 - Investigators West	483.8125R
INV 4 - Investigators South	483.6125R
INV 5 - Investigators Southeast	483.7375R
INV 6 - Investigators Central	483.6625R
INV 7 - Investigators East	483.5125R
LA Interagency Task Force	482.6125R
LA Interagency Task Force - Mobiles	460.050
LA Interagency Task Force - Mobiles	460.100
LA Interagency Task Force - Mobiles	482.150

MUTUAL AID CHANNELS

CHANNEL	FREQUENCY
M-AID 1 - Mutual Aid North / San Fernando Valley	483.5875R
M-AID 2 - Mutual Aid West Metropolitan	484.0875R
M-AID 3 - Mutual Aid East Metropolitan	483.7875R
M AID 4 - Mutual Aid San Gabriel Valley	484.1375R
M-AID 5 - Mutual Aid Central Metropolitan	484.0625R

LOS ANGELES COUNTY

SHERIFF'S OFFICE

SIMPLEX & SPECIAL UNIT CHANNELS

CHANNEL	FREQUENCY
CUST- Prisoner Custody	**483.7125**
CUST- Prisoner Custody	**486.7125**
CRT SVC - Court Services	**483.9375R**
CRT SVC - Court Services	**486.9375**
EMERG - Emergency Trigger	**483.2875R**
PHR F1 - Pitchess Honor Rancho Tactical	**483.9375R**
PHR F2 - Pitchess Honor Rancho North	**483.1125R**
PHR F3 - Pitchess Honor Rancho South	**483.9625**
PHR F4 - Pitchess Honor Rancho East	**484.0125**
SRC ACC - Sheriff Comm Center Access	**483.5625R**
SPC UNIT - Special Unit Dispatch (helicopters)	**483.5375R**
SIMPLEX A - Simplex Operations A	**486.0375**
SIMPLEX B - Simplex Operations B	**483.0375**
SIMPLEX C - Simplex Operations C	**486.1875**
SIMPLEX D - Simplex Operations D	**483.1875**
SIMPLEX E - Simplex Operations E	**486.4125**
SIMPLEX F - Simplex Operations F	**483.4125**
Portable Repeater A (Command Posts)	**483.6375R**
Portable Repeater B (Command Posts)	**483.8375R**
Sheriff Helicopters (AM)	**123.025**

SHERIFF MOBILE DATA TERMINALS

CHANNEL	FREQUENCY
DATA 1 - Data Communications Countywide	470.9375R
DATA 2 - Data Communications West 1	470.6875R
DATA 3 - Data Communications West 2	470.7375R
DATA 4 - Data Communications East 1	470.8625R
DATA 5 - Data Communications East 2	470.8375R
DATA 6 - Data Communications South 1	470.8875R
DATA 7 - Data Communications South 2	470.7625R
DATA 8 - Data Communications North	470.8125R
SRC ACC - Sheriff Radio Center Access	470.9125R

SHERIFF INTERAGENCY CHANNELS

CHANNEL	FREQUENCY
CLEMARS GOLD 1	**154.920**
CLEMARS GOLD 2	154.935
CLEMARS GOLD 6	**487.2375**
Small City Interagency	**484.2125**
Small City Interagency	**484.2375**

LOS ANGELES COUNTY

SHERIFF'S OFFICE

UNIT IDENTIFICATION SYSTEM

1. The number digits indicate the station:

10-19	Firestone	160-169	Carson
20-29	East Los Angeles	170-179	Lomita
30-39	Lennox	180-189	Avalon
40-49	Norwalk	190-199	Future Use
50-59	Temple	200-209	Future Use
60-69	Santa Clarita	210-219	Transit Services Bureau
70-79	Altadena / Crescenta Valley	220-229	Lost Hills
80-89	Future Use	230-239	Future Use
90-99	West Hollywood	240-249	Future Use
100-109	Malibu (substation)	250-259	Lynwood
110-119	Antelope Valley	260-269	Future Use
120-129	Future Use	270-279	Marina Del Rey
130-139	Lakewood	280-289	Future Use
140-149	City of Industry	290-299	San Dimas / Walnut
150-159	Pico Rivera		

2. The letter digit indicates the type of unit:

A,B,D,F	Two man unit	M	Motorcycle
C	Captain	N	Narcotic
E	Community Service	P	Station Posse
G	Gang unit	R	Reserve
H	Special Enforcement	S	Supervisor
I	Investigative	T	Traffic
J	Juvenile	V	Vice
K	One man unit	WS	Special Enforcement
K9	Canine unit	X	Out of Service
L	Lieutenants (WC's)		

3. The last number digits indicate the unit number.

4. Administrative Units:

A1	Chief of Administration	G1	Custody Chief
F1	Chief of Technical Services	D1	Detective Chief
I1	Chief of Court Services		

5. Special Units

240H	Horse Patrol	WL	SEB Watch Commander
240K-9	Canine unit	WS	SEB Field Sergeant
240R	Rescue unit	W1-W30	SEB Field Unit
240-250	Special Enforcement (SEB)		

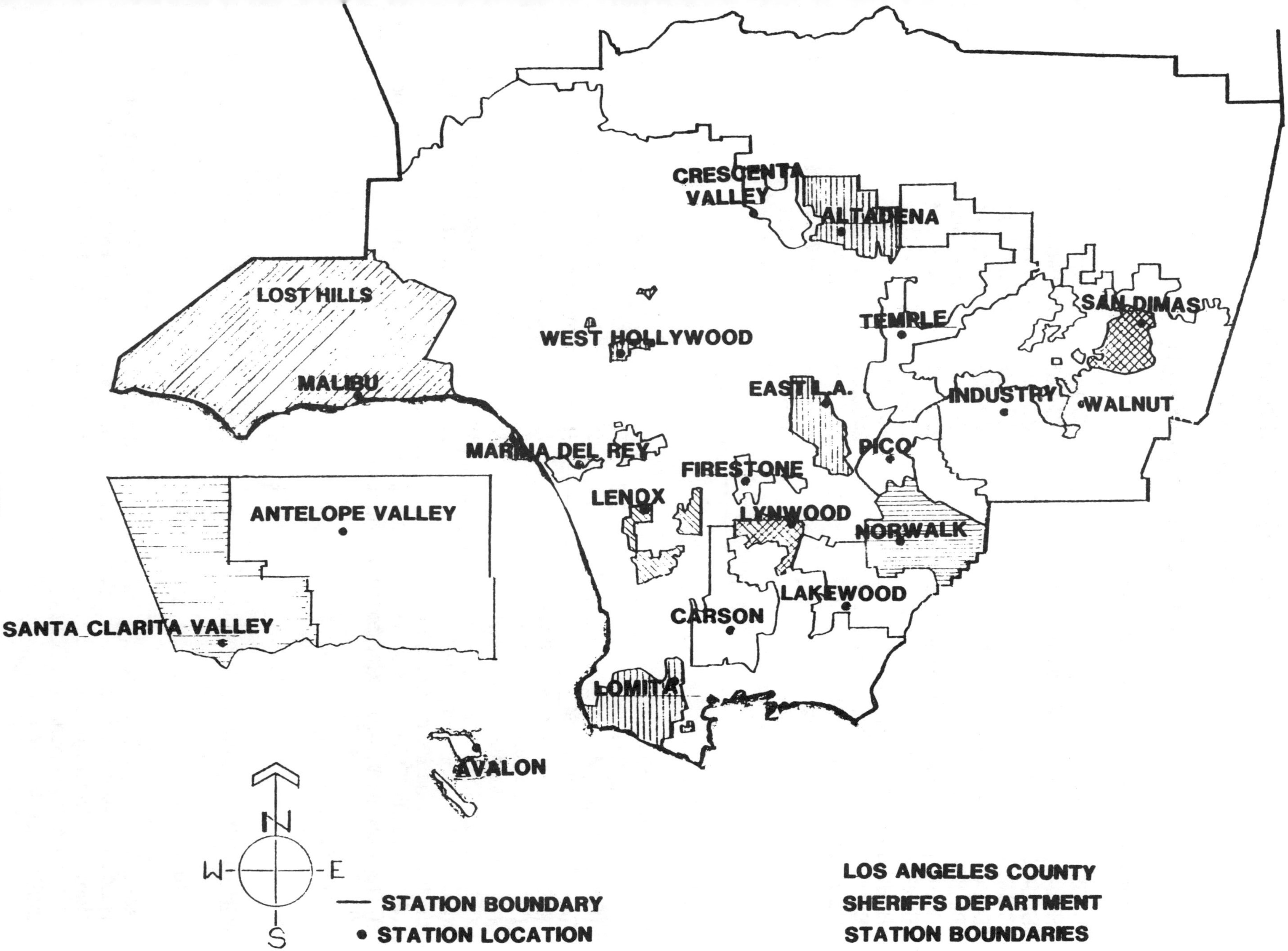

LOS ANGELES COUNTY
SHERIFFS DEPARTMENT
STATION BOUNDARIES

LOS ANGELES COUNTY

LACo FIRE DEPARTMENT

BLUE SYSTEM - UHF

CHANNEL	FREQUENCY	TONE*
BLUE 1 - Battalions 7, 8, 9, 14	**470.5625R**	151.4
BLUE 2 - L.A. Basin / Malibu Fireground	**470.4375R**	151.4
BLUE 3 - Battalions 1, 3, 5, 13	**470.6125R**	151.4
BLUE 4 - Countywide Command & Camps	**470.5125R**	151.4
BLUE 5 - San Gabriel Valley Fireground	**470.6625R**	151.4
BLUE 6 - Battalions 2, 4, 10, 12	**470.4125R**	151.4
BLUE 7 - San Gabriel Valley Fireground	**470.4625R**	151.4
BLUE 8 - Countywide Dispatch	**470.5375R**	151.4
BLUE 9 - Antelope Valley Fireground	**470.6375R**	151.4
BLUE 10 - Antelope Valley Fireground	**470.3625R**	151.4
BLUE 11 - Antelope Valley / Malibu Fireground	**470.5875R**	151.4
BLUE 12 - Battalions 6, 11	**470.4875R**	151.4

*Input tone

WHITE SYSTEM - VHF

CHANNEL	FREQUENCY	TONE
(F-1) LA County TAC 1	**154.430**	none
(F-2) LA County TAC 2	**154.340**	none
(F-3) LA County TAC 3	**154.400**	none
(F-4) LA County TAC 4	**153.830**	none
(F-5) Statewide Mutual Aid WHITE 1	**154.280**	none
(F-6) Statewide Mutual Aid WHITE 2	**154.265**	none
(F-7) Statewide Mutual Aid WHITE 3	**154.295**	none
(F-8) Office of Emergency Services TAC 1	**154.160**	none
(F-9) Ventura County Fire TAC 1	**154.010**	100.0
(F-10) Inglewood/San Marino FD	**154.175**	114.8
(F-11) Arcadia/El Monte/Monrovia/Sierra Madre FD	**154.370**	none
(F-12) Downey/Lynwood/Compton FD	**154.250**	none
(F-13) Long Beach FD	**153.950**	131.8
(F-14) Angeles National Forest	**171.575**	none

VHF-AIR OPERATIONS

CHANNEL	FREQUENCY	TONE
Air-to-Air	122.900	none
Air-to-Air	122.925	none
Air-to-Air	123.100	none
Helicopter Air-to-Air	118.925	none
Helicopter Air-to-Ground	118.950	none
Heliport Air-to-Ground	119.975	none
Interagency Air-to-Air	122.850	none

LOS ANGELES COUNTY

LACoFD Stations & Equipment

Battalion 1

Station	Location	Equipment
7	West Hollywood	E7
8	West Hollywood	E8, T8, SQ8, E208
14	South L.A	E14, E214, SQ14
18	Inglewood	E18, E505
38	Windsor Hills	E38, E538
58	Ladera Heights (HQ)	E58, E558, SQ58, U1
60	Universal City	E60, C60
110	Marina Del Rey	E110, T110, Fireboat 110

Battalion 2

Station	Location	Equipment
32	Azusa (HQ)	E32, SQ32, SQ509
"	"	WT32
62	Claremont	E62, Patrol 62
64	San Dimas	E64, SQ64, E264, U2
85	Glendora	E85, E585
86	Glendora	E86, T86, Patrol 86
97	Azusa	E97, Patrol 97
101	Claremont	E101, SQ101, E562
102	Claremont	E102, Patrol 102
141	San Dimas	E141, OES228, Patrol 502

Battalion 3

Station	Location	Equipment
1	East L.A. (City Terrace)	E1, F1, FD1, USAR1
3	East L.A.	E3, T3, SQ3, U3
22	City of Commerce	E22
25	Pico Rivera	E25
27	City of Commerce (HQ)	E27, T27, SQ27, U3
50	City of Commerce	E50, E250
103	Pico Rivera	E103, L103, E407

Battalion 4

Station	Location	Equipment
11	Altadena	E11, SQ11, E511, SQ511
12	Altadena	E12, Patrol 508
19	La Canada	E19, E505
63	La Crescenta	E63
66	Pasadena	E66, Patrol 66
74	San Fernando	E74, Patrol 74
82 (HQ)	Flintridge	E82, E282, T82, E582
"	"	Patrol 82, U4

Battalion 5

Station	Location	Equipment
65	Agoura	E65, E265, Patrol 65
67	Calabasas	E67
68	Calabasas	E68, E568, Patrol 68
69	Topanga	E69, E269, Patrol 69 & 269
70	Malibu (HQ)	E70, E570, T505, Patrol 511, U5
71	Malibu	E71, SQ71, Patrol 71
72	Malibu	E72, Patrol 509
88	Malibu	E88, SQ88, Squirt 88
99	Malibu	E99, E599
125	Calabasas	E125, E568, T125, SQ125/508
144	Westlake Village	E144, E571, T144, Patrol 144

Battalion 6

Station	Location	Equipment
73	Newhall (HQ)	E73, E273, T73, WT73
"	"	F73, Patrol 73, U73
75	Chatsworth	E75, Patrol 75
76	Valencia	E76, Hazmat 76
77	Gorman	E77, Patrol 77
81	Agua Dulce	E81, E573, E576, E581
"	"	E581A, Patrol 81
107	Canyon Co.	E107, E307, SQ107
"	"	SQ514, Patrol 107
111	Saugus	E111, SQ111, Patrol 111/510
123	Canyon Co.	E123, E576, MAU123
124	Stevenson R.	E124, SQ124
149	Castiac	E149, E575, WT149
"	"	Patrol 149/504

Battalion 7

Station	Location	Equipment
10	Carson (HQ)	E10, E510, F10, F50'
36	Carson	E36, E536, T36, SQ36
95	Gardena	E95
105	Compton Area	E105, E536, Hazmat & Deluge 105
116	Carson	E116, T116, SQ116
127	Carson (HQ)	E127, U7, B.C. Car

Battalion 8

Station	Location	Equipment
15	Whittier	E15, E515, T503, SQ502
17	Whittier	E17, Patrol 503
20	Norwalk	E20, T20, SQ20, U8
28	Whittier (HQ)	E28, T28, E404, SQ28
"	"	Mobile Aid 2, B.C. Car
49	La Mirada	E49, T49, SQ49
59	Whittier	E59, T502
96	Whittier	E96
115	Norwalk	E115, MAU115

BC = Battalion Chief
C = Cart (brush fires)
E = Engine Co.
F = Foam (airports)
FD = Fuel Disposal
MAU = Mobile Air Unit
L = Lightforce
SQ = Squad / Hazmat
T = Truck Co.
U = Utility Vehicle
USAR = Urban Search & Rescue
WT = Water Tender

LOS ANGELES COUNTY

LACoFD Stations & Equipment

Battalion 9

Station	Location	Equipment
23	Bellflower	E23, E523
30	Artesia (HQ)	E30, T30, SQ30, B.C. Car
31	Paramount	E31, T31, SQ31, SQ504
34	Hawaiian Gardens	E34
35	Cerritos	E35, E535
45	Lakewood	E45, T45, SQ45, U9
94	Lakewood	E94, E594
98	Bellflower	E98, SQ98
122	Lakewood	E122

Battalion 10

Station	Location	Equipment
4	Rosemead	E4, T501
5	San Gabriel	E5
29	Baldwin Park	E29, T29, E405, SQ29
42	Rosemead	E42, T42
44	Duarte	E44, E244, E544
"	"	Helitender 6, Patrol 44
47 (HQ)	Temple City	E47, SQ47, U10
48	Irwindale	E48, E548, SQ503, Patrol 506
90	S. El Monte	E90, E401, SQ90

Battalion 11

Station	Location	Equipment
24	Palmdale	E24, T24, E524, Patrol 502/512
33	Lancaster (HQ)	E33, SQ33, U33, WT33, B.C. Car
37	Palmdale	E37, E537, SQ37, SQ537
78	Palmdale	E78, E578, Patrol 78
79	Pearblossom	E79, E279, Patrol 79
80	Acton	E80, E580, OES229, Patrol 80
84	Quartz Hill	E84, SQ84, Patrol 84
92	Littlerock	E92, SQ92, Patrol 92
93	New Station	E93
112	Lancaster	E112
114	Palmdale	E114, Patrol 114
117	Lancaster	E117, E317, E517, Patrol 513
129	Lancaster	E129
130	Palmdale	E130
131	Palmdale	E131
140	Leona Valley	E140
157	Green Valley	E157

Battalion 12

Station	Location	Equipment
26	La Puente	E26, SQ26, SQ526
43	Industry	E43, Hazmat 43/543
61	Walnut	E61, SQ61
57	South Gate	E57, SQ501
87	Industry	E87, Deluge 87, SQ526
91	Hacienda Hts.	E91, Patrol 91
118	Industry	E118, T118, SQ118
119	Walnut	E119, SQ119
120	Diamond Bar	E120, T506, Patrol 120/506
145 (HQ)	Rowland Hts.	E145, E543, U12, B.C. Car
146	Walnut	E146

Battalion 13

Station	Location	Equipment
9	Los Angeles (Florence)	E9, E509, T504
16	South L.A.	E16, SQ16
39	Bell Gardens	E39, E539, SQ39
40	Pico Rivera	E40, SQ40, SQ512
41	Los Angeles	E41, E241, SQ41
54	South Gate	E54
57	South Gate	E57, SQ501
163	Bell	E163, E557, SQ163
164	Huntington Park (HQ)	E164, T164, SQ164
"	"	U164, B.C. Car
165	Huntington Park	E165, L165, U13

Battalion 14

Station	Location	Equipment
2	Palos Verdes	E2, E502, SQ2
6	Lomita	E6, SQ6
21	Lawndale	E21, SQ21
53	R. Palos Verd.	E53
55	Avalon	E55, E155, U55
"	"	Patrol 55/255
56	Rolling Hills	E56, Patrol 501
83	R. Palos Verd.	E83, E583, SQ513
"	"	Patrol 83
106 (HQ)	Rolling Hills E.	E106, E402, T106
"	"	SQ106, U14, B.C. Car

Fire Units:

Construction - Units used in the building of firebreaks and other emergency structures.
Deluge - Pickup truck with a giant water gun on back, carries foam, and follows engine on calls.
Lightforce - Consists of one truck and one engine. The engine is the main unit.
Mobile Air Unit - Carries oxygen supply and refills bottles during structure fires.
Patrol - One man unit, only in brush areas. Main job is to know topography and backroads.
School - Units relate to interdepartment training.
Squirt - Truck company with pump, used in place of an engine.
TaskForce - At least five engines and a chief; used for brush fires, mutual aids, etc.
USAR - Volunteer unit trained for water rescue, collapsed buildings, earthquakes, and other special incidents.

LOS ANGELES COUNTY FIRE DIVISIONS

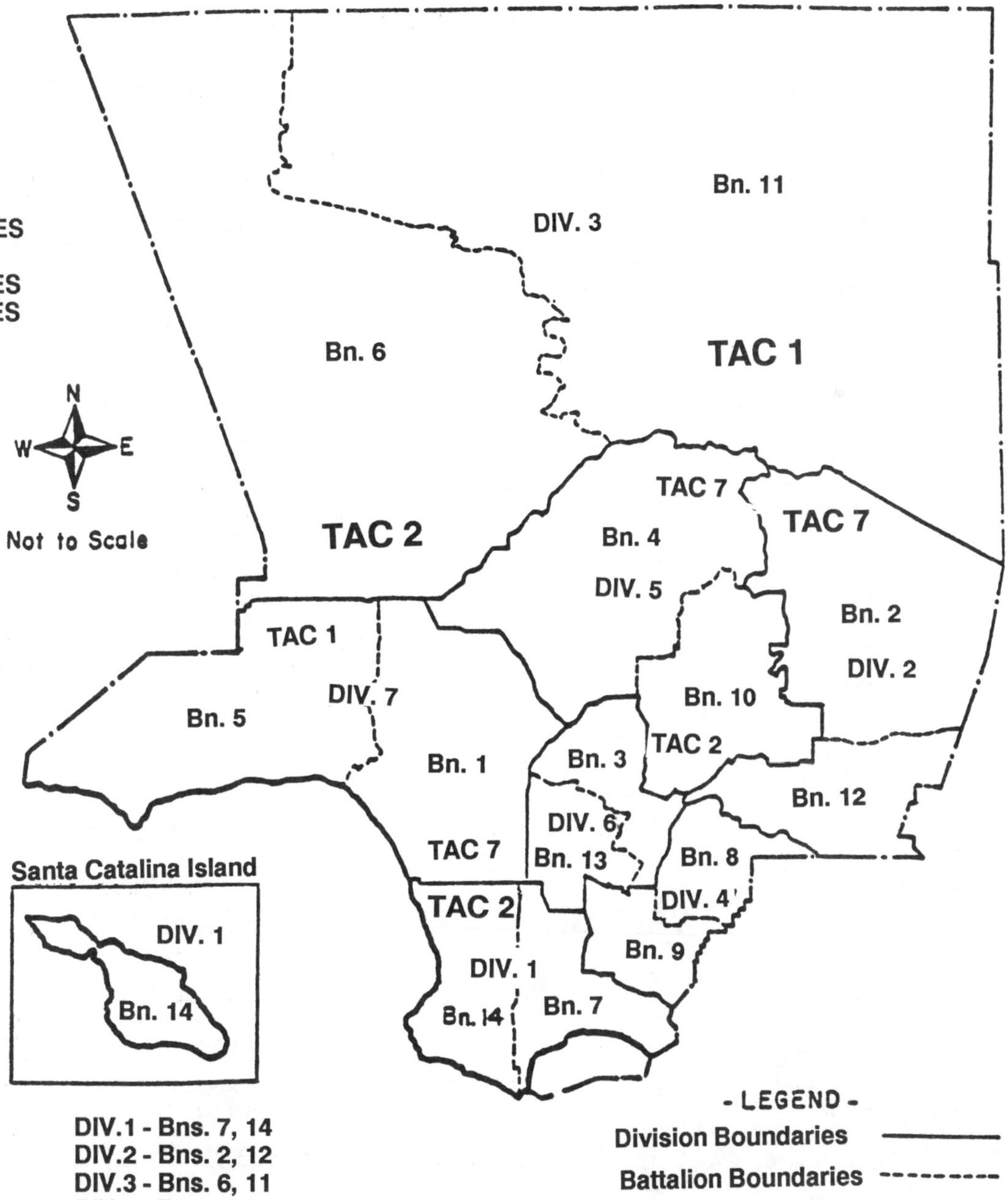

DIV.1 - Bns. 7, 14
DIV.2 - Bns. 2, 12
DIV.3 - Bns. 6, 11
DIV.4 - Bns. 8, 9
DIV.5 - Bns. 4, 10
DIV.6 - Bns. 3, 13
DIV.7 - Bns. 1, 5

DIVISION 1

(Battalions 7,14)
CARSON
LAWNDALE
LOMITA
PALOS VERDES ESTATES
ROLLING HILLS
ROLLING HILLS ESTATES
RANCHO PALOS VERDES
SOUTHGATE

DIVISION 2

(Battalions 2,12)
AZUSA
CLAREMONT
DIAMOND BAR
GLENDORA
INDUSTRY
LA PUENTE
SAN DIMAS
WALNUT

DIVISION 3

(Battalions 6,11)
LANCASTER
PALMDALE
SANTA CLARITA

DIVISION 4

(Battalions 8, 9)
ARTESIA
BELLFLOWER
CERRITOS
HAWAIIAN GARDENS
LAKEWOOD
LA MIRADA
NORWALK
PARAMOUNT
WHITTIER

DIVISION 5

(Battalions 4,10)
BALDWIN PARK
BRADBURY
DUARTE
IRWINDALE
LA CANADA-FLINTRIDGE
ROSEMEAD
SOUTH EL MONTE
TEMPLE CITY

DIVISION 6

(Battalions 3,13)
BELL
BELL GARDENS
COMMERCE
CUDAHY
HUNTINGTON PARK
MAYWOOD
PICO RIVERA

DIVISION 7

(Battalions 1,5)
AGOURA HILLS
CALABASAS
HIDDEN HILLS
MALIBU
WEST HOLLYWOOD
WESTLAKE VILLAGE

LOS ANGELES COUNTY

SOUTH BAY REGIONAL PUBLIC COMMUNICATIONS AUTHORITY

CHANNEL	INPUT	OUTPUT	TONE	CALL SIGN
Police Dispatch 1	473.6375	**470.6375R**	203.5	KDL-483
El Segundo (3)				
Hawthorne (5)				
Manhattan Beach (2)				
Police Dispatch 2	473.3875	**470.3875R**	203.5	KDL-483
Hawthorne (5)				
Police Dispatch 3	473.8125	**470.8125R**	203.5	KDL-483
El Segundo (3)				
Gardena (8)				
Police Tactical 1	474.1125	**471.1125R**	203.5	KDL-883
Police Tactical 2	473.3125	**470.3125R**	203.5	KDL-483
Police CLEMARS		154.920	none	N/A
Police Mutual Aid		153.860	167.9	N/A
Fire South Bay		**154.355**	103.5	KDW-493
El Segundo (30)				
Gardena (80)				
Hawthorne (50)				
Hermosa Beach (10)				
Manhattan Beach (20)				
Fire Mutual Aid WHITE 1		154.280	none	N/A
Fire MDT's	810.2375	855.2375	none	N/A

VERDUGO FIRE COMMUNICATIONS

CHANNEL	INPUT	OUTPUT	TONE	CALL SIGN
RED 1 - Burbank FD (Fireground & Tactical)	473.1375	470.1375R	156.7	WIK-334
RED 2 - Glendale FD (Fireground & Tactical)	473.2625	470.2625R	151.4	WIJ-366
RED 3 - Pasadena FD (Fireground & Tactical)	485.2375	482.2375R	107.2	WIL-459
RED 4 - Verdugo Dispatch	473.2125	**470.2125R**	107.2	WIL-592
RED 5 - Verdugo Command	485.3875	**482.3875R**	151.4	WIK-334
RED 6 - Burbank Fireground		460.3875	107.2	N/A
RED 7 - Glendale Fireground		460.5625	151.4	N/A
RED 8 - Pasadena Fireground		460.5875	107.2	N/A
RED 9 - Statewide FIREMARS		487.2375	179.9	WIJ-645
Burbank Incident Command	473.1875	470.1875R	151.4	WIK-334
Glendale Incident Command	473.0875	470.0875R	107.2	WIJ-366
Pasadena Incident Command	485.5375	482.5375R	151.4	WIJ-442
Mobile Data Terminals	814.2375	859.2375R	none	KNFU-724

LOS ANGELES COUNTY

EMERGENCY MEDICAL SERVICES

AMBULANCE & RESCUE

NAME	LOCATION	FREQUENCY	CALL SIGN
Adams	South Gate	155.220	KMF-819
Adams	South Gate	855.5875R	WNQL-544
Adams	La Crescenta	938.175 R	WNIQ-621
Allen's Professional	Los Angeles	42.960	KYA-913
Allen's Professional	Los Angeles	155.220	KLY-908
AME Ambulance	Whittier	155.295	N/A
AME Ambulance	Whittier	155.110	N/A
Bowers	Long Beach	47.620	KMD-456
California Search & Res.	Los Angeles	33.080	KA-69235
California Search & Res.	Los Angeles	155.160	KA-69235
Crippen Ambulance	Covina	47.620	KMG-790
Crippen Ambulance	Covina	155.280	KMG-790
Doctor's	Los Angeles	155.205	N/A
Emergency Rescue	Monrovia	155.220	WNKW-787
Event Medical Service	Cerritos	151.625	KA-69129
Event Medical Service	Cerritos	151.955	KA-69130
Event Medical Service	Los Angeles	151.865	N/A
Event Medical Service	Los Angeles	853.4875R	WNSK-719
Goodhew - Dispatch	Los Angeles	46.040	KMB-867
Goodhew	Los Angeles	155.355	N/A
Greybor	Los Angeles	463.575 R	WNGM-769
Greybor	Los Angeles	850 MHz	WNJG-694
Guardian Ambulance	Los Angeles	935 MHz	WNVY-205
High Desert Rangers	Lancaster	47.620	KRV-205
High Desert Rangers	Lancaster	155.160	KRV-205
High Desert Rangers	Lancaster	464.575 R	WNFN-957
Malibu EMS	Malibu	155.160	KNHK-428
Malibu EMS	Malibu	155.280	KNHK-428
Mauran	Sylmar	461.600 R	KNEQ-528
Mauran	Sylmar	850 MHz	WNNU-677
Medic	Los Angeles	155.400	KD-50437
Medic 1 - Dispatch	San Dimas	33.040	WNGB-871
Medic 1	San Dimas	463.0375R	WNGB-871
Medivac	El Monte	47.660	KMD-728
Medivac	El Monte	463.0625R	KMD-728
Medivac	El Monte	463.1875R	KMD-728
Mercy Air Ambulance	Los Angeles	155.220	WNSN-323
McCormick	Inglewood	155.220	KNCR-233
Newhall Ambulance	Chatsworth	33.160	KSL-604
Oxnard	Oxnard	155.235	KNGD-261

LOS ANGELES COUNTY

EMERGENCY MEDICAL SERVICES

AMBULANCE & RESCUE

NAME	LOCATION	FREQUENCY	CALL SIGN
Professional	Glendale	45.960	WSW-701
Professional - Dispatch	Glendale	46.000	WSW-701
Professional	Glendale	155.280	KNDT-857
Rescue 1	Santa Fe Springs	33.080	WNCL-503
Risher	Montebello	851.5875R	KNHY-651
Risher	Montebello	852.8875R	WNMK-826
Santa Monica-Malibu	Malibu	47.460	KNFB-450
Scheafer's	La Crescenta	461.950 R	WNGN-545
Scheafer's	La Crescenta	464.625 R	WNGN-545
Scheafer's	Southern Calif.	43.000	WZJ-657
Scheafer's	Southern Calif.	47.580	KMB-566
Service Search & Rescue	Glendale	33.100	KWQ-655
Southwest Desert Rescue	Palmdale	155.160	KA-75967
Southwest Desert Rescue	Palmdale	155.220	KA-75967
Southwest Desert Rescue	Palmdale	155.400	KA-75967
Tri-Valley EMS	Burbank-Glendale	155.235	KDB-316
Universal Health	Westlake	155.280	WNMK-667
Wilson Ambulance	Palmdale	47.540	KAY-959
Wilson Ambulance	Palmdale	460.550 R	WNPK-926

AMERICAN RED CROSS

CHANNEL	FREQUENCY
CH-1 National Disaster Net	47.420
CH-2 Southern California	47.500
CH-3 Southern California	47.460
CH-4 Southern California	47.540
CH-5 Southern California	47.580
Operations Repeaters	463.475 R
Operations Repeaters	**463.500 R**
Operations Repeaters	463.975 R
Operations Repeaters	507.8625R
Portable Repeaters	**462.1875R**
Portable Repeaters	463.3375R
Portable Repeaters	463.4125R
Portable Repeaters	463.4875R
Portable Repeaters	463.6375R
Portable Repeaters	464.0625R
Portables	154.515
Portables	154.540
Portables	154.600

LOS ANGELES COUNTY

EMERGENCY MEDICAL SERVICE

AMERICAN RED CROSS

CHANNEL	FREQUENCY
REACT Liaison	462.675 R
Sitor-A Mode	3815.5 KHz
Sitor-A Mode	6998.5 KHz
Sitor-A Mode	13569.5 KHz
Sitor-A Mode	13915.0 KHz
Sitor-A Mode	13950.0 KHz
Sitor-A Mode	13965.0 KHz

The Red Cross now rarely uses its low band frequencies. The UHF repeaters, Sitor, and cellular & satellite telephones handle most of the traffic during a major disaster.

HOSPITALS

NAME	LOCATION	FREQUENCY	CALL SIGN
Alsa Medical Center	Hollywood	151.925	WST-209
Bay Harbor Hospital	Bay Harbor	461.225 R	WNFJ-909
Bay Harbor Hospital	Bay Harbor	462.850 R	KNAT-417
Beverly Hospital	Montebello	152.0075	KVL-465
Beverly Hospital	Montebello	155.205	WXB-719
Beverly Hospital	Montebello	453.175 R	WNKB-631
Beverly Medical Center	Beverly Hills	154.570	KB-20881
Beverly Medical Center	Beverly Hills	461.100	KB-43573
Brotman Medical Center	Culver City	151.655	KNDJ-972
Brotman Medical Center	Culver City	151.955	KNDJ-972
Burbank Community Hosp.	Burbank	151.925	KNDP-633
California Hospital	Los Angeles	464.325 R	WZA-492
Casa Colina Hospital	Pomona	151.895	WZJ-913
Casa Colina Hospital	Pomona	151.955	WZJ-913
Cedars Sinai Medical	Los Angeles	463.650 R	KNIG-703
Centinela Hospital	Inglewood	155.265	KNBJ-902
Centinela Hospital	Inglewood	155.280	KNBJ-902
Centinela Hospital	Inglewood	453.025 R	KNBJ-902
Centinela Hospital	Inglewood	462.800 R	KSH-321
Centinela Hospital	Inglewood	464.975 R	KZN-896
Centinela Hospital	Los Angeles	462.800 R	KIC-587
Charter Community	Hawaiian Gardens	154.540	KNGW-384
City of Hope	Duarte	152.0075	KTB-232
City of Hope	Duarte	453.075	WNLK-544
Corona Community	Corona	461.0625	KB-79405

LOS ANGELES COUNTY

EMERGENCY MEDICAL SERVICE

HOSPITALS

NAME	LOCATION	FREQUENCY	CALL SIGN
Coronado Medical Center	Montrose	461.975 R	KNBC-683
Daniel Freeman Hospital	Inglewood	151.955	KVL-806
Daniel Freeman Hospital	Inglewood	158.460	KVL-806
Foothill Presbyterian	Glendora	35.680	N/A
Foothill Presbyterian	Glendora	151.625	KA-94009
Gardena Memorial	Gardena	461.125 R	KNFZ-807
Glendale Adventist	Glendale	464.325 R	KUO-318
Glendale Memorial	Glendale	461.1125	KB-82990
H.R. Medical Center	Santa Paula	461.075 R	KNAS-649
Hawthorne Community	Hawthorne	461.225 R	KNDS-744
Hoag Memorial Hospital	Newport Beach	462.800 R	KKY-846
Hollywood Presbyterian	Los Angeles	151.955	KNGC-827
Hollywood Presbyterian	Los Angeles	461.025 R	WNFA-769
Huntington Memorial	Pasadena	461.050 R	WNCG-735
Kaiser Foundation	Bellflower	461.475 R	KNAR-237
Kaiser Foundation	Los Angeles	462.0125	KT-3801
Kaiser Foundation	Los Angeles	462.0375	KB-24801
Kaiser Foundation	Panorama City	461.675 R	WNBW-214
Kaiser Hospital - Security	Harbor City	464.350 R	KTA-961
Las Encinas Hospital	Pasadena	463.775 R	WQW-278
Long Beach Comm. Hosp.	Long Beach	151.925	KYG-827
Los Alamitos General	Los Alamitos	151.895	WSY-249
Los Angeles Doctors	Los Angeles	463.5625	KA-68406
Methodist Hospital	Arcadia	462.825 R	WRB-875
Midway Hospital	Los Angeles	151.955	KLO-876
Northridge Hospital	Northridge	461.750 R	WNAS-257
Pleasant Valley Hospital	Santa Paula	151.775	WRJ-868
Pleasant Valley Hospital	Santa Paula	461.450 R	WNGB-348
Presbyterian Hospital	Whittier	151.895	KUD-544
Queen of Angels	Los Angeles	151.865	KNHT-740
Robert F. Kennedy	Hawthorne	151.775	WNDX-257
Robert F. Kennedy	Hawthorne	464.675 R	WNDX-257
Santa Fe Memorial	Los Angeles	33.040	N/A
Santa Monica Hospital	Santa Monica	461.050 R	WNFR-561
Santa Teresita	Duarte	157.450	WNCX-992
St. Francis Hospital	Lynwood	453.075 R	KUX-246
St. Francis Hospital	Lynwood	462.600 R	KAD-3147
St. Francis Hospital	Lynwood	463.375 R	KWK-571

LOS ANGELES COUNTY

EMERGENCY MEDICAL SERVICES

HOSPITALS

NAME	LOCATION	FREQUENCY	CALL SIGN
St. John Hospital	Santa Monica	151.715	WNDJ-957
St. Joseph Medical	Burbank	35.080	KNHW-960
St. Joseph Medical	Burbank	462.825 R	KNJF-704
St. Joseph Medical	Burbank	464.975 R	WNKT-365
St. Mary Hospital	Lancaster	151.835	WNDR-499
Tarzana Medical Center	Tarzana	463.925	KB-69999
U.C.L.A. Medical Center	Indio	463.375 R	KSN-401
White Memorial Med. Ctr.	Los Angeles	43.640	N/A
White Memorial Med. Ctr.	Los Angeles	461.675 R	KSE-208
Whittier Hospital	Whittier	462.850 R	WNAY-307
Valley Presbyterian	Van Nuys	461.550 R	WNBA-559
Valley Presbyterian	Van Nuys	461.675 R	WNBA-559

PARAMEDICS

CHANNEL	BASE FREQUENCY	MOBILE FREQUENCY
MED-1	463.000	468.000
MED-2	463.025	468.025
MED-3	463.050	468.050
MED-4	463.075	468.075
MED-5	463.100	468.100
MED-6	463.125	468.125
MED-7	463.150	468.150
MED-8	463.175	468.175

PARAMEDIC ASSIGNMENTS

HOSPITAL	LOCATION	BASE	MOBILE
Antelope Valley Community	Lancaster	463.150	468.150
Brotman Medical Center	Culver City	463.075	468.075
Cedars Sinai Med Ctr.	Los Angeles	463.000	468.000
Centinela Hospital	Inglewood	463.150	468.150
Charter Suburban Hospital	Paramount	463.125	468.125
Daniel Freeman Memorial	Inglewood	463.125	468.125
Downey Community Hospital	Downey	463.025	468.025

There are six tones used on the MED channels in Los Angeles County:

141.3, 146.2, 151.4, 156.7, 162.2, 203.5 Htz

LOS ANGELES COUNTY

EMERGENCY MEDICAL SERVICES

PARAMEDIC ASSIGNMENTS

NAME	LOCATION	BASE	MOBILE
Glendale Adventist Medical Ctr	Glendale	463.000	468.000
Glendora Community Hospital	Glendora	463.125	468.125
Henry Mayo Medical Center	Santa Clarita	463.075	468.075
Hollywood Community	Los Angeles	463.050	468.050
Holy Cross Medical	Mission Hills	463.125	468.125
Huntington Memorial Hospital	Pasadena	463.100	468.100
Martin Luther King Hospital	Los Angeles	463.175	468.175
Methodist Hospital	Arcadia	463.175	468.175
Northridge Medical Center	Northridge	463.000	468.000
Pomona Valley Hospital	Pomona	463.100	468.100
Presbyterian Inter-Community	Whittier	463.050	468.050
Queen of Angels Hospital	Hollywood	463.150	468.150
Queen of the Valley Hospital	West Covina	463.075	468.075
Robert F. Kennedy Memorial	Hawthorne	463.175	468.175
San Dimas Community	San Dimas	463.150	468.150
San Pedro Peninsula	San Pedro	463.175	468.175
Saint Joseph Medical	Burbank	463.100	468.100
Saint Mary Medical Center	Long Beach	463.100	468.100
Santa Monica Medical Center	Santa Monica	463.100	468.100
Santa Teresita	Duarte	463.000	468.000
South Bay Hospital	Redondo Beach	463.000	468.000
Torrance Memorial Hospital	Torrance	463.050	468.050
UCLA Medical Center	Los Angeles	463.025	468.025
USC Medical Center	Los Angeles	463.075	468.075
Verdugo Hills Hospital	Glendale	463.025	468.025
Westlake Medical Center	Westlake Village	463.150	468.150

HOSPITAL EMERGENCY NETWORKS

CHANNEL	FREQUENCY
Hospital Calling - Los Angeles County	**462.975 R**
Hospital Calling - Orange County	**462.950 R**
Hospital HEAR - Nationwide	**155.340**
Hospital HEAR - Regional	**155.280**

HEAR = Hospital Emergency Administrative Network

PL Tones: 103.5, 203.5

LOS ANGELES COUNTY

AIR QUALITY MANAGEMENT DISTRICT

CHANNEL	FREQUENCY
Air Pollution Alerts	**39.980**
Operations Primary	37.220
Operations Secondary	37.300
Operations Tactical	37.120
Operations Tactical	37.160

Call Sign Identification:

Hauser Peak KMF-964; Palos Verdes KMF-964; Santiago Peak KMF-964; Verdugo Peak KMF-964

PL Tone: 151.4 Htz

BEACHES & HARBORS DEPARTMENT

CHANNEL	FREQUENCY
Lifeguards - Avalon	37.060
Lifeguards - Avalon	37.080
Lifeguards - Hermosa Beach	37.060
Lifeguards - Marina Del Rey	47.540
Lifeguards - Marina Del Rey	155.160
Lifeguards - Point Dume	37.080
Lifeguards - Santa Monica	37.200
Lifeguards - Santa Monica	37.360
Lifeguards - Yuma Beach	37.080
Lifeguards - Yuma Beach	155.160
Lifeguards Secondary BROWN	453.150
Lifeguards Countywide	37.020
Lifeguards Countywide	37.200
Lifeguards Countywide	37.240
Lifeguards Countywide	37.280
Lifeguards Countywide	37.320
Lifeguards Countywide	37.340
Lifeguards Countywide	37.380
Maintenance Division	153.980
Marina Del Rey Patrol	156.425
Marina Del Rey Patrol	156.575

Call Sign Identification:

Avalon WNUS-602, Hermosa Beach KMC-219, Los Angeles KMF-288; Palos Verdes KMF-287, Santa Monica KMF-295, Zuma Beach WXF-746

PL Tones: 88.5, 103.5 Htz

LOS ANGELES COUNTY

LOS ANGELES COUNTY TRUNKING SYSTEM

A new 800 MHz General Electric trunking system was recently installed which provides communications for most LA County agencies including the Animal Control, District Attorney, Marshal's Office, General Services Department, Health Department, Road Department, Public Works, and Department of Parks & Recreation. The Southern California AQMD (Air Quality Management District), the Department of Beaches and Harbors, the Sheriff's Office, and County Fire are remaining on their current systems.

Countywide:
866.0875, 866.250, 866.275, 866.300, 866.5875, 866.775, 866.800, 866.9375, 867.225, 867.250, 867.275, 867.725, 867.775, 867.800, 868.0875, 868.250, 868.275, 868.725, 868.775, 868.800

Los Angeles Basin:
866.0625, 866.225, 866.4375, 866.725, 866.750, 866.9625, 867.0625, 867.0875, 867.300, 867.4375, 867.5875, 867.750, 867.9375, 867.9625, 868.0625, 868.225, 868.300, 868.4375, 868.5875, 868.750

METROPOLITAN TRANSIT AUTHORITY

CHANNEL	INPUT	OUTPUT	TONE	CALL SIGN
Bus Operations				
CH-1 Valley Area	475.5375	**472.5375RB**	136.5	KUP-236
CH-2A Valley Area	475.6375	**472.6375RB**	136.5	KUP-236
CH-2B Eastern Area	475.6375	**472.6375RB**	136.5	KUP-236
CH-3 South / Harbor Area	475.5625	**472.5625RB**	136.5	KUP-236
CH-4 South / Harbor Area	475.6625	**472.6625RB**	136.5	KUP-236
CH-5 Downtown Area	475.5875	**472.5875RB**	136.5	KUP-236
CH-6 Central Area	475.6875	**472.6875RB**	136.5	KUP-236
CH-7 Central Area	475.7125	**472.7125RB**	136.5	KUP-236
CH-8 Central Area	475.6125	**472.6125RB**	136.5	KUP-236
CH-9A Central / Valley Area	474.3625	**471.3625RB**	136.5	KUP-236
CH-9B Central / Valley Area	474.3625	**471.3625RB**	136.5	KUP-236
CH-10A Valley Area	474.3375	**471.3325RB**	136.5	KUP-236
CH-10B East Area	474.3375	**471.3375RB**	136.5	KUP-236
CH-10C South / Harbor Area	474.3375	**471.3375RB**	136.5	KUP-236
CH-11 South / Harbor Area	474.2125	**471.2125RB**	136.5	KUP-236
CH-12 Wide Area	474.9875	**471.9875R**	136.5	KUP-236
CH-13 East / Valley Area	475.0375	**472.0375RB**	136.5	KUP-236
CH-14 Downtown Area	475.0625	**472.0625RB**	136.5	KUP-236
CH-15 Western Area	475.1375	**472.1375RB**	136.5	KUP-236
Bus Operations - Maintenance	458.275	453.275 R	131.8	KRE-249
Bus Operations - Maintenance		43.780 R	114.8	KMA-454
Bus Staff & Supervisors	458.325	453.325 R	131.8	KSO-950

RB = Remote Base
Multiple input tones

LOS ANGELES COUNTY

METROPOLITAN TRANSIT AUTHORITY

CHANNEL	INPUT	OUTPUT	TONE	CALL SIGN
Bus Staff & Supervisors	458.475	453.475 R	131.8	KRW-210
Transit Police	458.625	**453.625 R**	131.8	KLE-920
Transit Police 1	465.200	460.200 R	131.8	WNXJ-405
Transit Police 2	465.425	460.425 R	131.8	WNXJ-405
MetroLink Rail Radio System				
CH1 - Road Channel		**161.415**	91.5	N/A
(Mission Tower-Claremont)				
CH2 - Yard Operations		160.815	91.5	N/A
CH3 - Road Channel		**160.545**	91.5	N/A
(Mission Tower-Burbank Junction)				
CH4 - Maintenance of Way		160.995	91.5	N/A
S.P. Railroad Liaison		161.550	none	N/A
San Bernardino Line		161.115	91.5	N/A
San Bernardino Line		161.175	91.5	N/A
MetroRail Radio System				
MetroRail 1 - Maintenance	160.635	**160.425 R**	136.5	WNNX-335
MetroRail 2 - Red Line Ops	160.755	**160.695 R**	136.5	WNNX-335
MetroRail 3 - Red Line Yard		161.505	136.5	WNNX-335
MetroRail 4 - Red Line Data	160.935	161.265 R	136.5	WNNX-335
MetroRail 5 - Red Line		161.145	136.5	WNNX-335
MetroRail 6 - Red LIne		160.725	136.5	WNBY-212
Metro Blue Line - LASO	487.1625	484.1625	186.2	N/A
Metro Blue Line - Operations	472.4875	472.4875R	136.5	N/A
Metro Blue Line - Yard	474.9875	471.9875R	136.5	N/A

OTHER SERVICES

CHANNEL	INPUT	OUTPUT	TONE	CALLSIGN
County MDT's	807.3125	852.3125R	none	KNER-448
County Medical Examiner	815.2625	860.2625R	107.2	KNER-447
District Attorney	458.575	453.575 R	107.2	WXP-574
District Attorney	458.775	453.775 R	107.2	WRG-879
Parks & Recreation Dept.	458.150	453.150 R		KWL-396
Public Works - Dams		44.840	none	KMB-285
Public Works - Dams		44.960	none	KMB-285
Southeast Mosquito District		37.260	none	KIG-770
Southeast Mosquito District		39.900	none	KIG-770

Los Angeles County is also implementing new radio systems on the following output frequencies.

470.175 / 470.200 / 470.225 / 470.250 / 470.475 / 470.975
482.100 / 482.125 / 482.175 / 482.200 / 482.300 / 482.475

LOS ANGELES COUNTY

AGOURA HILLS

CHANNEL	INPUT	OUTPUT	TONE	CALL SIGN
Police - L.A.S.O. #22	485.9125	**482.9125R**	186.2	WII-794
Fire - LACoFD Dispatch	473.5375	**470.5375R**	151.4	WGT-876
Fire - LACoFD Battalion 5	473.6125	**470.6125R**	151.4	WGH-876
Fire - LA County TAC 1		154.430	none	KMB-352
Local Government		153.905	127.3	WNWC-291
Las Virgenes Municipal Water	455.100	451.100 R		WNKA-633
Unified Schools	467.550	462.550 R		KAD-7802

ALHAMBRA

CHANNEL	INPUT	OUTPUT	TONE	CALL SIGN
Police Primary (F-1)	474.1375	**471.1375R**	127.3	KWT-671
Police Secondary (F-2)	474.0125	**471.0125R**	127.3	KZR-633
Police Direct (F-3)		471.1375	127.3	KWT-671
Police Direct (F-4)		471.0125	127.3	KZR-633
Police CLEMARS		154.920	none	WNFE-775
Fire & Medics (70)		**153.995**	136.5	KME-405
Fire San Marino FD		154.175	123.0	KME-405
Fire Mutual Aid WHITE 1		154.280	none	WNDS-204
Fire Mutual Aid WHITE 2		154.265	none	WNDS-204
Fire Mutual Aid WHITE 3		154.295	none	WNDS-204
Fire Tactical		153.890	136.5	WNDS-204
Fire Tactical		153.770	136.5	WNDS-204
Fire Tactical Reported		46.200	none	N/A
Fire Tactical Reported		46.500	none	N/A
Local Government	458.450	453.450 R	127.3	KML-741
Unified School District - Security		151.685	203.5	KNIV-439
Unified School District - Buses	466.075	461.075 R	118.8	WNAR-449

ALTADENA

CHANNEL	INPUT	OUTPUT	TONE	CALL SIGN
Police - L.A.S.O. #7	486.9875	**483.9875R**	186.2	WII-798
Fire - LACoFD	473.4125	**470.4125R**	151.4	KMB-355
Fire - LACoFD Dispatch	473.5375	**470.5375R**	151.4	KMB-355
Fireground		154.295	none	KMB-355
Lincoln Avenue Water District		37.520	141.3	KEJ-449

LOS ANGELES COUNTY

ARCADIA

CHANNEL	INPUT	OUTPUT	TONE	CALL SIGN
Police Primary (F-1)	510.0625	**507.0625R**	110.9	KXN-419
Police Direct (F-2)		507.0625	110.9	KXN-419
Police Secondary (F-3)	509.5625	**506.5625R**	110.9	KCX-723
Police Direct (F-4)		506.5625	110.9	KCX-723
Police CLEMARS		154.920	none	KYU-877
Police Low Band		45.420	none	WNQU-630
Police MDT's	823.6375	868.6375R	none	WNZB-824
Fire & Medic Dispatch		**154.370**	none	KFM-662
Fire Portable Repeater	153.830	154.370 R	218.1	KFM-662
Fireground		153.890	136.5	WNQE-913
Fire Mutual Aid WHITE 1		154.280	none	KMF-662
Fire El Monte F.D.		154.085	136.5	KNGQ-555
Public Works		155.115		WNZU-543
Public Works		155.880	179.9	KMK-951
Unified School District		30.920	none	KNGM-808

ARTESIA

CHANNEL	INPUT	OUTPUT	TONE	CALL SIGN
Police - L.A.S.O #13	486.1375	**483.1375R**	186.2	WII-795
Fire - LACoFD Battalion 9	473.5625	**470.5625R**	151.4	KME-463
Fire - LACoFD Dispatch	473.5375	**470.5375R**	151.4	KME-463
Fire - LA County TAC 1		154.430	none	KME-463
Fire Mutual Aid WHITE 1		154.280	none	KME-463

AVALON

CHANNEL	INPUT	OUTPUT	TONE	CALL SIGN
Police - L.A.S.O. #18	487.0375	**484.0375**R	186.2	WII-790
Fire - LACoFD Battalion 14	473.5625	**470.5625R**	151.4	KNIP-331
Fire - LACoFD Dispatch	473.5375	**470.5375R**	151.4	KNP-331
Fire - LA County TAC 2		154.340	none	KQL-546
Fire Low Band		46.360	103.5	KQL-546
Fire Mutual Aid WHITE 1		154.280	none	WNLW-582
Fire Mutual Aid WHITE 2		154.265	none	WNLW-582
Fire Mutual Aid WHITE 3		154.295	none	WNLW-582
Harbor Department		45.540	none	WNKX-957
Harbor Department		46.540	none	WNKX-957
Harbor Department VHF 9		156.450	none	KUF-763
Harbor Department VHF 12		156.600	none	KUF-763
Island Conservatory	159.450	151.145 R	110.9	WNNY-641

LOS ANGELES COUNTY

AZUSA

CHANNEL	INPUT	OUTPUT	TONE	CALL SIGN
Police Primary (F-1)		**154.845**	none	KMA-238
Police Secondary (F-2)		**155.490**	none	KMA-238
Police CLEMARS (F-3)		154.920	none	KRW-426
Police CLERS	155.430	158.790 R	186.2	KNFU-805
Fire - LACoFD Battalion 2	473.4125	**470.4125R**	151.4	WGH-876
Fire - LACoFD Dispatch	473.5375	**470.5375R**	151.4	WGH-876
Fireground		154.295	none	KNFU-805
Public Works		39.580	none	KNBV-316
Public Works		39.960	none	WPBQ-964
Unified School District		151.655	146.2	KNBT-767
Unified School District	469.600	464.600 R		WNHT-738
Utility Department	456.100	451.100 R	D243	KME-591
Valley Water Department		47.700		KMH-753

BALDWIN PARK

CHANNEL	INPUT	OUTPUT	TONE	CALL SIGN
Police Dispatch (F-1)	509.175	**506.175 R**	131.8	WIL-209
Police Secondary (F-2)	509.275	**506.275 R**	131.8	WIL-209
Police Direct (F-3)		506.175	131.8	WIL-209
Police Direct (F-4)		506.275	131.8	WIL-209
Police CLEMARS		154.920	none	KRS-232
Fire - LACoFD Battalion 10	473.4125	**470.4125R**	151.4	N/A
Fire - LACoFD Dispatch	473.5375	**470.5375R**	151.4	N/A
Fire - LA County TAC 2		154.340	none	KVP-622
Public Works		39.500	none	N/A
Public Works		45.360	none	KVR-825
Unified School District - Paging		30.760	none	KRC-365
Unified School District - PD	465.550	460.550 R		WNCP-757
Uniifed School District	469.425	464.425 R		WNPV-773

BELL

CHANNEL	INPUT	OUTPUT	TONE	CALL SIGN
Police Primary (F-1)		**155.190**	107.2	KNGT-891
Police CLEMARS (F-2)		154.920	none	WNJU-262
Police Tactical (F-3)		154.040	none	KDB-449
Fire - LACoFD Battalion 13	473.6125	**470.6125R**	151.4	KTX-604
Fire - LACoFD Dispatch	473.5375	**470.5375R**	151.4	KTX-604
Fire - LA County TAC 2		154.340	none	KTX-604
Public Works		154.040	none	KDB-449

LOS ANGELES COUNTY

BELLFLOWER

CHANNEL	INPUT	OUTPUT	TONE	CALL SIGN
Police - L.A.S.O. #13	487.1375	**483.1375R**	186.2	WII-794
Fire - LACoFD Battalion 9	473.5625	**470.5625R**	151.4	WIK-430
Fire - LACoFD Dispatch	473.5375	**470.5375R**	151.4	WIK-430
Fire - LA County TAC 1		154.430	none	KME-476
Public Works		158.775	88.5	KNFX-492
Unified School District	467.000	462.000 R		WSA-654
Unified School District	469.975	464.975 R		WSZ-849
Water District		47.900	none	KMG-656

BELL GARDENS

CHANNEL	INPUT	OUTPUT	TONE	CALL SIGN
Police Primary (F-1)	458.925	**453.925 R**	127.3	KRS-304
Police Direct (F-2)		453.925	127.3	KRS-304
Fire - LACoFD Battalion 13	473.6125	**470.6125R**	151.4	WIK-446
Fire - LACoFD Dispatch	473.5375	**470.5375R**	151.4	WIK-446
Fire - LA County TAC 2		154.340	none	KCS-516
Street Department		156.105		KNGL-838

BEVERLY HILLS

CHANNEL	INPUT	OUTPUT	TONE	CALL SIGN
Police Primary (F-1)	458.525	**453.525 R**	131.8	WNFT-622
Police Car-to-Car (F-2)		453.525	131.8	WNFT-622
Police Tactical (F-3)	458.650	**453.650 R**	131.8	WNFT-622
Police Tactical (F-4)		453.650	131.8	WNFT-622
Police Traffic Units (F-5)	458.225	**453.225 R**	131.8	WNFT-622
Police Traffic Units (F-6)		453.225	131.8	WNFT-622
Police CLEMARS		154.920	none	KNHW-255
Police CLERS	155.430	158.790 R	186.2	KNHW-255
Police NLEMARS		155.475	none	KNHW-255
Fire & Medics	465.625	**460.625 R**	156.7	KUX-244
Fire Command	465.575	**460.575 R**	156.7	KUX-244
Fire Mutual Aid WHITE 1		154.280	none	KNIU-672
Emergency Services		155.280	103.5	KNIU-672
Parks & Recreation	159.285	151.460 R	131.8	WNHM-757
Street Department	156.120	151.070 R	131.8	WNED-329
Unified School District	469.875	464.875 R		KEO-874
Water Department		153.590	131.8	KMG-655

Police Identification

The first digit indicates the shift: 1 Graveyard; 2 Day; 3 Evening
The second digit represents the type of unit:
A = Two-man unit L = One Man Unit M = Motorcycle
The last digits indicate the beat area.

LOS ANGELES COUNTY

BRADBURY

CHANNEL	INPUT	OUTPUT	TONE	CALL SIGN
Police - L.A.S.O. #5	485.9875	**482.9875R**	186.2	N/A
Fire - LACoFD Battalions	473.4125	**470.4125R**	151.4	WGH-875
Fire - LACoFD Dispatch	473.5375	**470.5375R**	151.4	WGH-875
Fire - LA County TAC 2		154.340	none	N/A
Fire Mutual Aid WHITE 1		154.280	none	N/A

BURBANK

CHANNEL	INPUT	OUTPUT	TONE	CALL SIGN
BLUE 1 - Police Dispatch	473.2375	**470.2375R**	141.3	WIK-334
BLUE 2 - Police Inquiry	473.2875	**470.2875R**	146.2	WIK-334
BLUE 3 - Police Tactical	473.1625	**470.1625R**	118.8	WIK-334
BLUE 4 - Police Tactical	473.0625	**470.0625R**	82.5	WIK-334
Police CLEMARS		154.920	136.5	KQM-693
Police CLERS	155.430	158.790 R	186.2	KNEC-954
Verdugo Fire Protection District				
Fire & Medics	473.2125	**470.2125R**	107.2	KME-279
Fire Incident Command	473.1875	**470.1875R**	151.4	WIK-334
Fire Mutual Aid WHITE 1		154.280	none	WNFQ-615
Airport Police		155.655	156.7	KNEC-954
Airport Police & Fire		153.935	156.7	WZZ268
Airport Ops		155.025	156.7	WZZ-268
Airport Maintenance		173.3375		WNCQ-293
Animal Control (F-1)	485.025	482.025 R	85.4	WIK-344
Animal Control (F-2)	485.050	482.050 R	162.2	WIK-344
Local Government (Sewage & Solid Waste)	473.325	470.325 R	167.9	WIK-334
Local Government	474.075	471.075 R	179.9	WIK-334
Parks & Recreation	473.1125	470.1125R	114.8	WIK-334
Public Works	473.0125	470.0125R	127.3	WIK-334
Street Department		151.100		KMG-288
Unified School District		154.540	D151	KNAK-434
Water Department	456.050	451.050 R	192.8	KMJ-941

CALABASAS

CHANNEL	INPUT	OUTPUT	TONE	CALL SIGN
Police - L.A.S.O. #22	485.9125	**482.9125R**	186.2	WII-813
Fire - LACoFD Battalions	473.6125	**470.6125R**	151.4	WIK-472
Fire - LACoFD Dispatch	473.5375	**470.5375R**	151.4	WIK-472
Fire - LA County TAC 1		154.430	none	WIK-472
Unified Schools		151.685		KNIL-414
Unified Schools	468.2125	463.2125R	D631	KNIL-414
Water District	456.100	451.100 R		WNKA-633

LOS ANGELES COUNTY

CARSON

CHANNEL	INPUT	OUTPUT	TONE	CALL SIGN
Police - L.A.S.O. #16	487.1625	**484.1625R**	186.2	WII-808
Fire - LACoFD Battalion 7	473.5625	**470.5625R**	151.4	WIK-418
Fire - LACoFD Dispatch	473.5375	**470.5375R**	151.4	WIK-418
Fireground		154.295	none	KTX-605
Fire Mutual Aid WHITE 1		154.280	none	WNCX-598
Public Works	458.450	453.450 R	82.5	KTP-895

CERRITOS

CHANNEL	INPUT	OUTPUT	TONE	CALL SIGN
Police - L.A.S.O. #13	486.1375	**483.1375R**	186.2	WII-811
Fire - LACoFD Battalion 9	473.5625	**470.5625R**	151.4	WIK-437
Fire - LACoFD Dispatch	473.5375	**470.5375R**	151.4	WIK-437
Local Government	458.5125	453.5125R		WPCD-404
Local Government	458.600	453.600 R		KJE-339
Cerritos Comm. College	469.425	464.425 R	131.8	WNJH-756

CLAREMONT

CHANNEL	INPUT	OUTPUT	TONE	CALL SIGN
Police Primary (F-1)	509.150	**506.150 R**	131.8	KXC-905
Police Secondary (F-2)	509.250	**506.250 R**	131.8	KXC-905
Police CLEMARS		154.920	none	KVG-728
Police CLERS	155.430	158.790 R	186.2	KXC-905
Fire - LACoFD Battalion 2	473.4125	**470.4125R**	151.4	N/A
Fire - LACoFD Dispatch	473.5375	**470.5375R**	151.4	N/A
Fireground		154.295	none	N/A
Local Government	509.3875	506.3875R	103.5	KXC-905
School District - High Schools		151.655		WSC-839
School District - Buses	467.550	462.550 R		KAE-5126
School District - Security	469.150	464.150 R		WNIC-675
School District - Security	469.275	464.275 R		KNJD-863
Claremont University	469.475	464.475 R		KRO-934
Municipal Water District		173.300		KNBC-722

COMMERCE

CHANNEL	INPUT	OUTPUT	TONE	CALL SIGN
Police - L.A.S.O. #2	486.2125	**483.2125R**	186.2	WII-809
Fire - LACoFD Battalion 3	473.6125	**470.6125R**	151.4	WIK-429
Fire - LACoFD Dispatch	473.5375	**470.5375R**	151.4	WIK-429
Fire - LA County TAC 1		154.430	none	KWH-7
Public Works	458.750	453.750 R	162.2	KFT-566

LOS ANGELES COUNTY

COMPTON

CHANNEL	INPUT	OUTPUT	TONE	CALL SIGN
Police Primary (F-1)		**155.850**	91.5	KNAR-940
Police Car to Car (F-2)		**155.745**	91.5	KVV-655
Police CLEMARS (F-3)		154.920	none	N/A
Police New System				
Police Primary	473.0625	470.0625R	123.0	N/A
Police Secondary	473.2875	470.2875R	123.0	N/A
Fire & Medics	154.740	**154.205 R**	103.5	KMF-467
Fire Secondary		154.250	none	KMF-467
Fire Mutual Aid WHITE 1		154.280	none	KMF-467
Code Enforcement		151.115	none	KMK-404
Public Works		158.760	none	KMK-404
Unified School District - Police	474.075	471.075 R	123.0	N/A
Unified School District - Security	469.375	464.375 R	131.8	WNGE-440
Unified School District - Security	469.675	464.675 R		KNJV-399
Unified School District - Buses	473.1625	470.1625R	186.2	WIL-353
Unified School District - Buses	809.9125	854.9125R		WFE-232

COVINA

CHANNEL	INPUT	OUTPUT	TONE	CALL SIGN
Police Dispatch (F-1)		**154.725**	167.9	KMA-507
Police Car to Car (F-2)		**155.490**	none	KMA-507
Police CLEMARS (F-3)		154.920	none	N/A
Police Tactical (F-4)		153.860	167.9	KBD-699
Fire Dispatch (90)		**154.070**	141.3	KAY-878
Fireground		154.310	192.8	KAY-878
Fire Mutual Aid WHITE 1		154.280	none	KAY-878
Fire Mutual Aid WHITE 3		154.295	none	KAY-870
Public Works		153.860	156.7	KBD-699
Unified Schools - Security		154.515	141.3	WNKX-778
Unified Schools - Security		154.540	141.3	WNKX-778
Unified Schools - Security		154.570	141.3	WNKX-778
Water District		153.635	162.2	KJD-851

CUDAHY

CHANNEL	INPUT	OUTPUT	TONE	CALL SIGN
Police - L.A.S.O. #2	486.2125	**483.2125R**	186.2	WII-811
Bell P.D.		155.190	107.2	KNGT-891
Fire - LACoFD Battalion 9	473.5625	**470.5625R**	151.4	WGH-877
Fire - LACoFD Dispatch	473.5375	**470.5375R**	151.4	WGH-877
Fire - LA County TAC 2		154.340	none	N/A
Public Works		45.320	none	N/A

LOS ANGELES COUNTY

CULVER CITY

CHANNEL	INPUT	OUTPUT	TONE	CALL SIGN
Police Primary (F-1)	485.2625	**482.2625R**	141.3	WIJ-364
Police Secondary (F-2)	485.4625	**482.4625R**	173.8	WIJ-364
Police CLEMARS		154.920	none	KLI-326
Police CLERS	155.430	158.790 R	186.2	KZR-79
Fire Dispatch (F-1)	485.3625	**482.3625R**	156.7	WIJ-364
Fire Secondary (F-2)	485.5625	**482.5625R**	192.8	WIJ-364
Fire Mutual Aid WHITE 1		154.280	none	KYF-286
Fire Mutual Aid WHITE 2		154.265	none	KYF-286
Fire Mutual Aid WHITE 3		154.295	none	KYF-286
Local Government	485.0625	**482.0625R**	114.8	WIJ-364
Los Angeles Community College		154.650		KNJY-515
Transit District	485.1625	**482.1625R**	88.5	WIJ-286
Unified School District - Security		151.745		WNWS-834
Unified School District - Buses	467.000	462.000 R		WDA-776
West Los Angeles College		154.515		WQV-919
West Los Angeles College		154.540		WQV-919

DIAMOND BAR

CHANNEL	INPUT	OUTPUT	TONE	CALL SIGN
Police - L.A.S.O. #29	485.9375	**482.9375R**	186.2	WII-812
Fire - LACoFD Battalion 12	473.4125	**470.4125R**	151.4	WGH-876
Fire - LACoFD Dispatch	473.5375	**470.5375R**	151.4	WGH-876
Fire - LA County TAC 1		154.430	none	N/A
Local Government		935 MHz	none	WNWK-280
Country Estates Security	469.325	464.325 R	D115	WNUQ-288
High Schools - Security		154.540		KNDX-576

DOWNEY

CHANNEL	INPUT	OUTPUT	TONE	CALL SIGN
Police Primary (F-1)	155.130	**155.895 R**	156.7	WNDX-988
Police Car-to-Car (F-2)		155.640	131.8	WNDX-988
Police Direct (F-3)		155.895	156.7	WNDX-988
Police CLEMARS (F-4)		154.920	156.7	KNBY-319
Fire & Medics	154.740	**154.205 R**	103.5	WPBH-204
Fire Secondary		154.250	none	WQN-823
Fire Mutual Aid WHITE 1		154.280	none	WQN-823
Fire Mutual Aid WHITE 2		154.265	none	WQN-823
Animal Control	37.420	37.100 R	none	KAG-238
Local Government MDT's	808.4375	853.4875R	none	WNWQ-617
Street Department	458.600	453.600 R		KMJ-616
Transit District		44.500	none	WNCK-316
Unified School District	466.225	461.225 R		KNHV-637
Water Department		153.680	103.5	WNBM-772

LOS ANGELES COUNTY

DUARTE

CHANNEL	INPUT	OUTPUT	TONE	CALL SIGN
Police - L.A.S.O. #5	485.9875	**482.9875R**	186.2	WII-812
Fire - LACoFD Battalion 10	473.4125	**470.4125R**	151.4	WIK-451
Fire - LACoFD Dispatch	473.5375	**470.5375R**	151.4	WIK-451
Fire - LA County TAC 2		154.340	none	N/A
Duarte R.E.A.C.T.	467.675	462.675 R		KAD-8993
Public Works		156.015		KWE-380
Unified School District - Security		154.540		KCD-232

EL MONTE

CHANNEL	INPUT	OUTPUT	TONE	CALL SIGN
Police Primary (F-1)	473.7125	**470.7125R**	127.3	KBY-750
Police Secondary (F-2)	473.9625	**470.9625R**	127.3	KBY-750
Police Tactical (F-3)		470.7125	127.3	KBY-750
Police Tactical (F-4)		470.9625	127.3	KBY-750
Police CLEMARS		154.920	none	KRT-726
Fire Dispatch BLUE (20-29)		**154.085**	136.5	KCE-468
Fire Secondary RED		154.370	131.8	KMK-861
Fire Mutual Aid WHITE 1		154.280	none	KMK-861
Parks & Recreation		45.360		WNCY-441
Parks & Recreation		45.380		KCZ-919
Transit District		44.500		KNFU-885
Mountain View Schools - Buses		154.515		KNAZ-675
Unified School District		151.715		KUP-559
Unified High School District	466.875	461.875 R		WNMX-365

EL SEGUNDO

CHANNEL	INPUT	OUTPUT	TONE	CALL SIGN
South Bay Regional Public Communications Authority				
Police Dispatch	473.6375	470.6375R	203.5	KDL-483
Police Tactical	474.1125	471.1125R	203.5	KDL-483
Police Tactical	473.3125	470.3125R	203.5	KDL-483
Police Mutual Aid		153.860		KNCJ-394
Fire Dispatch		**154.355**	103.5	KMF-391
Fireground		154.130		KMF-391
Public Works		155.055		KNCJ-394
Unified School District	469.575	464.575 R		KNIZ-574

LOS ANGELES COUNTY

GARDENA

CHANNEL	INPUT	OUTPUT	TONE	CALL SIGN
South Bay Regional Public Communications Authority				
Police Dispatch	473.6375	470.6375R	203.5	KDL-483
Police Tactical	474.1125	471.1125R	203.5	KDL-483
Fire Dispatch		**154.355**	103.5	KMF-526
Transit District	458.425	453.425 R		WAS-573

GLENDALE

CHANNEL	INPUT	OUTPUT	TONE	CALL SIGN
Police Primary	485.2875	**482.2875R**	107.2	WIJ-366
Police Tactical 1	485.1875	**482.1875R**	151.4	WIJ-366
Police Tactical 2	485.0874	**482.0875R**	151.4	WIJ-366
Police Tactical 3	485.1375	**482.1375R**	151.4	WIJ-366
Police Tactical 4	487.1875	**484.1875R**	107.2	WIJ-366
Galleria Mall	485.050	482.050 R	156.7	WIJ-366
Police CLEMARS		154.920	none	KSK-437
Police CLERS	155.430	158.790 R	186.2	KMB-79
Fire & Medics Verdugo	473.2125	**470.2125R**	107.2	KME-279
Fire Incident Command	473.0875	**470.0875R**	107.2	WIJ-366
Fire Mutual Aid WHITE 1		154.280	none	KME-279
Fire Mutual Aid WHITE 2		154.265	none	KME-279
Fire Mutual Aid WHITE 3		154.295	none	KME-279
Fire MDT's	814.2375	859.2375R	none	N/A
Local Government 1 (Animal Control) (Sanitation)	485.0875	**482.0875R**	151.4	WIJ-366
Local Government 2 (Maintenance) (Parks & Recreation)	487.2625	**484.2625R**	107.2	WIJ-366
Glendale Comm. College		151.745		KNCW-899
Glendale Comm. College	466.0125	461.0125R		WNXY-322
Electronic Department		47.980	103.5	KME-313
Public Works (Streets & Engineering)	473.3625	470.3625R		WIJ-366
Public Works VHF		154.055	103.5	KDB-518
Public Works VHF		158.955	103.5	KDB-518
Power Department		48.460	103.5	KME-313
School District Buses	809.3375	854.3375R		WXM-577
School District Security	469.1375	464.1375		WNLP-978
School District Security	469.2375	464.2375R		WNLP-978
Water & Power Dept.	485.4875	482.4875R	151.4	WIJ-366
Water & Power Dept.	485.4875	482.5875R	107.2	WIJ-366
Water & Power Dept.	487.1875	484.1875R	151.4	WIJ-366
Water Department - Portables		158.235		KK-8622

LOS ANGELES COUNTY

GLENDORA

CHANNEL	INPUT	OUTPUT	TONE	CALL SIGN
Police Primary (F-1)		**154.845**	none	KMH-354
Police CLEMARS (F-2)		154.920	none	KQW-344
Police Tactical (F-3)		153.860	none	KAY-232
Police Tactical (F-4)		155.730*	none	N/A
Police CLERS	155.430	158.790 R	186.2	KCS-74
Fire - LACoFD Battalion 2	473.4125	**470.4125R**	151.4	WGH-876
Fire - LACoFD Dispatch	473.5375	**470.5375R**	151.4	WGH-876
Fireground		154.295	none	N/A
Public Works		153.860	none	KAY-232
Unified School District		151.715	D371	WNSA-837

*Reported

HAWAIIAN GARDENS

CHANNEL	INPUT	OUTPUT	TONE	CALL SIGN
Police - L.A.S.O. #13	486.1375	**483.1375R**	186.2	WII-813
Fire - LACoFD Battalion 9	473.5625	**470.5625R**	151.4	WGH-876
Fire - LACoFD Dispatch	473.5375	**470.5375R**	151.4	WGH-876
Fire - LA County TAC 1		154.430	none	N/A
Fire Mutual Aid WHITE 1		154.280	none	N/A
Public Works		155.085		N/A

HAWTHORNE

CHANNEL	INPUT	OUTPUT	TONE	CALL SIGN
South Bay Regional Public Communications Authority				
Police Dispatch	473.3875	470.3875 R	203.5	KDL-483
Police Tactical	473.3125	470.3125 R	203.5	KDL-483
Police Tactical	474.1125	471.1125 R	203.5	KDL-483
Fire Dispatch		**154.355**	103.5	KMF-526
Fire Tactical		154.130		KMF-526
Public Works		158.805	123.0	KTV-887
School District - Security		151.685		KJJ-736
School District - Security		154.600		KJJ-736

HERMOSA BEACH

CHANNEL	INPUT	OUTPUT	TONE	CALL SIGN
Police Dispatch	473.1875	**470.1875R**	94.8	KDA-240
Police CLEMARS		154.920	none	KWF-708
Fire & Medics (10)		**154.355**	103.5	WNMG-920
Fire Mutual Aid WHITE 1		154.280	none	WNMG-920
Fire South Bay		154.130	127.3	WNMG-920
Public Works		153.860	123.0	KKV-552

LOS ANGELES COUNTY

HACIENDA HEIGHTS

CHANNEL	INPUT	OUTPUT	TONE	CALL SIGN
Police - L.A.S.O. #14	486.6875	**483.6875R**	186.2	WII-810
Fire - LACoFD Battalion 12	473.4125	**470.4125R**	151.4	WIK-494
Fire - LACoFD Dispatch	473.5375	**470.5375R**	151.4	WIK-494
Fire - LA County TAC 1		154.430	none	KVP-631

HIDDEN HILLS

CHANNEL	INPUT	OUTPUT	TONE	CALL SIGN
Police - L.A.S.O. #10	485.9125	**482.9125R**	186.2	WII-812
Fire - LACoFD Battalion 5	473.6125	**470.6125R**	151.4	WGH-876
Fire - LACoFD Dispatch	473.5375	**470.5375R**	151.4	WGH-876
Fire - LA County TAC 1		154.430	none	N/A

HUNTINGTON PARK

CHANNEL	INPUT	OUTPUT	TONE	CALL SIGN
Police Primary (F-1)	153.785	**154.815 R**	107.2	KNGV-851
Police Car to Car (F-2)		**154.040**	107.2	KML-951
Police CLEMARS (F-3)		154.920	none	KQY-292
Police CLERS	155.430	158.790 R	186.2	WNIM-956
Fire - LACoFD Battalion 13	473.6125	**470.6125R**	151.4	WGH-876
Fire - LACoFD Dispatch	473.5375	**470.5375R**	151.4	WGH-876
Fire - LA County TAC 2		154.340	none	KSU-514
Public Works		154.040	107.2	KML-951

INDUSTRY

CHANNEL	INPUT	OUTPUT	TONE	CALL SIGN
Police - L.A.S.O. #14	486.6875	**483.6875R**	186.2	WII-809
Fire - LACoFD Battalion 12	473.4125	**470.4125R**	151.4	WIK-450
Fire - LACoFD Dispatch	473.5375	**470.5375R**	151.4	WIK-450
Fire - LA County TAC 1		154.430	none	KVP-632
Fire Mutual Aid WHITE 1		154.280	none	KVP-632
Recreation Center	465.275	**460.275 R**	118.8	WNMP-618
Public Works	458.200	453.200 R	100.0	KWX-530
Street Department*	509.4125	506.4125R		KNS-455
Water Department	456.200	451.200 R		WNRF-637

*No activity heard.

LOS ANGELES COUNTY

INGLEWOOD

CHANNEL	INPUT	OUTPUT	TONE	CALL SIGN
Police Dispatch (F-1)	465.150	**460.150 R**	192.8	KMA-309
Police Tactical (F-2)	465.275	**460.275 R**	192.8	KNHR-792
Police Tactical (F-3)	465.050	**460.050 R**	192.8	KNHR-792
Police Tactical (F-4)	465.175	**460.175 R**	192.8	KNHR-792
Police Special Ops (F-5)	465.250	460.250 R	192.8	KNHR-792
Police Narcotic Ops (F-6)	465.325	460.325 R	192.8	KNHR-792
Police South Bay Liaison		153.860	167.9	KML-730
Police Tactical		155.820	103.5	KML-730
Police MDT's	810.9875	855.9875R	103.5	WNMA-848
Fire & Medics (40)		**154.175**	114.8	KMF-670
Fire Secondary		154.130	127.3	KMF-670
Local Government Trunked		850 MHz	none	SMR
Unified School District	466.350	461.350 R		KUL-540
Unified School District	469.825	464.825 R		WNAL-927

IRWINDALE

CHANNEL	INPUT	OUTPUT	TONE	CALL SIGN
Police Primary (F-1)		**155.790**	none	WZG-911
Police CLEMARS (F-2)		154.920	none	KWF-709
Police Tactical (F-3)		155.055	none	KAV-704
Fire - LACoFD Battalion 10	473.4125	**470.4125R**	151.4	WIK-455
Fire - LACoFD Dispatch	473.5375	**470.5375R**	151.4	WIK-455
Fire - LA County TAC 2		154.340	none	N/A
Public Works		155.055		KAV-704

LA CANADA - FLINTRIDGE

CHANNEL	INPUT	OUTPUT	TONE	CALL SIGN
Police - L.A.S.O. #12	486.9875	**483.9875R**	186.2	WII-482
Fire - LACoFD Battalion 4	473.4125	**470.4125R**	151.4	WGH-876
Fire - LACoFD Dispatch	473.5375	**470.5375R**	151.4	WGH-876
Fireground		154.295	none	KVP-635
Irrigation District		153.590		KBY-761

LA HABRA HEIGHTS

CHANNEL	INPUT	OUTPUT	TONE	CALL SIGN
Police - L.A.S.O. #14	486.6875	**483.6875R**	186.2	N/A
Fire - LACoFD Dispatch	473.5375	**470.5375R**	151.4	N/A
Volunteer Fire Department	33.360	33.740 R	79.9	KDN-945
Water District		47.880	none	WNAK-215

LOS ANGELES COUNTY

LAKEWOOD

CHANNEL	INPUT	OUTPUT	TONE	CALL SIGN
Police - L.A.S.O. #13	486.1375	**483.1375R**	186.2	WII-822
Fire - LACoFD Battalion 9	473.5625	**470.5625R**	151.4	WGH-876
Fire - LACoFD Dispatch	473.5375	**470.5375R**	151.4	WGH-876
Fire - LA County TAC 1		154.430	none	KME-460
Public Works		153.860	156.7	KML-641

LA MIRADA

CHANNEL	INPUT	OUTPUT	TONE	CALL SIGN
Police - L.A.S.O. #4	486.7625	**483.7625R**	186.2	N/A
Fire - LACoFD Battalion 8	473.5625	**470.5625R**	151.4	WIK-456
Fire - LACoFD Dispatch	473.5375	**470.5375R**	151.4	WIK-456
Fireground		154.295	none	KME-458
Public Works	474.0375	471.0375R		KUU-520
Unified School District	466.775	461.775 R		KLK-966

LANCASTER

CHANNEL	INPUT	OUTPUT	TONE	CALL SIGN
Police - L.A.S.O. #11	485.8625	**482.8625R**	186.2	WII-843
Fire - LACoFD Battalion 11	473.4875	**470.4875R**	151.4	WII-226
Fire - LACoFD Dispatch	473.5375	**470.5375R**	151.4	WII-226
Fire - LA County TAC 1		154.430	none	KBG-521
Local Government	458.650	**453.650 R**		WNXE-246
Local Government	155.865	153.920 R	88.5	WNLZ-731
Local Government		155.715	88.5	WNBE-532
Local Government-Telemetry	458.2875	453.2875R	none	WNZQ-837
Local Government-Telemetry	458.4625	453.4625R	none	KD-48994
Local Government-Telemetry	458.6375	453.6375R	none	KD-52392
Local Government-Telemetry	458.9375	453.9375R	none	WNVX-672
Mosquito Control	458.850	453.850 R		WNPE-496
Parks & Recreation	155.100	154.100 R		WNBE-432
Antelope Valley Comm College	466.175	461.175 R	131.8	WNCH-918
Antelope Valley Fairgrounds	458.425	453.425 R		WNMP-665
Antelope Valley High School	806.0625	851.0625R	103.5	WNSZ-930
Antelope Valley Schools		155.175	100.0	KNCC-511
Antelope Valley Schools		155.235	100.0	KNCC-511
Antelope Valley Schools	809.9875	854.9875R	none	WNJU-495
Antelope Valley Schools	814.2375	859.2375R		WNJU-495
Antelope Valley Water District	456.150	451.150 R		WBQ-843
Eastside High Schools		155.205		KSJ-229
Lancaster School District	467.0875	462.0875R		WNLQ-690
Lancaster School District	469.375	464.375 R		WNLQ-690
Westside Union High School		155.205		WNMZ-588
Wilsona School District		155.295		WNUJ-455

LOS ANGELES COUNTY

LA PUENTE

CHANNEL	INPUT	OUTPUT	TONE	CALL SIGN
Police - L.A.S.O. #14	486.6875	**483.6875R**	186.2	N/A
Fire & Medics (12)	473.5375	**470.5375R**	151.4	WIK-433
Fire - LA County TAC 1		154.430	none	KVP-639
Local Government		153.860		KBW-831
Basset USD - Police		**154.650**		WNBE-374
Basset USD - Security		151.655	179.9	WNBL-685
Basset USD - Security		151.895		KUU-955
Rowland USD	469.350	464.350 R		KRW-648
Rowland USD	469.425	464.425 R		WNHV-544
Water District		47.800		KNBN-495

LA VERNE

CHANNEL	INPUT	OUTPUT	TONE	CALL SIGN
Police Dispatch (F-1)		**155.490**	107.2	KDB-406
Police Secondary (F-2)		**155.550**	107.2	KDB-406
Police Tactical (F-3)		153.815	107.2	KDB-406
Police CLEMARS (F-4)		154.920	none	KTO-426
Police NLEMARS (F-5)		155.475	none	KTO-426
Fire & Medics (F-1)	485.575	**482.575 R**		WIL-293
Fire & Medics (F-2)		482.575		WIL-293
Public Works		153.815	107.2	KDB-406

LAWNDALE

CHANNEL	INPUT	OUTPUT	TONE	CALL SIGN
Police - L.A.S.O. #3	486.4375	**483.4375R**	186.2	WII-812
Fire - LACoFD Battalion 14	473.5625	**470.5625R**	151.4	WIK-428
Fire - LACoFD Dispatch	473.5375	**470.5375**R	151.4	WIK-428
Fire - LA County TAC 1		154.430	none	KMG-360
Public Works		158.775	none	KZM-843

LOMITA

CHANNEL	INPUT	OUTPUT	TONE	CALL SIGN
Police - L.A.S.O. #17	487.0375	**484.0375R**	186.2	WII-820
Fire - LACoFD Battalion 14	473.5625	**470.5625R**	151.4	WGH-876
Fire - LACoFD Dispatch	473.5375	**470.5375R**	151.4	WGH-876
Fire - LA County TAC 2		154.340	none	KME-539
Public Works		158.775		KTY-926

LOS ANGELES COUNTY

LONG BEACH

POLICE SERVICES

CHANNEL	INPUT	OUTPUT	TONE	CALL SIGN
Police Westside (F-1)	465.125	**460.125 R**	141.3	KMA-651
Police Eastside (F-2)	465.225	**460.225 R**	141.3	KMA-651
Police Inquiry (F-3)	465.350	**460.350 R**	141.3	KMA-651
Police Tactical (F-4)	465.450	**460.450 R**	141.3	KMA-651
Police Special Ops (F-5)	458.100	**453.100 R**	141.3	WNMQ-356
Police Special Ops (F-6)	458.350	**453.350 R**	141.3	WNMQ-356
Police Tactical*	465.400	**460.400 R**	141.3	KMA-651
Police UHF CLEMARS	465.025	**460.025**	103.5	KMA-651
Police CLEMARS		154.920	none	KVG-948
Police CLERS	155.430	158.760 R	186.2	KMC-54
Police Surveillance		155.040	none	KDO-236
Police Surveillance		155.550	none	KDO-236
Police Surveillance		453.125	none	N/A
Police MDT's	815.2375	860.2375R	none	KNCR-530

*Will replace 460.450.

Police Identification

1. The first digit indicates the shift:
   1 = Graveyard   2 = Day   3 = Evening
2. The second digit represents the type of unit:
   A = Two-man   L = One-Man   S = Supervisor
   D = Detective   M = Motorcycle   V = Vice

FIRE DEPARTMENT

CHANNEL	INPUT	OUTPUT	TONE	CALL SIGN
CH-1 Fire & Medics Dispatch	156.195	**153.950 R**	131.8	KMA-715
CH-2 Information & Response		**153.770**	131.8	KMA-715
CH-3 Administration		**155.925**	131.8	KMA-715
Fireground		**153.830**	131.8	WNYC-796
Fire Tactical		156.195	131.8	KMA-715
Fire Mutual Aid WHITE 1		154.280	none	KMA-715
Fire Mutual Aid WHITE 2		154.265	none	KMA-715
Fire Mutual Aid WHITE 3		154.295	none	KMA-715
Fire MDT's	823.6625	868.6625R	none	WNXZ-996

Channel three is also used as a fire tactical during major incidents.

LOS ANGELES COUNTY

LONG BEACH CITY FIRE STATIONS

19 Fire Engines (pumpers)
8 Paramedic Rescue Ambulances
4 Ladder Trucks, "Jaws of Life" Equipped
3 Fire Boats
1 Foam Application Apparatus

5 Reserve Engines (pumpers)
4 Reserve Paramedic Rescue
2 Reserve Ladder Trucks
3 Airport Firefighting Apparatus
1 Squirt Apparatus

Long Beach, with its 23 fire stations and 436 uniformed personnel, is one of the most respected fire departments in the nation. All firefighters are required to be certified as Emergency Medical Technicians.

LONG BEACH CITY SERVICES

CHANNEL	INPUT	OUTPUT	TONE	CALL SIGN
Airport Security	458.500	**453.500 R**	131.8	KDF-372
Animal Control	458.700	**453.700 R**	167.9	KDF-372
Beaches & Parks (F-1)	458.225	453.225 R	131.8	KTP-996
Beaches & Parks (F-2)	458.850	453.850 R	123.0	KTP-996
Building Security		151.865	123.0	KIR-538
Community College PD		154.650	123.0	KNDY-988
Community College		155.115	123.0	WNHP-965
Community College PD Parking	45.540	45.220 R		KNDY-988
Convention Center (F-1)	469.8125	464.8125R		N/A
Convention Center (F-2)	469.3125	464.3125		N/A
Local Government (Building & Engineering) (Parks & Beaches) (Streets & Sanitation)	458.850	**453.850 R**	123.0	KMA-324
Lifeguards (F-1)		155.925		KMM-310
Lifeguards (F-2)		155.160		KYU-541
Marine & Boat Patrol	458.550	**453.550 R**	131.8	KMF-644
Port Operations		153.230	123.0	KMB-593
Port Operations VHF		156.450	none	WXY-957
Port Operations VHF		156.600	none	WHU-336
Public Transit	509.4125	506.4125R	156.7	KWM-931
Queen Mary - Security	469.775	464.775 R		N/A
Queen Mary - Reported	469.375	464.375 R		N/A
Queen Mary - Reported		464.550		N/A
Unified School District		154.540		KNHY-735
Unified School District	468.400	463.400 R		WNBL-684
Unified School District	468.4375	463.4375R		WPAM-875
Unified School District	469.4375	464.4375R		WPAM-875
Unified School District	900.225	935.225 R	none	WNIT-941
Street Department		155.040	123.0	KDO-236
Water Department		173.3625		WNFG-658
Water Department	456.150	451.150 R		KME-993
Water Department	456.5625	451.5625		N/A
Water Department		952.8375	none	WNET-952

LOS ANGELES COUNTY

LOS ANGELES POLICE DEPARTMENT

BUREAU	AREA	CHANNEL	FREQUENCY	TONE
1. Central	Central Area (1)	C-1	**506.7375R**	110.9
	Central Traffic (24)	C-1	**506.7375R**	110.9
	Rampart (2)	C-2	**506.9375R**	110.9
	Northeast (11)	C-2	**506.9375R**	110.9
	Hollenbeck (4)	C-3	**507.1875R**	110.9
	Newton (13)	C-3	**507.1875R**	110.9
	Tactical "ROVER 4"		**506.5875R**	110.9
2. Valley	Van Nuys (9)	V-1	**506.7125R**	110.9
	West Valley (10)	V-1	**506.7125R**	110.9
	Valley Traffic (35)	V-1	**506.7125R**	110.9
	Devonshire (17)	V-2	**506.8875R**	110.9
	North Hollywood (15)	V-3	**507.1625R**	110.9
	Foothill (16)	V-3	**507.1625R**	110.9
	Tactical "ROVER 5"	-	**507.2625R**	110.9
3. South	LA Harbor (5)	S-1	**506.7875R**	110.9
	Southeast (18)	S-1	**506.7875R**	110.9
	Southwest (3)	S-2	**506.9875R**	110.9
	77th Street (12)	S-3	**507.2375R**	110.9
	South Traffic (25)	S-3	**507.2375R**	110.9
	Tactical "ROVER 7"	-	**507.0375R**	110.9
4. West	West Los Angeles (8)	W-1	**506.7625R**	110.9
	West Traffic (34)	W-1	**506.7625R**	110.9
	Wilshire (7)	W-2	**506.9625R**	110.9
	Pacific (14)	W-2	**506.9625R**	110.9
	Hollywood (6)	W-3	**507.2125R**	110.9
	Tactical "ROVER 6"	-	**506.8125R**	110.9
Citywide	Metropolitan Division	-	**506.8375R**	110.9
Citywide	Emergency "ROVER 8"	-	**507.0875R**	110.9

The Metro Division is for Special Operations, Tactical, and SWAT.
"ROVER 8" is the citywide Emergency Trigger channel.

ROVER: Remote Out of Vehicle Emergency Radio
CRASH: Community Resources Against Street Hoodlums

Transmitter Sites (KJC-625)
Central Bureau: City Hall Building, Mount Lee.
South Bureau: Baldwin Hills, San Pedro Hill.
West Bureau: Baldwin Hills, GTE Building, San Vicente Peak.
Valley Bureau: Oat Mountain, Verdugo Peak.

LOS ANGELES COUNTY

LOS ANGELES POLICE DEPARTMENT

UHF TACTICAL SYSTEM

CHANNEL	INPUT	OUTPUT	TONE	CALL SIGN
Crash & Gang Detail	487.6875	484.6875R	173.8	WIJ-803
Crash & Gang Detail	487.9875	484.9875R	173.8	WIL-499
Criminal Intelligence	487.4875	484.4875R	100.0	WIJ-803
Criminal Intelligence & Vice	487.5125	484.5125R	100.0	WIJ-803
LA Interagency Task Force	485.6125	482.6125R	203.5	N/A
Narcotics Bureau	458.100	453.100 R	none	KGR-382
Narcotics Bureau (Tactical Simplex)	458.125	453.125	none	KD-5913
Police CLEMARS (Interagency Tactical)		460.025	none	N/A
Special Operations (Anti-terrorists) (Narcotic Bureau) (Organized Criminal Intelligence) (SWAT Operations)	458.875	453.875 R	none	KTU-271
Special Operations (Surveillance & Tactical Simplex)	458.975	453.975 R	none	KD-5913
Vice Squad (6Y)	458.350	453.350 R	none	KIZ-232

VHF TACTICAL SYSTEM

CHANNEL	INPUT	OUTPUT	TONE	CALL SIGN
Tactical 1 (Hot Stops, Special Enforcement)		**154.830**	none*	KMA-357
Tactical 2 (Special Enforcement)		**154.770**	none*	KMA-787
Tactical 3 (Surveillance & VIP Escort)		154.995	none	WNBE-338
Tactical 4 (CLEMARS)		154.920	none	N/A
Anti-terrorist		154.965	none	WRA-551
Burglary Auto Detail BAD		154.755	none	KMA-367
LoJack Vehicle Tracking		173.075	none	WNSZ-607
Police CLERS	155.430	158.790 R	186.2	N/A
Police NLEMARS		155.475	none	N/A
Surveillance Operations		154.875	none	N/A
Surveillance Operations		155.040	none	N/A
Surveillance Operations		155.190	none	N/A
Surveillance Operations		155.310	none	N/A
Surveillance Operations		155.655	none	N/A
Surveillance Operations		156.825	none	N/A

*The tone 110.9 Htz has been reported.

LOS ANGELES COUNTY

LOS ANGELES POLICE DEPARTMENT

AIR SUPPORT DIVISION

CHANNEL	INPUT	OUTPUT	TONE	CALL SIGN
Air to Air, Primary		**122.750**	none	KSD-3
Air to Air, Secondary		122.250	none	KNN-3
Air to Air, Interagency		122.850	none	N/A
Air to Air, Multicom		122.900	none	KA-97205
Helicopter Air to Air		**123.025**	none	KMG-7
Heliports - Airports		**123.050**	none	KMG-7
Heliports - Hospitals		123.075	none	KSD-3
L.A.S.O. Helicopter Common	486.5375	483.5375R	186.2	N/A
Tactical		118.900	none	N/A
Tactical		118.950	none	N/A
Tactical		119.800	none	N/A
Tactical		119.950	none	N/A
Tactical		119.975	none	N/A

MOBILE DATA TERMINALS (MDT)

CHANNEL	INPUT	OUTPUT	TONE	CALL SIGN
Central Bureau	155.520	159.180 R	none	WRL-685
South Bureau	155.010	158.910 R	none	WRL-685
Valley Bureau	155.730	159.150 R	none	WLR-685
West Bureau	155.070	158.865 R	none	WLR-685
Citywide	155.580	159.030 R	none	WLR-685

LAPD FREQUENCY EXPANSION

Los Angeles PD is expanding its UHF radio system. The department is purchasing Astro Saber radios manufactured by Motorola which have both digital and analog voice capability. According to the head of LAPD Communications, these portable radios will employ the multi-level digital modulation technique known as Frequency Division Multiple Access (FDMA). The FDMA technique is a proposal of APCO Project 25 and is favored due to its compatibility with existing analog FM systems. Thus LAPD may use the digital modulation on its new 484 MHz frequencies in conjunction with the current 506 MHz voice channels. All LAPD communications will be equipped with digital radios in the future as funding and resources permit. Digital modulation will provide the department with increased privacy, improved voice quality, and better spectrum efficiency. It appears LAPD will also keep its myriad of VHF and 453 MHz channels.

Regarding the assignments of the new frequencies: there are several plans and the exact configuration will not be known until the Astro radios have been programmed and testing has proceeded. We suggest you search these frequencies which are arranged numerically and report what you hear. A lot of interesting activity has been reported in recent months. LAPD units such as Crash, Organized Criminal Intelligence, Burglary Auto Theft, Detectives and others have been observed using plain voice (analog transmission) on the frequencies that are bold faced.

LOS ANGELES COUNTY

LOS ANGELES POLICE DEPARTMENT

EXPANDED RADIO SYSTEM

Frequency	Channel Use*	Tone
484.2875R	Harbor Area F5	173.8
484.3125R	South Bureau Tac 2	173.8
484.3375R	South Traffic Division F22	173.8
484.3625R	West Traffic Division F23	173.8
484.3875R	Pacific Area F14	173.8
484.4125R	West Bureau Tac 2	173.8
484.4375R	Central Traffic Division F19	173.8
484.4625R	Citywide Tactical 4	173.8
484.4875R	Organized Criminal Intelligence F43	100.0 / 173.8
484.5125R	Organized Criminal Intelligence F42	100.0 / 173.8
484.5375R	Burglary Auto Theft F35	173.8
484.5625R	Bunco-Forgery F34	173.8
484.5875R	Detective Support F40	173.8
484.6125R	Robbery-Homicide F37	173.8
484.6375R	Citywide Tac 5	173.8
484.6875R	Crash / Gang Section F36	173.8 / 186.2
484.7125R	Special Investigations F38	173.8
484.7375R	Detectives Headquarters F33	173.8
484.7625R	Citywide Tac 3	173.8
484.8125R	Central Traffic Division F21	173.8
484.8375R	Northeast Area F11	173.8
484.8625R	Newton Area F13	173.8
484.8875R	Valley Bureau Tac 2	173.8
484.9125R	Valley Traffic Division F24	173.8
484.9375R	Pacific Traffic Division	173.8
484.9625R	North Hollywood Area F15	173.8
484.9875R	Crash / Gang Section	167.9 / 173.8
506.3125R	MDT's	100.0
506.3625R	Special Enforcement	N/A
506.4375R	Administrative Vice	91.5
506.4625R	Communications	N/A
506.5375R	Undetermined	N/A
506.6125R	MDT's	100.0
506.6375R	MDT's	100.0
506.6875R	Special Enforcement	N/A
506.8625R	Special Enforcement	N/A
506.9125R	MDT's	100.0
507.0125R	MDT's / System Control	100.0
507.1125R	Citywide Tac 1	N/A
507.1375R	MDT's	100.0

*Channel assignments are only proposed and will be altered or possibily scrapped.

Call Sign Identification:

Los Angeles (WIJ-798 / WIJ-800 / WIJ-803 / WIJ-804 / WIK-744 / WIL-499), Oat Mountain WIJ-792, Rolling Hills WIJ-499, San Dimas WIJ-801, Santa Monica WIJ-802, Verdugo Peak WIJ-796

LOS ANGELES COUNTY

LOS ANGELES POLICE DEPARTMENT

LAPD UNIT IDENTIFICATION SYSTEM

1. The first one or two number digits represent the service area:

1	Central	7	Wilshire	13	Newton Street	24	Central Traffic
2	Rampart	8	West LA	14	Pacific	25	South Traffic
3	Southwest	9	Van Nuys	15	North Hollywood	34	West Traffic
4	Hollenbeck	10	West Valley	16	Foothill	35	Valley Traffic
5	Harbor	11	Northeast	17	Devonshire		
6	Hollywood	12	77th Street	18	Southeast		

2. The second or third digit, which is a letter, indicates the type of unit:

A	Two-man Patrol	L	One Officer Patrol	UC	UCLA Police
B	Two-Man	M	Motorcycle	V	Vice
C	Bicycle Patrol	MQ	Special Enforcement	VS	Visitor Services
E	Traffic	OP	Observation Post	W	Detectives
F	Felony	P	Special Events	X	Extra Patrol
FB	Foot Patrol	Q	Special Detail	Z	Special, Reserve
J	Juvenile	T	Traffic Investigation		
JW	Female Juvenile	U	Reports		

3. The last number digits indicate the master beat or individual unit. Some units have last digits designating special units:

10 Watch Commander
20, 30, 40, 50, 60, 70, 80 Supervisors
90 Station Desk

4. "David" Support Units:

1D	Field Support Services	2D	Aircraft Support Services

5. "Henry" Administrative Units:

1H	Detentions	8H	Fiscal Services
2H	Crime Laboratory	9H	Property Management
3H	Construction	10H	Communications
4H	Transportation	11H	Unassigned
5H	Personnel	12H	Records & Identification
6H	Developmental Services	13H	Supplies
7H	Unassigned	14H	Training

6. "King" Investigative Units:

2K = Burglary Auto Theft   3K = Bunco-Forgery   4K = Homicide-Robbery

7. "I" Internal Affairs "Y" Narcotics

3I	Internal Affairs Division	7Y/8Y/9Y	Narcotics Task Force
4I	Organized Criminal Intelligence		
5I	Anti-Terrorist Division		
6I	Vice Administrative		

LOS ANGELES POLICE DEPARTMENT

AREA BOUNDARIES

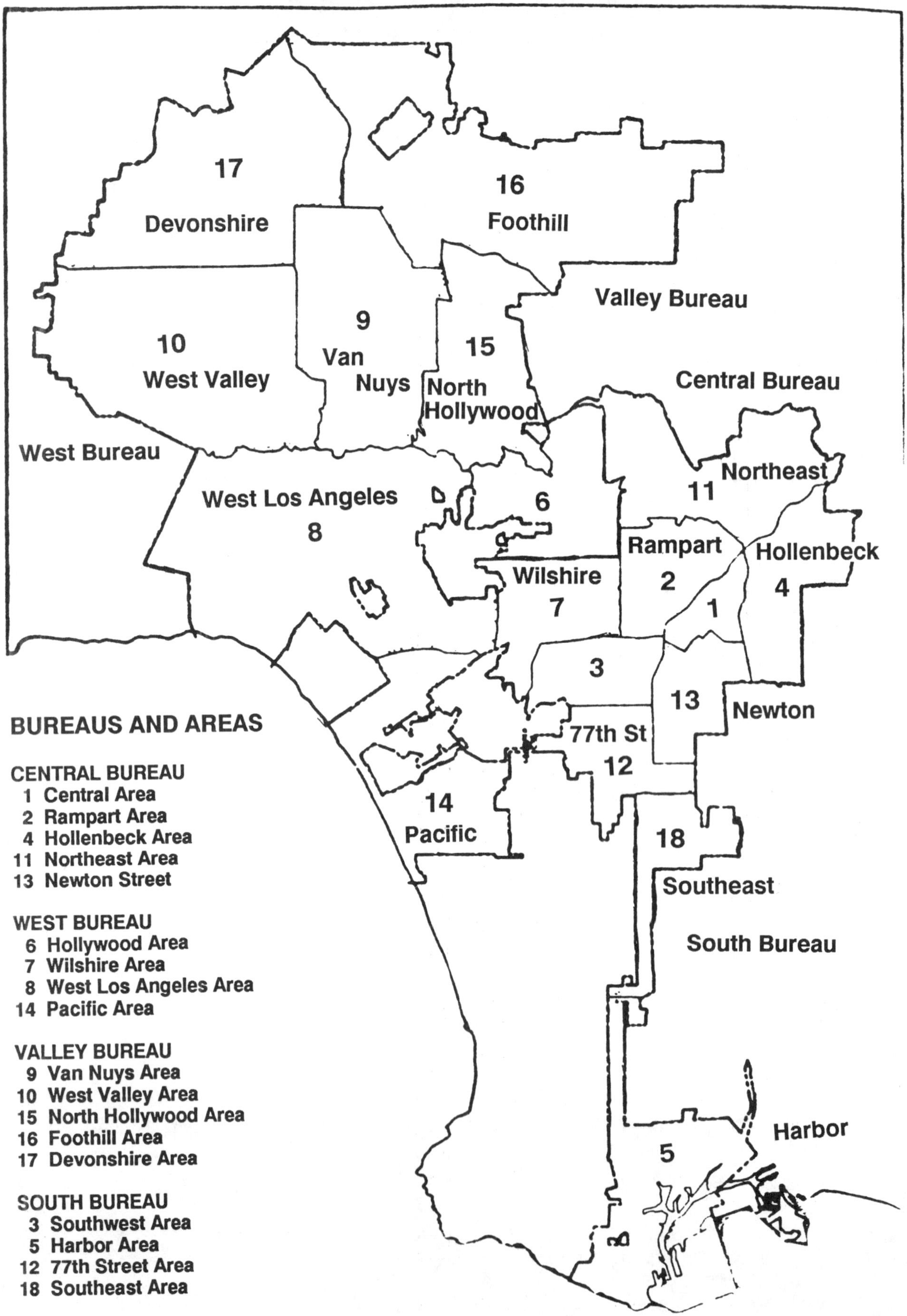

LOS ANGELES COUNTY

LA CITY FIRE DEPARTMENT

CHANNEL	INPUT	OUTPUT	TONE*	CALL SIGN
(F-1) Division 1	815.9375	**860.9375R**	85.4	KNFV-977
(F-2) Division 2	814.9375	**859.9375R**	88.5	KNFV-977
(F-3) Division 3	813.9375	**858.9375R**	91.5	KNFV-977
(F-4) Rescue - Div. 1 & 2	812.9375	**857.9375R**	94.8	KNFV-977
(F-5) Fire Prevention & Arson	811.9375	**856.9375R**	91.5	KNFV-977
(F-6) Fireground	815.4375	**860.4375R**	186.2	KNFV-977
(F-7) Fire Dispatch - Div. 1 & 2	814.4375	**859.4375R**	192.8	KNFV-977
(F-8) Fire Dispatch - Div. 3	813.4375	**858.4375R**	103.5	KNFV-977
(F-9) Major Incident Dispatch	812.2375	**857.2375R**	107.2	KNFV-977
(F-10) Rescue Units	811.2375	**856.2375R**	123.0	KNFV-977
(F-11) Command	815.7625	**860.7625R**	127.3	KNFV-977
(F-12) Fireground	813.2375	**858.2375R**	131.8	KNFV-977
(F-13) Fireground	812.4375	**857.4375R**	141.3	KNFV-977
(F-14) Fireground	811.4375	**856.4375R**	146.2	KNFV-977
(F-15) Fireground	814.7625	**859.7625R**	151.4	KNFV-977
(F-16) Fireground	813.7625	**858.7625R**	162.2	KNFV-977
(F-17) Fireground	812.7625	**857.7625R**	167.9	KNFV-977
(F-18) Fireground	811.7625	**856.7625R**	173.8	KNFV-977
(F-19) Statewide Mutual Aid		**866.0125**	156.7	N/A
(F-20) Statewide Mutual Aid		**866.5125**	156.7	N/A
(F-21) Statewide Mutual Aid		**867.0125**	156.7	N/A
(F-22) Statewide Mutual Aid		**867.5125**	156.7	N/A
(F-23) Statewide Mutual Aid		**868.0125**	156.7	N/A
(F-24) Statewide Mutual Aid		**868.5125**	156.7	N/A

*Output tones.

LA CITY FIRE STATIONS & EQUIPMENT

Division 1

Battalion 1

Station	Location	Equipment
3	108 N. Fremont	TF3, RA3, RA803, Bus, LU, HMRU
4	800 N. Main	TF4, Squad4, CPU
9	430 E. 7th St. (HQ)	9TF, RA9
17	1601 S. Santa Fe	17TF, EMS5, HMRU

Battalion 2

Station	Location	Equipment
12	5921 N. Figueroa	TF12
42	2021 Colorado Bld.	E42
44	1410 Cypress Ave.	E44
50	3036 Fletcher Dr.	TF50, BC2
55	4455 E. York Bld.	E55, RA55
56	2759 Rowens Ave.	E56, HU56

Battalion 5

Station	Location	Equipment
27	1327 N. Cole Ave.	TF27, Squad 27, RA27
35	1601 N. Hillhurst Ave.	TF35, RA35
41	1439 N. Gardner St.	E41
52	4957 Melrose Ave.	E52, RA52, EMS1
76	3111 Cahuenga Ave.	E76
82	1800 N. Bronson Ave.	E82, RA82

Battalion 7

Station	Location	Equipment
1	2230 Pasadena Ave.	TF1, RA1, FS1
2	1962 E. Brooklyn	TF2, RA2
16	2011 N. Eastern Ave.	E16
25	2927 Whittier Blvd.	E25
47	4575 Huntington Dr.	TF47

LOS ANGELES COUNTY

LA CITY FIRE STATIONS & EQUIPMENT

Battalion 9

Station	Location	Equipment
19	12229 Sunset Blvd.	E19, RA19
23	17281 Sunset Blvd.	E23
37	1090 Vernon Ave. (HQ)	TF37, RA37
59	15045 Olympic Blvd.	E59
69	15045 Sunset Blvd	TF69, RA69
71	107 S. Beverly Glen	E71

Battalion 11

Station	Location	Equipment
6	326 N. Virgil Ave.	E6
11	1819 W. 7th St.	TF11, RA11
13	1206 S. Vermont	E13, RA13
20	2144 Sunset Blvd.	TF20, RA20
29	158 S. Western	TF29

Division 2

Battalion 3

Station	Location	Equipment
14	3401 S. Central Ave.	TF14, RA14
15	915 W. Jefferson Blvd.	TF15, RA15, RA815
21	1187 E. 52nd St.	E21, RA21
26	2009 S. Western Ave.	TF26, RA26
34	3661 7th Ave.	E34, RA34, RA834, FC
46	4370 S. Hoover St.	E46, RA46, EMS2

Battalion 4

Station	Location	Equipment
5	6621 Manchester Ave.	5TF, RA5
51	10435 Sepulveda Bl.	E51
62	3631 Centinela Ave.	E62
63	1930 Shell Ave.	TF63, RA63
80	6911 World Way West	E80, CR80
95	10010 Intl. Rd.	TF95, RA95

Battalion 6

Station	Location	Equipment
38	124 E. "I" St.	TF38, RA38
40	330 Ferry St.	E40, Snorkel
48	1601 S. Grand Ave.	TF48, Squad 48
49	400 Yacht St. (HQ)	E49, BT3, BT4, Fireboat
53	438 N. Mesa	E53, RA53, FC
85	1331 W. 253rd St.	TF85, RA85
101	1414 25th St.	E101
110	2945 Miner St.	BT5, Fireboat
111	945 S. Seaside St.	BT1, Fireboat
112	550 Sampson Way	BT2, Fireboat

Battalion 13

Station	Location	Equipment
33	6406 S. Main St.	TF33, RA33, RA833
57	7800 S. Vermont (HQ)	E57, RA57, RA857
64	118 W. 108th St.	TF64, RA64
65	1525 E. 103rd St.	E65, RA65, RA865
66	1909 W. Slauson Bl.	TF66, RA66, RA866, FS
79	18030 S. Vermont	PE79

Battalion 18

Station	Location	Equipment
58	1556 Robertson Blvd.	TF58
61	5821 W. 3rd St.	TF61, RA61
68	5023 Washington Blvd.	E68, RA68
92	10556 W. Pico Blvd.	TF92, ESM4
94	4470 Coliseum St.	TF94

Division 3

Battalion 10

Station	Location	Equipment
39	14415 Sylvan St.	TF39, Squad 39, RA39
83	5001 Balboa Ave.	E83, LU83
88	5101 N. Sepulveda	TF88, AU88, HMRU88, Fuel Tank
90	7921 Woodley Ave.	TF90, Crash 90, Helitender 90
99	14415 Mulholland Dr.	E99, RA99
109	16500 Mulholland Dr.	E109

Battalion 12

Station	Location	Equipment
24	9411 Wentworth St.	E24
74	7777 Foothill Bl.	TF74, RA74
75	15345 San Fernando	TF75, RA275
77	8943 Glenoaks Bl.	E77
81	14123 Nordhoff St.	E81, RA81
91	14430 Polk St.	E91, RA91
98	13035 Van Nuys	TF98, RA98

Battalion 14

Station	Location	Equipment
60	5320 Tujunga Ave.	TF60, RA60, FS60
78	4230 Coldwater Cyn.	PE78
86	4305 Vineland Ave.	E86, FC86, EMS6
89	7063 Laurel Cyn.	TF89, RA89, Fuel Tanker
97	8021 Mulholland Dr.	E97
102	13200 Burbank Bl.	TF102
108	15250 Mulholland Dr.	E108

Battalion 15

Station	Location	Equipment
8	11351 Tampa Ave.	E8
18	12050 Balboa Ave.	E18
70	9861 Reseda Bl.	TF70
87	10241 Balboa Bl.	E87, RA87
96	21800 Marilla St.	TF96
103	18143 Parthenia St.	E103
104	8349 Winnetka Ave.	E104, RA104
107	20225 Devonsfire St.	E107, RA107

Battalion 17

Station	Location	Equipment
72	6811 DeSoto Ave.	E72
73	7419 Reseda Bl.	TF74, RA74
84	5340 Canoga Ave.	E84
93	19059 Ventura Bl.	TF93, RA93
100	6751 Louise Ave.	PE100, RA100, EMS3
105	6345 Fallbrook St.	TF105, RA105
106	23004 Roscoe Bl.	E106

LOS ANGELES COUNTY

LA CITY FIRE EQUIPMENT

BC = Battalion Chief

E = Engine
Called a "triple" when it is equipped with a high-capacity water pump.

EA = Emergency Air
A truck that carries extra oxygen cylinders for use during internal fires.

EMS = Emergency Medical Service
The captain or supervisor of a paramedic or rescue ambulance

FOAM = Crash Foam Rescue
A specially equipped unit usually located at stations near airports or major runways. It carries water and chemical foam to extinguish or help prevent a major fire.

FS = Food Service
Mobile canteen during major incidents

HMRU = Hazardous Materials Response Unit
This unit responds to hazardous materials calls of all types.

HU = Heavy Utility Unit
One unit designed for transporting, lifting, and cutting heavy objects.

LF = Light Force
Consists of one truck (fully manned) and one engine.

LU = Electric Light Utility

PB = Utility Vehicle

PE = Paramedic Engine

RA = Rescue Ambulance
Responds to medical calls and transports the sick or injured.

TF = Task Force
Consists of one truck (fully manned) and two engines.

Squad
Unit designed to handle firefighting and hazardous materials calls.

USAR = Urban Search & Rescue Unit
Coordinates search & rescue operations in the city limits

Bus
A twenty-four passenger bus for transporting personnel to the scene or the staging area for a major incident.

LOS ANGELES COUNTY

LA CITY SERVICES*

NAME	INPUT	OUTPUT	TONE	CALL SIGN
Airport Police	465.550	**460.550 R**	156.7	WNLB-660
(Los Angeles Intl. & Ontario)				
(Palmdale & Van Nuys)				
Airport Police	465.525	**460.525 R**	110.9	WNNS-629
(Los Angeles Intl. & Ontario)				
Airport Maintenance	465.100	**460.100 R**	118.8	WNKX-789
(Ontario, Palmdale & Van Nuys)				
Airport Operations	458.375	**453.375 R**	118.8	KTU-271
(Los Angeles Intl. & Ontario)				
(Palmdale & Van Nuys)				
Airport Maintenance		153.800	100.0	KSU-564
(Los Angeles Intl. & Ontario)				
(Palmdale & Van Nuys)				
Civil Defense Command		37.900		WNNA-608
Civil Defense Operations		47.500	none	KC-4103
City College PD (F-1)		154.740	none	KNID-882
City College PD (F-2)		154.725	none	KNID-882
City College PD		154.625		KNHD-559
Community Colleges PD		**154.650**	141.3	KNJY-515
Community Colleges PD		155.685	141.3	KNJY-515
Community Colleges PD		156.090	141.3	KP-2569
Community College Maintenance		151.685		N/A
Convention Center Ops	458.225	453.225 R		N/A
Convention Center Ops	458.600	453.600 R		N/A
Convention Center Security	458.500	453.500 R		N/A
Convention Center Security	465.550	460.550 R	110.9	WNLB-660
Electronic Department		37.140	none	N/A
Emergency Services		39.900		WNHE-685
Emergency Services		155.280		KFG-564
Emergency Services	169.200	170.200 R	156.7	WNKU-928
General Services Dept.		153.800		KSU-564
Harbor Department	809.9625	**854.9625R**	103.5	WNSS-857
Harbor Department (F-1)		45.620	192.8	KMA-367
Harbor Department (F-2)		45.740	192.8	KMA-367
Local Government 1				
(Animal Control & Housing)	453.250	**453.250 R**	186.2	KMF-546
(Street Lighting)	458.250	**453.250 R**	186.2	KMF-546
Local Government 2				
(Sanitation & Sold Waste)	458.050	**453.050 R**	none	KXL-317
Parks & Recreation				
Park Rangers		153.800	88.5	KET-364
Park Rangers		153.920	88.5	KET-364
Park Rangers		153.935	88.5	KET-364
School District - Police	458.800	**453.800 R**		KMK-877
School District - Security		151.775		KZD-675
School District - Security	469.250	464.250 R		KWS-798

*A 40 channel trunked 800 MHz system is planned for most city services.

LOS ANGELES COUNTY

LA CITY SERVICES

NAME	INPUT	OUTPUT	TONE	CALL SIGN
School District Transportation	808.3125	853.3125R		KXD-940
School District Transportation	808.3375	853.3375R		KXD-940
School District Transportation	808.3625	853.3625R		KXD-940
School District Transportation	808.3875	853.3875R		KXD-940
School District Transportation	808.4125	853.4125R		KXD-940
Street Maintenance Bureau				
Downtown Area		33.060	110.9	KGT-542
Valley Area		33.020	110.9	KGT-542
Traffic Signal Repair (TSR)	458.400	453.400 R	100.0	KNDV-915
Water & Power Department (trunked):				

935.3875, 935.400, 935.4125, 935.425, 935.4375, 935.450, 935.4625, 936.475, 935.4875, 935.500, 936.3875, 936.400, 936.4125, 936.425, 936.4375, 936.450, 936.4625, 936.475, 936.4875, 936.500

LYNWOOD

CHANNEL	INPUT	OUTPUT	TONE	CALL SIGN
Police - L.A.S.O. #25	486.2625	**483.2625R**	186.2	WII-827
Fire Primary		154.205		WQN-826
Fire Mutual Aid WHITE 1		154.280	none	WQN-826
Public Works		45.280		WZU-647
Unified School District		155.280		KNJX-285
Unified School District	469.825	464.825 R		KXK-452

MALIBU

CHANNEL	INPUT	OUTPUT	TONE	CALL SIGN
Police - L.A.S.O. #10	485.9125	**482.9125R**	186.2	WII-826
Fire - LACoFD Battalion 5	473.6125	**470.6125R**	151.4	WGH-876
Fire - LACoFD Dispatch	473.5375	**470.5375R**	151.4	WGH-876
Fireground		154.430	none	KVP-645
Emergency Services		155.160		KNHK-428
Emergency Services		155.280		KNHK-428

MANHATTAN BEACH

CHANNEL	INPUT	OUTPUT	TONE	CALL SIGN
South Bay Regional Public Communications Authority				
Police - Dispatch	473.6375	470.6375R	203.5	KDL-483
Fire - South Bay Dispatch		154.355	103.5	KMF-434
Public Works		39.580	none	WZM-644
Unified School District		151.835		KNBS-709
Undetermined	469.475	464.975 R		KAJ-327

LOS ANGELES COUNTY

MAYWOOD

CHANNEL	INPUT	OUTPUT	TONE	CALL SIGN
Police Primary (F-1)		**156.030**	107.2	WNAN-857
Police CLEMARS (F-2)		154.920	none	WNJU-262
Fire - LACoFD Dispatch	473.5375	**470.5375R**	151.4	WGH-876
Fire - LACoFD	473.6125	**470.6125R**	151.4	WGH-876
Fire - LA County TAC 2		154.340	none	KTX-604
Public Works		37.180	none	KNDR-708

MONROVIA

CHANNEL	INPUT	OUTPUT	TONE	CALL SIGN
Police Dispatch (F-1)		**158.730**	131.8	KNFP-459
Police Repeater (F-2)	155.970	158.730 R	131.8	KNFP-459
Police CLEMARS (F-3)		154.920	none	KYU-717
Police Tactical (F-4)		154.115	131.8	WNGI-435
Police Tactical (F-5)		155.880	131.8	KBM-676
Fire Dispatch		154.370	none	WNUT-823
Public Works		154.115		WNGI-435
Public Works		155.880		KBM-676

MONTEBELLO

CHANNEL	INPUT	OUTPUT	TONE	CALL SIGN
Police Primary	485.2125	**482.2125R**	D343	WIJ-365
Police Tactical	485.3125	**482.3125R**	D465	WIJ-365
Police Tactical & Interagency	485.6125	**482.6125R**	203.5	WIJ-365
Police Detectives	485.1125	**482.1125R**	D223	WIJ-365
Police CLEMARS		154.920	none	KJS-750
Fire & Medics (50-59)	485.4125	**482.4125R**	D532	WIJ-365
Fire Mutual Aid WHITE 1		154.280	none	KMF-655
Fireground WHITE 3		154.295	none	KMF-655
Fireground WHITE 2		154.265	none	KMF-655
Fire MDT's	814.2625	859.2625R	none	KNGD-850
Local Government	485.0375	482.0375R	D351	WIJ-365
Local Government (Parks & Recreation) (Streets & Sanitation)	485.5125	482.5125R	D654	WIJ-365
Local Government (Telephone Interconnect)	485.1125	**482.1125R**	D223	WIJ-365
Transit District	808.5375	853.4875R		WNXE-677
Unified School District	468.450	463.450 R		KTY-556

LOS ANGELES COUNTY

MONTEREY PARK

CHANNEL	INPUT	OUTPUT	TONE	CALL SIGN
Police Primary (F-1)	154.950	**155.565 R**	136.5	KNAK-313
Police Car to Car (F-2)		**155.415**	136.5	KNAK-313
Police CLEMARS (F-3)		154.920	136.5	KNFU-620
Police Detectives (F-4)		154.890	136.5	KNAK-313
Police CLERS (F-5)	155.430	158.790 R	186.2	KNFM-279
Fire Primary (60)		**153.995**		KMK-857
Fire Mutual Aid WHITE 1		154.280	none	KNCW-994
Fire Mutual Aid WHITE 3		154.295	none	KNCW-994
Public Works		45.740		KNFP-549
Public Works		46.580		KIZ-348

NORWALK

CHANNEL	INPUT	OUTPUT	TONE	CALL SIGN
Police - L.A.S.O. #4	486.7625	**483.7625R**	186.2	WII-824
Police CLEMARS		154.920	none	KNAO-471
Fire - LACoFD Battalion 8	473.5625	**470.5625R**	151.4	WII-373
Fire - LACoFD Dispatch	473.5375	**470.5375R**	151.4	WII-373
Fireground		154.295	none	KMJ-547
Local Government	808.4875	853.4875R		WNXE-677
Public Works		46.540	131.8	KFO-844
Unified School District	458.950	453.950 R	131.8	KNAV-725
Unified School District	466.775	461.775 R		KLK-966
Unified School District	896.150	935.150 R		WNIT-940
Water Department		952.4125	none	WNEM-825

PALMDALE

CHANNEL	INPUT	OUTPUT	TONE	CALL SIGN
Police - L.A.S.O. #11	485.8625	**482.8625R**	186.2	WII-798
Fire - LACoFD Battalion 11	473.4875	**470.4875R**	151.4	WGH-876
Fire - LACoFD Dispatch	473.5375	**470.5375R**	151.4	WGH-876
Fire - LA County TAC 1		154.430	none	KBG-521
Local Government 1 (Engineering & Streets) (Parks & Recreation) (Solid Waste & Sanitation)	474.0375	**471.0375R**	103.5	WIG-515
Local Government 2	473.3375	**470.3375R**	103.5	WIG-515
Mosquito Control	458.850	453.850 R		WNPE-496
Palmdale School District - Security		151.895		WNGL-538
Palmdale School District - Buses	810.2125	855.2125R		WNSK-604
Palmdale School District - Buses	810.7125	855.7125R		WNVU-657
Water District		173.3125		WNJT-987
Water District	456.100	451.100 R		KNAY-345

LOS ANGELES COUNTY

PALOS VERDES ESTATES

CHANNEL	INPUT	OUTPUT	TONE	CALL SIGN
Police Primary (F-1)	45.340	**45.660 R**	D074	WSL-641
Police Car to Car (F-2)		**45.660**	D074	WSL-641
Police Tactical (F-3)		**45.340**	D074	WSL-641
Police CLEMARS		154.920	none	N/A
Fire - LACoFD Battalion 14	473.5625	**470.5625R**	151.4	WGH-876
Fire - LACoFD Dispatch	473.5375	**470.5375R**	151.4	WGH-876
Fire - LA County TAC 2		154.340	none	N/A
Lifeguards		37.120	none	KMF-964
Lifeguards		37.160	none	KMF-964
Lifeguards		37.220	none	KMF-964
Lifeguards		37.300	none	KMF-964
Public Works		46.580		KDY-435

PARAMOUNT

CHANNEL	INPUT	OUTPUT	TONE	CALL SIGN
Police - L.A.S.O. #13	486.1375	**483.1375R**	186.2	WII-792
Fire - LACoFD Battalion 9	473.5625	**470.5625R**	151.4	WIK-438
Fire - LACoFD Dispatch	473.5375	**470.5375R**	151.4	WIK-438
Fire - LA County TAC 1		154.430	none	KJA-932
Unified High School		35.080	none	KNBG-892
Unified School District		45.960		KJI-718
Unified School District		43.000		KJI-718
Unified School District		151.865		KJI-718
Public Works	45.600	45.120 R		KUL-798
Public Works		45.280		WNFE-762

PASADENA

CHANNEL	INPUT	OUTPUT	TONE	CALL SIGN
Police Primary (F-1)	485.3375	**482.3375R**	118.8	WIJ-442
Police Tactical (F-2)	485.4375	**482.4375R**	94.8	WIJ-442
Police Tactical (F-3)	485.5375	**482.5375R**	151.4	WIJ-442
Police Detectives (F-4)	485.6375	**482.6375R**	82.5	WIJ-442
Police CLEMARS		154.920	none	KNEJ-534
Police CLERS	155.430	158.760 R	186.2	KMC-58
Fire & Medics Verdugo	473.2125	**470.2125R**	107.2	KME-279
Fire Incident Command	485.5375	**482.5375R**	151.4	WIJ-442
Fireground	485.2375	**482.2375R**	107.2	WIJ-442
Fire MDT's	821.5625	866.5625R	none	WNVF-960
Local Government		158.820	136.5	KMM-517
Local Government 1 (Animal Control)	509.200	506.200 R	162.2	WIL-549
Local Government 2	509.225	506.225 R	162.2	WIL-549
City College PD	467.675	462.675 R		KAA-7406

LOS ANGELES COUNTY

PASADENA

CHANNEL	INPUT	OUTPUT	TONE	CALL SIGN
Community College Police		154.650	141.3	KNGQ-557
Community College Police		154.920	none	WNLP-339
Community College Police	465.050	460.050 R		WNRL-778
Unified School District	466.925	461.925 R		KXR-395
Unified School District (HT's)		154.570	D134	KA-61273
Street Department	458.950	453.950 R	162.2	KMF-241
Water Department		153.650	162.2	KMA-767
Water Department (HT's)		158.235	162.2	KMA-767

PICO RIVERA

CHANNEL	INPUT	OUTPUT	TONE	CALL SIGN
Police - Montebello PD at night				
Police - L.A.S.O. #15	486.7625	**483.7625R**	186.2	WII-834
Fire - LACoFD Battalion 3	473.6125	**470.6125R**	151.4	WGH-876
Fire - LACoFD Dispatch	473.5375	**470.5375R**	151.4	WGH-876
Fire - LA County TAC 1		154.430	none	KMG-216
Fire Mutual Aid WHITE 1		154.280	none	KMG-216
Public Works		45.200		KFT-620
Special Emergency		155.340		KOP-949
Water District	510.3125	507.3125R		WIF-276

POMONA

CHANNEL	INPUT	OUTPUT	TONE	CALL SIGN
BLUE - Police Dispatch		**155.550**	none	KMA-384
BROWN - Police Inquiry		**153.935**	none	KAY-838
ORANGE - Police Car-to-Car		**153.860**	none	KAY-838
Police CLEMARS GOLD		154.920	none	WQP-808
Police Detectives		155.250	none	KMA-384
Police CLERS	155.430	158.790 R	186.2	KMB-78
Police NLEMARS		155.475	none	WQP-808
Police PRECOM		155.595	none	WQP-808
Police Mobile (rare use)		39.060	none	KMA-384
Fire Primary RED (80-89)		**154.070**	141.3	KAY-933
Fireground		**154.415**	none	KAY-933
Community College		154.540		WNCS-836
Humane Society	469.325	464.325 R		KNGR-834
Public Works		153.860		KAY-883
Unified School District - Security	469.7625	464.7625R	D431	WNUJ-404
Unified School District*	474.8375	471.8375R		WII-384
Water District	456.250	451.250 R		KMF-735
Water District	456.250	451.525 R		KMF-735

*Not heard

RANCHO PALOS VERDES

CHANNEL	INPUT	OUTPUT	TONE	CALL SIGN
Police - L.A.S.O. #17	484.0375	**484.0375R**	186.2	WIJ-619
Fire - LACoFD Battalion 14	473.5625	**470.5625R**	151.4	WGH-876
Fire - LACo FD Dispatch	473.5375	**470.5375R**	151.4	WGH-876
Fire - LA County TAC 2		154.340	none	KMB-650
Fire Mutual Aid WHITE 1		154.280	none	KMB-650

REDONDO BEACH

CHANNEL	INPUT	OUTPUT	TONE	CALL SIGN
Police Dispatch (F-1)	465.075	**460.075 R**	203.5	KLR-373
Police Secondary (F-2)	465.475	**460.475 R**	203.5	KNIF-652
Police CLEMARS (A)		154.920	none	KWF-911
Police CLEMARS (B)		154.935	none	KWF-911
Police Tactical (C)		153.860	none	KMA-926
Police NLEMARS (D)		155.475	none	WNHI-284
Fire Primary (F-1)	810.7375	**855.7375R**	156.7*	KNCD-499
Fire Secondary (F-2)	810.9625	**855.9625R**	156.7	KNCD-499
Fire Mutual Aid WHITE 1		154.280	none	KMF-416
Port Operations VHF 12		156.600	none	KWB-433
Public Works	458.200	453.200 R	103.5	KMM-563
School District Buses	469.275	464.275 R		KNJH-249
School District High School		151.955		KNDC-911

*Tone not confirmed.

ROLLING HILLS

CHANNEL	INPUT	OUTPUT	TONE	CALL SIGN
Police - L.A.S.O. #17	487.0375	**484.0375R**	186.2	N/A
Fire - LACoFD Battalion 14	473.5625	**470.5625R**	151.4	WGH-876
Fire - LACoFD Dispatch	473.5375	**470.5375R**	151.4	WGH-876
Fire - LA County TAC 2		154.340	none	KMG-941
Public Works		158.775		KBG-262

ROSEMEAD

CHANNEL	INPUT	OUTPUT	TONE	CALL SIGN
Police - L.A.S.O. #5	485.9875	**482.9875R**	186.2	N/A
Fire - LACoFD Battalion 10	473.4125	**470.4125R**	151.4	WIK-449
Fire - LACoFD Dispatch	473.5375	**470.5375R**	151.4	WIK-449
Fire - LA County TAC 2		154.340	none	KVP-650
Public Works		158.895		KFX-343

LOS ANGELES COUNTY

SAN DIMAS

CHANNEL	INPUT	OUTPUT	TONE	CALL SIGN
Police - L.A.S.O. #8	485.9375	**482.9375R**	186.2	WII-829
Fire - LACoFD Battalion 2	473.4125	**470.4125R**	151.4	WGH-876
Fire - LACoFD Dispatch	473.5375	**470.5375R**	151.4	WGH-876
Fire - LA County TAC 2		154.340	none	KMG-941
Fire Mutual Aid WHITE 1		154.280	none	KMG-941
Fire Mutual Aid WHITE 2		154.265	none	KMG-941
Fire Mutual Aid WHITE 3		154.295	none	KMG-941
Unified Schools		154.515		KAY-411
Unified Schools - Buses		155.235		KIK-786
Public Works Primary		45.160		KIZ-655
Public Works		45.080		KIZ-655

SAN FERNANDO

CHANNEL	INPUT	OUTPUT	TONE	CALL SIGN
Police Primary (F-1)	465.350	**460.350 R**	131.8	KNCL-521
Police Tactical (F-2)	465.225	**460.225 R**	131.8	KNCL-521
Police Tactical (F-3)	465.125	**460.125 R**	131.8	KNCL-521
Police CLEMARS		154.920	none	KNDA-357
Fire - L.A. City FD	813.4375	**858.4375R**	103.5	KNFV-977
Fire Mutual Aid WHITE 1		154.280	none	KSS-308
Fire Rescue	46.440	46.080 R		N/A
Public Works		45.600		KMK-918

SAN GABRIEL

CHANNEL	INPUT	OUTPUT	TONE	CALL SIGN
Police Primary (F-1)	465.175	**460.175 R**	131.8	KMA-440
Police Tactical (F-2)	465.075	**460.075 R**	131.8	KMA-440
Police CLEMARS		154.920	none	KXA-296
Fire Primary (50-59)		**153.995**		KNED-334
Fire - LACoFD Battalion 10	473.4125	**470.4125R**	151.4	WGH-876
Fire - LACoFD Dispatch	473.5375	**470.5375R**	151.4	WGH-876
Fire - LA County TAC 2		154.340	none	KVP-653
Public Works		45.200		KBA-858
School District - Buses	467.025	462.025 R		WNMV-859

SAN MARINO

CHANNEL	INPUT	OUTPUT	TONE	CALL SIGN
Police Primary (F-1)	155.100	**156.330 R**	107.2	KMA-235
Police CLEMARS (F-2)		154.920	none	N/A
Fire Primary (40-49)		**154.175**	114.8	WSX-361
Fire Mutual Aid WHITE 1		154.280	none	WSX-361
High Schools - Buses		33.400		WNSY-234

LOS ANGELES COUNTY

SANTA CLARITA

CHANNEL	INPUT	OUTPUT	TONE	CALL SIGN
Police - L.A.S.O. #5	485.8625	**482.8625R**	186.2	N/A
Fire - L.A.CoFD Dispatch	473.5375	**470.5375R**	151.4	N/A
Local Government	474.000	471.000 R		N/A
Local Government	474.025	471.025 R		N/A
Local Government	474.100	471.100 R		N/A

SANTA FE SPRINGS

CHANNEL	INPUT	OUTPUT	TONE	CALL SIGN
Police - L.A.S.O. #4	486.7625	**483.7625R**	186.2	WII-782
Fire Primary RED (80-89)		**154.205**	103.5	KMJ-581
Fire Mutual Aid WHITE 3		154.295	none	KMJ-581
Public Works	46.220	44.720 R	none	N/A
Public Works		45.240		KLK-698

SANTA MONICA

CHANNEL	INPUT	OUTPUT	TONE	CALL SIGN
Police Primary (F-1)	474.0625	**471.0625R**	110.9	KVJ-951
Police Secondary (F-2)	474.0875	**471.0875R**	110.9	KVJ-951
Police Tactical (F-3)	474.0125	**471.0125R**	110.9	KVJ-951
Police Tactical (F-4)	473.125	**470.125 R**	110.9	KVJ-951
Police Tactical (F-5)	474.000	**471.000 R**	110.9	KVJ-951
Police CLERS	155.430	158.790 R	186.2	WGM-639
Police CLEMARS		154.920	none	KNFU-618
Fire & Medic Dispatch	485.525	**482.525 R**	none	WIJ-878
Fire TAC-1 - Response	485.0125	**482.0125R**	100.0	WIJ-878
Fire TAC-2	485.325	**482.325 R**	none	WIJ-878
Local Government	458.750	453.750 R	110.9	KMF-370
Local Government	458.850	453.850 R	110.9	KMF-370
Airport Operations	474.0375	471.0375R	136.5	WIL-258
Santa Monica College Police		154.650		KNCV-305
Santa Monica College Security		151.835		WNMH-879
Santa Monica College Security		154.570		WNMH-879
Transit District	509.6625	506.6625R	none	KZD-414

SIERRA MADRE

CHANNEL	INPUT	OUTPUT	TONE	CALL SIGN
Police Dispatch		**158.730**	131.8	KMA-302
Police CLEMARS		154.920	none	KNHS-545
Fire Primary (40)		**154.370**	131.8	KAY-727
Fire Mutual Aid WHITE 1		154.280	none	KAY-727
Search & Rescue		155.160		KSV-856
Public Works		155.880		N/A

LOS ANGELES COUNTY

SIGNAL HILL

CHANNEL	INPUT	OUTPUT	TONE	CALL SIGN
Police Trunked	811.2875	**856.2875T**	none	WNAR-366
Police Trunked	812.2875	**857.2875T**	none	WNAR-366
Police Trunked	813.2875	**858.2875T**	none	WNAR-366
Police Trunked	823.6125	**868.6125T**	none	WNVG-202
Police Trunked	823.9375	**868.9375T**	none	WNVG-202
Fire - Long Beach FD	156.195	**153.950 R**	131.8	KLI-207

SOUTH EL MONTE

CHANNEL	INPUT	OUTPUT	TONE	CALL SIGN
Police - L.A.S.O. #5	485.9875	**482.9875R**	186.2	WII-832
Fire - LACoFD Battalion 10	473.4125	**470.4125R**	151.4	WGH-876
Fire - LACoFD Dispatch	473.5375	**470.5375R**	151.4	WGH-876
Fire - LA County TAC 2		154.340	none	KVP-655
Public Works	474.0375	471.0375R	136.5	WIF-696

SOUTH GATE

CHANNEL	INPUT	OUTPUT	TONE	CALL SIGN
Police Dispatch (F-1)		**155.670**	none	KMA-966
Police CLEMARS (F-2)		154.920	none	KXA-295
Police CLERS	155.430	158.790 R	186.2	KCU-38
Fire - LACoFD Battalion 13	473.6125	**470.6125R**	151.4	WGH-876
Fire - LACoFD Dispatch	473.5375	**470.5375R**	151.4	WGH-875
Fire - LA County TAC 2		154.340	none	N/A
Public Works		158.760		KMK-250

SOUTH PASADENA

CHANNEL	INPUT	OUTPUT	TONE	CALL SIGN
Police Dispatch (F-1)	473.9875	**470.9875R**	91.5	KVF-692
Fire Primary (80-89)		**153.995**	136.5	KNDN-751
Fire Mutual Aid WHITE 1		154.280	none	KWE-410
Public Works		155.835		WQU-994

LOS ANGELES COUNTY

TEMPLE CITY

CHANNEL	INPUT	OUTPUT	TONE	CALL SIGN
Police - L.A.S.O. #5	485.9875	**482.9875R**	186.2	WII-832
Fire - LACoFD Battalion 10	473.4125	**470.4125R**	151.4	WGH-876
Fire - LACoFD Dispatch	473.5375	**470.5375R**	151.4	WGH-876
Fire - LA County TAC 2		154.340	none	N/A
Public Works		39.180		KJB-949
Public Works		158.925		KNAH-328
Unified School District	467.000	462.000 R		WNND-336

TORRANCE

CHANNEL	INPUT	OUTPUT	TONE	CALL SIGN
Police Dispatch	509.1625	**506.1625R**	123.0	WIK-591
Police Tactical / Inquiry	509.2125	**506.2125R**	146.2	WIK-591
Police Tactical	485.500	482.500	85.4	WIK-591
Police Tactical	485.600	482.600	91.5	WIK-591
Police CLEMARS		154.920	none	KLO-454
Police CLERS	155.430	158.760 R	186.2	KHQ-35
Police MDT's	812.2625	857.2625R	none	WNRE-795
Fire & Medics (F-1)	509.2625	**506.2625R**	167.9	WIK-591
Fire & Medics (F-2)	509.2875	**506.2875R**	179.9	WIK-591
Lifeguards	485.425	482.425 R	110.9	WIK-591
Local Government				
Citywide Common	509.500	506.500 R	192.8	WIK-591
Planning & Streets	509.1375	506.1375R	118.8	WIK-591
Water & Power	509.2375	506.2375R	156.7	WIK-591
Local Government - Paging		929.0125R	none	WNIX-993
Transit District	808.5125	**853.5125R**	D072	WXD-691
Unified School District - Security		151.895	114.8	KLZ-587
Unified School District - Buses		462.750 R		KTE-898
Unified High School Security		462.075		KNFE-304
Water District		48.060		KMG-971

VERNON

CHANNEL	INPUT	OUTPUT	TONE	CALL SIGN
Police Primary (F-1)	156.135	**158.745 R**	179.9	KNIV-369
Police CLEMARS (F-2)		154.920	none	KNIN-874
Police Vehicle Location		45.580		KMC-230
Fire & Medics (1-9)		**154.130**	none	KMC-214
Fire RED		154.250	103.5	KMC-214
Fire Mutual Aid WHITE 1 (F-2)		154.280	none	KMC-214
Fire Fireground WHITE 3 (F-3)		154.295	none	KMC-214
Local Government		158.760		KAZ-873

LOS ANGELES COUNTY

WALNUT

CHANNEL	INPUT	OUTPUT	TONE	CALL SIGN
Police - L.A.S.O. #29	485.9375	**482.9375R**	186.2	WII-819
Fire - LACoFD Battalion 12	473.4125	**470.4125R**	151.4	N/A
Fire - LACoFD Dispatch	473.5375	**470.5375R**	151.4	N/A
Fire - LA County TAC 1		154.430	none	KQT-906
Local Government		856-860.0375	none	SMR
Local Government		861-865.4875	none	SMR
Unified School District - Security		154.515		KNAP-468
Unified School District - Buses		155.235		WQN-383

WEST COVINA

CHANNEL	INPUT	OUTPUT	TONE	CALL SIGN
Police Primary (F-1)	509.3375	**506.3375R**	103.5	KYO-871
Police Tactical (F-2)	509.5125	**506.5125R**	103.5	KYO-871
Police Secondary (F-3)	509.4875	**506.4875R**	103.5	KYO-871
Police Tactical (A)		155.715	103.5	KMA-449
Police Tactical (B)		155.790	1-3.5	KMA-449
Police CLERS	155.430	158.790 R	186.2	KQB-52
Police CLEMARS		154.920	none	KLO-478
Fire Primary (70)		**156.165**	192.8	KAZ-313
Fireground		154.310	none	KAZ-313
Fire Mutual Aid WHITE 1		154.280	none	KAZ-313
Fire Mutual Aid WHITE 3		154.295	none	KAZ-313
Local Government	509.4875	506.4875R	103.5	KYO-872
Unified School District	469.175	464.175 R		WNCS-790
Unified School District	469.825	464.825 R		WNCS-790
Unified School District - Security		154.570		KB-63664
West Covina Disposal		151.895		KML-464

WESTLAKE VILLAGE

CHANNEL	INPUT	OUTPUT	TONE	CALL SIGN
Police - L.A.S.O. #22	485.9125	**482.9125R**	186.2	WII-872
Fire - LACoFD Battalion 5	473.6125	**470.6125R**	151.4	WGH-876
Fire - LACoFD Dispatch	473.5375	**470.5375R**	151.4	WGH-876
Fire - LA County TAC 1		154.430	none	KRO-240
Westlake High School	469.475	464.475 R		KNBR-214

WEST HOLLYWOOD

CHANNEL	INPUT	OUTPUT	TONE	CALL SIGN
Police - L.A.S.O. #9	486.3625	**483.3625R**	186.2	WII-818
Fire - LACoFD Battalion 1	473.6125	**470.6125R**	151.4	WGH-876
Fire - LACoFD Dispatch	473.5375	**470.5375R**	151.4	WGH-876
Fireground		154.295	none	N/A

WHITTIER

CHANNEL	INPUT	OUTPUT	TONE	CALL SIGN
Police Primary (F-1)	473.3375	**470.3375R**	162.2	KBQ-770
Police Direct (F-2)		470.3375	162.2	KBQ-770
Police Tactical 1	473.1125	**470.1125R**	162.2	KBQ-770
Police Tactical 2		470.1125	162.2	KBQ-770
Police CLEMARS		154.920	none	KLW-287
Fire - LACoFD Battalion 8	473.5625	**470.5625R**	151.4	WGH-876
Fire - LACoFD Dispatch	473.5375	**470.5375R**	151.4	WGH-876
Fireground		154.295	none	N/A
Local Government (Building & Engineering) (Parks & Recreation) (Street Department)	458.300	**453.300 R**	D346	KMM-610
Local Government	33.840	33.640 R		WNYB-775
Whittier High Schools	466.050	461.050 R		KKM-423
Whittier High Schools	466.550	461.550 R		WNMQ-513
Whittier Schools - Security		463.550		KD-51836
Whittier College - Security	469.825	464.825 R		KYV-454
Whittier College		154.540		KVY-454

ORANGE COUNTY

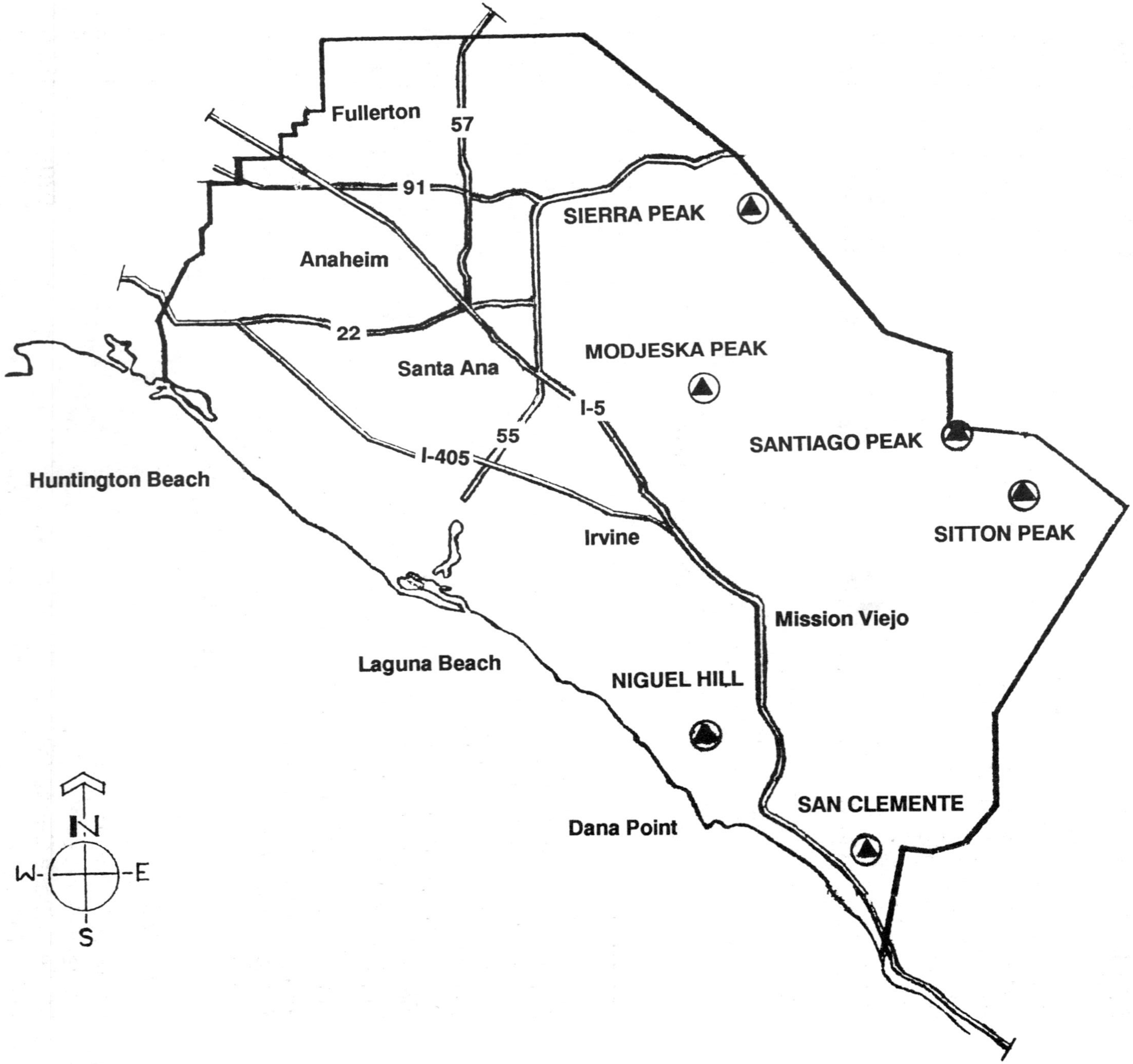

Fullerton
57
91
Anaheim
22
Santa Ana
I-5
55
I-405
Huntington Beach
Irvine
SIERRA PEAK
MODJESKA PEAK
SANTIAGO PEAK
SITTON PEAK
Mission Viejo
Laguna Beach
NIGUEL HILL
Dana Point
SAN CLEMENTE
N
W
E
S

ORANGE COUNTY

SHERIFF'S OFFICE

DISPATCH CHANNELS

CHANNEL	FREQUENCY	TONE
Sheriff Dispatch "Main"	**460.425 R**	103.5
Sheriff Dispatch "South"	**460.175 R**	103.5
Police Common ORANGE North	**460.400 R**	103.5
Police Common ORANGE South	**460.200 R**	103.5
Police Task Force BLUE	**460.500 R**	103.5
Police Tactical WHITE	**460.150**	103.5
Police Countywide RED	**460.025 R**	103.5
Police Inquiry PURPLE	**460.300 R**	103.5
Police Tactical GRAY	**460.550**	103.5

Dispatch "South" is for the City of San Clemente
Countywide RED is for traffic Sig Alerts, major disasters, pursuits, etc.
Each police department in Orange County is assigned one Green channel.

CITY POLICE CHANNELS

CHANNEL	FREQUENCY	TONE
City Police Channel GREEN 1	**460.325 R**	vary
City Police Channel GREEN 2	**460.050 R**	vary
City Police Channel GREEN 3	**460.275 R**	vary
City Police Channel GREEN 4	**460.250 R**	vary
City Police Channel GREEN 5	**460.475 R**	vary
City Police Channel GREEN 6	**460.100 R**	vary
City Police Channel GREEN 7	**460.175 R**	vary
City Police Channel GREEN 8	**460.075 R**	vary
Comm Center CLEMARS	154.920	103.5
Comm Center CLERS (input 155.430)	158.790 R	186.2
Air-to-Air Coordination	45.300	103.5
Tactical - NLEMARS	155.475	none

SHERIFF DETENTIONS

CHANNEL	FREQUENCY	TONE
Detention & Jail Paging	37.040	103.5
Intake Release Center (IRC)	154.800	136.5
Intake Release Center (IRC)	158.970	136.5
Irvine Honor Camp	154.800	136.5
James Musick Jail	154.800	136.5
Theo Lacey County Jail	154.965	136.5
Theo Lacey County Jail	158.775	136.5
Theo Lacey County Jail	158.970	136.5

ORANGE COUNTY

SHERIFF'S OFFICE

SPECIAL OPERATIONS

CHANNEL	FREQUENCY	TONE
Surveillance Operations "Gold"	465.150	none
Surveillance Operations	460.500	103.5
Surveillance Operations "Black"	**460.525 R**	103.5
Surveillance Operations "Pink"	**460.550**	103.5
Surveillance Operations	465.550	none
Surveillance Operations-Reported only	457.550	none
Surveillance Operations-Reported only	467.575	none

AREAS SERVED

Ana Aliso Viejo, Anaheim (unincorporated), Black Star, Coto De Caza, Cowan Heights, Dana Point, East Tustin, El Toro, Irvine Subdivisions, Laguna Hills, Laguna Niguel, Lake Forest, Lemon Heights, Midway City, Mission Viejo, Modjeska Canyon, Olinda, Rancho Santa Margarita, Rossmoor, Red Hill, San Juan Capistrano, San Juan Hot Springs, Santa Ana Heights, Silverado Canyon, Stanton (unincorporated), Sunset Beach, Tonner Canyon, Trabuco Canyon, Villa Park, Yorba Linda (unincorporated)

UNIT IDENTIFICATION SYSTEM

The first number is 18 which represents "Station 18" for the Sheriff's Department. See the countywide numbering system on page 109 for further details.

The first digit following the station number indicates the shift:

1. Graveyard Shift
2. Day Shift
3. Evening Shift

The last two digits generally represent the patrol area:

01	City of Villa Park
02-03	Unincorporated (Atwood, East Lake, Olinda, Yorba Linda Areas)
04-11	Unincorporated (Western Beats)
12-19	Unincorporated (Central Beats)
20	Unincorporated (Dove & Modjeska Canyon Areas)
21-29	City of Stanton
30-33	City of San Juan Capistrano
34-39	City of Dana Point
40-43	City of Laguna Hills
44-49	Unincorporated Areas
50-59	City of Mission Viejo
60-69	Unincorporated (Lake Forest, Rancho Santa Margarita, Silverado Canyon)
70-79	City of San Clemente (South Dispatch)

Example: A typical unit that might be heard on South Dispatch is **18-273** which is a **Day Shift Patrol** Unit in the **City of San Clemente**.

ORANGE COUNTY

SHERIFF'S OFFICE

UNIT IDENTIFICATION SYSTEM

The last three digits, if 400 or above, indicate the type of unit:

18-	400-499	Transportation
18-	500-599	Coroner
18-	600-699	Special Units
18-	700-799	Traffic Units
18-	800-899	Mixed Operations
18-	900-999	Field Portable Units

3100 = Special
3200 = Special
3300 = Special

Many units employ letter designations:

18-	ALPHA	2-Man Patrol
18-	BRAVO	Bomb Squad
18-	CHARLIE	Commercial Division
18-	DELTA	Special Detail
18	DUKE	Sheriff Helicopter
18-	FOXTROT	Special Events
18-	HOTEL	Homicide Investigator
18-	KING	K-Nine Dog
18-	LIMA	Lieutenants
18-	MIKE	Motorcycle Patrol
18-	NOV.	Narcotics
18-	PAPA	Parking
18-	RN	Regional Narcotics
18-	ROMEO	Investigators
18-	SAM	Sergeant
18-	SIERRA	Investigators
18-	TANGO	Traffic
18-	WHISKEY	Warrant
18-	YANKEE	Sex Crimes
18-	ZULU	Lieutenants

An alternate form of Sheriff identification is the shift number immediately followed by a letter designations from above. Examples include **2-SAM-30, 3-TANGO-50**.

COUNTYWIDE IDENTIFICATION

Every public safety unit in the county has a unique four or five digit identification number. The first two digits indicate the department:

07	Parks & Recreation
08	Road Department
09	Unassigned
10	Animal Control
11	UC Irvine PD
13	Cal State Fullerton PD
15	Marshal's Office
16	Military Police
17	Animal Control?
18	Sheriff's Office
19	District Attorney
21	La Habra PD
22	Brea PD
23	Buena Park PD
24	Fullerton PD
25	Placentia PD
24	Fullerton PD
25	Placentia PD
26	Anaheim PD
30	Airport Security
31	Santa Ana PD
32	Garden Grove PD
33	Orange PD
34	Tustin PD
35	Anassigned
36	Cypress PD
37	La Palma PD
38	Unassigned
39	Irvine PD
40	Fountain Valley PD
41	Seal Beach PD
42	Huntington Beach PD
43	Newport Beach PD
44	Costa Mesa PD
45	Laguna Beach PD
46	Unassigned
47	Westminster PD
48	Los Alamitos PD
49	Unassigned
50	Unassigned
51	Seal Beach PD
52	Huntington B. Lifeguard
53	Newport B. Lifeguard
54	Harbor Department
55	Laguna B. Lifeguard
56	San Clemente Lifeguard
57	Unassigned
58	Dana Point Lifeguard

ORANGE COUNTY

FIRE RADIO SYSTEM

The Orange County 800 MHz Fire system consists of three trunking groups. Each trunked group employs four to five rotating frequencies plus one or two dedicated channels. The dedicated frequencies are assigned as control and talk-around channels. Talk-around channels bypass the normal repeater mode and are used for direct communications at the fire scene; they are not part of the trunking system.

Central Net uses Sierra Peak and has seven frequencies: five for repeated voice, one for control data, and a dedicated talk-around channel. North Net follows the same type of frequency format as Central Net and the primary radio site is Signal Hill. Countywide Net also consists of five trunked channels all duplicated at four main repeater sites: San Clemente Reservoir, Santiago Peak, Sierra Peak, and Signal Hill. Laguna Beach Fire uses the dedicated frequency of 860.9625 for dispatch and response; its primary radio site is at the Moorhead Reservior. Orange County Fire also has several fill-in frequencies for poor coverage areas. Changes to the basic frequency configuration occur from time to time.

Listening to the system is not difficult as most signals are fairly strong. The data channels from the site controller rotate randomly once every twelve hours (usually at noon and midnight). Remember to lock out these channels after each rotation and unlock the frequencies that were previously being used for data.

CENTRAL NET

CHANNEL	INPUT	OUTPUT	TONE	CALL SIGN
CH-1 Control Channel (ID's)	811.4625	**856.4625 T**	none	WQZ-938
CH-2 Trunked System	815.7125	**860.7125 T**	none	WQZ-938
CH-3 Trunked System	814.7125	**859.7125 T**	none	WQZ-938
CH-4 Trunked System	813.9625	**858.9625 T**	none	WQZ-938
CH-5 Trunked System	813.7125	**858.7125 T**	none	WQZ-938
Direct Talk-around		**857.9625**	none	WQZ-938

COVERAGE AREA: Central Net includes all fire departments in Central Orange County: Costa Mesa, Fountain Valley, Huntington Beach, Newport Beach, Santa Ana, and Westminster.

NORTH NET

CHANNEL	INPUT	OUTPUT	TONE	CALL SIGN
CH-1 Control Channel (ID's)	811.9625	**856.9625 T**	none	WQZ-938
CH-1A Trunked System	812.2125	**857.2125 T**	none	WQZ-938
CH-3 Trunked System	815.4625	**860.4625 T**	none	WQZ-938
CH-4 Trunked System	814.9625	**859.9625 T**	none	WQZ-938
CH-5 Trunked System	814.4625	**859.4625 T**	none	WQZ-938
CH-6 Trunked System	813.4625	**858.4625 T**	none	WQZ-938
Direct Talk-around		**857.4625**	none	WQZ-938

COVERAGE AREA: North Net includes all fire departments in North Orange County: Anaheim, Brea, Buena Park, Fullerton, Garden Grove, La Habra, Orange, and Stanton.

ORANGE COUNTY

FIRE RADIO SYSTEM

COUNTYWIDE NET

CHANNEL	INPUT	OUTPUT	TONE	CALL SIGN
CH-1 Control Channel (ID's)	811.2125	**856.2125 R**	none	WQZ-938
CH-2 Trunked System	815.2125	**860.2125 R**	none	WQZ-938
CH-3 Trunked System	814.2125	**859.2125 R**	none	WQZ-938
CH-4 Trunked System	813.2125	**858.2125 R**	none	WQZ-938
CH-5 Trunked System	812.7125	**857.7125 R**	none	WQZ-938
CH-6 Trunked System	811.7125	**856.7125 R**	none	WQZ-938
Conventional Channels				
Laguna Beach Dispatch	815.9625	**860.9625 R**	none	WQZ-938
Orange County Failsoft	809.9875	854.9875 R	none	WNSF-261
Orange County Failsoft	811.7125	856.7125 R	none	WNSF-261
Orange County Fill-in	810.4625	855.4625 R	none	WNFH-922
Orange County MDT's	810.7125	855.7125 R	none	WNHV-479
Temporary Repeater	814.7125	859.7125 R	D331	WQZ-938
Temporary Repeater	814.9625	859.9625 R	D331	EQZ-938

COVERAGE AREA: Countywide Net includes the dispatch for Laguna Beach, Laguna Niguel, Lake Forest, Irvine, San Clemente, Tustin, Villa Linda, and Yorba Linda. Laguna Beach Fire is outside the coverage of the Countywide Net and uses its own frequency and radio site at the Moorhead Reservoir.

DIRECT TALK-AROUND

CHANNEL	INPUT	OUTPUT	TONE	CALL SIGN
"Central Net" Talk-around		**857.9625**	none	WQZ-938
"Countywide Net" Talk-around		**858.2125**	none	WQZ-938
"Laguna Fire" Talk-around		**860.9625**	D331	WQZ-938
"North Net" Talk-around		**857.4625**	none	WQZ-938

MUTUAL AID CHANNELS

CHANNEL	INPUT	OUTPUT	TONE	CALL SIGN
(A) Fire Mutual Aid WHITE 1		154.280	none	WNIE-837
(B) Fire Mutual Aid WHITE 2		154.265	none	WNIE-837
(C) Fire Mutual Aid WHITE 3		154.295	none	WNIE-837
Statewide Mutual Aid		866.0125	156.7	WPAT-870
Statewide Mutual Aid		866.5125	156.7	WPAT-870
Statewide Mutual Aid		867.0125	156.7	WPAT-870
Statewide Mutual Aid		867.5125	156.7	WPAT-870
Statewide Mutual Aid		868.0125	156.7	WPAT-870
Statewide CLEMARS	823.5125	868.5125R	156.7	WPAT-870
Statewide FIREMARS	823.9875	868.9875R	156.7	WPAT-870

ORANGE COUNTY FIRE DEPARTMENT

STATION LOCATIONS

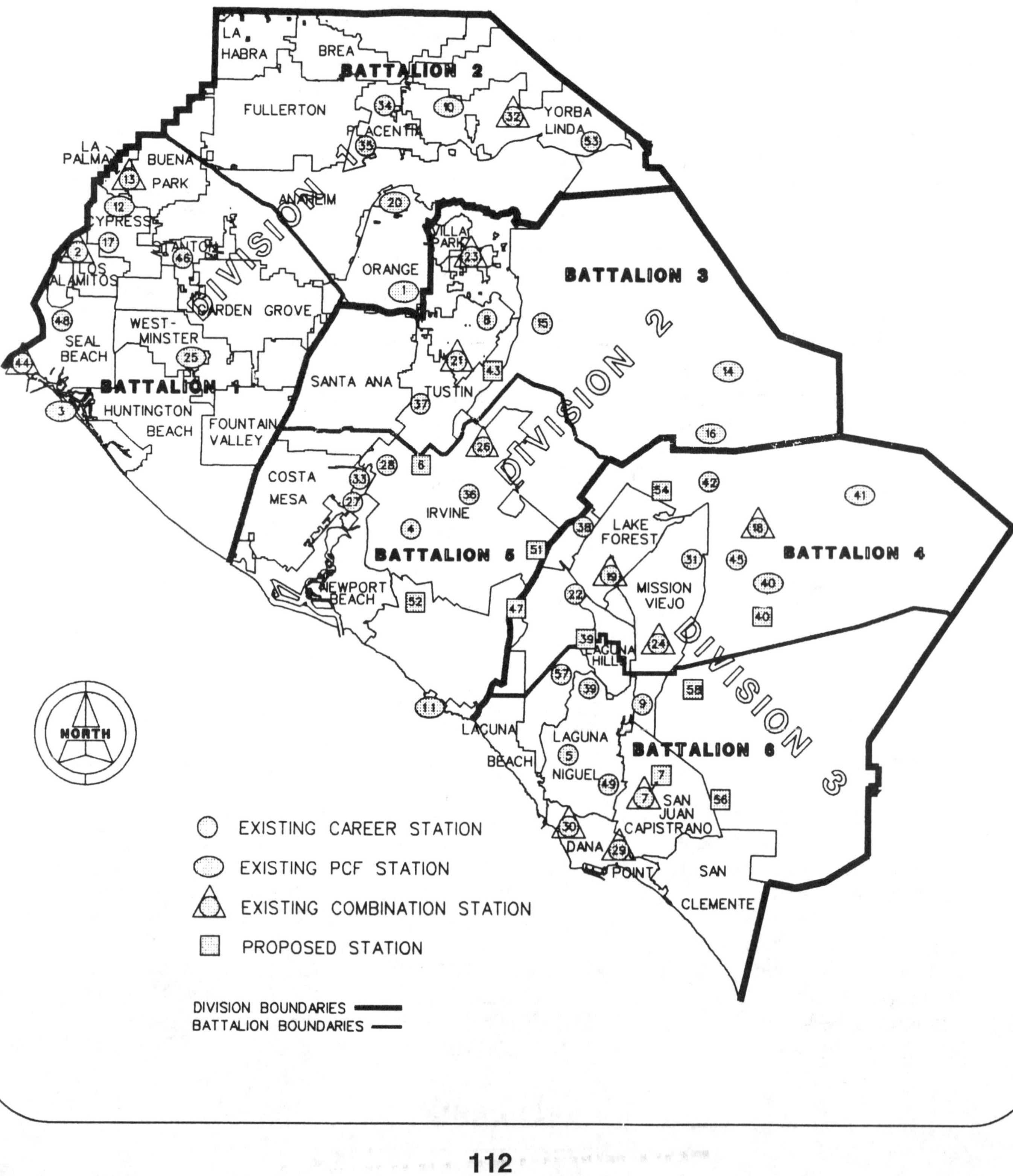

ORANGE COUNTY

FIRE STATIONS & EQUIPMENT

Station No.	Location	Equipment
01	Orange (HQ)	E1, E301, AU1, DOZ1
02	Los Alamitos	E2, E202, AU2, WT2
03	Sunset Beach	E3, P3
04	Irvine (U.C.)	E4, T4, HM4, M4
05	Laguna Niguel	E5, M5
07	San Juan Capistrano	E7, E207, E307, M7, WT7
08	Skyline	E8, RE931
09	S. Mission Viejo	E9, T9
10	Yorba Linda	E10, E210, WT10
11	Emerald Bay	E11, P11
12	Cypress	E12, TAU12
13	La Palma	E13, E213
14	Silverado Canyon	E14, E214, P14
15	Irvine Lake	E15, E315, RE935
16	Modjeska Canyon	E16, P16, WT16
17	Tri-Cities	E17, T17, M17
18	Trabuco Canyon	E18, E218, E318, P18, RE941
19	El Toro	E19, E219, AU19, DOZ19
20	Olive	E20, E220
21	Red Hill (Tustin)	E21, E221, T21, M21
22	Laguna Hills	E22, E222, T22, M22
23	Villa Park	ME23, E223, WT23
24	Mission Viejo	E24, E224, E324, M24
25	Midway City	E25, E225, RE911
26	Valencia (Irvine)	E26, E226, E326, M26
27	Airport (Santa Ana)	E27, RE952, RE953, Foam Truck
28	Irvine Industrial	E28, T28, RE951, FT28
29	Doheny	E29, E229
30	Niguel (Dana Point)	E30, E230, E330, AU30, RE961
31	N. Mission Viejo	E31, Parade 5101
32	E. Yorba Linda	E32, E323, E332, M32
33	Airport (Costa Mesa)	Crash 1-5
34	Placentia	E34, T34
35	Placentia	E35, RE921
36	Irvine	E36, RT951
37	Tustin	E37
38	Lake Forest	E38, M38
39	N. Laguna Niguel	E39
40	Cota De Caza	E40, WT40
41	Holy Jim Canyon	Surplus Orco Engine
42	Portola Hills	E41, RE933, RE943, FC1
43	Tustin	Future Station
44	Seal Beach	E44, E244, T44
45	Santa Margarita	E45, T45, M45, FC2, RE942
46	Stanton	E46, T46, RE911, RE912
47	Irvine, Laguna Hills	Future Station
48	Seal Beach	E48, E248, M48, RE913
49	Bear Brand (Laguna Niguel)	Quint49, RE962
53	E. Yorba Linda	E53, RE922, RE923, RT921, Parade
54	Foothill Ranch	E54, RE944, RT941
57	Aliso Viejo	E57, RE962, RT951

AU = Air Utility
DOZ = Dozer
E = Engine
FC = Communications
HM = Hazmat
M = Medic
RE = Crown Trucks
T = Truck (Also referred to as Quints)
WT = Water Tender

ORANGE COUNTY

COUNTYWIDE TRUNKING SYSTEM

The Sheriff's Office, Marshal's Office, Harbor Department, and all municipal law enforcement agencies will change to a recently awarded Motorola trunking system on the following 800 MHz frequencies. Testing is being conducted and surveillance traffic has been reported on the new 61 frequencies. The transition will take several years depending on the availability of funding. Orange County Fire and other county services will remain on their current systems.

The new frequencies are as follows:

866.100
866.125
866.150
866.175
866.200
866.325
866.350
866.375
866.400
866.425
866.450
866.600
866.625
866.650
866.675
866.700
866.825
866.850
866.875
866.900
866.925
867.100
867.125
867.150
867.175
867.200
867.325
867.350
867.375
867.400
867.425
867.450
867.600
867.625
867.650
867.675
867.700
867.825
867.850
867.875
867.900
867.925
868.100
868.125
868.150
868.175
868.200
868.325
868.350
868.375
868.400
868.425
868.600
868.625
868.650
868.675
868.700
868.825
868.850
868.875
868.925

Call Sign Data

Anaheim WNSS-312 / WNSS-313, Orange WNSS-311

ORANGE COUNTY

EMERGENCY MEDICAL SERVICES

AMBULANCE & RESCUE

NAME	LOCATION	FREQUENCY	TONE	CALL SIGN
Doctor's Ambulance	Laguna Hills	155.205	114.8	N/A
Goodhew Ambulance	Orange	46.040	none	KMB-867
Goodhew Ambulance	Los Alamitos	155.220		KMB-867
FHP Ambulance	El Toro	852.6375R		WNPM-306
Lifefleet Ambulance	Laguna Hills	47.460	100.0	KUJ-655
Lifefleet Ambulance	Laguna Hills	45.960	94.8	KCZ-886
Lifefleet Ambulance	Laguna Hills	46.040	94.8	KCZ-886
Medix Ambulance	Tustin	935 MHz	none	SMR
Morgan Ambulance	Santa Ana	47.580	none	KFG-460
Saddleback S. & R.	Orange County	155.160	none	N/A
Saddleback S & R.	Orange County	155.235	none	N/A
Sealson, Inc.	Costa Mesa	155.160	82.5	KTN-426

AMERICAN RED CROSS

NAME	FREQUENCY
CH-1 National Disaster Net	47.420
CH-2 Southern California	47.500
CH-3 Southern California	47.460
CH-4 Southern California	47.540
CH-5 Southern California	47.580
Operations Repeaters	463.475 R
Operations Repeaters	463.500 R
Portable Repeaters	462.1875R
REACT Liaison	462.675 R

HOSPITALS

NAME	LOCATION	FREQUENCY	CALL SIGN
Anaheim Memorial	Anaheim	151.895	KVX-478
Anaheim Memorial	Anaheim	154.515	WNAJ-591
Canyon General Hospital	Anaheim	157.450	WSQ-492
Childrens Hospital	Orange	35.000	WNFU-261
Choc Hospital	Orange	464.575 R	WNQO-326
Choc Hospital	Orange	464.675 R	WNQO-326
City of God St. Joseph	Brea	151.955	KIF-5880
Fountain Valley Medical	Tustin	507.7375R	KGE-448
Friendly Hills Regional	La Habra	154.625	KVB-935
Hoag Memorial - Security	Newport Beach	154.515	KJN-627
Humana Hospital West	Westminster	158.460	KSX-959
Kaiser Foundation	Anaheim	464.875 R	WNMX-657

ORANGE COUNTY

EMERGENCY MEDICAL SERVICES

HOSPITALS

NAME	LOCATION	FREQUENCY	CALL SIGN
Kaiser Permanente	Anaheim	154.515	KNIS-281
Kaiser Permanente	Anaheim	157.450	WRX-768
Los Alamitos General	Los Alamitos	151.895	WSY-249
Mission Community	Mission Viejo	464.875	WNJZ-930
Mission Regional	Mission Viejo	35.680	WNAB-308
Pacifica Hospital	Huntington Beach	35.680	N/A
Pacifica Hospital	Huntington Beach	37.980	N/A
Pacificare Health	Cypress	463.800 R	WNXM-438
Pacificare Health	Cypress	464.700 R	WNXA-363
Saddleback Community	Laguna Hills	35.640	N/A
Saint Joseph	Orange	35.640	KGY-318
Saint Joseph	Orange	464.325 R	KCT-424
Saint Joseph	Orange	464.775 R	KCT-424
Saint Joseph	Orange	464.825 R	KCT-424
Saint Joseph	Orange	464.875 R	KCT-424
Saint Joseph	Orange	464.925 R	KCT-424
Western Medical Center	Santa Ana	464.525 R	KKY-342
Western Medical Center	Santa Ana	464.825 R	WNNL-692

HOSPITAL EMERGENCY NETWORKS

CHANNEL	FREQUENCY	TONE
Hospital-Medic Assignment "Control 1"	**462.950 R**	103.5
Hospital-Medic Assignment - OCFD	**850 MHz**	none
HEAR - Nationwide	**155.340**	103.5
HEAR - Regional	**155.280**	103.5

HEAR = Hospital Emergency Administrative Network

PARAMEDICS

CHANNEL	BASE FREQUENCY	MOBILE FREQUENCY
MED-1	463.000	468.000
MED-2	463.025	468.025
MED-3	463.050	468.050
MED-4	463.075	468.075
MED-5	463.100	468.100
MED-6	463.125	468.125
MED-7	463.150	468.150
MED-8	463.175	468.175

PL Tone: 162.2 Htz

ORANGE COUNTY

ANIMAL REGULATIONS OFFICE

CHANNEL	INPUT	OUTPUT	TONE	CALL SIGN
(F-1) Dispatch (10)	159.105	**153.920 R**	103.5	KMK-457
(F-2) Operations Direct		153.920	103.5	KMK-457

EMS COMMUNICATIONS SYSTEM

CHANNEL	INPUT	OUTPUT	TONE	CALL SIGN
California State OES Region 4	44.920	**46.320 R**	103.5	N/A
Orange County EMS Primary		**46.560**	103.5	N/A

ENVIRONMENTAL MANAGEMENT AGENCY

CHANNEL	INPUT	OUTPUT	TONE	CALL SIGN
(F-1) Flood Control (6)	159.375	**151.160 R**	136.5	KMK-358
(F-2) Flood Control Direct		151.160	136.5	KMK-358
(F-3) Road Maintenance	159.000	**151.085 R**	136.5	KMK-950
(F-4) Road Maintenance		151.085	136.5	KMK-950
Operations Portables		458.2625	none	KB-65185
Operations Portables		458.4625	none	KB-65185

GENERAL SERVICES ADMINISTRATION

CHANNEL	INPUT	OUTPUT	TONE	CALL SIGN
(F-1) Operations Primary	807.0625	**851.0625R**	103.5	WNRM-448
(F-2) Operations Direct		851.0625	103.5	WNRM-448
Airport Security	159.105	153.980 R	103.5	N/A
Microwave Link	409.725	412.725 R	none	WNRI-622
Mobile Data Terminal (A)	810.2125	855.2125R	none	WNHV-480
Mobile Data Terminal (B)	810.7125	855.7125R	none	WNHV-480
Paging Services		929.2375	none	WNMV-987

HARBOR DEPARTMENT

CHANNEL	INPUT	OUTPUT	TONE	CALL SIGN
(F-1) Primary (54)		**45.140**	none	KMA-752
(F-2) Secondary		45.400	none	KMA-752
(F-3) Harbor Lifeguard Liaison		45.920	none	KMA-752
(F-4) Interagency & Tactical		45.300	none	KMA-752
Reported Activity		45.160	none	N/A
VHF Maritime 12 (Port Operations)		156.600	none	KTD-448
VHF Maritime 16 (Safety-Calling)		156.800	none	KTD-448
VHF Maritime 22 (Coast Guard)		157.100	none	KTD-448

ORANGE COUNTY

MARSHAL'S OFFICE

CHANNEL	INPUT	OUTPUT	TONE	CALL SIGN
(F-1) Dispatch (15)	509.8625	**506.8625R**	103.5	KBA-850
(F-2) Tactical	509.3625	506.3625R	103.5	KBA-850
(F-3) Tactical		506.6125	103.5	KBA-650
(F-4) Tactical		509.6125	103.5	KBA-850
(F-5) Tactical		510.1125	103.5	KBA-850

PROBATIONS DEPARTMENT

CHANNEL	INPUT	OUTPUT	TONE	CALL SIGN
Joplin Boys Ranch		151.235	103.5	WNJM-714
Los Pinos Boys Ranch		151.235	103.5	WNJM-714

PARKS & RECREATION DEPARTMENT

CHANNEL	INPUT	OUTPUT	TONE	CALL SIGN
Regional Parks (Caspers & O'Neill Regional Parks)	159.375	151.160 R	103.5	WNJM-714

TRANSIT DISTRICT

CHANNEL	INPUT	OUTPUT	TONE	CALL SIGN
(F-1) Dial A Ride	511.4625	508.4625R	88.5	WIJ-588
(F-2) Dial A Ride	511.5125	508.5125R	88.5	WIJ-588
(F-3) Dial A Ride	511.6125	508.6125R	88.5	WIJ-588
(F-4) Dial A Ride	511.6625	508.6625R	88.5	WIJ-588
(F-5) Dial A Ride	511.7125	508.7125R	88.5	WIJ-588
Bus Operations 1	811.4875	**856.4875R**	none	KNCM-802
Bus Operations 2	812.4875	**857.4875R**	none	KNCM-802
Bus Operations 3	813.4875	**858.4875R**	none	KNCM-802
Bus Operations 4	814.4875	**859.4875R**	none	KNCM-802
Security Detail	815.4875	**860.4875R**	none	KNCM-802

WATER DEPARTMENT

CHANNEL	INPUT	OUTPUT	TONE	CALL SIGN
Operations Yorba Linda		49.460	141.3	WNHB-735
Operations Yorba Linda		49.160		WNHB-735

ORANGE COUNTY

ANAHEIM

CHANNEL	INPUT	OUTPUT	TONE	CALL SIGN
Police Dispatch (26)	465.050	**460.050 R**	103.5	KUN-399
Police Tactical	465.400	**460.400 R**	103.5	KUN-399
Police Tactical	465.550	460.550	103.5	WNHN-744
Police MDT's	810.4875	855.4875R	none	WNFH-924
Police Air Units		45.300	103.5	KMA-751
Fire - OCFD North Net		850 MHz	none	WNIB-733
Public Works		45.440	123.0	KML-963
Public Works - Data	458.9625	453.9625R	none	KML-963
Convention Center Security	458.950	**453.950 R**	123.0	KML-963
Convention Center reported	466.275	461.275 R		N/A
Convention Center reported	466.5125	461.5125		N/A
Union High School Security		154.540		KWG-712
Public Works		47.940	none	KMD-621
Utilities		37.580	none	KMD-621
Water & Power Dept.		173.210		WNKT-508
Water & Power Dept.	456.475	451.475 R	123.0	KMD-621

BREA

CHANNEL	INPUT	OUTPUT	TONE	CALL SIGN
Police Dispatch (22)	465.100	**460.100 R**	156.7	KUQ-632
Police Tactical	465.400	**460.400 R**	103.5	KUQ-632
Police MDT's	810.2375	855.2375R	none	WNFH-923
Fire - OCFD Countywide Net "Station 62"				
Emergency Ambulance		850 MHz	none	WNLS-724
Public Works		45.520	107.2	KAR-840

BUENA PARK

CHANNEL	INPUT	OUTPUT	TONE	CALL SIGN
Police Dispatch (23)	465.250	**460.250 R**	127.3	KUP-204
Police Tactical	465.400	**460.400 R**	103.5	KUP-204
Fire - OCFD North Net "Station 63"				
Public Works		158.880	123.0	KAQ-271
School District	469.425	464.425 R		WNLI-536

ORANGE COUNTY

COSTA MESA

CHANNEL	INPUT	OUTPUT	TONE	CALL SIGN
Police Dispatch (44)	465.075	**460.075 R**	173.8	KVM-235
Police Tactical	465.200	**460.200 R**	103.5	KVM-235
Fire - OCFD Central Net "Station 84"				
Parks & Recreation	812.2625	857.2625R	103.5	KNEC-203
Mobile Printers & Voice	810.9875	855.9875R	none	KNBF-692
Public Works		45.080	107.2	KMJ-849

CYPRESS

CHANNEL	INPUT	OUTPUT	TONE	CALL SIGN
Police Dispatch (36)	465.075	**460.075 R**	141.3	KUN-537
Police Tactical	465.400	**460.400 R**	103.5	KUN-537
Fire - OCFD Countywide Net "Station 66"				
Public Works		45.320	114.8	KRZ-238
Cypress Comm College	469.575	464.575 R		WNVG-411
School District	806.9375	851.9375R	D132	WZZ-492

DANA POINT

CHANNEL	INPUT	OUTPUT	TONE	CALL SIGN
Sheriff South Dispatch	465.425	**460.425 R**	103.5	KMA-752
Fire - OCFD Countywide Net "Station 68"				
Local Government		856-860.800 T	none	SMR
Lifeguards Primary		**45.920**	none	KOP-903
Lifeguards Secondary		45.400	none	KUN-457
Lifeguards & Harbor Dept.		45.140	none	KMJ-540
Sanitary Services		158.835	103.5	KNBD-631
Southeast Reclamation Authority		154.115	123.0	WNFA-728

FOUNTAIN VALLEY

CHANNEL	INPUT	OUTPUT	TONE	CALL SIGN
Police Primary (40)	465.325	**460.325 R**	173.8	KUP-279
Police Tactical	465.200	**460.200 R**	103.5	KUP-279
Fire - OCFD Central Net "Station 40"				
Public Works		45.520	107.2	KCY-226
School District		151.805		KGB-492

ORANGE COUNTY

FULLERTON

CHANNEL	INPUT	OUTPUT	TONE	CALL SIGN
Police Primary (24)	465.175	**460.175 R**	173.8	KUL-828
Police Tactical	465.400	**460.400 R**	103.5	KUL-828
Fire - OCFD North Net "Station 24"				
Local Government	473.3625	470.3625R	173.8	KML-734
Local Government	473.3125	470.3125R	173.8	KML-734
Fullerton Elementary	807.1125	852.1125R	D152	KNCD-382
Fullerton High School - Buses	468.625	463.625 R	74.4	KFJ-481
Fullerton High School	469.375	464.375 R	118.8	WCL-728
Fullerton High School	900.9375	939.9375		WNLU-594
Parks & Recreation		154.600	D632	N/A

GARDEN GROVE

CHANNEL	INPUT	OUTPUT	TONE	CALL SIGN
Police Primary (32)	465.275	**460.275 R**	103.5	KUQ-685
Police Tactical	465.400	**460.400 R**	103.5	KUQ-685
Police MDT's	813.3375	858.3375R	none	WNCE-549
Police Vehicle Location		45.180	none	N/A
Fire - OCFD North Net		850 MHz	none	WNIB-733
Public Works		154.040	77.0	KMM-309
Unified School District	815.3875	860.3875R		WNDG-522
Water Department		153.710	118.8	KMJ-537

HUNTINGTON BEACH

CHANNEL	INPUT	OUTPUT	TONE	CALL SIGN
Police Primary (42)	465.100	**460.100 R**	103.5	KUL-829
Police Tactical	465.400	**460.400 R**	103.5	KUL-829
Police MDT's	811.2625	856.2626R	none	KNDN-332
Fire - OCFD Central Net "Station 82"				
Building Security	469.625	464.625 R		WQG-568
Lifeguards Primary (52)		45.920	none	KEU-992
Lifeguards Secondary		45.400	none	KEU-992
Local Government		158.805		KMK-914
Local Government		173.3375		KMK-914
Utilities		45.480		KMK-914
Unified Schools - Security		151.685		KJJ-736
Unified Schools - Buses		155.265		KSJ-598
Unified Schools - Buses	511.0875	508.0875R	82.5	WAK-399
High Schools	899.2125	938.2125R		WNNH-630

ORANGE COUNTY

IRVINE

CHANNEL	INPUT	OUTPUT	TONE	CALL SIGN
Police Primary (39)	465.600	**460.600 R**	173.8	KWI-746
Police Tactical	465.200	**460.200 R**	103.5	KWI-746
Police Vehicle Location		45.580	none	WNDP-478
Fire - OCFD Countywide Net "Station 79"				
Fire MDT's	821.075	866.075 R	none	WNVF-992
Local Government	509.4375	506.4375R	103.5	KYM-818
Local Government SMR		850 MHz	none	WNKK-594
Building Security		154.600	D734	KD-40301
Unified Schools - Security F1		154.540		KA-20618
Unified Schools - Security F2		151.625	D306	KNDK-292

LAGUNA BEACH

CHANNEL	INPUT	OUTPUT	TONE	CALL SIGN
Police Primary (45)	465.475	**460.475 R**	173.8	KUN-538
Police Tactical	465.200	**460.200 R**	103.5	KUN-538
Fire - "Laguna Fire"	815.9625	**860.9625R**	none	WNIB-734
Fire - Talk-around		860.9625	none	WNIB-734
Lifeguards Primary (52)		45.920	none	KTA-932
Lifeguards Secondary		45.400	none	KTA-932
Public Works		45.640	107.2	N/A
Water District		173.3125		WNFC-306

LAGUNA NIGUEL

CHANNEL	INPUT	OUTPUT	TONE	CALL SIGN
Sheriff North	465.425	**460.425 R**	103.5	KMA-752
Fire - OCFD Countywide Net "Station 68"				

LA HABRA

CHANNEL	INPUT	OUTPUT	TONE	CALL SIGN
Police Primary (21)	465.475	**460.475 R**	103.5	KUN-539
Police Tactical	465.400	**460.400 R**	103.5	KUN-539
Fire - OCFD North Net (61) "La Habra Fire"		850 MHz	none	WNIB-734
Public Works		45.640	146.2	KDC-265
School District - Security		154.600	118.2	KD-25924
School District - Buses	813.0375	858.0375R		WNCU-209

ORANGE COUNTY

LA PALMA

CHANNEL	INPUT	OUTPUT	TONE	CALL SIGN
Police Primary (37)	465.325	**460.325 R**	110.9	KUQ-521
Police Tactical	465.400	**460.400 R**	103.5	KUQ-521
Fire - OCFD Countywide Net				
Local Government		155.745	141.3	KSQ-339

LOS ALAMITOS

CHANNEL	INPUT	OUTPUT	TONE	CALL SIGN
Police Primary	465.325	**460.325 R**	156.7	KUP-203
Police Tactical	465.400	**460.400 R**	103.5	KUP-203
Fire - OCFD Countywide Net "Station 91"				
Local Government		158.805	141.3	WNLG-671

MISSION VIEJO

CHANNEL	INPUT	OUTPUT	TONE	CALL SIGN
Orange County Sheriff	465.425	**460.425 R**	103.5	KMA-752
Fire - OCFD Countywide Net "Station 69"				
Local Government (M)	806.0625	851.0625 R	103.5	N/A
Saddleback Comm. College				
Campus Police (F-1)	154.815	154.650 R	167.9	KCE-631
Student Escorts (F-2)		154.515	167.9	KXB-220
Campus Police (F-3)		155.475	none	N/A
Maintenance (F-4)		154.600	141.3	N/A
Telephone Interconnect	458.5375	453.5375R		KCE-631
Athletics Department	469.425	464.425 R		WNJZ-934

NEWPORT BEACH

CHANNEL	INPUT	OUTPUT	TONE	CALL SIGN
Police Primary (43)	465.275	**460.275 R**	173.8	KUS-544
Police Tactical	465.200	**460.200 R**	103.5	KUS-544
Fire - OCFD Central Net "Station 83"				
Lifeguards Primary (53)		45.920	none	KLE-686
Lifeguards Secondary		45.400	none	KLE-686
Local Government		158.880	107.2	KMM-444
Local Government		158.760	156.7	KMM-444

ORANGE COUNTY

ORANGE

CHANNEL	INPUT	OUTPUT	TONE	CALL SIGN
Police Primary (33)	465.250	**460.250 R**	103.5	KUS-447
Police Tactical	465.400	460.400 R	103.5	KUS-447
Police MDT's	822.450	867.450 R	none	WNSJ-654
Fire - OCFD North Net "Station 73"				
Fire Mutual Aid WHITE 1		154.280	none	KD-9905
Fire MDT's	810.4625	855.4625R	none	WNFH-992
Local Government		158.760		KDT-341
Public Works		46.520	none	KDT-341
Sanitation District (AVL)		44.760	none	N/A
School District Security		154.570	D162	KB-60864
School District - Buses	806.6375	851.6375R		KLJ-265

PLACENTIA

CHANNEL	INPUT	OUTPUT	TONE	CALL SIGN
Police Primary (25)	465.325	**460.325 R**	127.3	KUQ-684
Police Tactical	465.400	**460.400 R**	103.5	KUQ-684
Fire - OCFD Countywide Net "Station 65"				
Public Works		153.785	114.8	KIZ-312
School District Security		151.655	D624	WNCL-214
School District Security		151.775		WNCL-214
School District Security		151.865		WNCL-214
School District Transportation	897.6375	936.6375R	none	WNNC-964

SAN CLEMENTE

CHANNEL	INPUT	OUTPUT	TONE	CALL SIGN
Sheriff South Dispatch	465.175	**460.175 R**	103.5	KSO-637
Sheriff Tactical	465.200	**460.200 R**	103.5	KSO-637
Border Patrol Check Point	162.925	**163.725 R**	100.0	KBY-352
Fire - OCFD Countywide Net "Station 86"				
Lifeguards Primary (56)		**45.920**	none	KFG-592
Lifeguards Secondary		45.400	none	KFG-592
Public Works Primary		45.360	none	KFG-592
Public Works & Lifeguards		45.440	none	KFG-592
Public Works Telemetry	458.9625	453.9625R	none	KMJ-649
Public Works Telemetry	458.9875	453.9875R	none	KMJ-649

ORANGE COUNTY

SAN JUAN CAPISTRANO

CHANNEL	INPUT	OUTPUT	TONE	CALL SIGN
Sheriff North Dispatch (18)	465.425	**460.425 R**	103.5	KZT-450
Fire - OCFD Countywide Net "Station 69"		850 MHz	none	WNIB-735
Local Government		155.805	179.9	KRA-420
Capistrano USD-Buses		46.000		KRB-453
Capistrano USD-Maintenance	808.2625	853.2625R	D131	KRT-844
Capistrano USD-Security		154.570		KA-91259
Capistrano USD-Security		154.600		KA-91259
Capistrano USD-Security		151.715		KAY-416

SANTA ANA

CHANNEL	INPUT	OUTPUT	TONE	CALL SIGN
Police Primary (31)	465.475	**460.475 R**	103.5	KUN-540
Police Tactical	465.200	**460.200 R**	103.5	KMG-796
Police Sheriff Liaison	465.300	460.300 R	103.5	KMG-796
Police MDT's	810.2125	855.2125R	none	WNHV-479
Fire - OCFD Central Net "Station 71"				
Fire MDT's	823.450	868.450 R	none	WNSS-307
Local Government	458.700	453.700 R	173.8	KMJ-950
Bookmobile		37.320	none	KMJ-950
Public Works - Low Power	458.4125	453.4125R	none	KMJ-950
Public Works - Low Power	458.9125	453.9125R	none	KMJ-950
School District Buses		151.805		KNDU-601
School District Security		155.175	none	KQW-330
Street Department	509.3875	506.3875R		N/A
Water District	456.200	451.200 R	118.8	KMF-829
Zoo Maintenance		151.160	127.3	N/A

SEAL BEACH

CHANNEL	INPUT	OUTPUT	TONE	CALL SIGN
Police Primary (41)	465.175	**460.175 R**	127.3	KUN-536
Police Tactical	465.400	**460.400 R**	103.5	KUN-536
Fire - OCFD Countywide Net "Station 81"				
Lifeguard Primary (51)		45.920	none	KNBG-287
Lifeguard Secondary		45.400	none	KNBV-934
Local Government		155.805		KFG-554

ORANGE COUNTY

STANTON

CHANNEL	INPUT	OUTPUT	TONE	CALL SIGN
Police - Sheriff (18)	465.425	**460.425 R**	103.5	KMA-752
Police Sheriff	465.200	460.200 R	103.5	KUP-205
Fire - OCFD North Net				
Local Government		154.115	123.0	KWQ-494

TUSTIN

CHANNEL	INPUT	OUTPUT	TONE	CALL SIGN
Police Primary (34)	465.325	**460.325 R**	103.5	KUQ-522
Police Tactical	465.200	**460.200 R**	103.5	KUQ-522
Fire - OCFD Countywide Net				
Local Government		153.785	167.9	KGK-606
School District		154.540		KWH-922
School District		464.4875		KD-28801

VILLA PARK

CHANNEL	INPUT	OUTPUT	TONE	CALL SIGN
Police - Sheriff (18)	465.425	**460.425 R**	103.5	KMA-752
Fire - OCFD Countywide Net				
Public Works		155.955	151.4	KNIF-765

WESTMINSTER

CHANNEL	INPUT	OUTPUT	TONE	CALL SIGN
Police Primary (47)	465.250	**460.250 R**	173.8	KUP-206
Police Tactical	465.200	**460.200 R**	103.5	KUP-206
Fire - OCFD Central Net "Station 87"				
Local Government		158.835	151.4	KEL-456
Sanitation District		154.085		KNFF-494
School District	856.4125	851.4125R		WNGC-825

YORBA LINDA

CHANNEL	INPUT	OUTPUT	TONE	CALL SIGN
Police - Brea PD	465.100	**460.100 R**	156.7	WNHN-744
Police Tactical	465.400	**460.400 R**	103.5	WNHN-744
Fire - OCFD Countywide Net				
Public Works	509.3875	506.3875R	127.3	KNO-597
School District	806.4875	851.4875R		WQH-658
Water District	814.2625	859.2625R		KAY-281

IMPERIAL COUNTY

Imperial County

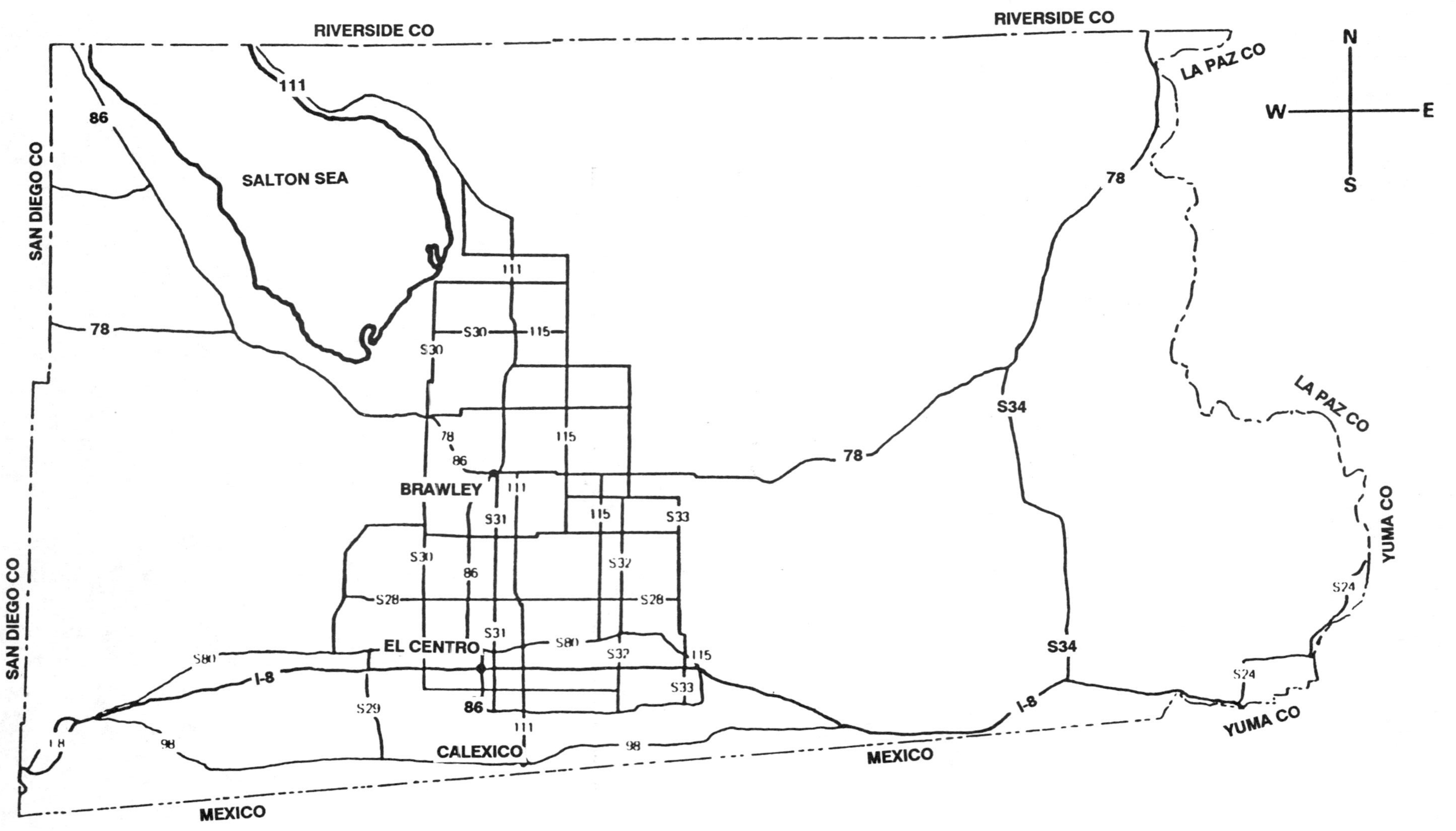

IMPERIAL COUNTY

SHERIFF'S OFFICE

CHANNEL	INPUT	OUTPUT	TONE	COLOR
(F-1) Central Dispatch	154.890	**156.030 R**	110.9	BLUE
(F-2) Central Direct		156.030	110.9	BLUE
(F-3) East Dispatch	155.370	**154.845 R**	110.9	BLACK
(F-4) East Direct		154.845	110.9	BLACK
(F-5) West Dispatch	155.415	**154.995 R**	110.9	WHITE
(F-6) West Direct		154.995	110.9	WHITE
(F-7) Countywide Tactical	155.100	**155.745 R**	110.9	RED
(F-8) Police CLEMARS		154.920	none	GOLD
(F-9) Police Tactical		155.520	110.9	none
(F-10) Police CWMA		**155.130**	110.9	none
(F-11) Police NLEMARS		155.475	none	none
(F-12) Interagency CALCORD		156.075	none	none

CALL SIGN IDENTIFICATION

Brawley	KNDQ-930
Brawley	KVN-844
Brawley	WGJ-510
El Centro	KVL-911
El Centro	WGH-635
Imperial	KNDQ-867
Imperial	KNDQ-891
Imperial	KNFM-205
Mobile I.D.	KA-35828
Salton City	KVL-912
Salton City	KNDQ-887
Winterhaven	KNDQ-895
Winterhaven	KNFU-853

UNIT IDENTIFICATION SYSTEM

1. The first three digits define the type of unit or the patrol area:

200-299	Administration
300-399	North Stations
400-499	Reserve Units
500-599	Central Stations
600-699	Salton City
700-799	Westmoreland
800-899	District Attorney
900-999	Winterhaven

2. Letters are used to represent special units:

C = Chiefs
K = Canine units
L = Lieutenants
S = Sergeants

IMPERIAL COUNTY

FIRE DEPARTMENTS

CHANNEL	INPUT	OUTPUT	TONE	COLOR
(F-1) Central Dispatch		**154.370**	110.9	ORANGE
(F-2) North Dispatch		**154.010**	110.9	GREEN
(F-3) East Dispatch		**154.340**	110.9	BLUE
Fireground		153.830	110.9	none
Statewide Fire Mutual Aid		154.280	none	WHITE 1
Statewide Fire Mutual Aid		154.265	none	WHITE 2
Statewide Fire Mutual Aid		154.295	none	WHITE 3

CALL SIGN IDENTIFICATION

Brawley	KML-733
Brawley	WRX-681
Brawley	KNBF-430
Calexico	KNJA-502
Calipatria	KGQ-346
El Centro	KDF-428
El Centro	KG-7901
El Centro	KNCU-866
El Centro	KNFZ-529
El Centro	KWX-687
Holtville	KDG-236
Imperial	KNCU-866
Imperial	KWX-687
Ocotillo	KXC-493
Ogilby	WZM-846
Palo Verde	KIT-920
Winterhaven	KGT-519

FIRE UNIT IDENTIFICATION

600	California Parks & Recreation
700	Unassigned
800	Calexico FD
900	Brawley FD
2700	Animal Control
2600	Ambulances & Paramedics
2800	Imperial County FD
2900	Unassigned
3000	Unassigned
3100	El Centro FD
3300	Unassigned
3400	Imperial FD
3500	Unassigned
3600	Holtville FD
4200	Ocotillo FD
5100	Salton City CSD
5200	Calipatria FD
5300	Niland FD
5400	Salton Sea Beach FD
6100	Westmoreland FD
8200	Winterhaven FD

IMPERIAL COUNTY

EMERGENCY MEDICAL SERVICES

AMBULANCE SERVICE

NAME	LOCATION	FREQUENCY	CALL SIGN
Gold Cross Ambulance*	El Centro	155.325	N/A
Imperial Ambulance	Imperial	461.275 R	WNBJ-844
Imperial Ambulance	Imperial	863.8375R	WNBS-682
Imperial Ambulance	Imperial	864.8375R	WNBS-682
Life Flight Helicopters	Mt. Laguna	462.975 R	KNCG-493
Yuma Ambulance	Winterhaven	464.275 R	KUT-974

***Gold Cross Ambulances**

2640	El Centro
2645	Brawley
2650	Backup
2655	Unassigned
2660	Salton Sea
2665	Unassigned
2670	Calexico

COUNTY EMERGENCY MEDICAL SYSTEM

CHANNEL	USE	FREQUENCY
1	Hospital - Medics	**155.325**
2	Hospital - Medics	**155.175**
3	Hospital Administration	155.385
4	Hospital Emergency	155.340

Call Sign Identification:

Brawley KYQ-463; Calexico KYQ-464; El Centro KYQ-466; Imperial KNDM-589; Mobile I.D. KV-6493; Ogilby KNAG-318

HOSPITALS

NAME	LOCATION	FREQUENCY	CALL SIGN
El Centro Regional Med.	El Centro	152.300	KNAS-634
Pioneer Memorial	Brawley	31.000	WNCC-210
Pioneer Memorial	Brawley	163.250	WRY-271

IMPERIAL COUNTY

EMERGENCY MEDICAL SERVICES

PARAMEDIC SERVICE

CHANNEL	BASE FREQUENCY	MOBILE FREQUENCY
MED-1	463.000	468.000
MED-2	463.025	468.000
MED-3	463.050	468.050
MED-4	463.075	468.075
MED-5	463.100	468.100
MED-6	463.125	468.125
MED-7	463.150	468.150
MED-8	463.175	468.175

Call Sign Identification:

Calexico Hospital KYW-809;
El Centro Memorial KYW-811;
Pioneers Memorial KYW-810

SEARCH & RESCUE

NAME	LOCATION	FREQUENCY	CALL SIGN
County of Imperial	El Centro	155.160	WRG-623
De Anza Rescue Units	Brawley	155.160	KZR-411
De Anza Rescue Units	El Centro	155.160	KZR-412
Los Explorados	El Centro	155.160	WZM-938

IMPERIAL COUNTY

UTILITY SYSTEMS

COUNTYWIDE MUTUAL AID

CHANNEL	FREQUENCY
Imperial Countywide Mutual Aid	155.130

Call Sign Identification:

Brawley PD KMA-781; Calexico PD KMA-826; Calipatria PD KMA-781; El Centro PD KMA-790; Holtville PD KMB-263; Imperial PD KQT-239; Sheriff's Office KNDQ-867

Tone: 110.9 Htz

EDUCATION DEPARTMENT

CHANNEL	FREQUENCY
Bus Operations Primary	155.235
Bus Operations	155.265

Call Sign Identification:

Imperial KNBU-847; Mobile I.D. KV-6493

EMS COMMUNICATIONS SYSTEM

CHANNEL	FREQUENCY
California State Region 4	**46.320 R**
Coordination with San Diego County	46.340 R
Countywide EMS Communications	46.140
Countywide EMS Communications	46.280
Countywide EMS Communications	46.440

Call Sign Identification:

WNYT-799

Tone: 103.5, 110.9 Htz

IMPERIAL COUNTY

UTILITY SYSTEMS

GENERAL SERVICES DEPARTMENT

CHANNEL	INPUT	OUTPUT	TONE	CALL SIGN
Community Services				
(Salton Sea)		154.085	110.9	KNDQ-887
Local Government 1				
Agriculture Department	153.920	155.880 R	110.9	see below
Road Department	153.920	155.880 R	110.9	see below
Local Government 2				
Animal Control	155.040	155.760 R	110.9	see below
Health Department	155.040	155.760 R	110.9	see below

Call Sign Identification:

El Centro KNFH-768/KNFH-769/WGH-635/WGJ-983/WGJ-985; Imperial KNDQ-891; Mobile I.D. KD-9484; Ogilby KMK-359; Salton City KNDQ-887; Winterhaven KNDQ-895

NEW COUNTYWIDE 800 MHz SYSTEM

The Sheriff's Office, General Services Department, and most municipalities plan to change to an 800 MHz trunked radio system when funding becomes available. The new frequencies will be part of a regional radio system shared with San Diego County. More than sixty radio sites or cells will be used to provide coverage for the two counties.

Police & Fire Services:
856.7625, 857.7625, 858.7625, 859.7625, 860.7625
864.050, 865.050

Local Government Services:
866.100, 866.600, 862.100, 868.100, 868.375, 868.600, 868.925

Local Government Services:
866.3875, 866.8875, 867.3875, 867.6125, 867.875

IMPERIAL COUNTY

BRAWLEY

CHANNEL	INPUT	OUTPUT	TONE	CALL SIGN
Police Primary (F-1)		**155.010**	110.9	KMA-781
Police Encrypted (F-2)		155.010	110.9	KMA-781
Police CWMA (F-3)		155.130	110.9	KMA-781
Police CLEMARS (F-4)		154.920	none	WQJ-512
Fire & Medics (900)		**154.010**	110.9	KML-733
Fire Mutual Aid WHITE 1		154.280	none	KML-733
Local Government		154.115	110.9	KNEK-282
School District - Buses		155.235	110.9	WNKX-960

CALEXICO

CHANNEL	INPUT	OUTPUT	TONE	CALL SIGN
Police Primary (F-1)		**154.725**	110.9	KMA-826
Police CWMA (F-2)		155.130	110.9	KMA-826
Police CLEMARS (F-3)		154.920	none	WRL-476
Fire & Medics (800)		**154.370**	110.9	KMJA-502
Fire Mutual Aid WHITE 1		154.280	none	KMA-826

CALIPATRIA*

CHANNEL	INPUT	OUTPUT	TONE	CALL SIGN
Police Primary (F-1)		**155.010**	110.9	KMA-781
Police CWMA (F-2)		155.130	110.9	KMA-781
Police CLEMARS (F-3)		154.920	none	WYZ-323
Fire & Medics (5200)		**154.010**	110.9	KGQ-346
Fire - Central County		154.370	110.9	KGQ-346
Fireground BROWN		154.250	110.9	KGQ-346
Fire Mutual Aid WHITE 1		154.280	none	KGQ-346

*Dispatched by Brawley PD.

IMPERIAL COUNTY

EL CENTRO

CHANNEL	INPUT	OUTPUT	TONE	CALL SIGN
Police Primary (F-1)		**154.815**	110.9	KMA-790
Police Car-to-Car (F-2)		**155.250**	110.9	KMA-790
Police CWMA (F-3)		155.130	110.9	KMA-790
Police CLEMARS (F-4)		154.920	none	KXO-974
Police NLEMARS (F-5)		155.475	none	KMA-790
Fire & Medics (3100)		**154.370**	110.9	KDF-428
Fireground YELLOW		153.890	110.9	WRU-992
Fire Mutual Aid WHITE 1		154.280	none	KDF-428
Local Government		154.040	110.9	WQN-932
Union High School - Security		151.655		WNUH-503
Union High School		151.775		WNUH-503

HOLTVILLE

CHANNEL	INPUT	OUTPUT	TONE	CALL SIGN
Police Primary		**155.130**	110.9	KMB-263
Police CLEMARS		154.920	none	KOM-865
Fire & Medics (3600)		**154.370**	110.9	KDG-236
Fire Mutual Aid WHITE 1		154.280	none	KDG-236
Local Government		158.955	110.9	WNWQ-757
Unified School District		155.295		WNMS-864

IMPERIAL

CHANNEL	INPUT	OUTPUT	TONE	CALL SIGN
Police Primary (F-1)		**154.815**	110.9	KMA-790
Police CWMA (F-2)		155.130	110.9	KQT-239
Police CLEMARS (F-3)		154.920	none	KMD-934
Fire & Medics (3400)		**154.370**	110.9	KNCU-866
Fire Secondary		154.010	110.9	KNCU-866
Fire Mutual Aid WHITE 1		154.280	none	KNCU-866
Local Government		45.160	none	KBW-825
Unified School District - Buses	811.525	856.525 R		WNSK-650

WESTMORELAND

CHANNEL	INPUT	OUTPUT	TONE	CALL SIGN
Sheriff East (700)		**154.845**	110.9	KNDQ-895
Fire Dispatch		154.010	110.9	KGT-519
Local Government		158.895	110.9	WNGV-445

RIVERSIDE & SAN BERNARDINO COUNTIES

RIVERSIDE COUNTY

SHERIFF RADIO SYSTEM

Cell Site 1

866.2125
866.2625
866.7125
866.7625
867.2125
867.7125
868.2125

Cell Site 2

866.3125
866.8125
867.3125
867.7625
867.8125
868.2625
868.3125

Cell Site 3

866.2375
866.7375
867.2375
867.7375
868.2375
868.7375

Cell Site 4

866.2375
866.7375
867.2375
867.7375

Cell Site 5

866.2675
867.7875
868.2875
868.7875

MDT System

866.7875
867.2875

Mutual Aid

154.920
155.475
156.075
460.025

Palm Springs

866.2125
866.7125
867.2125
867.7125
868.2125
868.7125

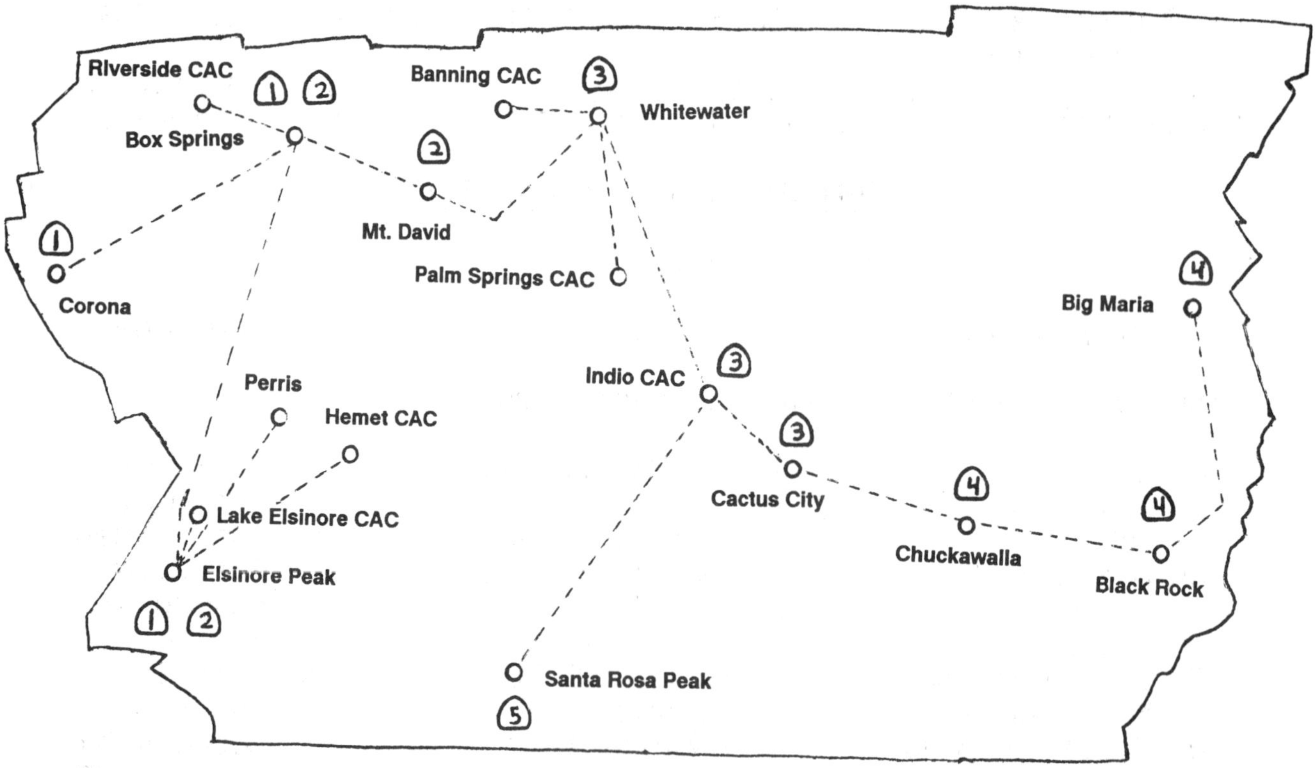

Radio Cell Sites

Cell Site 1: Box Springs, Blue Mtn, Elsinore Peak, Pleasant Peak
Cell Site 2: Box Springs, Elsinore Peak, Mt. David
Cell Site 3: Cactus City, Indio, Palm Springs, Whitewater
Cell Site 4: Big Maria, Black Rock, Chuckwalla
Cell Site 5: Santa Rosa Peak

RIVERSIDE COUNTY

SHERIFF VHF SYSTEM

CHANNEL	INPUT	OUTPUT	TONE	CALL SIGN
(F-1) Riverside & Norco Area (Box Springs Mountain)		**158.850**	110.9	KMA-880
(F-2) Blythe Area (Chuckwalla & Big Maria Peaks)		**159.090**	110.9	WNYD-718
(F-3) Car-to-Car		**158.970**	110.9	KMA-583
(F-4) Banning, Beaumont, Elsinore (Elsinore Peak)		**158.820**	110.9	KNIW-496
(Mt David)		**158.820**	110.9	KXG-846
(F-5) Coachella, Indio Area (Black Butte & Whitewater Hill)		**158.760**	110.9	KGJ-782
(F-6) Inquiry, Mutual Aid		**158.925**	110.9	KMB-350
(F-7) CLEMARS		**154.920**	110.9	KUB-791
(F-8) Tactical		**159.450**	110.9	KE-6932
(F-9) Tactical		**158.775**	110.9	KE-6932
(F-10) Temecula Dispatch		**154.100**	110.9	WNFJ-919
Comm Center Box Springs Link	158.850	453.0875R	none	KB-78753
County Courthouse Security (F-1)	465.350	460.350 R	131.8	WNHN-920
Security (F-2)	465.225	460.225 R	131.8	WNHN-920

UNIT IDENTIFICATION SYSTEM

1. The first number digit indicates the shift:

1. Graveyard Shift
2. Day Shift
3. Evening Shift

2. The second digit, which is a letter, represents the station:

A	Blythe SO	J	Indio Wells Contract	RE	Crime Unit
B	Banning SO	K	Corona PD	RW	Crime Unit
C	Indio SO	L	Lake Elsinore SO	S	Rancho Mirage
CC	Cathedral City	LA	La Quinta	T	Palm Desert
D	Hemet	M	Perris PD	U	Banning PD
E	Riverside	MV	Moreno Valley	V	Palm Desert
F	Lake Elsinore SO	N	UC Riverside PD	W	Indio PD
G	Norco Contract	O	Unassigned	X	Blythe PD
H	Desert Hot Springs	P	Riverside PD	Y	Coachella PD
I	San Jacinto PD	Q	Hemet PD	Z	Beaumont PD

3. The Last digits indicate the type of unit:

1-99	Patrol Units	200-299	Administrative	700-799	Coroner
100-199	Supervisors	300-399	Special Enforcement	Star 80 -	Helicopter

Example: **254-BOY-15** = Day shift patrol unit out of the Banning Station.

RIVERSIDE COUNTY

COUNTY FIRE DEPARTMENT

CHANNEL	INPUT	OUTPUT	TONE	CALL SIGN
Perris Operations CDF NET 3 - Dispatch West	159.360	**151.385 R**	110.9	KVZ-63
Perris Operations CDF NET 4 - East & Central	159.285	**151.175 R**	none	KRP-770
CDF-1 Statewide	159.300	**151.355 R**	none	N/A
CDF-2 Statewide	159.330	**151.265 R**	none	N/A
CDF Administration Net	159.345	**151.340 R**	none	N/A
CDF Air Operations YELLOW		151.310	none	KA-3134
CDF Red Net		**151.220**	none	KA-3134
Corona & Norco FD		**154.235**	192.8	KGJ-726
Fire Command Net		151.460	none	KA-3134
Fire Support Net 5		154.145	167.9	WNNV-408
Fire Support Net 6		**154.445**	167.9	WNNV-408
Fire Support Net 7		154.130	none	WNML-669
Fire Support Net 8		153.770	none	KD-21834
Fire Support Net 9		154.175	none	WNVE-264
Fire Tactical Net 2		151.160	none	KA-3134
Fire Tactical Net 5		151.250	none	KA-3134
Fire Tactical Net 10		151.400	none	KA-3134
Fire Tactical Net 13		151.475	none	KA-3134
Fire Tactical Net 15		159.270	none	KA-3134
Fire Tactical Net 18		159.345	none	KA-3134
Fire Tactical Net 20		159.375	none	KA-3134
Fire Tactical Net 22		159.405	none	KA-3134
Fire Tactical Net 23		159.450	none	KA-3134

Please see the Department of Forestry in the State of California section for additional frequency and tone data.

RIVERSIDE COUNTY

EMERGENCY MEDICAL SERVICES

AMBULANCE & RESCUE

NAME	LOCATION	FREQUENCY	TONE	CALL SIGN
Blythe Ambulance	Blythe	155.295	110.9	N/A
Competition Ambulance	Riverside	155.220	110.9	KNJB-407
Goodhew Ambulance	Riverside	155.355	110.9	WPHA-612
Goodhew & Hemet Valley	Riverside	155.295	110.9	WPHA-612
Goodhew & Hemet Valley	Hemet	155.265	110.9	WNIZ-766
Hemet Valley Ambulance	Hemet	463.1125R	71.9	WNIZ-766
Hemet Valley Rescue	Hemet	155.160		N/A
Life Care Ambulance	Banning	46.000	none	WZW-963
Mercy Air Ambulance	Beaumont	155.220		WPBG-407
Springs Ambulance	Palm Springs	155.205	110.9	KLK-710

HOSPITAL EMERGENCY MEDICAL NETWORK

SERVICE	FREQUENCY	TONE
EMS-1 Ambulance-Hospital	**155.265**	110.9
EMS-2 Ambulance-Hospital	**155.295**	110.9
EMS-3 Ambulance-Hospital	155.175	110.9
EMS-4 Ambulance-Hospital	155.205	110.9
EMS Tactical CALCORD	156.075	none
Hospital Emergency HEAR	155.340	167.9
Valley Emergency Services VESCOMM	154.130	none
Valley Emergency Services VESCOMM	154.515	

UTILITY SYSTEMS

CHANNEL	INPUT	OUTPUT	TONE	CALL SIGN
Coachella Valley Parks & Rec.		153.815	110.9	WNDP-878
Coachella Valley Parks & Rec.	458.9875	453.9875R	D162	WNDP-878
Conventional 800 MHz	809.9625	854.9675R	156.7	WNJE-985
Conventional 800 MHz	810.9875	855.9875R	156.7	WNJE-985
County Fairground	458.875	453.875 R	141.3	WNIE-438
Flood Control District	458.225	453.225 R	179.9	WHF-95
General Hospital	465.275	460.275 R	107.2	WNHN-919
Riverside County*				
Central Riverside County	158.760	159.090 R	110.9	WNYD-718
Eastern Riverside County	159.450	158.850 R	110.9	WPAC-862
Western Riverside County	159.400	158.850 R	110.9	WPAC-862
Road Department	151.160	151.055 R	110.9	KMK-716
Road Department		159.135	110.9	WNGN-683

*The County of Riverside is offering the VHF channels from the Sheriff for use to local cities and other agencies not associated with the new 800 MHz trunking system.

RIVERSIDE COUNTY

MUNICIPALITIES

CITY	INPUT	OUTPUT	TONE	CALL SIGN
Banning, City of				
Sheriff - Cell Site 2				
Police Primary "Station U"		158.820	110.9	KXG-864
Police Tactical		158.745	110.9	WSR-225
Fire Dispatch (CDF)	159.360	151.385 R	110.9	KVZ-63
Fire Secondary		154.445	110.9	N/A
Local Government		46.580	none	KGU-739
Water & Power		47.900		KMG-251
Beaumont, City of				
Sheriff - Cell Site 2				
Police Primary "Station Z"		158.820	110.9	KXG-864
Fire Dispatch (CDF)	159.360	151.385 R	110.9	KVZ-63
Fire Secondary		154.415	110.9	WNML-669
Blythe, City of				
Sheriff - Cell Site 4				
Sheriff VHF	159.450	159.090 R	110.9	KNJC-928
Police Primary "Station X"		158.760	110.9	KVH-312
Fire Dispatch		154.340	118.8	KCN-819
Fire Interagency		155.220	127.3	KCN-819
Cathedral City				
Police Dispatch	465.100	460.100 R	162.2	KNIX-969
Police Tactical	465.050	460.050 R	162.2	KNIX-969
Fire Dispatch		154.415	110.9	WNKW-561
Local Government	458.775	453.775 R	123.0	KNIH-673
Coachella, City of				
Sheriff - Cell Site 4				
Police Primary "Station Y"		158.910	110.9	KWF-757
Police Tactical		155.610	110.9	N/A
Fire Dispatch		154.130	none	KDG-856
Agriculture Ops	458.350	453.350 R	146.2	KNHK-405
Corona, City of				
Police Dispatch	156.045	154.755 R	94.8	KMA-855
Police CLEMARS		154.920	none	KTG-607
Fire Dispatch		154.235	192.8	KGJ-726
Desert Hot Springs				
Sheriff - Cell Site 3				
Police Primary "Station H"	155.370	159.030 R	110.9	KNGL-592
Police Tactical		155.190	110.9	KNGL-592
Fire Dispatch (CDF)	159.285	151.175 R	none	KVP-501
Local Government		155.745	none	WQV-954
Hemet, City of				
Police Primary	155.880	155.055 R	110.9	KVV-714
Police Secondary		154.995	110.9	KVV-714
Police Inquiry		154.995	110.9	KVV-714
Police CLEMARS		154.920	none	N/A
Fire Dispatch		154.415	110.9	KUS-588
Street Department		156.135	173.8	KNJD-743

RIVERSIDE COUNTY

MUNICIPALITIES

CITY	INPUT	OUTPUT	TONE	CALL SIGN
Indian Wells, City of				
Sheriff Dispatch	158.760	159.090 R	110.9	WNYD-718
Fire Dispatch (CDF)	159.285	151.175 R	none	KVP-501
Local Government		155.055	110.9	WNHE-801
Indio, City of				
Sheriff - Cell Site 5				
Police Dispatch	156.090	158.730 R	110.9	KNBN-870
Police Inquiry	155.640	158.880 R	110.9	KMA-583
Fire Dispatch	153.950	154.400 R	110.9	KNEQ-451
Local Government		155.880	110.9	KNAD-287
Lake Elsinore, City of				
Sheriff - Cell Sites 1&2				
Sheriff Dispatch		158.820	110.9	KNIW-495
Fire Dispatch (CDF)	159.360	151.385 R	110.9	KVZ-63
Local Government		155.865	110.9	KDN-433
La Quinta, City of				
Sheriff - Cell Site 5				
Sheriff - Dispatch	158.760	159.090 R	110.9	KMA-880
Fire Dispatch (CDF)	159.285	151.175 R	none	KRP-770
Moreno Valley, City of				
Sheriff Dispatch		158.925	110.9	KMB-350
Fire Dispatch	159.360	151.385 R	110.9	KVZ-63
Emergency Services	467.675	462.675 R		KAE-5242
Murrieta, City of				
Police - 800 MHz SMR				
Fire Dispatch		154.400	none	WNQE-752
Fire Services	810.4875	855.4875R		WPAJ-865
Norco, City of				
Sheriff - Cell Site 1				
Sheriff Dispatch		158.850	110.9	KMA-880
Fire Dispatch		154.235	192.8	KSJ-208
Fire Secondary		154.145	192.8	KSJ-208
Street Department		156.225	127.3	WNJE-436
Palm Desert, City of				
Sheriff - Cell Site 3				
Sheriff Dispatch	158.760	159.090 R	110.9	WNYD-718
Fire Dispatch (CDF)	159.285	151.175 R	none	KVP-501
Local Government	155.100	155.985 R	123.0	KNGZ-319
Local Government	458.050	453.050 R		WNKR-293
Palm Springs, City of				
Airport Operations		155.865		KNIC-611
Police Trunked	812.4875	857.4875 T	none	WNZH-770
Police Trunked	813.9625	858.9625 T	none	WNZH-770
Police Trunked	814.4625	859.4625 T	none	WNZH-770
Police Trunked	815.7125	860.7125 T	none	WNZH-770
Fire Dispatch		154.355	71.9	WNNO-763
Fire Secondary		154.280	71.9	WNNO-763

RIVERSIDE COUNTY

MUNICIPALITIES

CITY	INPUT	OUTPUT	TONE	CALL SIGN
Perris, City of				
Sheriff - Cell Sites 1&2				
Police Dispatch		158.955	110.9	WNAK-241
Fire Dispatch (CDF)	159.360	151.385 R	110.9	KVZ-63
Fire Secondary		153.770	none	KD-21834
Fire Tactical		154.145	192.8	WNNV-408
Local Government		154.040		KFO-495
Rancho Mirage, City of				
Sheriff - Cell Site 3				
Sheriff Dispatch	158.760	159.090 R	110.9	WNYD-718
Fire Dispatch (CDF)	159.285	151.175 R	none	KVP-501
Local Government	458.525	453.525 R	123.0	KRG-880
Riverside, City of				
Police Main (F-1)	465.175	**460.175 R**	131.8	KQR-573
Police Inquiry (F-2)	465.300	**460.300 R**	131.8	KQR-573
Police Alternate (F-3)	465.325	460.325 R	131.8	KQR-573
Police Car-to-Car (F-4)	465.475	460.475 R	131.8	KQR-573
Police Direct (F-5)		460.475	131.8	KQR-573
Police Direct (F-6)		460.300	131.8	KQR-573
Police Detectives (F-7)		460.250	131.8	KQR-573
Police Direct (F-8)		460.250	131.8	KQR-573
Police Special Enforcement		460.050	131.8	KQR-573
Fire Dispatch (F-1)	465.575	**460.575 R**	123.0	KQR-610
Fire Alternate (F-2)	465.600	460.600 R	123.0	KQR-610
Fire Call Back (F-3)	465.625	460.625 R	123.0	KQR-610
Fire Command (F-4)		460.125 R	123.0	KQR-610
Fireground (F-5)		465.1125	123.0	KQR-610
Fireground (F-6)		465.0875	123.0	KQR-610
Fireground (F-7)		460.0875	123.0	KQR-610
Fire Portable Repeater (F-8)	465.0875	460.0875R	123.0	KQR-610
Fire & Police Common (F-9)		453.925	123.0	WNWX-710
Fire & Police Common (F-10)	458.925	453.925 R	123.0	WNWX-710
Fire Tactical Direct (F-11)		460.575	123.0	KQR-610
Fire Tactical Direct (F-12)		460.600	123.0	KQR-610
Fire Tactical Direct (F-13)		460.625	123.0	KQR-610
Local Government	458.425	453.425 R		KWL-310
Transit District	457.800	452.800 R	141.3	KZF-847
Water District		48.020		KNIR-231
San Jacinto, City of				
Sheriff - Cell Site 2				
Police Dispatch	155.580	158.895 R	110.9	KDV-442
Fire Dispatch	159.360	151.385 R	110.9	KVZ-63
Fire Mutual Aid		154.280	none	KEO-271
Local Government		158.805	110.9	WQN-903
Temecula, City of				
Sheriff - Cell Sites 1&2				
Police Dispatch		154.100	110.9	WNFJ-919
Fire Dispatch	159.360	151.385 R	110.9	KVZ-63

SAN BERNARDINO COUNTY

TRUNKED RADIO SYSTEM

Valley System

852.175	Trunked
852.200	"
853.075	
853.450	
854.000	
854.775	
854.800	
854.825	
854.925	
855.975	
856.000	
856.275	
856.425	
866.1375	Trunked
866.3375	"
866.3625	
866.3875	
866.4125	
866.6125	
866.6375	
866.8375	
866.8625	
866.8875	Failsoft
866.9125	Trunked
867.1375	"
867.1625	
867.1875	
867.3625	
867.3875	
867.4125	
867.6125	
867.6375	
867.6625	
867.6875	
867.8625	
867.8875	
867.9125	
868.1375	
868.1625	
868.1875	
868.3375	
868.3625	
868.3875	
868.4125	
868.6125	
868.8375	

West-end System

851.500	Trunked
852.200	"
853.450	
854.975	
855.375	
856.425	
856.825	
858.525	
858.775	
859.825	
859.925	
860.325	
860.825	
861.200	
861.850	
862.200	
862.350	
862.850	
863.850	
864.850	
865.850	Failsoft
868.6375	Trunked
868.6625	"
868.6875	
868.8875	
868.9125	

Needles System

855.225	Trunked
856.450	"
857.975	
858.250	
858.950	
859.975	
860.250	

Moronga Basin

856.475	Trunked
857.475	"
858.225	
860.225	
860.250	
860.450	
860.950	

Victorville System

854.975	Trunked
855.475	"
856.225	
856.725	
857.225	
857.950	
858.250	
858.450	
858.750	
859.450	
859.950	Failsoft
860.250	Trunked

Desert System

856.475	Trunked
857.475	"
858.225	
858.250	
858.725	
859.725	
860.225	
860.250	
860.475	Failsoft
860.950	Trunked

Baker & Trona

855.725	(D331)
857.450	
858.250	
858.475	
860.250	
860.450	
868.8625	
868.9375	

Mutual Aid

866.0125	
866.5125	
867.0125	
867.5125	
868.0125	
868.5125	
868.9375	

Barstow System

855.225	Trunked
855.725	"
856.450	
857.450	
857.975	
858.250	
858.475	
858.950	
859.475	
859.975	
860.250	
860.450	

Mountain System

866.1875	Trunked
866.6875	"
867.1125	
867.3375	
867.8375	Failsoft
868.1125	Trunked
868.6375	"
868.6875	

County Jails

866.1625	
866.6625	

Park Rangers

866.4125	(D131)
866.9125	
867.6875	
868.6375	
868.6875	
868.9375	

Phone Patches

858.525	
858.775	
867.1125	

Air-to-Ground

868.8625	

SAN BERNARDINO COUNTY

TRUNKED RADIO SYSTEM

One of the most ambitious projects in public safety radio was undertaken in 1989, the design and construction of an 800 MHz radio system to cover the nation's largest county, an area more than three times the size of Rhode Island. Now the core of the system is complete and it has become a model for other agencies throughout the world.

There are seven trunking systems, 41 radio sites, and 118 output frequencies. Over 235 base stations and 5300 mobile units are in service. Frequencies are reused up to four times throughout the county and changes occur often. Some frequencies have been set aside for expansion or will be placed in a conventional format to serve as an emergency backup system. A portable trunking system is available which uses the same frequencies as the Needles System.

Every city in San Bernardino County except Barstow and Adelanto are committed to changing to the new system, a transition that will take perhaps 5-10 years to fully complete. The agencies on line (Sheriff, Roads, Animal Control, etc.) are very pleased with the performance. County Park units use conventional repeaters when north of Barstow.

San Bernardino took a different approach than normal when it designed and constructed this system – it tailored the radios to the needs of the everyday user and left the county radio technicians and engineers the responsibility of coordinating all construction. As a result, costs have been significantly reduced (less than 40 million has been spent) and communications have greatly been enhanced.

Disaster and interagency communications have also been a major priority from the start of the project. All radios are similarly programmed so that one user can talk to another anywhere in the county and outside agencies on VHF or UHF can be patched through during a major disaster. Alternate power is available at many sites and satellite phones will soon be employed as a backup to the 911 telephone system.

Desert System
Twentynine Palms area

Moronga Basin
Joshua Tree, Moronga Valley, and Yucca Valley.

Mountain System
Big Bear City, Big Bear Valley, Big Bear Lake, Fawnskin, Forest Falls, Lake Arrowhead, Moonridge, Running Springs, Twin Peaks, and Yucaipa City.

Needles System
Bullhead City, Colorado River, Essex, Ludlow, Needles, Parker, Topock, and Vidal.

Valley System
Central SO; Cities of Colton, Fontana, Highland, Loma Linda, Ontario, Redlands, Rialto, and San Bernardino. The most difficult system to monitor.

Victorville Area
Apple Valley, Barstow, Dagget, Hesperia, Lucerne Valley, and Victorville.

West-end Cities
Cities of Chino, Montclair, Ontario, Rancho Cucamonga and Upland; Cucamonga County Water District, Rancho Cucamonga Fire Protection District, and Chino Valley Independent Fire District.

SAN BERNARDINO COUNTY

FIRE DEPARTMENTS

CHANNEL	FREQUENCY	TONE
Barstow-Victorville		
(Desert Gray)	**154.205**	167.9
CDF - San Bernardino West & Mountain	**151.445 R**	none
CDF - San Bernardino High Desert	**151.325 R**	none
CFD Tactical Common	151.250	none
Central & Mountain Cites BLUE	**154.190**	167.9
Desert Protection District		
(Desert Yellow)	**154.070**	167.9
Mountain Protection District		
(Mountain Yellow)	33.640	none
(Mountain Blue-Mutual Aid)	33.440	none
(Mountain Red-Mutual Aid)	33.760	none
(Fireground)	33.420	none
Parker Dam VFD	164.575 R	none
West Valley Cities BLUE	**154.025**	167.9
Countywide Mutual Aid	**154.325**	167.9
Colorado River Area	**154.070**	167.9

FIRE UNIT IDENTIFICATION

61-69 = Chino FD
71-85 = Central Valley FD
91-95 = Lake Arrowhead FD
101-105 = Wrightwood FD
111-115 = Lucerne Valley
121-124 = Yucca Valley
125 = Hinkley
127 = Searles Valley
128 = Forest Falls
129 = Green Valley Lake
131-149 = Ontario FD
151-159 = Montclair FD
161-169 = Upland FD
171-189 = Rancho Cucamonga FD
191-199 = Mount Baldy
201-210 = Rialto FD
211-220 = Colton FD
221-240 = San Bernardino FD
241-250 = Unassigned
251-259 = Loma Linda FD
261-269 = Redlands FD
271-279 = Lake Arrowhead
281-290 = Big Bear Lake
291-299 = Big Bear City FD
301-309 = Hesperia FD
311-319 = Victorville FD
321-329 = Adelanto FD
331-349 = Apple Valley
351-359 = Unassigned
361-369 = Barstow FD
371-379 = Daggett
381-389 = Yermo
391-399 = Newberry Springs
401-409 = Barstow USMC
411-419 = Fort Irwin FD
421-429 = Twentynine Palms
451-459 = Twentynine Palms
461-469 = Baker

HOSPITAL EMERGENCY MEDICAL NETWORK

CHANNEL	FREQUENCY	TONE
Ambulance-Hospital	**155.400**	167.9
Hospital Emergency HEAR	155.340	167.9

SAN BERNARDINO COUNTY

EMERGENCY MEDICAL SERVICE

AMBULANCE & RESCUE

NAME	LOCATION	FREQUENCY	TONE	CALL SIGN
Bear Valley Community	Big Bear Lake	155.205		WNMR-973
Courtesy Ambulance	San Bernardino	155.235	131.8	KNCC-502
Courtesy Ambulance	San Bernardino	155.280	131.8	KNCC-502
Courtesy Ambulance	San Bernardino	154.515		WNDR-814
Desert Ambulance	Barstow	155.280	none	KJI-601
Desert Rescue Squad	Barstow	47.460	none	KNIT-460
Howard Ambulance	Redlands	46.000	123.0	WZW-963
Howard Ambulance	Redlands	46.040	123.0	WZW-963
Joshua Tree Ambulance	Joshua Tree	155.205		KNBD-947
Mercy Air Ambulance	Fontana	155.220	167.9	WNSN-323
Rescue 3	Barstow	155.160	167.9	KQY-802
Rescue 3	Barstow	155.175	167.9	KQY-802
Rescue 3	Barstow	155.295	167.9	KQY-802
Valley Medical	Victorville	155.280	167.9	WNBG-873

PARAMEDICS

CHANNEL	BASE FREQUENCY	MOBILE FREQUENCY
MED-1	463.000	468.000
MED-2	463.025	468.025
MED-3	463.050	468.050
MED-4	463.075	468.075
MED-5	463.100	468.100
MED-6	463.125	468.125
MED-7	463.150	468.150
MED-8	463.175	468.175

HOSPITAL ASSIGNMENTS

HOSPITAL	CHANNEL	TONE
Barstow Community Hospital	MED-3	167.9
Doctors Hospital	MED-5	167.9
Hi Desert Memorial Center	MED-4	167.9
Loma Linda Medical Center	MED-1	167.9
Ontario Community Hospital	MED-4	167.9
San Bernardino Community	MED-6	167.9
Redlands Community Hospital	MED-2	167.9
Victor Valley Community Hospital	MED-1	167.9

SAN BERNARDINO COUNTY

MUNICIPALITIES

CITY	INPUT	OUTPUT	TONE	CALL SIGN
Adelanto, City of				
Police Dispatch		155.685	167.9	KNIP-802
Police CLEMARS		154.920	none	KNIP-802
Police Car-to-Car		155.520	167.9	KNIP-802
Police New	465.050	460.050 R		WNWT-948
Local Government		155.745	167.9	KMA-917
Unified Schools		155.205	none	WYU-881
Apple Valley				
Sheriff - Victorville System				
Fire Dispatch		154.205	167.9	WNCP-524
Unified Schools		155.205		KNAB-908
Barstow, City of				
Police Dispatch		155.595	100.0	KLE-738
Police Car-to-Car		155.730	100.0	WNHQ-519
Police CLEMARS		154.920	none	KLE-738
Fire Dispatch		154.070	91.5	KYB-659
Local Government	458.625	453.625 R		WNDU-906
Public Works		153.860	100.0	KMM-564
Big Bear Lake, City of				
Sheriff - Mountain System				
Fire Dispatch		154.190	91.5	KLR-456
Lake Patrol		155.835	91.5	KNBQ-201
Public Works	153.905	155.775 R	91.5	WNHR-608
Chino, City of				
Police - West-end System				
Fire Dispatch		154.025	167.9	KNFW-530
School District Buses		155.235		KCS-319
School District PD	458.500	453.500 R		KNFB-977
Colton, City of				
Police - Valley System				
Police Dispatch		154.815	91.5	KRF-325
Fire Dispatch		154.190	91.5	KMF-642
Local Government		155.025	91.5	KCR-581
Fontana, City of				
Sheriff - Valley System				
Fire Dispatch		154.190	91.5	KMF-643
Hesperia, City of				
Sheriff - Victorville System				
Fire Dispatch (CDF)	159.390	151.445 R	none	KMF-238
Parks & Recreation	458.050	453.050 R		WNYY-762
Highland, City of				
Sheriff - Valley System				
Fire Dispatch (CDF)	159.360	151.445 R	none	KMF-238
Loma Linda, City of				
Sheriff - Valley System				
Fire Dispatch		154.190	91.5	KJF-927

SAN BERNARDINO COUNTY

MUNICIPALITIES

CITY	INPUT	INPUT	TONE	CALL SIGN
Montclair, City of				
Police - West-end System				
Fire Dispatch		154.025	167.9	KMB-350
Needles, City of				
Sheriff - Needles System				
Fire Dispatch (CDF)	159.360	151.445 R	none	KMF-238
Fire - Mojave County, AZ	153.890	154.355 R		N/A
Ontario, City of				
Police - Valley System				
Fire Dispatch BLUE		154.025	167.9	KUI-675
Fire Tactical		154.830	167.9	WRH-346
Ontario Montclair School PD		154.650		WNQC-671
Rancho Cucamonga, City of				
Sheriff - West-end System				
Fire Dispatch BLUE		154.025	91.5	KUO-675
Fireground	458.550	453.550 R		KB-73167
Redlands, City of				
Police - Valley System				
Police Dispatch	155.535	154.800 R	91.5	KMF-833
Fire Dispatch		154.190	91.5	KDV-647
Fire Secondary		154.385	91.5	KDV-647
Rialto, City of				
Police - Valley System				
Police Dispatch		156.030	91.5	KIB-839
Fire Dispatch		154.190	91.5	KNCB-852
Fire Secondary		154.385	91.5	KMB-346
San Bernardino, City of				
Police - Moving to Valley System				
Police Dispatch		155.655	91.5	KEX-310
Police Secondary		156.210	91.5	KEX-310
Police Car-to-Car		154.740	91.5	KEX-310
Fire Dispatch		154.385	91.5	KTZ-349
San Bern. Comm. College	465.150	460.150 R	107.2	WFR-388
School District PD	465.075	460.075 R	110.9	WFR-388
Twentynine Palms, City of				
Sheriff - Desert System				
Fire Dispatch		154.070	167.9	KYB-658
Upland, City of				
Police - West-end System				
Fire Dispatch		154.190	91.5	KMB-349
Local Government	458.900	453.900 R	67.0	KMM-521
Victorville, City of				
Sheriff - Victorville System				
Fire Dispatch		154.205	91.5	WNCP-524
Fire Tactical		153.830	91.5	N/A
Local Government	458.550	453.550 R	107.2	WNEB-252

VENTURA COUNTY

VENTURA COUNTY

SHERIFF'S OFFICE

CHANNEL	INPUT	OUTPUT	TONE	CALL SIGN
A-1 West County Dispatch		**159.210 R**	100.0	vary
A-2 Inquiry		155.535	100.0	WNZN-461
A-3 East Valley Dispatch		**156.150**	100.0	vary
A-4 Car-to-Car West		158.730	100.0	WNAQ-813
A-5 Countywide Mutual Aid		156.015	100.0	KK-2948
A-6 Car-to-Car East		155.145	100.0	KSZ-932
(A-7 thru A-12 are local police channels)				
A-13 Police CLEMARS		154.920	none	KDC-275
A-14 Police CALCORD		156.075	100.0	N/A
A-16 Police NLEMARS		155.475	100.0	N/A
B-6 Los Padres National Forest		170.550	none	N/A
Sheriff Search & Rescue		155.160	100.0	KNBZ-695
Sheriff Tactical		155.745	100.0	WYX-316

COUNTY FIRE DEPARTMENT

CHANNEL	INPUT	OUTPUT	TONE	CALL SIGN
YELLOW - Countywide Dispatch		**154.010**	100.0	vary
BLUE 1 - Command 1		154.325	100.0	KMB-552
BLUE 1R - Command 1	155.835	154.325 R	100.0	KMB-552
BLUE 2 - Command 2		154.100	100.0	WNVA-304
BLUE 2R - Command 2	155.100	154.100 R	100.0	WNVA-304
BROWN 1 - Fireground		154.235	100.0	KA-9883
BROWN 2 - Fireground		153.860	100.0	none found
BROWN 3 - Fireground		153.950	100.0	KA-9883
BROWN 4 - Fireground		153.830	100.0	KA-9883
PURPLE - Paramedics		**155.385**	100.0	WNLC-583
Emergency Medical		155.205	100.0	vary

EMERGENCY MEDICAL SERVICES

AMBULANCE SERVICE

NAME	LOCATION	FREQUENCY	CALL SIGN
Ojai Ambulance	Ojai	155.280	N/A
Oxnard Ambulance	Oxnard	155.220	KNGD-261
Oxnard Ambulance	Oxnard	155.235	KNGD-261
Oxnard Ambulance	Oxnard	464.125 R	KNAO-378
Simi Valley Ambulance	Simi Valley	155.175	N/A

HOSPITAL-AMBULANCE NETWORK

CHANNEL	FREQUENCY
Ambulance-Hospital	155.205

VENTURA COUNTY

UTILITY SYSTEMS

SERVICE	INPUT	OUTPUT	TONE	CALL SIGN
Airports & Harbor Department		155.145	100.0	KSZ-932
Animal Control & Parks	158.775	153.785 R	141.3	KXQ-695
Emergency Services	158.940	153.845 R		KNGJ-809
Facilities Security	458.550	453.550 R	141.3	KNDM-592
Health, Sanitation & Water	158.745	153.815 R	123.0	KMM-469
Road Department	156.240	151.025 R		KMJ-498

MUNICIPALITIES

CITY	INPUT	OUTPUT	TONE	CALL SIGN
Camarillo, City of				
Police - Sheriff (8)		**159.210**	100.0	WNAQ-813
Fire Dispatch		154.010	100.0	KDT-971
Local Government		154.965	100.0	WNDP-431
Local Government	458.225	453.225 R	100.0	WNHT-896
Water Department		47.780		KMH-737
Water Department	456.1125	451.1125R		KB-79555
Fillmore, City of				
Police - Sheriff (6)		**159.210**	100.0	KQS-768
Fire Dispatch		154.205	100.0	KOM-441
Fire Support		154.010	100.0	KMD-360
Fire		154.055	100.0	WNDP-956
Local Government		158.835		KTE-619
Street Department		159.105	100.0	KMB-238
Moorpark, City of				
Police - Sheriff (2)		**156.150**	100.0	WZX-690
Fire Dispatch		154.010	100.0	KMD-356
School District	467.625	462.625 R		KAC-9400
Ojai, City of				
Police - Sheriff (1)		**159.210**	100.0	KMF-707
Fire Dispatch		154.010	100.0	KMD-359
Local Government		158.835	100.0	KTE-679
Street Department		159.105	100.0	KXM-711
Oxnard, City of				
Police "Station 3"		**155.655**	100.0	KUU-511
Police Inquiry		155.745	100.0	KKD-497
Police Car-to-Car		156.090	100.0	WNRU-881
Police Tactical		155.625	100.0	KMA-905
Police Ventura	156.210	155.310 R	100.0	KMA-905
Fire Dispatch		154.145	100.0	KMF-789
Fire Secondary		154.370	100.0	KMB-429
Local Government		158.880	100.0	KMK-885
Transit District	458.425	453.425 R	123.0	KBJ-919

VENTURA COUNTY

MUNICIPALITIES

CITY	INPUT	OUTPUT	TONE	CALL SIGN
Port Hueneme, City of				
Police "Station 8"		**155.655**	100.0	KMF-302
Police Inquiry		155.745	100.0	WSR-209
Police Mutual Aid		156.015	100.0	KMF-302
Fire Dispatch		154.145	100.0	KMA-671
Fire Support		154.370	100.0	KMA-671
Local Government		153.965	100.0	KNIB-560
Local Government		158.880	100.0	KJU-963
San Buenaventura, City of				
Police "Station 2"	156.210	**155.310 R**	100.0	KMF-302
Police Inquiry		155.625	100.0	KMF-302
Police Car-to-Car		156.090	100.0	WNRU-881
Police Mutual Aid		156.015	100.0	KK-2948
Fire Dispatch RED		**154.370**	100.0	KMA-677
Fire Support		154.145	100.0	KMA-677
Fire Support		154.010	100.0	WYC-728
Local Government-Housing		153.935	141.3	KAN-503
Local Government		155.955	100.0	KCP-553
Transit District	458.425	453.425 R	123.0	KVN-569
Santa Paula, City of				
Police "Station 4"		**158.835**	100.0	KTE-678
Fire Dispatch		154.370	100.0	KMB-535
Fire Support		154.415	100.0	WNHE-836
Local Government		159.105	100.0	KVI-726
School District	469.425	464.425 R		WNLI-535
Union High School		157.480		N/A
Simi Valley, City of				
Police "Station 9"		**160.785**	100.0	WNVI-253
Police Secondary	158.805	153.875 R	100.0	WNDG-332
Police Mutual Aid		156.015	100.0	KK-2948
Police Santa Paula PD		158.835	100.0	KTE-619
Fire Dispatch		**154.010**	100.0	KAS-604
Local Government	458.700	453.700 R		KNFB-476
Local Government	458.925	453.925 R	71.9	KNFB-476
Parks & Recreation	458.200	453.200 R		KWI-960
Thousand Oaks, City of				
Police - Sheriff		**156.150**	100.0	KLI-941
Fire Dispatch		**154.010**	100.0	KMD-363
Conejo Parks & Recreation	159.375	151.460 R		KSN-318
Public Works		47.920 / 48.240		KNHP-464
Public Works		154.115	127.3	KJU-978

FUTURE COUNTYWIDE SYSTEM

Licensed for Sheriff's Office, County Fire, etc.
866.175 / 866.700 / 867.200 / 867.700 / 868.200 / 868.800 / 868.9375 / 868.9625

OTHER AREAS

BAJA CALIFORNIA NORTE

DIRECTORATE OF NATIONAL SECURITY

CHANNEL	FREQUENCY
Operations Tijuana	154.475 R
Operations Tijuana	154.775 R

FEDERAL HIGHWAY PATROL

CHANNEL	FREQUENCY
Ensenada Area	160.750 R
Tijuana & Mexicali	154.700 R
Links for Remote Bases	167.360 R
Links for Remote Bases	167.400 R
Federal Highway Maintenance	167.300 R

No coverage from El Rosario to Guerrero Negro.

FEDERAL JUDICIAL POLICE

CHANNEL	FREQUENCY
Operations Tijuana	172.000 R

MEXICAN CUSTOM'S

CHANNEL	FREQUENCY
Operations Tijuana	165.325 R

STATE GOVERNOR'S OFFICE

CHANNEL	FREQUENCY
Statewide Repeaters (rare use)	450.700 R

STATE HIGHWAY PATROL

CHANNEL	FREQUENCY
Operations - Statewide Repeaters	151.825 R
Operations - Tijuana	31.850 R

BAJA CALIFORNIA NORTE

STATE JUDICAL POLICE

CHANNEL	FREQUENCY
(F-1) Local Use (Tijuana, Ensenada, Tecate)	161.300 R
(F-2) Mexicali	160.225 R
(F-3) Statewide Use	159.075 R

STATE PUBLIC WORKS (SAHOPE)

CHANNEL	FREQUENCY
Operations - Tijuana & Mexicali	148.175 R
Operations - Tijuana	148.225

EMERGENCY SERVICES

NAME	FREQUENCY
Mexican Red Cross	
Emergency Room	163.400
Ensenada	147.450
Interagency	148.125
Tijuana	163.475 R
Rosarito Beach	163.275
Rescue Teams	
Cuerpo de Rescates	144.800
Delta 7 Rescue Team	170.775 R
Halcones Rescue Team	159.450 R
Radio Brigada de Auxilio	147.000
Rescates Rescue Team	165.400
Toll Road Emergency Services	
Ensenada Toll Road	149.150 R
Tijuana-Mexicali Toll Road	149.275 R

Many volunteer ambulance & rescue teams use CB and amateur radio.

ENSENADA

CHANNEL	FREQUENCY
(F-1) Operations Primary	166.875
(F-2) Traffic Units	169.725
Police Rural Areas	166.750 R
Police Rural Areas	148.225
Traffic Police - Old System	148.300
Traffic Police - Old System	148.500

BAJA CALIFORNIA NORTE

MEXICALI

CHANNEL	FREQUENCY
Police (F-1)	166.500 R
Police (F-2)	166.700 R
Fire Dispatch	166.075 R
Local Government	166.175 R
Local Government	166.475 R

TECATE

CHANNEL	FREQUENCY
Police Primary	171.900
Police Repeater	171.900 R
Police Tactical (rare use)	172.550
Fire Dispatch	171.950

TIJUANA

CHANNEL	FREQUENCY
Police Department	
Commercial Division (Banks, etc.)	166.100 R
Downtown & Motorcycle Traffic	166.125 R
Downtown & Patrol Cars	166.650 R
La Mesa	166.300 R
La Presa / East	166.625 R
Playas / Southwest	166.525 R
Rosarito	166.725 R
San Antonio / South	166.425 R
Tijuana Jail	166.605
Fire Dispatch	151.400 R
Red Cross Interagency	148.125
State Penitentiary - Security	166.200
Public Works	
Construction - La Mesa Sector	150.450 R
Downtown - Beaches Sector	151.460 R
East Sector	156.750 R
Water Department	148.900
Water Department	168.700 R
Water Department	171.100 R
Water Department	171.150

ARIZONA BORDER AREA

AGENCY NAME	FREQUENCY
Arizona Department of Fish & Game	
Southwest Area (Yuma)	**151.475 R**
Statewide Common	151.460 R
Arizona Department of Public Safety (DPS)	
Criminal Investigations - Telegraph Pass	156.090
Criminal Investigations - Telegraph Pass	460.275 R
Criminal Investigations - Telegraph Pass	460.500 R
District 4, Yuma to Dateland	**460.400 R**
District 19, Parker Area	**460.300 R**
Statewide UHF	460.225 R
Statewide Aircraft Coordination	154.935
Statewide Repeaters	155.445 R
Arizona Department of Transportation (DOT)	
District 1 - Phoenix	156.120 R
District 2 - Tucson	156.105 R
District 3 - Prescott	156.225 R
District 4 - Flagstaff	156.135 R
Arizona Medical Comms (Southwest)	463.000
Arizona Medical Comms (Statewide)	463.100
Bullhead City Police	155.715 R
Lake Havasu Police Primary	154.875
Lake Havasu Police Tactical	154.770
Lake Havasu Fire Dispatch	154.355
Law Enforcement - Interagency	155.475
Law Enforcement - Interagency	460.375 R
Mojave County Sheriff (F-1)	**155.640 R**
Mojave County Sheriff (F-2)	**156.150 R**
Mojave County Fire - Dispatch	**154.355 R**
Mojave County Fire	154.190
Mojave County Fire	154.325
San Luis Police	156.210
Somerton Police	154.740
Yuma County Sheriff (F-1)	**158.730 R**
Yuma County Sheriff (F-2)	**158.805**
Yuma County Sheriff - Northern Link	154.710
Yuma County Sheriff - Eastern Link	154.800
Yuma County Fire (F-1)	154.325
Yuma County Fire (F-2)	154.400
Yuma Fire Dept. (F-1)	154.175
Yuma Fire Dept. (F-2)	154.205
Yuma Police (F-1)	**460.350 R**
Yuma Police (F-2)	**460.450 R**
Yuma Police (F-3)	**460.100 R**
Yuma Proving Ground - Military Police	141.000 / 163.535
Yuma Proving Ground - Fire & Rescue	140.875

Yuma Proving Ground - Range Control:
140.950 (repeater), 274.000 / 290.100 / 358.600 (aircraft)

LAS VEGAS, NEVADA

AGENCY NAME	FREQUENCY
LAS VEGAS POLICE DEPARTMENT	
(F-1) Inquiry / Detectives	159.150 R
(F-2) Central / Downtown Strip South	159.090 R
(F-3) Car-to-Car / Secondary Dispatch	158.970 R
(F-4) East / Downtown Strip North	158.745 R
(F-5) Car-to-Car	159.210 R
(F-6) Western Area	159.030 R
(F-7) Administrative (various units)	158.970 R
(F-8) Car-to-Car	156.210 R
(F-9) Blue Diamond / Sandy Valley Area	158.850 R
(F-10) Glendale, Moapa, Logandale Area	158.850
(F-11) Laughlin Area	158.850
(F-12) Car-to-Car	156.210
(F-13) Search & Rescue, Mt. Potosi	156.030 R
LOCAL FIRE DEPARTMENTS	
Clark County FD - Eastern County	453.700
Clark County FD - Western County	453.100
Las Vegas FD - Downtown & East	453.150
Las Vegas FD - North & West	453.400
Rural Areas FD	154.400 R
Rural Areas FD (Laughlin)	154.430 R
LAS VEGAS CITY SERVICES	
Animal Control	158.685 R
Parks & Recreation	155.820
Road Department	159.000 R
Transit Department	453.950 R
Undetermined	158.925 R
COUNTY AGENCIES	
Animal Control	158.820 R
Building Department	153.905
District Attorney	453.925 R
Housing Department	154.115
Road Department	154.100 R
Sanitation Department	453.600 R
EMERGENCY MEDICAL SERVICES	
EMS Dispatch	462.975
Fight For Life	158.940 R
Intl. Air Ambulance	155.160
Mercy Ambulance (F-1)	155.205
Mercy Ambulance (F-2)	155.175
NEVADA HIGHWAY PATROL (NHP)	
Nevada Highway Patrol - Checkpoints	42.560
Nevada Highway Patrol - Stateline to Las Vegas	**42.880**
Nevada Highway Patrol - Statewide	42.780
Nevada Highway Patrol - Statewide	42.940

STATE OF CALIFORNIA

STATE OF CALIFORNIA

ALCOHOLIC BEVERAGE CONTROL

CHANNEL	INPUT	OUTPUT	TONE	CALL SIGN
(F-1) Southern California	465.375	**460.375 R**	127.3	KYJ-255
(F-2) Central California	465.450	460.450 R	127.3	KEG-395
(F-3) Northern California	458.825	453.825 R	127.3	KYJ-333
(F-4) Operations Direct		460.375	127.3	KV-2969
(F-5) Operations Direct		460.450	127.3	KV-2969
(F-6) Operations Direct		453.825	127.3	KV-2969
(F-7) UHF CLEMARS		460.025	none	KA-4993

CORRECTIONS DEPARTMENT

CALIFORNIA INSTITUTION FOR MEN - CHINO

CHANNEL	FREQUENCY	CALL SIGN
Security Operations	866.4625R	WNSS-304
Security Operations	867.4625R	WNSS-304
Security Operations	868.4625R	WNSS-304

CALIFORNIA REHABILITATION CENTER - CORONA

CHANNEL	FREQUENCY	CALL SIGN
Security Trunked	866.9875R	WNSS-303
Security Trunked	867.9875R	WNSS-303
Security Trunked	868.9875R	WNSS-303

CALIFORNIA REHABILITATION CENTER - NORCO

CHANNEL	FREQUENCY	CALL SIGN
Security Operations	866.4625R	WNVR-878
Security Operations	866.5625R	WNVR-878
Security Operations	867.5625R	WNVR-878
Security Operations	868.5625R	WNVR-878

CALIFORNIA STATE PRISON - BLYTHE

CHANNEL	FREQUENCY	CALL SIGN
Security Trunked	861.200 R	WNCM-653
Security Trunked	862.200 R	WNCM-653
Security Trunked	863.200 R	WNCM-653
Security Trunked	864.200 R	WNCM-653
Security Trunked	865.200 R	WNCM-653

STATE OF CALIFORNIA

CORRECTIONS DEPARTMENT

CALIFORNIA STATE PRISON - CALIPATRIA

CHANNEL	FREQUENCY	CALL SIGN
Security Operations	863.200 R	WNCM-654
Security Operations	864.200 R	WNCM-654
Security Operations	865.200 R	WNCM-654

CALIFORNIA STATE PRISON - EAST LOS ANGELES

CHANNEL	FREQUENCY	CALL SIGN
Security Trunked	866.4625R	N/A
Security Trunked	867.4625R	N/A
Security Trunked	868.4625R	N/A

CALIFORNIA STATE PRISON - LANCASTER

CHANNEL	FREQUENCY	CALL SIGN
Security Trunked	866.4625R	WNSS-305
Security Trunked	867.4625R	WNSS-305
Security Trunked	868.4625R	WNSS-305

CALIFORNIA STATE PRISON - OTAY MESA

CHANNEL	FREQUENCY	CALL SIGN
Security Trunked	863.200 R	WNCM-655
Security Trunked	864.200 R	WNCM-655
Security Trunked	865.200 R	WNCM-655

STATE CONSERVATION CAMPS & YOUTH AUTHORITY

CHANNEL	FREQUENCY
El Centro Youth School - Security	155.445
Nelles Youth School - Whittier	39.100
Pilot Rock Conservation Camp - Crestline	153.755
Puerta La Cruz Conservation Camp - Warner Springs	153.755
Rainbow Conservation Camp (female)	153.755
Southern Youth Center - Los Angeles	39.100
Southern Youth Center - Norwalk	39.120

STATE OF CALIFORNIA

EMERGENCY & MUTUAL AID

CALIFORNIA LAW ENFORCEMENT MUTUAL AID RADIO SYSTEM
C.L.E.M.A.R.S.

CHANNEL	FREQUENCY
Police Mutual Aid GOLD 1	**154.920**
Police Mutual Aid GOLD 2	154.935
Police Mutual Aid GOLD 3*	**155.475**
Police Mutual Aid GOLD 4	460.025
Police Mutual Aid GOLD 5	460.025R
Police Mutual Aid GOLD 6	39.460
Police Mutual Aid GOLD 7	39.460R
Police Mutual Aid GOLD 8	868.5125
Police Mutual Aid GOLD 9	868.5125R
Police Mutual Aid GOLD 20	868.9875
Police Mutual Aid GOLD 21	868.9875R

*National Law Enforcement Mutual Aid Radio System (NLEMARS)

CALIFORNIA LAW ENFORCEMENT RADIO SYSTEM
C.L.E.R.S.

CHANNEL	FREQUENCY
Los Angeles Co.	158.790 R
San Luis Obispo, San Diego, Ventura Co's.	453.675 R

CALIFORNIA MULTI-AGENCY RADIO SYSTEM
C.M.A.R.S.

CHANNEL	FREQUENCY
(F-1) Law Enforcement Services	856.2625
(F-2) Office of Emergency Services	857.2626
Multiple-user	860.100
Multiple-user	860.125
Multiple-user Trunked A	858.2626
Multiple-user Trunked A	859.2626
Multiple-user Trunked A	860.2626
Multiple-user Trunked B	861.200
Multiple-user Trunked B	862.200
Multiple-user Trunked B	863.200
Multiple-user Trunked B	864.200
Multiple-user Trunked B	865.200

STATE OF CALIFORNIA

EMERGENCY & MUTUAL AID

CALIFORNIA MULTI-AGENCY RADIO SYSTEM
C.M.A.R.S.

CHANNEL	FREQUENCY
Multiple-user Trunked C	866.3875
Multiple-user Trunked C	866.4625
Multiple-user Trunked C	867.4625
Multiple-user Trunked C	868.4625
Multiple-user Trunked D	866.5375
Multiple-user Trunked D	867.5375
Multiple-user Trunked D	868.5375
Multiple-user Trunked E	866.5625
Multiple-user Trunked E	867.5625
Multiple-user Trunked E	868.5625
Multiple-user Trunked F	866.9875
Multiple-user Trunked F	867.9875
Multiple-user Trunked F	868.575

State CMARS Users:
Air Resources Board, Caltrans, Corrections Dept., Disaster Services, General Services Dept., Forestry Dept., Office of Emergency Services (OSI), Parks & Recreation Dept., State Fire Marshal, State Universities, and others.

CALIFORNIA ON-SCENE EMERGENCY COORDINATION
CALCORD

CHANNEL	FREQUENCY
CALCORD - Interagency	**156.075**

FIRE MUTUAL AID RADIO SYSTEM
FIREMARS

CHANNEL	FREQUENCY
(F-1) OES Tactical	154.160
(F-2) OES Tactical	154.220
(F-3) OES Primary San Diego	**154.160 R**
(F-4) OES Primary Orange	154.220 R
Statewide Fireground	153.830
Statewide Mutual Aid WHITE 1	154.280
Statewide Mutual Aid WHITE 2	154.265
Statewide Mutual Aid WHITE 3	154.295

CTCSS Tones: 179.9, 192.8

STATE OF CALIFORNIA

EMERGENCY & MUTUAL AID

OFFICE OF EMERGENCY SERVICES

CHANNEL	FREQUENCY	TONE
Operations Primary (input 154.980)	**153.755 R**	192.8
Operations Direct	153.755	192.8
Statewide	153.740	none
Statewide FIREMARS (input 33.980)	154.160 R	multi
Statewide FIREMARS (input 33.660)	154.220 R	multi
Statewide Packet (X.25 Protocal)	37.020	none
Statewide Packet (X.25 Protocal)	37.380	none
Statewide Mutual Aid WHITE 1	154.280	none
Statewide Mutual Aid WHITE 2	154.265	none
Statewide Mutual Aid WHITE 3	154.295	none

STATE MUTUAL AID RADIO SYSTEM
S.M.A.R.S.

CHANNEL	FREQUENCY	TONE
(F-1) National Common 1	866.0125	156.7
(F-2) National Common 2	866.5125	156.7
(F-3) National Common 3	867.0125	156.7
(F-4) National Common 4	867.5125	156.7
(F-5) National Common 5	868.0125	156.7
(F-6) California Common 6	868.5125	156.7
(F-7) California Common 7	868.9875	156.7
(F-8) California Common 8	866.200	156.7
(F-9) California Common 9	866.9125	156.7
(F-10) California Common 10	868.7875	156.7

STATE EMERGENCY COMMUNICATIONS USING RADIO EFFECTIVELY
S.E.C.U.R.E.

CHANNEL	FREQUENCY
(F-1) Fixed Day Operations USB	7.4800
(F-2) Fixed Day Operations USB	7.8020
(F-3) Day & Night Fixed USB	5.1400
(F-4) Night Operations USB	2.4190
(F-5) Night Operations USB	2.4220
(F-6) Night Operations USB	2.8120
(F-7) Night Operations USB	2.8040
(F-8) Interstate Coordination Fixed USB	2.3260
(F-9) Interstate Coordination Fixed USB	5.1950
(F-10) Interstate Coordination Fixed USB	7.8050

STATE OF CALIFORNIA

FISH & GAME DEPARTMENT

CHANNEL	INPUT	OUTPUT	TONE	CALL SIGN
Region 5 Southern California				
Operations Repeaters	159.420	151.430 R	multi	Radio Sites
Operations Direct		151.430	multi	KD-3640
Region 6 Long Beach Harbor				
Operations Repeaters	159.420	151.430 R	114.8	N/A
Operations Direct		151.430	114.8	KD-3640
Operations Repeaters	159.435	151.415 R	114.8	Radio Sites
Operations Direct		151.415	114.8	KD-3640
Operations Repeaters	813.2625	858.2625R	179.9	KNER-450
Statewide Mobiles		31.260	none	N/A

Radio Sites:

Avalon WRW-698, Big Bear Lake KSU-541, Black Mountain WPAF-837, Hauser Peak KCS-995, Long Beach KMB-446, Lyons Peak KMA-898, Monument Peak KME-547, Mt. Lukens KMF-418, Mt. Palomar KMB-226, Mt. Santiago KMG-603, Ojai KBQ-390, Santa Ynez Peak KME-261.

Statewide Tone Plan

A: 94.8	D: 107.2	G: 127.3	J: 156.7	M: 173.8	P: 192.8
B: 100.0	E: 114.8	H: 141.3	K: 162.2	N: 179.9	
C: 103.5	F: 118.8	I: 151.4	L: 167.9	O: 186.2	

FORESTRY DEPARTMENT

RIVERSIDE RANGER DISTRICT (3100)

CHANNEL	INPUT	OUTPUT	TONE	CALL SIGN
CDF Statewide NET 1	159.300	**151.355 R**	multi	KRP-770
CDF Statewide NET 2	159.330	**151.265 R**	multi	KRP-770
Perris Dispatch NET 3				
Western & Central Area	159.360	**151.385 R**	multi	KRP-770
Perris Dispatch NET 4				
Eastern (Palm Springs-Blythe)	159.285	**151.175 R**	multi	KRP-770

Riverside West (NET 3)

Tone 1: 110.9 <Elsinore Peak>
Tone 2: 123.0 <Thomas Mtn.>
Tone 3: 131.8 <Box Springs>
Tone 4: 136.5 <Mt. David>

Riverside East (NET 4)

Tone 1: 110.9 <Chuckwalla>
Tone 2: 123.0 <Black Rock>
Tone 3: 131.8 <Cactus City>
Tone 4: 136.5 <Snow Peak>

multi = multiple tones

STATE OF CALIFORNIA

FORESTRY DEPARTMENT

RIVERSIDE RANGER DISTRICT

CHANNEL	INPUT	OUTPUT	TONE	CALL SIGN
Administration Net	159.345	**151.340 R**	none	KRP-770
Air Operations YELLOW		151.310	none	KA-3134
Air Operations RED NET		151.220	none	KA-3134
Corona & Norco FD		154.235	192.8	KGJ-726
Fire Command Net		151.460	none	KA-3134
Fire Support Net 5		154.145	none	WNNV-408
Fire Support Net 6		154.445	none	WNNV-408
Fire Support Net 7		154.130	none	WNML-669
Fire Support Net 8		153.770	none	KD-21834
Fire Support Net 9		154.175	103.5	WNVE-264
Fire Tactical Net 2		151.160	none	KA-3134
Fire Tactical Net 5		151.250	none	KA-3134
Fire Tactical Net 7		151.370	none	KA-3134
Fire Tactical Net 10		151.400	none	KA-3134
Fire Tactical Net 13		151.475	none	KA-3134
Fire Tactical Net 15		159.270	none	KA-3134
Fire Tactical Net 18		159.345	none	KA-3134
Fire Tactical Net 20		159.375	none	KA-3134
VHF Air Operations USFS		168.625	none	N/A

Call Sign I.D.
Blue Mountain KFN-612; Box Springs KRP-770; Cactus City KMF-338; Chuckwalla Mountain KVP-501; Elsinore Peak KSP-243; Mount David WNAF-200; Snow Peak KNIF-714; Thomas Mountain KDV-649.

RIVERSIDE CDF FIRE STATIONS*

STATION	LOCATION	EQUIPMENT
1	Perris HQ	Batt. 3111, E1
"	"	E3184, E3185,
"	"	SQ1, D3141,
"	"	Transport 3141, Truck 1
2	Sunnymead	Batt. 9, E2, E2A, SQ2
3	Nuview	E3, E3R, SQ3
4	Cajalco	E4, SQ4
5	Quail Valley	E5, SQ5
6	Edgemont	E6, E6A
7	Sun City	Batt. 7, E7, E7A, SQ7
8	Woodcrest	E8, SQ8
9	Good Meadow	E9, WT9
10	Elsinore	Batt. 3112, E10, E10A,
"	"	E3179, E3186, SQ10, U3138

*Most CDF Stations are shared with Riverside County Fire.

STATE OF CALIFORNIA

FORESTRY DEPARTMENT

RIVERSIDE CDF FIRE STATIONS

STATION	LOCATION	EQUIPMENT
11	Lakewood Village	E11, E11A, SQ11
12	Temecula (to relocate)	E12, E12R, E3161,
"	"	E3182, SQ12, U3138
13	Home Gardens	E13, SQ13
14	Corona	Batt. 3114, E14,
"	"	E3178, U3135
15	El Centro	E15, SQ15
16	Pedly	E16, E16A, SQ16
17	Glen Avon	E17, E17A, SQ17, WT17
18	West Riverside	E3175, SQ18
19	Highgrove	E19, SQ19
20	Beaumont	E3160, E3180, D3142
21	Calimesa	E21, E21A, R21, SQ21
22	Cherry Valley	E22, E22A, SQ22
23	Pine Cove	Batt 11, E23
24	Cabazon	E24, SQ24
25	San Jacinto	Batt. 3115, E3171, E3183,
"	"	SQ25, U3137
26	Little Lake	E26, E26A, E26B, SQ26
27	Ryan Field	Batt. 3108, D3140, E27, E27A,
"	"	Transport 3140
28	Sage	E3163, SQ28, WT28
29	Anza	E29, E3162, SQ29, WT29
30	Pinyon	E30, WT30
31	Bermuda Dunes	E31, E31R, SQ31
32	La Quinta	E32, E32A, SQ32
33	Palm Desert	E33, M33, T33, SQ33
35	Thousand Palms	Batt. 6, E35, SQ35
36	N. Palm Springs	E36, E36R
37	Desert Hot Springs	Batt. 10, E37, E37A, SQ37
38	Indio	E38, E38R, SQ38
39	Thermal	E39, E39R, SQ39, WT39
40	Mecca	Batt. 7, E40, E40A,
"	"	SQ40, WT40
41	North Shore	E41, SQ41
42	Oasis	E42, E42R, SQ42
43	Blythe	Batt. 8, E43, E43A
"	"	SQ43, WT43
44	Ripley	E44
45	Blythe Air Base	E45, E45R
46	River Bend	E46
47	Lost Lake	A47, E47
48	North County	E48
49	Lake Tamarisk	E49, R49, WT49
50	Rancho Mirage	E50, M50, M50R, R50, T50

STATE OF CALIFORNIA

FORESTRY DEPARTMENT

RIVERSIDE CDF FIRE STATIONS

STATION	LOCATION	EQUIPMENT
51	El Cariso	E51, SQ51
52	Cottonwood	E52
53	Garner Valley	E53, SQ53
54	Homeland	E54
55	Indian Wells	Batt. 12, E55, M55, R55, T55
56	Sky Valley	E56, SQ56
57	Indio Hills	A57, E57
58	Moreno	E58, SQ58
59	Mead Valley	E59, E59R
60	Canyon Lake	E60, SQ60
61	Wildomar	E61, E61A, SQ61
62	Rancho Carrillo (v)	E62
63	Poppet Flats (v)	A63, E63
64	Juniper Flats (v)	E64, SQ64
65	Kennedy Park	E65, S65, SQ65
66	Beaumont	A66, Batt. 3113, E66, E66A
"	"	E66R, R66, U3186, WT66
67	Mesa View	E67, E67R
68	Menifee	E68, SQ68
69	N. Rancho Mirage	E69
70	La Quinta	E70
71	Palm Springs	E71, E71R
72	Valle Vista	E72, E72R
73	Temecula	Batt. 15A, E73, E73R
74	Rancho Capistrano (v)	E74, R74
75	Bear Creek	E75
76	Terwilliger (v)	E76
77	Lake Riverside (v)	E77
78	Mountain Center (v)	E78
79	Coachella City	E79, E79R
80	Eagle Mountain	E80, E80R, R80, SQ80
81	Gavalen Hills	E81

(v) = Volunteer FD

AIR ATTACK BASE

STATION	LOCATION	EQUIPMENT
Air Attack Base	Ryan Field	Air Attack 310,
"	"	T72, T80, Tanker 2,
"	"	Tanker 19, Copter 301

Equipment assignments vary from day to day as equipment is placed in and out of service. Listen to the morning quickcall at 8:00 AM for further updates.

STATE OF CALIFORNIA

FORESTRY DEPARTMENT

SAN DIEGO RANGER DISTRICT (3300)

CHANNEL	INPUT	OUTPUT	TONE	CALL SIGN
"Monte Vista" Dispatch				
Local 1 <Red Mountain>	159.225	**151.190 R**	110.9	KVE-80
Local 2 <Whitestar>	159.225	**151.190 R**	123.0	KVE-80
Local 3 <Cuyamaca>	159.225	**151.190 R**	131.8	KVE-80
Local 4 <Boucher Hill>	159.225	**151.190 R**	136.5	KVE-80
Local 5 <Tecate Peak>	159.225	**151.190 R**	146.2	KVE-80
Local 6 <Lyons Peak>	159.225	**151.190 R**	156.7	KVE-80
Local 7	159.225	**151.190 R**	167.9	KVE-80
(F-8) Car-to-Car		151.190	none	KVE-80
(F-9) CDF Tactical 5		151.250	none	KA-3134
(F-10) CDF Tactical 8		151.370	none	KA-3134
(F-11) CDF Tactical 11		151.460	none	KA-3134
(F-12) Fire Mutual Aid WHITE 1		154.280	none	KC-5112
(F-13) Fire Mutual Aid WHITE 3		154.295	none	KC-5112
(F-14) San Diego County Fire RED		155.085	103.5	KM-8817
(F-15) Volunteer Inland Fire		154.175	103.5	KM-8396
(F-16) Reservation Tactical		151.145	none	KA-3134
(F-17) CDF Tactical Net 1	159.300	151.355 R	vary	KVE-80
(F-18) CDF Tactical Net 2	159.330	151.265 R	vary	KVE-80
(F-19) CDF Air Tactical RED NET		151.220	none	KA-3134
(F-20) CDF Riverside	159.360	151.385 R	vary	KRP-770
CDF Tactical "74"		151.475	none	KA-3134
CDF Tactical "75"		159.270	none	KA-3134
CDF Tactical "76"		159.345	none	KA-3134
CDF Tactical "77"		159.375	none	KA-3134
CDF Tactical "78"		159.405	none	KA-3134
VHF Air Operations GREEN		**151.295**	none	KA-3134
VHF Air Operations YELLOW		**151.310**	none	KA-3134

Call Sign Identification:

Boucher Hill KMB-226; Cuyamaca Peak KMB-229; El Cajon KVE-80; Red Mountain KDJ-491; Tecate Peak KDJ-492; Warner Springs KSP-244; White Star KMA-989

SAN DIEGO CDF FIRE STATIONS

STATION	EQUIPMENT
Campo	Batt. 3314, E3374,
"	E3394, U3334
Cuyamaca	E3375
De Luz	E3378
Dulzura	Batt. 3313, E3393
Flinn Springs	E3382, E3388

STATE OF CALIFORNIA

FORESTRY DEPARTMENT

SAN DIEGO CDF FIRE STATIONS

STATION NAME	EQUIPMENT
Grossmont-Mt, Helix #21	E21
Grossmont-Mt. Helix #22	E22, SQ22
Grossmont-Mt. Helix #23	B24, E23, R23,
Julian	Batt. 3315, D3345, E3365,
"	E3385, Transport 3345,
"	U3357, WT3345
La Jolla Reservation	R6191
Lyons Peak	E3380
Miller	E3381
Monte Vista	Batt. 3312, D3342, E3372, E3379
"	Transport 3342, U3332, WT3342
Pala Reservation	E6611, E6688, WT6665
Potrero	E3363, E3383
Ramona	Batt. 3316, E3367, E3376,
"	E3392, E3396, U3336
Red Mountain	Batt. 3311, E3371, E3381, U3332
Rincon Reservation	E3377, E3379
San Marcos	E3387
Sycuan	E6411, R6491, U6401
Station 72	E3372, E272, SQ72
Station 73	E3373, SQ73
Station 80	E80, E82, Medic 80
Station 81	Batt 3315, E81, E82, Medic 81, Rescue 82
Station 82	E3382, E282, R82, U3338, EMS8
Valley Center #1	Batt. 3317, E3389, E3397, U3337
Valley Center #2	E1, E2, E3, E80, SQ1, Prevention 3327
Valley Center #3	E3, SQ2
Warner Springs	E3366, E3388
White Star	E3384
Witch Creek	E3388

CDF FIRE LOOKOUTS

Boucher Hill
Mount Woodson
Red Mountain

AIR ATTACK BASE

STATION LOCATION	EQUIPMENT*
Ramona Air Field	Air Attack 330
"	Air Tanker 14
"	Air Tanker 70
"	Air Tanker 71

*Equipment assignments vary during each fire season. Listen to the morning quickcall at 8:00 AM on 151.190 MHz for further updates.

STATE OF CALIFORNIA

FORESTRY DEPARTMENT

SAN BERNARDINO RANGER DISTRICT (3500)

CHANNEL	INPUT	OUTPUT	TONE	CALL SIGN
San Bernardino Local West	159.390	**151.445 R**	multi	KVE-74
Tone 1: 110.9 <Quartzite Peak>		Tone 5: 146.2 <Butler Peak>		
Tone 2: 123.0 <Cajon Pass>		Tone 6: 156.7 <Onyx Peak>		
Tone 3: 131.8 <Strawberry Peak>		Tone 7: 167.9 <Toro Peak>		
Tone 4: 136.5 <Keller Peak>		Tone 8: 103.5 <Black Peak>		
San Bernardino Local East	159.315	**151.325 R**	multi	KVE-74
Tone 1: 110.9 <Government Peak>		Tone 3: 131.8 <Ord Mountain>		
Tone 2: 123.0 <Onyx Peak>		Tone 4: 136.5 <Yucca Valley>		
Statewide CDF-1	159.300	151.355 R	none	KVE-74
Statewide CDF-2	159.330	151.265 R	none	KVE-74
Riverside West	159.360	151.385 R	none	KRP-770
Riverside East	159.285	151.175 R	none	KRP-770
San Bernardino Co. RED		**154.325**	91.5	KTZ-349
San Bernardino Co. GREEN		154.190	167.9	KTZ-349
San Bernardino Co. GRAY		154.205	167.9	WNCP-524
San Bernardino Co. YELLOW		154.070	167.9	KTZ-349
San Bernardino Co. ORANGE		154.385	167.9	KTZ-349
San Bernardino Tactical East		**151.400**	none	KA-3134
San Bernardino Tactical West		**159.405**	none	KA-3134
Administration Net	159.345	151.340 R	multi	KVE-74
VHF Air Operations YELLOW		**151.310**	none	KA-3134

Call Sign Identification

Black Peak N/A; Bolero Peak KMF-238; Cajon Summit KMF-238; Gilman Peak KAT-542; Government Peak KMC-928; Keller Peak KMC-928; Little Mountain KFR-608; Onyx Peak KNAV-419; Ord Mountain KMF-780; Strawberry Peak KMA-991.

SAN BERNARDINO CDF FIRE STATIONS

STATION	LOCATION	EQUIPMENT
1	San Bernardino	D3541, E1, E3575,
"	"	SQ1, Transport 3541,
"	"	U3535, U3536, U3537
2	Devore	E2, E3581
3	East Highland	E3
4	Helendale	E4, SQ4
5	Hesperia	D3540, E3562, E3585
"	"	Transport 3540
6	Highland	E6, E6A, M6
7	Loma Linda	E3582
8	Lucerne Valley	E3578
9	Mentone	E9, E9A

STATE OF CALIFORNIA

FORESTRY DEPARTMENT

SAN BERNARDINO CDF FIRE STATIONS

STATION	LOCATION	EQUIPMENT
10	Phelan	E10, E3580, U10
11	El Mirage	E11
12	San Antonio	E12, E3586
13	Yucaipa	E13, E3576, E3583, SQ13
14	Yucca Valley	E3560
15	Angelus Oaks	E15, E15A, SQ15
16	Baldy Mesa	E16, R16
17	Big River	E17, E17A
18	Crafton Hills	E18, E18A, SQ18
19	Homestead Valley	E19, E19R, R19
20	Lytle Creek	E20, E20A
21	Parker Dam	E21, E221, Prevention 12
22	Spring Valley Lake	E22
23	Grand Terrace	E23, E23A, E218
24	Cedar Pines	E24
25	Crestline	E25, E25A, R25
26	Twin Peaks	E26, E26A, Loader 26,
"	"	Snow Cat 26
27	Blue Jay	E27, WT27
28	Valley of Enchantment	E28
29	Lake Gregory	E29R, E161
30	Rim Forest	E30, WT161
31	Needles	E31, E31A, WT31
32	Moronga - West	E32
33	Moronga - East	A33, WT33
34	Moabi Park	E34, WT34
35	Lake Havasu	E35, E35A
36	Joshua Tree	E36, E36A, Recon 36,
"	"	SQ36, WT36
37	Mountain View	E37, E37A
38	Pioneertown	E38
39	Oak Glenn	A39, E39
40	Oak Hills	E40, E40A, SQ40
41	Wonder Valley 1	E41, E41A, Recon 41,
"	"	SQ41, WT41
42	Parker Dam (v)	E42, WT42
43	Johnson Valley (v)	A43, E43, WT43
44	Copper Mountain	A44, E44, P44, WT44
45	Daggett	E45
46	Harvard (v)	E46, E46A
47	Chino Hills	N/A
48	Summit Valley (v)	A48, E48, P48, WT48
49	Fawn Skin	E49, E49A, R49, WT49
50	Running Springs #1	E50, Truck 50, WT50
51	Running Springs #2	Air Support 51,
"	"	Ambulance 51, E51, E51A,
"	"	Prevention 51,
"	"	Snow Cat 51, U51

STATE OF CALIFORNIA

FORESTRY DEPARTMENT

SAN BERNARDINO CDF FIRE STATIONS

STATION	LOCATION	EQUIPMENT
52	Baron Prison	E52, E52A, P52
53	Baker (v)	Ambulance 53, E53,
53	"	E53A, WT53
54	Red Mountain	E54, E54A
55	Black Meadow	E55, E55A
56	Las Flores (v)	E56, WT56
57	Closed	N/A
58	Owens Valley	D3542, E3561, E3563, E573,
"	"	Stakeside 3553, Transport 3542
59	Pilot Rock	E59, E59A
271	Arrowbear FPD	E271, SQ271, WT271
371	Daggett FPD	E371, E371A, E372, WT373
421	Palms FPD	E421, E421A, SQ421

Batt. = Battalion
D = Dozer
E = Engine
P = Prevention
R = Rescue
SQ = Squad
WT = Water Tender
U = Utility

CDF FIRE LOOKOUTS

Cajon Peak

AIR OPERATIONS

CHANNEL	FREQUENCY
VHF Aircraft Band (AM)	
Air Tactics	118.950
Air Tactics	122.850
Air Tactics	122.900
Federal & State	**122.925**
Helicopters Air-to-Air	123.025
Helicopter Operations	119.950
Helicopter Operations	119.975
Heliport Unicom	123.050
Heliport Unicom	123.075
VHF-High Band	
Air Operations BLUE	151.280
Air Operations GREEN	151.295
Air Operations YELLOW	151.310
RED NET - Air Drop Coordination	151.220

STATE OF CALIFORNIA

FORESTRY DEPARTMENT

FIRE COMMAND & OPERATIONS

CHANNEL	FREQUENCY
CDF 1 - Statewide	**151.355 R**
CDF 2 - Statewide	**151.265 R**
CDF Administration Net	**151.340 R**
California Traveler's Net - Statewide	**169.125 R**

STATEWIDE TACTICAL OPERATIONS

The following frequencies are excellent to have in your scanner during a major brush or forest fire involving CDF units.

CHANNEL	FREQUENCY
01	151.145
02	151.160
03	151.175
04	151.190
05	151.250
06	151.325
07	151.340
08	151.370
09	151.385
10	151.400
11	151.445
12	151.460
13	159.225
14	159.270
15	159.285
16	159.315
17	159.345
18	159.360
19	159.375
20	159.390
21	159.405
22	159.450

STATE OF CALIFORNIA

HIGHWAY PATROL DEPARTMENT

BORDER DIVISION - EL CENTRO COMMAND CENTER

EL CENTRO OFFICE (85)

LOCATION	CHANNEL	BASE	MOBILE	CALL SIGN	TONE
El Centro Office	Maroon	**42.920**	42.740	KMD-934	162.2
Boulder Peak	Maroon	42.920	42.740	KMD-934	162.2
Monument Peak	Maroon	42.920	42.740	KMD-934	162.2

WINTERHAVEN OFFICE (116)

LOCATION	CHANNEL	BASE	MOBILE	CALL SIGN	TONE
Winterhaven Office	Maroon	42.920	42.740	KRM-699	162.2
Black Mountain	Maroon	42.920	42.740	KRM-699	162.2

BORDER DIVISION - INDIO COMMAND CENTER

BLYTHE OFFICE (6)

LOCATION	CHANNEL	BASE	MOBILE	CALL SIGN	TONE
Blythe Office	Red	**42.440**	42.280	KUL-656	162.2
Black Mountain	Remote Link	75.640	72.120	KUL-656	none
Blythe	Red	42.440	42.280	KUL-656	162.2
Chuckwalla	Red	42.440	42.280	KGK-672	162.2

INDIO OFFICE (76)

LOCATION	CHANNEL	BASE	MOBILE	CALL SIGN	TONE
Indio Office	Red	42.440	42.280	KMD-892	162.2
Cactus City	Red	42.440	42.280	KJW-727	162.2

BANNING OFFICE (97)

LOCATION	CHANNEL	BASE	MOBILE	CALL SIGN	TONE
Banning Office	Red	42.440	42.280	KBQ-804	162.2
Snow Peak	Red	42.440	42.280	KBN-448	162.2

STATE OF CALIFORNIA

HIGHWAY PATROL DEPARTMENT

BORDER DIVISION - SAN DIEGO COMMAND CENTER

SAN DIEGO OFFICE (87)

LOCATION	CHANNEL	BASE	MOBILE	CALL SIGN	TONE
San Diego Office	Orange	**42.880**	42.660	KRO-424	162.2
Mt. Soledad	Orange	42.880	42.660	KMD-711	162.2
Mt. Woodson	Remote Link	75.780	72.460	KBT-579	none

The San Diego Office patrols all highways within the San Diego City limits.

OCEANSIDE OFFICE (92)

LOCATION	CHANNEL	BASE	MOBILE	CALL SIGN	TONE
Oceanside Office	Tan	42.420	42.840	KMH-722	162.2

The Oceanside Office patrols all highways north of the San Diego City limits up to the Orange County and Riverside County lines. Primary patrol areas include I-5 north of Miramar, I-15 north of Rancho Bernardo, HW 76, HW 78, S6, S12, S13, and S14.

EL CAJON OFFICE (108)

LOCATION	CHANNEL	BASE	MOBILE	CALL SIGN	TONE
El Cajon Office	Gold	**42.120**	42.200	KZJ-917	162.2
Cuyamaca Peak	Gold	42.120	42.200	KME-357	162.2
Monument Peak	Gold	42.120	42.200	KCR-934	162.2
Tecate Peak	Gold	42.120	42.200	KJF-567	162.2

The El Cajon Office patrols all highways east of the San Diego City limits. Primary patrol areas include I-8 east of SDSU, HW67, HW79, HW94, HW188, S1, S2, S3, S4, S5, S6, S7, and S22.

TEMECULA OFFICE (115)

LOCATION	CHANNEL	BASE	MOBILE	CALL SIGN	TONE
Rancho Calif. Office	Tan	42.420	42.840	WNJV-836	162.2
Boucher Hill	Tan	42.420	42.840	KCS-568	162.2
Toro Peak	Tan	42.420	42.840	WNVK-262	162.2

The Rancho California Office primary patrol is I-15 from the Riverside county line, I-215, HW74, HW79 (Riverside County), HW371, and R3.

STATE OF CALIFORNIA

HIGHWAY PATROL DEPARTMENT

BORDER DIVISION - SANTA ANA COMMAND CENTER

SANTA ANA OFFICE (86)

LOCATION	CHANNEL	BASE	MOBILE	CALL SIGN	TONE
Santa Ana Office	Purple	**42.400**	42.160	KMD-454	162.2
Freeway Service Patrol		**44.140**		KMD-454	D631
Modjeska Peak	Remote Link	954.700	958.300	KNIP-504	none
Santiago Peak	Purple	42.400	42.160	KDG-272	162.2
Sierra Peak	Remote Link	954.500	958.100	KNIP-505	none

WESTMINSTER OFFICE (55)

LOCATION	CHANNEL	BASE	MOBILE	CALL SIGN	TONE
Westminster Office	Gray	42.480	42.680	KRP-673	162.2

SAN JUAN CAPISTRANO OFFICE (113)

LOCATION	CHANNEL	BASE	MOBILE	CALL SIGN	TONE
San Juan Capistrano	Gray	42.480	42.680	WXB-735	162.2
Laguna Hills	Gray	42.480	42.680	KMJ-214	162.2
Modjeska Peak	Gray	42.480	42.680	KNIP-504	162.2
San Clemente Hill	Gray	42.480	42.680	KNIP-503	162.2
Sierra Peak	Gray	42.480	42.680	KNIP-505	162.2

COASTAL DIVISION - VENTURA COMMAND CENTER

VENTURA OFFICE (31)

LOCATION	CHANNEL	BASE	MOBILE	CALL SIGN	TONE
Ventura Office	Purple	42.400	42.160	KDC-275	167.9
Geary Hill	Purple	42.400	42.160	KWO-542	167.9
Oat Mountain	Extender	154.695	155.445	KCX-987	167.9
Red Mountain	Purple	42.400	42.160	KMB-521	167.9
South Mountain	Purple	42.400	42.160	KME-354	167.9
Torrey Hill	Purple	42.400	42.160	KNIP-564	167.9

SANTA BARBARA OFFICE (29)

LOCATION	CHANNEL	BASE	MOBILE	CALL SIGN	TONE
Goleta Office	Green	42.540	42.240	KDK-709	167.9

STATE OF CALIFORNIA

HIGHWAY PATROL DEPARTMENT

INLAND DIVISION - BARSTOW COMMAND CENTER

VICTORVILLE OFFICE (12)

LOCATION	CHANNEL	BASE	MOBILE	CALL SIGN	TONE
Victorville Office	White	**42.560**	42.720	KMH-962	118.8
Quartzite Peak	White	42.560	42.720	KNDA-386	118.8

MOJAVE DESERT OFFICE (59)

LOCATION	CHANNEL	BASE	MOBILE	CALL SIGN	TONE
Mojave Office	White	**42.560**	42.720	KMH-723	118.8
Government Peak	White	42.560	42.720	KMD-887	118.8
Oak Peak	White	42.560	42.720	KUI-764	118.8

BARSTOW OFFICE (73)

LOCATION	CHANNEL	BASE	MOBILE	CALL SIGN	TONE
Barstow Office	White	**42.560**	42.720	KME-264	118.8
Bristol	White	42.560	42.720	KUQ-264	118.8
Clark Mountain	Remote Link	72.260	75.980	KME-264	none
Rodman Mountain	White	42.560	42.720	KMG-903	118.8

MORONGO BASIN OFFICE (106)

LOCATION	COLOR	BASE	MOBILE	CALL SIGN	TONE
Twentynine Palms	White	42.560	42.720	KWE-486	118.8
Iron Mountain	White	75.920	72.020	WNMT-213	none
Yucca Valley	White	956.650	953.050	KIA-737	118.8

NEEDLES OFFICE (114)

LOCATION	CHANNEL	BASE	MOBILE	CALL SIGN	TONE
Needles Office	White	**42.560**	42.720	KML-956	118.8
Parker, Arizona	White	42.560	42.720	WNNA-885	118.8
Sacramento Mountain	Remote Link	75.980	72.020	KGV-301	none

STATE OF CALIFORNIA

HIGHWAY PATROL DEPARTMENT

INLAND DIVISION - SAN BERNARDINO COMMAND CENTER

RIVERSIDE OFFICE (74)

LOCATION	CHANNEL	BASE	MOBILE	CALL SIGN	TONE
Riverside Office	Green	**42.540**	42.240	KBQ-743	118.8
Box Springs	Green	42.540	42.240	KMD-453	118.8
Santiago Peak	Green	42.540	42.240	KDG-272	118.8

SAN BERNARDINO OFFICE (75)

LOCATION	CHANNEL	BASE	MOBILE	CALL SIGN	TONE
San Bernardino Office	Copper	**42.600**	42.740	KKR-346	118.8
Little Mountain	Copper	42.600	42.740	KMD-952	118.8
Onyx Peak	Copper	42.600	42.740	KVV-806	118.8

ARROWHEAD OFFICE (101)

LOCATION	CHANNEL	BASE	MOBILE	CALL SIGN	TONE
Running Springs	Copper	**42.600**	42.740	KBO-577	118.8
Quartzite Peak	Copper	42.600	42.740	KNDA-386	118.8
Strawberry Peak	Copper	42.600	42.740	KCX-980	118.8

ONTARIO OFFICE (123)

LOCATION	CHANNEL	BASE	MOBILE	CALL SIGN	TONE
Ontario Office	Green	**42.540**	42.240	KBO-577	118.8
Box Springs	Green	42.540	42.240	KMD-453	118.8
Onyx Peak	Green	42.540	42.240	KVV-806	118.8
Santiago Peak	Green	42.540	42.240	KDG-272	118.8

SOUTHERN DIVISION - LOS ANGELES COMMAND CENTER

CENTRAL LOS ANGELES OFFICE (15)

LOCATION	CHANNEL	BASE	MOBILE	CALL SIGN	TONE
Central Los Angeles	Black	**42.460**	42.700	KRB-412	192.8
Elysian Heights	Black	42.460	42.700	KRB-411	192.8
Mount Lee	Black	42.460	42.700	KRB-411	192.8

STATE OF CALIFORNIA

HIGHWAY PATROL DEPARTMENT

WEST VALLEY OFFICE (56)

LOCATION	CHANNEL	BASE	MOBILE	CALL SIGN	TONE
Van Nuys Office	Brown	**42.500**	42.820	KVG-865	192.8
Castro Peak	Brown	42.500	42.820	KJP-457	192.8

SOUTH LOS ANGELES OFFICE (77)

LOCATION	CHANNEL	BASE	MOBILE	CALL SIGN	TONE
Torrance Office	White	**42.560**	42.720	KBO-575	192.8
Baldwin Hills	White	42.560	42.720	KMA-801	192.8
Rolling Hills	White	42.560	42.720	KMA-801	192.8
San Pedro Hills	White	42.560	42.720	KMA-801	192.8

NEWHALL OFFICE (78)

LOCATION	CHANNEL	BASE	MOBILE	CALL SIGN	TONE
Newhall Office	Tan	**42.420**	42.840	KME-311	192.8
Hauser Peak	Tan	42.420	42.840	KMH-277	192.8
Los Pinos Peak	Tan	42.420	42.840	KMH-277	192.8

WEST LOS ANGELES OFFICE (79)

LOCATION	CHANNEL	BASE	MOBILE	CALL SIGN	TONE
Culver City Office	Pink	**42.440**	42.760	KBO-570	192.8
Baldwin Hills	Pink	42.440	42.760	KMB-443	192.8

BALDWIN PARK OFFICE (81)

LOCATION	CHANNEL	BASE	MOBILE	CALL SIGN	TONE
Baldwin Park Office	Orange	**42.880**	42.660	KBO-571	192.8
Johnstone Peak	Orange	42.880	42.660	KCQ-240	192.8
La Habra Peak	Orange	42.880	42.660	KCQ-240	192.8

EAST LOS ANGELES OFFICE (82)

LOCATION	CHANNEL	BASE	MOBILE	CALL SIGN	TONE
Montebello Office	Yellow	**42.520**	42.300	KBO-57	192.8
Commerce	Yellow	42.520	42.300	WQA-721	192.8
Elysian Heights	Yellow	42.520	42.300	WQA-721	192.8

STATE OF CALIFORNIA

HIGHWAY PATROL DEPARTMENT

SOUTHERN DIVISION - LOS ANGELES COMMAND CENTER

SANTA FE SPRINGS OFFICE (83)

LOCATION	CHANNEL	BASE	MOBILE	CALL SIGN	TONE
Santa Fe Springs Office	Gold	42.120	42.200	KBO-573	192.8
La Habra Hills	Gold	42.120	42.200	KTN-273	192.8
Rolling Hills	Gold	42.120	42.200	KTN-273	192.8

ANTELOPE VALLEY OFFICE (89)

LOCATION	CHANNEL	BASE	MOBILE	CALL SIGN	TONE
Lancaster Office	Tan	**42.420**	42.840	KMH-724	192.8
Hauser Peak	Tan	42.420	42.840	KMH-277	192.8

VERDUGO HILLS OFFICE (98)

LOCATION	CHANNEL	BASE	MOBILE	CALL SIGN	TONE
Verdugo Hills Office	Maroon 2	42.360	42.640	WPAE-678	192.8
Mt. Lukens	Maroon 2	42.360	42.640	WPAE-678	192.8
Verdugo Peak	Maroon 2	42.360	42.640	WPAE-678	192.8

MALIBU OFFICE (109)

LOCATION	CHANNEL	BASE	MOBILE	CALL SIGN	TONE
Malibu Office	Pink	42.440	42.760	KBO-574	192.8
Baldwin Hills	Pink	42.440	42.760	KMB-443	192.8
Castro Peak	Pink	42.440	42.760	KMB-443	192.8

CHP MOBILE RADIO EXTENDERS

CHANNEL	FREQUENCY	TONE
(F-1) Mobile Repeater	**154.905**	192.8*
(F-2) Mobile Repeater	**154.905**	192.8*
(F-3) Tactical	**154.905**	none
(F-4) Police CLEMARS	154.920	none
(F-5) California On-Scene CALCORD	156.075	none
(F-6) Police NLEMARS	155.475	none

*No tone is often used.

STATE OF CALIFORNIA

HIGHWAY PATROL DEPARTMENT

FREEWAY SERVICE PATROL

The CHP began the Freeway Service Patrol (FSP) program in 1992 to specifically alleviate the hazards and congestion caused by stalled and disabled vehicles. Private tow trucks, under a strict contract enforced by the CHP, scan the busy highways during peak commuter hours to spot potential problems that will impede the flow of traffic. A distressed motorist may be provided with a tow to a safe location, free gasoline, transportation, or assistance in changing a tire.

Towing companies:

Los Angeles Cty:	AAA of Southern Calif. and many others.
Orange County:	AAA of Southern Calif., Harbor Towing, Quality Motors Towing
Riverside County:	Program being developed
San Diego County:	AAA of Southern Calif. (I-8), Rancho Del Oro (HW78).

The towing companies transmit on their own licensed or SMR frequencies. Freeway Service Patrol Officers in Orange County are dispatched on the frequency 44.140 with the tone D631.

NEW CHP FREQUENCIES

The CHP has acquired the folllowing 39 MHz frequencies that formerly were licensed to the Los Angeles County Sheriff's Office. Most will be used as repeater inputs for the 42 MHz frequencies. The frequency 39.100 is already being used in Los Angeles County and the other channels will be used within the next two years.

39.100, 39.200, 39.220, 39.260, 39.340, 39.360, 39.400, 39.420, 39.440, 39.600, 39.680, 39.720, 39.800, 39.880, 39.920

STATEWIDE CHANNELS

CHANNEL	FREQUENCY
CHP Blue - Bases	42.340
CHP Blue - Mobiles	42.180
CHP Ivory - Air-to-Ground	**45.020**
CHP Ivory - MDT's	45.020
CHP Tactical (input 42.840)	39.100 R
CHP CLEMARS (input 45.860)	39.460 R

CHP Blue is used as a secondary, tactical, statewide common, emergency, and air-to-ground channel. CHP Ivory is also used for emergency, air-to-ground, and tactical communications. Both channels have been heavily used during recent major disasters in California.

STATE OF CALIFORNIA

HIGHWAY PATROL DEPARTMENT

STATEWIDE FREQUENCY PLAN

DESIG.	BASE	MOBILE	DESIG.	BASE	MOBILE
AMBER	42.080	42.280	ORANGE	42.880	42.660
AQUA	42.620	42.840	PINK	42.440	42.760
BLACK	42.460	42.700	PURPLE	42.400	42.160
BLUE	42.340	42.180	RED	42.440	42.280
BRONZE	42.120	42.200	RUBY	42.500	42.280
BROWN	42.500	42.820	SILVER	42.080	42.280
COPPER	42.600	42.740	TAN	42.420	42.840
GOLD	42.120	42.200	TURQUOISE	42.600	42.020
GRAY	42.480	42.680	VIOLET	42.160	42.640
GREEN	42.540	42.240	WHITE	42.560	42.760
IVORY	45.020	45.020			
MAROON	42.920	42.740			

UNIT INDENTIFICATION

1. The first one or two digits indicate the office number:

3. Buellton
6. Blythe
8. King City
9. Templeton
12. Victorville
14. Santa Maria
15. Central LA
23. Hollister-Gilroy
26. Santa Cruz
27. Monterey
28. San Luis Obispo
29. Santa Barbara
31. Ventura
55. Westminster
56. West Valley
59. Mojave Desert
72. Bishop
73. Barstow
74. Riverside
75. San Bernardino
77. South LA
78. Newhall
79. West LA
81. Baldwin Park
82. East LA
83. Santa Fe Springs
85. El Centro
86. Santa Ana
87. San Diego
89. Antelope Valley
92. Oceanside
97. Banning
98. Verdugo Hills
101. Lake Arrowhead
102. Fort Tejon
106. Moronga Basin
108. El Cajon
109. Malibu
113. Westminster
114. Needles
115. Rancho California
123. Ontario

2. The next two or three digits represent the beat number. There are thousands of CHP patrol beats in California. ScanFan is compiling a list for future publication.

3. Special units:

5000 Governor
5001 Lieutenant Governor
5010 Southern Calif. Representative
5049 Governor's Driver
5098 Lt. Governor's Driver
5100's Office of Emergency Services
5300's California Military Department
5700's California Attorney General

Many special unit identifiers exist, but most are rarely heard on the air. An example of a typical unit I.D. used everyday is **115-50**: a CHP unit out of the Rancho California Office in beat number 50.

CHP GEOGRAPHICAL ORGANIZATION

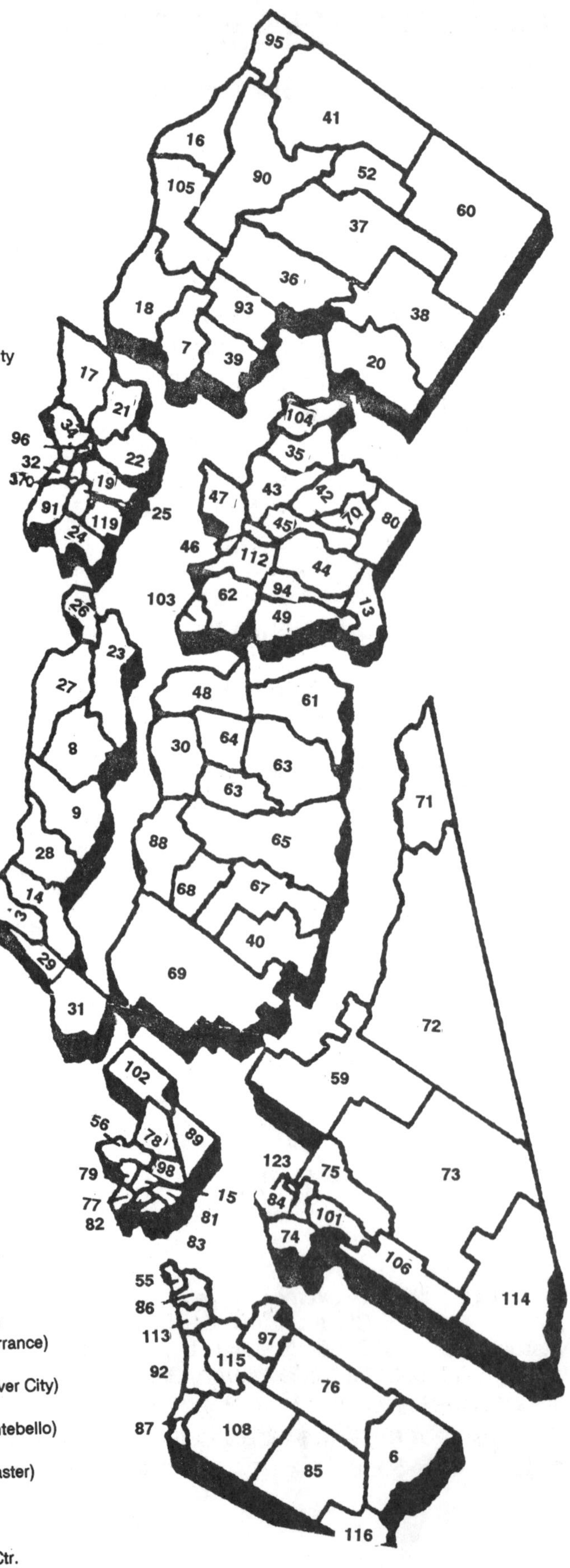

301 GOLDEN GATE DIVISION
Vallejo

17 Santa Rosa
19 Contra Costra (Martinez)
21 Napa
22 Solano (Fairfield)
24 San Jose
25 Hayward
32 San Francisco
34 Marin-Golden Gate
91 Redwood City
96 Oakland
118 Livermore
119 Dublin
346 Mission Grade I.F.
347 Nimitz I.F.
366 Cordelia Inspection Facility

401 CENTRAL DIVISION
Fresno

30 Los Banos
40 Porterville
61 Sonora
63 Mariposa
64 Merced
65 Fresno
66 Madera
67 Visalia
68 Hanford
69 Bakersfield
88 Coalinga
117 Buttonwillow
424 Wheeler Ridge I.F.

701 COASTAL DIVISION
San Luis Obispo

3 Buellton
8 King City
9 Templeton
14 Santa Maria
23 Hollister - Gilroy
26 Santa Cruz
27 Monterey (Salinas)
28 San Luis Obispo
29 Santa Barbara (Goleta)
31 Ventura
726 Gilroy Inspection F.
766 Cornejo Inspection F.

501 SOUTHERN DIVISION
Los Angeles

15 Central Los Angeles
56 West Valley (Van Nuys)
77 South Los Angeles (Torrance)
78 Newhall
79 West Los Angeles (Culver City)
81 Baldwin Park
82 East Los Angeles (Montebello)
83 Santa Fe Springs
89 Antelope Valley (Lancaster)
98 Verdugo Hills
102 Fort Tejon (Lebec)
109 Malibu
514 L.A Communication Ctr.
541 Castaic Inspection Facility

101 NORTHERN DIVISION
Redding

7 Lakeport
16 Humbolt
18 Ukiah
20 Quincy
36 Red Bluff
37 Redding
38 Susanville
39 Williams
41 Yreka
52 Mount Shasta
60 Alturas
90 Trinity River
93 Willows
95 Crescent City
105 Garberville
147 Mt. Shasta Inspection Facility
151 Clear Lake

201 VALLEY DIVISION
Sacramento

13 Lake Valley (Meyers)
35 Oroville
42 Grass Valley
43 Yuba-Sutter
44 Placerville
45 Auburn (New Castle)
46 North Sacramento
47 Woodland
48 Modesto
49 San Andreas
62 Stockton
70 Gold Run (Dutch Flat)
80 Truckee
94 Amador
103 Tracy
104 Chico
112 South Sacramento
214 Sacramento Comm. Center
223 Truckee Insp. Facility

801 INLAND DIVISION
San Bernardino

12 Victorville
59 Mojave
71 Bridgeport
72 Bishop
73 Barstow
74 Riverside
75 San Bernardino
101 Arrowhead (Running Springs)
106 Moronga Basin (29 Palms)
114 Needles
123 Ontario

601 BORDER DIVISION
San Diego

6 Blythe
55 Westminster
76 Indio
85 El Centro
86 Santa Ana
87 San Diego
92 Oceanside
97 Banning
108 El Cajon
113 San Juan Capistrano
116 Winterhaven
621 Winterhaven Insp. Facility
651 San Onofre I.F.
656 Banning Inpection F.

STATE OF CALIFORNIA

HOSPITALS, STATE

CHANNEL	FREQUENCY
Chino, Norco, Whittier	39.040
Norwalk	39.120
Norwalk	156.090
Pomona	156.090

JUSTICE DEPARTMENT

CHANNEL	INPUT	OUTPUT	TONE	CALL SIGN
(F-1) Repeaters	155.460	154.680 R	94.8	See below
(F-2) Repeaters	155.460	154.680 R	94.8	See below
(F-3) Repeaters	155.460	154.680 R	94.8	See below
(F-4) Repeaters	155.460	154.680 R	94.8	See below
(F-5) Repeaters	155.460	154.680 R	94.8	See below
(F-6) Repeaters	155.460	154.680 R	94.8	See below
(F-7) Direct		154.680	94.8	KF-2353
(F-8) Car-to-Car		154.665	94.8	KF-2353
(F-9) Car-to-Car		154.695	94.8	KF-2353
(F-10) CLEMARS		154.920	none	KF-2353
(F-11) NLEMARS		155.475	none	KF-2353
(F-12) CALCORD		156.075	none	KF-2353
Medical Fraud - San Diego	815.100	860.100 R	D115	WPBQ-450
Surveillance Operations		154.935	none	KF-2353
Surveillance Operations		155.040	none	KF-2353

Call Sign Identification:

Los Pinos KBT-203; Mt. Otay KXR-569; Mt. Palomar KFX-368; Mobile I.D. KF-2353; San Diego KBT-571; Santiago Peak KBE-464; Stevenson Peak KBT-579

Statewide Input Tones

Tone 1:	94.8	Tone 4:	146.2
Tone 2:	110.9	Tone 5:	167.9
Tone 3:	127.3	Tone 6:	192.8

Unit Identification

IDA = Investigator
MARY = Medi-Cal Fraud
NORA = Narcotics Task Force
PAUL = Prosecutor
ROBERT = Organized Crime Task Force

Units from the State Justice Department are often heard working with federal agencies and local police. Statewide repeaters conform to the above tone plan on channels one thru six, but some machines are currently not in operation. The output tone 94.8 Htz is used on all repeaters.

STATE OF CALIFORNIA

MILITARY DEPARTMENT

LAW ENFORCEMENT ASSISTANCE FORCE (LEAF)

CHANNEL	FREQUENCY
Convoys - Statewide	30.100
Convoys - Statewide	40.900
Convoys - Statewide	40.940
East Bay - WHITE	
(F-1) Operations Direct	399.650
(F-2) Operations Repeater	399.650 R
(F-3) Operations Direct	399.700
(F-4) Operations Repeater	399.700 R
Military Police	
Tactical 1	399.900
Tactical 2	399.950
National Guard	
LA Riots	138-140 MHz
West Bay - YELLOW	
(F-1) Operations Direct	399.650
(F-2) Operations Repeater	399.650 R
(F-3) Operations Direct	399.750
(F-4) Operations Repeater	399.750

PARKS & RECREATION DEPARTMENT

CHANNEL	FREQUENCY
Anza-Borrego Recreational Area	858.125 R
Channel Islands Coast District	855.7375R
Frontera District	
Lifeguards	866.4625R
Visitor Services	858.125 R
Visitor Services	859.125 R
Visitor Services reported	855.125
High Desert District	868.5375R
Lake Perris Recreational Area	855.7375R
Montane District	857.125 R
Mojave River District	855.7375R
Orange Coast District	
Lifeguards	867.4625R
Operations "Sand"	866.5625R
Operations "Yellow"	867.5625R
Operations (Santiago Peak)	867.5375R
Pendleton Coast District	
Lifeguards	867.4625R
Operations	858.125 R
Visitor Services	856.125 R

STATE OF CALIFORNIA

PARKS & RECREATION DEPARTMENT

CHANNEL	FREQUENCY
Picacho District	859.125 R
Salton Sea Recreational Area	858.125 R
Salton Sea Recreational Area	868.4625R
Santa Monica Mountains Conservatory	866.4625R
Santa Monica Mountains Conservatory	868.5625R
Statewide CMARS - Mt. Soledad	860.100 R
Statewide CMARS - Boulevard	860.125 R
Statewide CMARS - Santiago Peak	867.5375R
Statewide CMARS - Black Mountain	866.5375R
Statewide Mobiles 1	855.4875
Statewide Mobiles 2	855.7125
Statewide Mobiles 3	855.7375
Statewide Mobiles 4	855.9625

DISTRICTS

Channel Coast District: Carpinteria, Emma Wood Camp, McGrath, San Buenaventura
Frontera District: Carlsbad to Silver Strand
Montane District: Cuyamaca and Palomar Mountain State Parks
Mojave River District: Providence Mountains, Silverwood Lake, Summit Valley
Orange Coast District: Bolsa Chica, Crystal Cove, Doheny, Huntington Beach
Pendleton Coast District: San Clemente, San Onofre
Picacho District: Winterhaven

Tone: D115

PLANT INDUSTRY DIVISION

CHANNEL	INPUT	OUTPUT	TONE	CALL SIGN
Operations Repeaters	39.500	37.260 R	107.2	Vary
Operations San Diego		171.525	none	N/A

STATE FIRE MARSHAL

CHANNEL	FREQUENCY
(F-1) Operations Primary	**153.755 R**
(F-2) Operations Direct	**153.755**

REGIONS:

North Region Sacramento
Coastal Region San Leandro
South Region West Covina

STATE OF CALIFORNIA

STATE POLICE

CHANNEL	FREQUENCY
(F-1) Operations Primary	**460.375 R**
(F-2) Operations Tactical	460.375
(F-3) Police CLEMARS	460.025

Output Tone: 127.3 Htz

Southern California Offices:
Castiac, Long Beach, Los Angeles HQ, Pearblossom
San Bernardino, San Diego, Santa Ana.

Mutiple Law Enforcement Services
Alcoholic Beverage Control
Consumer Affairs Department
Dignitary Protection Services
Health Department
Motor Vehicle Compliance
State Insurance Commission
State Lottery Commission
State Police Officers

STATE UNIVERSITIES

CALIFORNIA STATE UNIVERSITY, DOMINGUEZ HILLS

CHANNEL	FREQUENCY	CALL SIGN
Campus Police	153.740	KQS-541
Police CLEMARS	154.920	WRL-652
CPPS Operations	154.540	KXG-398

CALIFORNIA STATE UNIVERSITY, FULLERTON

CHANNEL	FREQUENCY	CALL SIGN
Campus Police	868.575 R	KRG-679
CPPS Operations	151.715	KWH-950
CPPS Operations	152.480	KWH-950

CALIFORNIA STATE UNIVERSITY, LONG BEACH

CHANNEL	FREQUENCY	CALL SIGN
Campus Police	153.740	KLK-542
CPPS Operations & Paging	154.540	N/A
Maintenance	462.0125R	WNRE-759
Student Union Ops	464.525 R	N/A
Student Union Ops	464.550	N/A
Student Union Ops	464.775 R	WNRE-759

STATE OF CALIFORNIA

CALIFORNIA STATE UNIVERSITY, LOS ANGELES

CHANNEL	FREQUENCY	CALL SIGN
Campus Police	153.740	KZG-206
CPPS Operations	151.655	N/A
CPPS Operations	152.480	N/A

CALIFORNIA STATE UNIVERSITY, NORTHRIDGE

CHANNEL	FREQUENCY	CALL SIGN
Campus Police	153.740	KBA-859
Campus Police	154.540	N/A
Administrative Ops	154.515	N/A

CALIFORNIA STATE UNIVERSITY, POMONA

CHANNEL	FREQUENCY	CALL SIGN
Campus Police (F-1)	153.740	KFB-892
Police Mutual Aid (F-2)	154.920	KLW-288
Security & Telephone	464.1625R	N/A

CALIFORNIA STATE UNIVERSITY, SAN BERNARDINO

CHANNEL	FREQUENCY	CALL SIGN
Campus Patrol	153.740	KLS-653
CPPS Operations	151.655 / 154.570	N/A

CALIFORNIA STATE UNIVERSITY, SAN DIEGO*

CHANNEL	FREQUENCY	CALL SIGN
(F-1) Campus Police	866.3875R	N/A
(F-2) Campus Police	868.575 R	N/A
Administrative	151.835	WNGP-427
Aztec Center Security	151.745	WNGP-427
Student Escort	151.655	KYU-755

*The above 800 MHz frequencies are also used at California State University San Marcos and the SDSU extension campus in Calexico. The tone used for all radios at the three facilities is D152.

STATE OF CALIFORNIA

UNIVERSITY OF CALIFORNIA, IRVINE

CHANNEL	FREQUENCY	CALL SIGN
Campus Police	155.505 R	KUB-764
Campus Police	155.445	KUB-764
Health Services	156.015 R	KNCJ-871
Parking Control	155.055 R	KSM-438
Trunking System	866.4875 T	WNWX-380
Trunking System	866.9875 T	WNWX-380
Trunking System	867.4875 T	WNWX-380
Trunking System	867.9875 T	WNWX-380
Trunking System	868.4875 T	WNWX-380

UNIVERSITY OF CALIFORNIA, LOS ANGELES

CHANNEL	FREQUENCY	CALL SIGN
Trunking System	866.4875 T	WNWX-381
Trunking System	866.9875 T	WNWX-381
Trunking System	867.4875 T	WNWX-381
Trunking System	867.9875 T	WNWX-381
Trunking System	868.4875 T	WNWX-381

UNIVERSITY OF CALIFORNIA, RIVERSIDE

CHANNEL	FREQUENCY	CALL SIGN
Trunking System	866.4875 T	WNWX-381
Trunking System	867.4875 T	WNWX-381
Trunking System	868.4875 T	WNWX-382

UNIVERSITY OF CALIFORNIA, SAN DIEGO

CHANNEL	FREQUENCY	CALL SIGN
(F-1) Campus Police	155.505	KUB-765
(F-2) Police CLEMARS	154.920	KUB-765
(F-3) Police Secondary	155.445	KUB-765
CPPS Operations	856.425	N/A
Health Services	856.225 R	WNBY-265
Health Services	857.225 R	WNBY-265
Medical Center	464.675 R	KXO-250
Trunking System	866.4875 T	WPAH-209
Trunking System	866.9875 T	WPAH-209
Trunking System	867.4875 T	WPAH-209
Trunking System	867.9875 T	WPAH-209
Trunking System	868.4875 T	WPAH-209

STATE OF CALIFORNIA

TRANSPORTATION DEPARTMENT CALTRANS

DISTRICT 7 - LOS ANGELES

CHANNEL	FREQUENCY
Commuter Info & Sig Alerts	47.240
Maintenance - Regional Repeater	860.9875R
Maintenance - South Los Angeles	859.7375R
Maintenance - Ventura County	857.9875R
Maintenance - Ventura County	860.7375R
Operations Trunked	**856.7375R**
Operations Trunked	**857.7375R**
Operations Trunked	**857.9875R**
Operations Trunked	**858.7375R**
Operations Trunked	**859.9875R**
Operations Trunked	**860.7375R**
Vincent Thomas Bridge	453.900 R

Tones: 110.9 Htz, 123 Htz, D503

DISTRICT 8 - SAN BERNARDINO

CHANNEL	FREQUENCY
(F-1) Primary GRAY	47.160
Operations Trunked	**857.7375 R**
Operations Trunked	**859.7375 R**
Operations Trunked	**860.7375 R**
Operations Trunked	**860.9875 R**

DISTRICT 11 - SAN DIEGO

CHANNEL	FREQUENCY
(F-1) Primary YELLOW	**47.100**
(F-2) Alternate BLACK	47.040
Maintenance - Districtwide	856.100 R
Maintenance - Districtwide	857.100 R
Maintenance - Districtwide	858.100 R
Maintenance - Districtwide	859.100 R
Maintenance - Districtwide	860.100 R
Maintenance - Districtwide	860.125 R
Maintenance VHF	151.010

STATE OF CALIFORNIA

TRANSPORTATION DEPARTMENT
CALTRANS

DISTRICT 11 - SAN DIEGO

CHANNEL	FREQUENCY
Riverside Trunked	861.200 T
Riverside Trunked	862.200 T
Riverside Trunked	863.200 T
Riverside Trunked	864.200 T
Riverside Trunked	865.200 T
San Diego / El Centro Trunked	861.150 T
San Diego / El Centro Trunked	862.150 T
San Diego / El Centro Trunked	863.150 T
San Diego / El Centro Trunked	864.150 T
San Diego / El Centro Trunked	865.150 T
Toll Bridge - Coronado	453.900 R

The San Diego District is still on the VHF band.

DISTRICT 12 - ORANGE COUNTY

CHANNEL	FREQUENCY
Operations Trunked	**856.9875R**
Operations Trunked	857.9875R
Operations Trunked	858.7375R
Operations Trunked	860.7375R

WATER RESOURCES DIVISION

SOUTHERN DIVISION

CHANNEL	FREQUENCY
(F-1) Car-to-Car	151.205
(F-2) Repeater - Box Springs Mtn	151.205 R
(F-3) Repeater - Hauser Peak	151.205 R
(F-4) Repeater - Strawberry Peak	151.205 R
(F-5) Tactical	151.235

Radio Sites		Statewide Tone Plan	
Box Springs	Quartzite Peak	A: 110.9	D: 136.5
Hauser Peak	Strawberry Hill	B: 123.0	E: 146.2
Los Pinetos Peak	Tejon Peak	C: 131.8	F: 167.9

U.S. GOVERNMENT

U.S. AGRICULTURE DEPARTMENT

ANGELES NATIONAL FOREST

CHANNEL	FREQUENCY	CALL SIGN
(F-1) Forest Fire Net	171.575	KME-22
(F-2) Forest Fire Net	**171.575 R**	KME-22
(F-3) Administration Net	169.950	KME-22
(F-4) Fire Tactical	168.200	KME-22
(F-5) Region 5 Interagency	169.125	KME-22
(F-6) Air to Ground Tactical	170.000	KME-22
(F-7) Fire Tactical	168.600	KME-22
(F-8) Fire Tactical	168.125	KME-22
(F-9) Tactical	169.175	KME-22
(F-10) Air Tactical	169.200	KME-22
Law Enforcement	166.125 R	KME-22

RANGER DISTRICT CONTROL LINKS

LOCATION	FREQUENCY	CALL SIGN
La Canada - Mount Lukens	415.225 R	KME-22
Glendora - Johnstone Peak	415.325 R	KME-22
San Fernando - Mount Lukens	415.225 R	KME-22

CLEVELAND NATIONAL FOREST

CHANNEL	FREQUENCY	CALL SIGN
(F-1) Forest Fire Net	168.750	KME-23
(F-2) Forest Fire Net	**168.750 R**	KME-23
(F-3) Administration Net	168.150	KME-23
(F-4) Fire Camp Service Net	164.125	KME-23
(F-5) Fire Tactical	168.200	KME-23
(F-6) Air to Ground Tactical	170.000	KME-23
(F-7) Administration Net	168.125	KME-23
(F-8) Air Tactical	166.675	KME-23
(F-9) Air to Ground Tactical	168.050	KME-23
(F-10) South Zone Supply Net	168.625	KME-23
Fire Command 3	168.075	KME-23

RANGER DISTRICT CONTROL LINKS

LOCATION	FREQUENCY	CALL SIGN
Alpine - Cuyamaca	415.425 R	KME-23
Escondido - Boucher Hill	415.250 R	KME-23
San Diego - Boucher Hill	415.450 R	KME-23
Santa Ana - Santiago Pk	415.275 R	KME-23

U.S. AGRICULTURE DEPARTMENT

SAN BERNARDINO NATIONAL FOREST

CHANNEL	FREQUENCY	CALL SIGN
(F-1) Forest Fire Net	171.475	KME-20
(F-2) Forest Fire Net	**171.475 R**	KME-20
(F-3) Fire Camp Service Net	164.125 R	KME-20
(F-4) Region Command 4	169.875	KME-20
(F-5) Fire Tactical	172.575	KME-20
(F-6) Fire Tactical	169.925	KME-20
(F-7) Fire Project Net	168.650	KME-20
(F-8) Air Tactical Net	169.150	KME-20
(F-9) Fire Tactical	168.600	KME-20
(F-10) Air to Ground Tactical	170.000	KME-20
(F-11) Administration Net	172.225	KME-20
(F-12) Reclamations Common	172.725 R	KME-20

RANGER DISTRICT CONTROL LINKS

LOCATION	FREQUENCY	CALL SIGN
San Bernardino - Keller Peak	415.350 R	KME-20
San Bernardino - Keller Peak	415.475 R	KME-20
San Bernardino - Pine Cove	415.575 R	KME-20

SAN DIMAS FOREST PROJECT

CHANNEL	FREQUENCY	CALL SIGN
(F-1) Forest Direct	169.175	KML-899
(F-2) Forest Net	**169.175 R**	KML-899
(F-3) Tactical	171.500	KML-899
(F-4) Tactical	172.250	KML-899

ADMINISTRATIVE & LAW ENFORCEMENT

CHANNEL	FREQUENCY	CALL SIGN
Animal & Plant Inspectors	171.525*	KMM-747
Animal & Plant Inspectors	171.5625	KMM-747
Animal & Plant Inspectors	173.7625	KMM-747
Grain Inspectors Nationwide	415.300 R	N/A
Grain Inspectors Nationwide	415.300	N/A
Office of Inspector General	166.125 R	N/A
Office of Inspector General	166.175	N/A
Office of Inspector General	168.675	N/A

*Also used by State of California inspectors.

U.S. AGRICULTURE DEPARTMENT

AIR OPERATIONS

CHANNEL	FREQUENCY
Air Guard	**168.625**
Air Safety	172.600
Air Tactical 1	**166.675**
Air Tactical 2	169.150
Air Tactical 3	169.200
Air Tactical BLM	168.550
Air to Ground	167.950
VHF-1 AM - Air to Air Tactics	118.950
VHF-2 AM - Air to Ground Tactics	122.850
VHF-3 AM - Air to Air Tactics	122.900
VHF-4 AM - Air to Air Tactics	122.975
VHF-5 AM - Helicopter Unicom	123.050
VHF-6 AM - Helicopter Unicom	123.075
VHF-7 AM - Aircraft Tanker Base	122.925
VHF-8 AM - Helicopter Base	119.950
VHF-9 AM - Air to Ground	119.975
VHF-10 AM - Helicopter Air to Air	123.025
VHF AM - Backup	118.925
VHF AM - Backup	123.100
VHF AM - Backup	127.475
VHF AM - Backup	127.925
Air to Ground Net	170.000

FIRE CAMP SERVICE NETS

CHANNEL	FREQUENCY
Fire Camp Net	164.125 R
Fire Camp Net	164.100 R
Fire Camp Net	164.825 R
Fire Camp Net	164.800 R

RIVERSIDE FOREST FIRE LAB

CHANNEL	FREQUENCY
Firescope Project Net	168.650
Operations Net (F-1)	169.175
Operations Net (F-2)	169.175 R

U.S. AGRICULTURE DEPARTMENT

REGION 5 COMMUNICATIONS

CHANNEL	FREQUENCY	USE
Air Dispatch & Operations	**168.625**	Regional
Air to Ground Net	170.000	Regional
Cache Repeater	168.775 R	Regional
Cache Direct	171.3875	Regional
Cache Direct	171.500	Regional
Cache Direct	172.225	Regional
Cache Direct	172.575	Regional
California Interagency Mobilization Plan		
Tactical 1	168.050	Regional
Tactical 2	168.200	Regional
Tactical 3	168.600	Regional
Tactical 4	173.9125	Statewide
Tactical 5	173.9625	Statewide
Tactical 6	173.9875	Statewide
Command Net 1	168.700	Nationwide
Command Net 2	168.100	Nationwide
Command Net 2R	168.100 R	Nationwide
Command Net 3	168.075	Nationwide
Command Net 3R	168.075 R	Nationwide
Command Net 4	**169.975 R**	Regional
Command Net 5	169.175	Regional
Experimental Projects	169.175 R	Regional
Fire Dispatchers Net - North	166.5875R	Regional
Fire Dispatchers Net - South	164.9125R	Regional
High Frequency Net - USB	9.410	Regional
Hydrology	164.150	Regional
Hydrology	169.575	Regional
Hydrology	170.375	Regional
Interagency	168.350	Nationwide
Logistics 1	414.650	Nationwide
Logistics 2	415.400	Nationwide
Logistics 3	415.500	Nationwide
Logistics 4	417.300	Nationwide
Logistics 5	417.350	Nationwide
Logistics 6	417.800	Nationwide
Public Information Officers	36.670	Regional
Public Information Officers	36.370	Regional
Regional Calling	169.125 R	Regional
Regional Common	168.650	Regional
Region 5 - Tactical	**169.925 R**	Regional
Region 5 - Tactical	168.125 R	Regional
Region 5 - Tactical	169.175	Regional
Region 5 - Tactical	169.950	Regiona
Research & Development	173.9375	Statewide
Research & Development	173.9875	Statewide
Traveler's Net	**169.175 R**	Statewide

U.S. COMMERCE DEPARTMENT

INTERNATIONAL TRADE COMMISSION

CHANNEL	FREQUENCY
(F-1) Operations Repeaters	414.625 R
(F-2) Operations Direct	414.625
(F-3) Operations Repeaters	410.725 R
(F-4) Operations Direct	410.725
(F-5) Operations Tactical	410.850
(F-6) Operations Tactical	410.750
(F-7) Operations Tactical	410.875
(F-8) Operations Tactical	410.975
(F-9) Operations Tactical	410.900
(F-10) Operations Tactical	410.150

NATIONAL OCEANIC & ATMOSPHERIC ADMINISTRATION

CHANNEL	FREQUENCY
Weather Channel 1 - Los Angeles	**162.550**
Weather Channel 2 - San Diego	**162.400**
Weather Channel 3 - Special Broadcasts	162.475
Weather Channel 4 - Nationwide	162.425
Weather Channel 5 - Nationwide	162.450
Weather Channel 6 - Nationwide	162.500
Weather Channel 7 - Nationwide	162.525
Link to Mt. Woodson	416.375

NATIONAL OCEAN SERVICE

CHANNEL	FREQUENCY
(F-1) Coastal Operations	164.075
(F-2) Coastal Operations	164.025
National Assignment	162.100

NATIONAL MARINE FISHERIES SERVICE

CHANNEL	FREQUENCY
(F-1) Operations Primary	166.075 R
(F-2) Operations Direct	166.075
Nationwide Assignment	34.810
Nationwide Assignment	34.830

U.S. EXECUTIVE BRANCH

HEALTH AND HUMAN SERVICES DEPARTMENT

CHANNEL	FREQUENCY
Operations Nationwide (F-1)	413.425
Operations Nationwide (F-2)	407.700

HOUSING AND URBAN DEVELOPMENT DEPARTMENT

CHANNEL	FREQUENCY
Nationwide Assignment	164.700
Nationwide Assignment	169.050
Nationwide Assignment	169.700

LABOR DEPARTMENT

CHANNEL	FREQUENCY
(F-1) Operations Primary	406.200
(F-2) Operations	407.175
(F-3) Operations	408.025
Operations Los Angeles	172.300

STATE DEPARTMENT

CHANNEL	FREQUENCY
NET 1 - Los Angeles Operations	408.600 R
NET 2 - Los Angeles Operations	407.600 R
NET 3 - San Diego Operations	**407.200 R**
NET 4 - National Operations	409.625 R
Nationwide Assignments	164.700
Nationwide Assignments	169.050
Nationwide Assignments	169.100
Nationwide Assignments	169.800
Nationwide Assignments	169.925
Orange County Operations	169.625 R
San Diego Operations	169.6125
U.S. Consulate - Tijuana	163.575 R

U.S. INDEPENDENT AGENCIES

CENTRAL INTELLIGENCE AGENCY

CHANNEL	FREQUENCY
San Diego Office - Security	**417.200 R**
Multi-Frequency Portables	163-174 MHz
Multi-Frequency Portables	406-420 MHz
Security Operations	467.700
Security Operations	467.800

ENVIRONMENTAL PROTECTION AGENCY

CHANNEL	FREQUENCY
(F-1) Operations Los Angeles	**164.450** R
(F-2) Operations Los Angeles	167.1875
(F-3) Operations Los Angeles	167.900
Nationwide - Portables	406.100

FEDERAL COMMUNICATIONS COMMISSION

CHANNEL	FREQUENCY
(F-1) Operations San Diego	**167.050 R**
(F-2) Operations Direct	167.050
(F-3) Operations Tactical	167.075

FEDERAL EMERGENCY MANAGEMENT AGENCY

CHANNEL	FREQUENCY
Operations Repeaters	**138.225 R**
Operations Direct	138.225
Operations Repeaters	139.825 R
Operations Direct	139.825
Operations Tactical	142.975
Operations Tactical	142.350
Operations Tactical	142.375
Operations Tactical	142.425

FEDERAL EMERGENCY WARNING RADIO SYSTEM

CHANNEL	FREQUENCY
Operations Primary	**167.975 R**
Operations Direct	167.975
Operations Tactical	167.925
Los Angeles Emergency System	170.200 R

U.S. INDEPENDENT AGENCIES

FEDERAL MARITIME COMMISSION

CHANNEL	FREQUENCY
Operations Primary	166.150
Operations Secondary	166.175

FEDERAL RESERVE BANK*

CHANNEL	FREQUENCY
Building Intercomms - California	468.4125
Operations Los Angeles	409.525 R

*Private Corporation

GENERAL SERVICES ADMINISTRATION

CHANNEL	FREQUENCY
(F-1) Federal Protection Services	**415.200 R**
(F-2) Federal Protection Services	**417.200 R**
(F-3) Federal Protection Services - Direct	415.200
(F-4) Federal Protection Services - Direct	417.200
(F-5) Federal Protection Services - Tactical	417.425
(F-6) Federal Protection Services - Tactical	412.400
Personal Property Services	164.275

INTERSTATE COMMERCE COMMISSION

CHANNEL	FREQUENCY
Operations Nationwide	409.200

NATIONAL TRANSPORTATION SAFETY BOARD

CHANNEL	FREQUENCY
(F-1) Administration Direct	165.750
(F-2) Administration Repeaters	165.750
(F-3) Crash Investigations	166.175

U.S. INDEPENDENT AGENCIES

NUCLEAR REGULATORY COMMISSION

CHANNEL	FREQUENCY
Field Inspection Teams (F-1)	411.075
Field Inspection Teams (F-2)	411.200
National Assignment	165.6625

U.S. POSTAL SERVICE

CHANNEL	FREQUENCY
Postal Inspectors (F-1)	**414.750 R**
Postal Inspectors Direct (F-2)	414.750
Postal Inspectors (F-3)	**415.050 R**
Postal Inspectors Direct (F-4)	415.050
Postal Inspectors BLUE	169.650
Postal Inspectors RED	169.850
Postal Inspectors YELLOW	169.375
Building Security Direct (F-1)	418.300
Building Security (F-2)	418.300 R

VETERAN'S ADMINISTRATION

CHANNEL	FREQUENCY
Disaster Net (F-1)	406.325 R
Disaster Net (F-2)	406.325
Emergency / Transport - San Diego	168.575
Hospital Security	168.525
Hospital Security	168.575
Hospital Security	169.300
Paging Services - San Diego	163.000
Paging Services - Los Angeles	169.300
Paging Services - Sepulveda	171.3875

U.S. INTERIOR DEPARTMENT

FISH & WILDLIFE SERVICE

CHANNEL	FREQUENCY
(F-1) Wildlife Refuges	**34.810 R**
(F-2) Wildlife Refuges Direct	34.810
(F-3) Wildlife Refuges Car-to-Car	34.230
Law Enforcement (F-1)	408.675 R
Law Enforcement (F-2)	408.675
Lake Havasu Operations	164.625 R

Wildlife Refuges:

Calipatria Wildlife Refuge
Havasu Lake National Wildlife Refuge
Imperial National Wildlife Refuge
Tijuana Slough

INDIAN AFFAIRS BUREAU (BIA)

CHANNEL	FREQUENCY
(F-1) Administrative Repeaters	**163.125 R**
(F-2) Administrative Direct	163.125
(F-3) Fire Control	166.6125
(F-4) Car-to-Car	166.725
(F-5) Law Enforcement	164.8625R
Imperial County Reservation	171.750
Parker, Arizona	168.325
Yarupa Reservation	406.475

GEOLOGICAL SURVEY, U.S.

CHANNEL	FREQUENCY
Geothermal Research	166.375
Hydrology	164.150
Hydrology	170.300
Seismic Studies (portables)	162.590
Seismic Studies (portables)	163.150
Topographical Mapping	164.575
Water Resources Divison - Ventura	164.800

Seismic monitors are utilized at various fault locations. If one is placed in your area, it is easy to locate by searching the following ranges:

162.500-164.000 MHz,
164.800-168.000 MHz
171.100-171.500 MHz

U.S. INTERIOR DEPARTMENT

LAND MANAGEMENT BUREAU (BLM)

FIRE CONTROL AND SAFETY

CHANNEL	FREQUENCY
Air Operations, Primary	168.550
Air Safety	172.600
Air Tactical 1	166.675
Air Tactical 2	169.150
Air Tactical 3	169.200
Air-to-Ground Net	167.950
BLM Calling Common	168.550
BLM & Local Fire Tactical	166.6125
Fire Ground Net TAC 1	167.250 R
Fire Ground Net TAC 2	166.775
Fire Ground Net TAC 3	168.250
Fire Ground Net TAC 4	168.400
Fire Ground Net TAC 5	168.400 R
Fire Ground Net TAC 6	167.100
Fire Ground Net TAC 7	167.100 R
Fire Ground Net TAC 8	169.350

LAW ENFORCEMENT

CHANNEL	FREQUENCY
(F-1) Operations Direct	166.375
(F-2) Operations Repeaters	**166.375 R**
(F-3) Operations Direct	166.4875
(F-4) Operations Repeaters	166.4875R
(F-5) Ranger Net	172.575
(F-6) Ranger Tactical	166.2625
(F-7) Ranger Tactical	166.9375
(F-8) Car-to-Car	167.000
(F-9) Fire Tactical	166.675
(F-10) Administration Net	166.750
Riverside-San Bernardino Link	**172.725 R**
Yuma District (F-1)	167.175 R
Yuma District (F-2)	167.175

U.S. INTERIOR DEPARTMENT

NATIONAL PARK SERVICE

CABRILLO NATIONAL PARK

CHANNEL	FREQUENCY
(F-1) Rangers	**171.750**
(F-2) Operations	171.500

CHANNEL ISLANDS NATIONAL PARK

CHANNEL	FREQUENCY
(F-1) Operations Primary	**171.700 R**
(F-2) Operations Direct	171.700
(F-3) Operations Tactical	171.675

DEATH VALLEY NATIONAL PARK

CHANNEL	FREQUENCY
(F-1) Administration	**170.100 R**
(F-2) Administration Direct	170.100

JOSHUA TREE NATIONAL PARK

CHANNEL	FREQUENCY
(F-1) Administration Direct	171.675
(F-2) Administration Primary	**171.675 R**

SANTA MONICA MOUNTAINS NATIONAL PARK

CHANNEL	FREQUENCY
Park Rangers	172.525 R

RECLAMATIONS BUREAU

CHANNEL	FREQUENCY
Operations Repeaters	172.725 R
Operations Repeaters	172.775 R
Car to Car	167.075
Colorado River Operations (F-1)	164.475
Colorado River Operations (F-2)	164.575
Colorado River Operations (F-3)	167.125
Colorado River Operations (F-4)	166.200

U.S. JUSTICE DEPARTMENT

BORDER PATROL - IMMIGRATION

CHULA VISTA SECTOR (10)

CHANNEL	FREQUENCY
Alien Detentions & Buses	163.700
Alien Detentions	163.750
Alien Detentions (Spanish)	165.825
Anti-Smuggling Units, Primary	165.875 R
Anti-Smuggling Units	165.975
Border Crime Units	165.850
Brown Field Primary	**163.650 R**
Brown Field Direct	163.650
Brown Field Tactical	162.900
Chula Vista Primary	**163.625 R**
Chula Vista Direct	163.625
Chula Vista Tactical	163.950
El Cajon Area Primary	**163.775 R**
El Cajon Area Direct	163.775
El Cajon Area Tactical	168.450
North County & Temecula Primary	**163.675 R**
North County Temecula Direct	163.675
San Clemente Primary (Checkpoint)	**163.725 R**
San Clemente Direct (Checkpoint)	163.725
Sectorwide Tactical / Surveillance	163.050

EL CENTRO SECTOR (11)

CHANNEL	FREQUENCY
Alien Detentions	163.750
Calexico / El Centro	**163.725 R**
Calexico / El Centro Direct	163.725
Calexico / Indio	**163.775 R**
Calexico / Indio Direct	163.775
Sectorwide Tactical	170.675

YUMA SECTOR (12)

CHANNEL	FREQUENCY
Blythe & El Centro Area	**163.750 R**
Indio Area	**163.775 R**
Parker & Yuma, Arizona Area	**163.625 R**
Yuma North	**168.875 R**

U.S. JUSTICE DEPARTMENT

DRUG ENFORCEMENT ADMINISTRATION

REGION 14 LOS ANGELES (LA CENTER)

CHANNEL	FREQUENCY
(F-1) Operations Repeaters	**418.625 R**
(F-2) Operations Repeaters	**418.900 R**
(F-3) Car-to-Car, Regional	418.750
(F-4) Car-to-Car, Primary	418.675
(F-5) Operations Repeaters	**418.825 R**
(F-6) Operations Repeaters	**418.950 R**
(F-7) Operations Repeaters	418.975 R
(F-8) Operations Tactical & Air Surveillance FLINT	**418.175**
Interagency Common	418.075
Operations Repeaters - Descanso	419.250 R
Operations Repeaters - LA & Tecate	419.250 R
Operations Repeaters - San Diego	419.300 R
Operations Repeaters - Vista (I-15 Corridor)	419.225 R
Surveillance Operations	417.200
Surveillance Operations	417.275
Surveillance Operations	417.500
Surveillance Operations	418.050
Surveillance Operations	**418.500**
Surveillance Operations	**418.575**
Surveillance Operations	418.700
Surveillance Operations	**418.775**
Surveillance Operations	418.800
Surveillance Operations	418.875
Vehicle Tracking System?	1950 MHz
Wire Taps	72-76 MHz
Wire Taps	171-174 MHz

The channel arrangement of the surveillance frequencies varies.

San Diego Repeater Sites:	Tone
F-1: Mount Palomar	156.7
F-2: Black Mountain, Tecate Peak	156.7
F-5: Santiago Peak, San Ysidro Peak	156.7
F-6: Hi Pass, Mount Soledad, Mount Otay	156.7
F-7: Mount Woodson	156.7

Tecate Repeater: Tecate Peak
Vista Repeater: San Marcos Mountain

U.S. JUSTICE DEPARTMENT

FEDERAL BUREAU OF INVESTIGATIONS

LOS ANGELES DIVISION

CHANNEL	FREQUENCY
A-1 Input frequency	167.2875
A-2 Surveillance	167.475
A-3 Glendale & Pasadena	162.6375 R
A-4 Santa Barbara	165.5875 R
A-5 San Gabriel Valley	163.9125 R
A-6 Malibu	163.8875 R
A-7 Downey & Norwalk	165.7125 R
A-8 Long Beach	163.8375 R
Metro Bank Robbery Squad	
B-1 Los Angeles Metro	167.4125 R
B-2 Surveillance	167.600
B-3 Glendale & Pasadena	163.9375 R
B-4 Orange County	173.150 R
B-5 East Los Angeles	173.7375 R
B-6 Malibu	173.050 R
B-7 Sierra Peak	167.3125R
B-8 Police CLEMARS	154.920
Resident Offices	
C-1 Input frequency	167.2875
C-2 Surveillance	167.5375
C-3 Resident Agent Offices	162.7625 R
C-4 Resident Agent Offices	165.5875 R
C-5 Resident Agent Offices	163.8375 R
C-6 Resident Agent Offices	163.9125 R
C-7 Resident Agent Offices	162.7375 R
C-8 Nationwide Car-to-Car	167.5625
Surveillance	
D-1 Surveillance	167.650
D-2 Surveillance	167.7375
D-3 Surveillance	167.7875
D-4 Surveillance	167.4375
D-5 Surveillance	167.325
D-6 Nationwide Car-to-Car	167.5625
D-7 Surveillance Repeaters	163.9875 R
D-8 Surveillance Repeaters	164.050 R
Orange & Santa Ana	170.050 R
Organized Crime Task Force	165.925 R
Special Weapons & Tactics - Rare use	163.8625 R
Surveillance	167.6625

U.S. JUSTICE DEPARTMENT

FEDERAL BUREAU OF INVESTIGATIONS

SAN DIEGO DIVISION

CHANNEL	FREQUENCY
A-1 Surveillance	167.3375
A-2 Surveillance	167.6875
A-3 Surveillance	167.2125
A-4 Nationwide Car-to-Car	167.5625
A-5 Surveillance	167.2625
A-6 Surveillance	167.4625
A-7 Surveillance	167.5375
A-8 Surveillance	167.500
B-1 Operations Repeaters - Primary	167.3875R
B-2 Operations Repeaters - Mt. Woodson	167.2625R
B-3 Operations Repeaters - Mt. Soledad	167.5125R
B-4 Operations Repeaters - Santiago Peak	167.4125R
B-5 Operations Repeaters - San Diego Metro	167.7125R
B-6 Operations Repeaters - Cuyamaca Peak	167.6875R
B-7 Operations Repeaters - Mt. Laguna	170.900 R
B-8 Operations Repeaters - San Diego	163.9125R
C-1 Repeater Input	164.350
C-2 Bank Robbery Squads	167.7625R
C-3 Bank Robbery Squads	167.825 R
C-4 Bank Robbery Squads	167.5875R
C-5 San Diego Metro	170.825R
C-6 Operations Tactical	167.2375
C-7 Nationwide Car-to-Car	167.5625
C-8 Police CLEMARS	154.920
D-4 San Diego (input 162.6125)	171.2625R
Microwave Links	
Office-Mt. Laguna	414.125R
Office-Mt. Soledad	412.425
Tactical - Simplex	162.7875
Tactical - Simplex	163.9125
Tactical - Simplex	167.5875

Motorola Spectra Syntor 9000 radios are used by the FBI. These radios may be programmed remotely from any base station or equipped mobile and allow the FBI to change the system's frequencies, tones, and encryption codes at will. Therefore, obtaining accurate system data at any particular time is difficult.

Most repeater outputs in Southern California are now being heard in the 167.0-167.7875 and 170 MHz ranges with inputs falling between 162.6125-162.7375 and 164.050-164.8625 MHz. DES encryption is also more prevalent than in the past.

The common PL tone for the FBI is 167.9 Htz.

U.S. JUSTICE DEPARTMENT

MARSHAL, U.S.

CHANNEL	FREQUENCY
(F-1) Operations Primary	**163.200 R**
(F-2) Operations Direct	163.200
(F-4) Operations Tactical	164.600
(F-5) San Diego Metro	**162.7875R**
(F-6) Treasury Common	166.4625
Court Security	170.850
Los Angeles Metro	**162.7125R**

PRISONS BUREAU

CHANNEL	FREQUENCY
(F-1) Operations Primary	**170.875**
(F-2) Operations Secondary	**170.925**
(F-3) Operations Tactical	170.650
(F-4) Operations Tactical	170.825
(F-5) Operations Tactical	170.900

Local Facilities:
Federal Correctional Institute - Terminal Island
Metro Correctional Center - Los Angeles
Metro Correctional Center - San Diego

U.S. ATTORNEY

CHANNEL	FREQUENCY
(F-1) Operations Repeaters	410.875 R
(F-2) Operations Direct	410.875
(F-3) Operations Tactical	410.850
(F-4) Operations Tactical	416.175

The above frequencies are unconfirmed, except 416.175 which has not been heard in some time.

U.S. TRANSPORTATION DEPARTMENT

COAST GUARD

CHANNEL	FREQUENCY
Air Operations - UHF Primary	**381.800**
Air Operations - UHF Secondary	383.900
Air Operations - UHF Alternate	381.700
Air Operations - UHF Tactical	277.800
Air Operations - High Frequency	2.18200
Air Operations - High Frequency	5.69200
Air Operations - High Frequency	8.98000
Air Operations - High Frequency	8.98400
Air Operations - Narcotics Interdiction	353.900
Air Operations - Narcotics Interdiction	361.800
Air Operations - Narcotics Interdiction	387.800
Air Operations - Search & Rescue UHF	282.800
Air Operations - Search & Rescue UHF	237.900
Air Operations - Search & Rescue UHF	381.400
Air Operations - Search & Rescue VHF	121.500
Air Operations - Search & Rescue VHF	122.900
Air Operations - Search & Rescue VHF	123.100
Air Station Security	165.2625
Air Station Security	171.3375
Air Station Security	171.3625
Auxiliaries (F-1)	**148.725 R**
Auxiliaries (F-2)	148.825 R
Auxiliaries (F-3)	143.280
Auxiliaries (F-4)	148.875 R
Civilian Liason VHF 22	**157.100**
Command & Control (F-1/2)	**163.175**
Command & Control (F-3)	162.125
Command & Control (F-4)	162.325
Distress Calling VHF 16	**156.800**
Law Enforcement - Primary (F-1)	**165.3375R**
Law Enforcement - Direct (F-2)	165.3375
Law Enforcement - Tactical (F-3)	167.900
Law Enforcement - Tactical (F-4)	165.3125
Law Enforcement - Long Beach (F-5)	164.300 R
Law Enforcement - San Diego (F-6)	162.125
Law Enforcement - Tactical (F-7)	165.5625
Law Enforcement - Tactical (F-8)	164.9125
Law Enforcement - Tactical (F-9)	164.550
Oil Spills Tactical	165.3125
Primary Operations, San Diego	**157.150**
Special Enforcement	415.925
Special Enforcement	416.850

U.S. TRANSPORTATION DEPARTMENT

FEDERAL AVIATION ADMINISTRATION

LAW ENFORCEMENT

CHANNEL	FREQUENCY
Facilities Security Nationwide	166.175
Facilities Security Nationwide	165.500
Facilities Security Nationwide	165.6625

MAINTENANCE

CHANNEL	FREQUENCY
Maintenance Direct	172.825
Maintenance Direct	172.850
Maintenance Direct	172.875
Maintenance Direct	172.900
Maintenance Direct	172.925
Maintenance Direct	172.950
Maintenance Direct	172.975
Maintenance Repeater Operations 1	172.925 R
Maintenance Repeater Operations 2	172.950 R
Maintenance Repeater Operations 3	172.975 R
Maintenance Repeater Operations 4	172.850 R
Maintenance Repeater Operations 5	172.875 R
Maintenance Repeater Operations 6	172.900 R
Maintenance Repeater Operations 7	172.825 R
Maintenance & Flight Standards	172.125
Maintenance & Flight Standards	172.150
Maintenance & Flight Standards	172.175
Maintenance - Lindbergh Field	416.550

U.S. TREASURY DEPARTMENT

ALCOHOL TOBACCO & FIREARMS

CHANNEL	FREQUENCY
A-1 Operations Repeaters	**165.2875R**
A-2 Operations Direct	165.2875
A-3 Operations Tactical	166.5375
A-4 Treasury Common	166.4625
A-5 Operations Tactical	165.9125
A-6 Operations Tactical	173.8875
A-7 Operations Tactical	168.000
A-8 Operations Repeaters	173.8875R
B-1 ATF-Customs Tactical	166.5875
B-2 ATF-Customs Tactical	165.7375
C-1 Police CLEMARS	154.920
ATF Repeaters	418.225 R
ATF Repeaters	419.975 R
ATF Tactical	418.175
ATF Tactical	418.200
ATF-DEA Liaison	418.075

CUSTOMS SERVICE

CHANNEL	FREQUENCY
A-1 Primary Operations	165.2375
A-2 Operations Repeaters	**165.2375R**
A-3 Treasury Common	166.4625
A-4 ATF-Customs Tactical	166.5875
A-5 Tactical Direct	163.125
A-6 Tactical Repeated	**163.125 R**
A-7 Tactical Southern California	165.4375
A-8 Tactical Southern California	165.4375R
B-3 Air Operations "X-Ray"	165.7375
B-5 Tactical Southern California	165.7625
B-6 Tactical Southern California	165.7625R
B-7 Tactical San Diego	165.5125
B-8 Tactical San Diego	165.5125 R
C-1 Tactical Southern California	169.450
C-2 Tactical Southern California	169.450 R
D-8 Tactical Intersystem NLEMARS	155.475
Customs Surveillance	164.4625
Customs Surveillance	164.8625
Customs Surveillance	165.4875
Customs Surveillance	166.6625
Customs Surveillance	169.550
UNIFORM - Custom's Aircraft	282.425

U.S. TREASURY DEPARTMENT

INTERNAL REVENUE SERVICE

CHANNEL	FREQUENCY
(F-1) Operations San Diego	**418.225 R**
(F-2) Operations Direct	418.225
(F-3) Car-to-Car	418.175
(F-4) Tactical	418.250
VHF System	
Car to Car	165.950
Car to Base	167.000
Primary Operations	**165.950 R**
Treasury Common	166.4625

SECRET SERVICE

CHANNEL	FREQUENCY
Common Designators	
Able	169.925
Baker - Field Office Repeaters	**165.7875R**
Charlie - Field Office Repeaters	**165.375R**
Delta - Tactical	**165.925**
Foxtrot	166.575
Golf	166.400
Hotel	167.900
Kilo	167.825
Lima	168.7875
Mike - Presidental Protection	**165.2125**
November - Staff Protection Teams	**166.700**
Oscar - Staff Protection Teams	**164.8875**
Quebec	166.700
Romeo	164.400
Sierra - Staff Protection Teams	**166.5125**
Tango - Presidental Protection	**164.650**
Whiskey - Staff Protection Teams	**167.025**
X-Ray (Treasury Common)	166.4625
Yankee (Ground-to-Air)	162.6875
Zulu (Air-to-Ground)	171.2875
Special Assignments	
San Diego Office	**165.6875R**
High Frequency USB	6.6830
High Frequency USB	6.7155
High Frequency USB	11.1800
High Frequency USB	11.2490
High Frequency USB	13.2150
Operations UHF	**414.850**
Operations UHF	416.650
Police Liaison - LAPD	154.830
Police Liaison - SDPD	154.725

U.S. MILITARY

U.S. AIR FORCE

AIR NATIONAL GUARD

CHANNEL	FREQUENCY
Northern California Net	149.150
Southern California Net	**149.545**
Statewide Net	149.675
UHF Common - Nationwide	251.900 / 303.000

VAN NUYS AIR NATIONAL GUARD STATION

CHANNEL	FREQUENCY
Security Primary	149.600
Security Tactical	149.265
Fire & Crash	148.450
Commander's Net	149.150
Disaster Net	148.500
Airfield Operations	149.505
Air Terminal Information Service (ATIS)	118.450
Approach / Departure (VNY 160-280)	134.200 / 338.200
Approach / Departure (VNY 280)	120.400 / 360.600
Approach / Departure (BUR 150-VNY 160)	124.600 / 395.900
Approach / Departure (BUR 050-150)	135.050 / 317.500
Clearance Delivery	126.600 / 239.000
Ground Control	121.700
Tower - Helicopters	119.000
Tower	119.300
Tower	120.200 / 239.000

AIR REFUELING

266.500, 267.800, 276.100, 276.500, 279.800, 281.000, 283.900, 286.200, 286.900, 287.400, 287.425, 291.900, 295.400, 295.900, 296.200, 297.300, 297.600, 305.700, 314.300, 314.500, 318.000, 322.800, 322.950, 324.900, 336.100, 341.100, 352.700, 352.900, 354.200, 358.200, 359.100, 361.700, 366.300, 368.600, 369.200, 370.400, 372.300, 375.700, 376.000, 384.600, 385.800, 388.400, 391.000, 391.800, 394.600, 394.900, 396.200

CIVIL AIR PATROL (CAP)

CHANNEL	FREQUENCY
Operations Primary (input 143.900)	**148.150 R**
Operations San Diego (input 148.150)	**143.750 R**
Pacific Region Net - USB	4.585
Search & Rescue Missions	123.100
Search & Rescue, Air-to-Air	122.900
VHF Portables - rare use	26.620

U.S. AIR FORCE

MILITARY AFFILIATED RADIO SYSTEM

CHANNEL	FREQUENCY
Statewide Repeaters (input 142.150)	**143.450 R**

NORTH AMERICAN AIR DEFENSE COMMAND

CHANNEL	FREQUENCY
CH-1 "Sierra Pete"	228.900
CH-2 "Sierra Pete"	234.600
CH-3 "Sierra Pete"	281.600
CH-4 "Sierra Pete"	320.600
CH-5 "Sierra Pete"	254.200
CH-6 "Sierra Pete"	274.400
CH-7 "Sierra Pete"	336.600
CH-8 "Sierra Pete"	346.200
CH-9 "Sierra Pete"	309.400
CH-10 "Sierra Pete"	398.000
CH-11 "Sierra Pete"	238.400
CH-12 "Sierra Pete"	260.800
CH-13 "Sierra Pete"	341.800
CH-14 "Sierra Pete"	377.000
CH-15 "Sierra Pete"	387.800
CH-16 "Sierra Pete"	270.200
CH-17 "Sierra Pete"	316.300
CH-18 "Sierra Pete"	324.000
CH-19 "Sierra Pete"	327.900
CH-20 "Sierra Pete"	267.000
CH-21 "Sierra Pete"	**364.200**
Fighter Discrete	270.800
Fighter Discrete	278.400
Fighter Discrete	279.400
Fighter Discrete	287.700
Fighter Discrete	293.800
Fighter Discrete	299.000
Fighter Discrete	374.000
Fighter Discrete	386.000
Fighter Discrete	390.200
Fighter Discrete	394.200
Fighter Discrete	397.800
High Frequency Operations USB	6.761
High Frequency Operations USB	9.023
High Frequency Operations USB	11.243

U.S. AIR FORCE

OFFICE OF SPECIAL INVESTIGATIONS

CHANNEL	FREQUENCY
(F-1) Operations Nationwide	138.075
(F-2) Operations Nationwide	138.175
Tactical Operations	40.170
Tactical Operations	40.190
Tactical Operations	40.220

NATIONWIDE ASSIGNMENTS

CHANNEL	FREQUENCY
Airborne Warning & Control (AWACS)	
"Sentry"	376.200
"Sentry"	391.800
High Frequency Net USB	9.023
Military Aircraft Distress (MAD)	**243.000**
Military Airlift Command (MAC)	
Command Posts	297.000
Command Posts	349.400
High Frequency Net USB	5.197
Strategic Air Command (SAC)	311.000
Strategic Air Command (SAC)	321.000
Tactical Air Command (TAC)	
Command Posts	381.300
Command Posts	320.000
Tactical Net	272.200
High Frequency ROMEO	9.027

THUNDERBIRDS FLIGHT DEMONSTRATION SQUADRON

CHANNEL	FREQUENCY
Air-to-Air Primary	**322.950**
Air-to-Air Tactical 1	295.700
Air-to-Air Tactical 2	273.500
Air-to-Air Tactical 3	283.500
Air-to-Air Tactical 4	322.300
Air-to-Air Tactical 5	322.600
Air-to-Air Tactical 6	382.900
Air-to-Air Tactical 7	394.000
Air-to-Air VICTOR 1	**141.850**
Air-to-Air VICTOR 2	143.850
Ground Support - Low Power	413.075
Ground Support - Low Power	**413.100**

U.S. AIR FORCE

Note: Many U.S. military bases, including some Air Force installations are being closed in accordance with Public Law 87-297 and international agreements.

EDWARDS AIR FORCE BASE

CHANNEL	FREQUENCY
Base Police (input 148.675)	149.925 R
Base Police	148.870
Department of Energy Security	148.050
EOD Teams	163.5875
EOD Teams	148.4875
Engineering Service	149.5625
Fire & Crash	173.5875
Flight Research Center	165.5125R
Fuel & Lubrication	148.905
Fuel & Lubrication	149.900
Medical Services	173.4875
Paging Services	413.500
Public Works	149.275
Range Control "Dag Rag"	148.275
Rocket Propulsion Lab	407.250
Runway Maintenance	148.085
Transportation Services	149.650
Test Range	163.200
Test Range - Security	163.4875
Airfield	
Air Refueling	354.400
Air Terminal Information Service (ATIS)	116.400 / 269.900
Approach / Departure	133.650 / 348.700
Approach / Departure	126.100 / 290.300
Aviation Engineering Flight Activity	141.100 / 339.900
Command Post	304.000
Flight Service Station	269.200
Ground Control	121.800 / 390.100
Pilot to Metro Service	239.800 / 375.200
Tower VHF	120.700
Tower UHF	236.600 / 318.100

Air-to-Ground Comms:

257.000 / 259.900 / 260.700 / 264.600 / 266.300 / 267.900 / 272.000 / 274.200
275.900 / 284.100 / 286.400 / 287.200 / 289.100 / 290.600 / 291.800 / 294.600
297.400 / 308.700 / 311.200 / 314.400 / 315.900 / 319.600 / 322.700 / 325.900
337.000 / 341.600 / 349.300 / 351.400 / 358.400 / 359.200 / 378.100 / 379.700
383.000 / 384.400 / 385.900 / 395.100 / 398.100

Space Shuttle UHF System	
Astronaut-Orbitor	296.800
Astronaut Secondary	259.400
Astronaut Secondary	259.700
Astronaut Reported	286.800

U.S. AIR FORCE

FORT MACARTHUR

CHANNEL	FREQUENCY
Fire Department	143.375
Barracks Maintenance	143.415
National Guard Service 1	150.465
National Guard Service 2	150.525
Undetermined	139.325

LOS ANGELES AIR FORCE STATION

CHANNEL	FREQUENCY
Aerospace Corps Security	165.0125
Air Station Security (F-1)	149.200
Air Station Security (F-2)	142.225
Commander's Net	148.550
Contingency Communications 1 (Space & Missiles)	163.4625
Contingency Communications 2 (Space & Missiles)	163.4875

MARCH AIR FORCE BASE

CHANNEL	FREQUENCY
Aircraft Maintenance "Job Control"	407.475
Air Force Reserves	413.325
Base Police	163.4625
Base Police "Blue Control"	163.4875
Base Police - Airfield	164.700
Base Weapons Security "Alpha Control"	165.0625
Base Police & Security	164.500
Civil Engineers "Prime Beef"	413.200
Civil Engineers "Prime Beef"	413.300
Commander's Net	149.535
Fire & Crash Net	150.345
Flight Line Refueling	150.175
Maintenance A	149.175
Maintenance B	148.515
Maintenance C	148.175
Maintenance E	150.050
Medical Services	150.225
Motor Pool & Base Taxi	149.475
Munitions & SAC Arsenal	148.245
Paging Services	138.325
Ramp Control	413.025 / 413.475

U.S. AIR FORCE

MARCH AIR FORCE BASE

CHANNEL	FREQUENCY
Airfield	
163 TFG ANG (Grizzley Operations)	293.700
AFRES Operations	138.450
AFRES Operations (C-130 aircraft)	298.450
AFRES Operations (KC-135 aircraft)	252.100
Air Force Reserve Ops	252.100
Air Refueling	235.100 / 344.100
Air Refueling - Backup	319.500
Approach / Departure (east-south)	134.000 / 278.300
Approach / Departure (south-southwest)	135.400 / 295.700
Approach / Departure (north)	125.500 / 351.100
Approach / Departure (north-northeast)	127.250 / 318.200
Air Terminal Information Service (ATIS)	134.750 / 270.100
Ground Control	126.500 / 335.800
Pilot to Metro Service	239.800
Tower	127.650 / 253.500

Some services on March Air Force Base are moving to a UHF trunking system. Please see the following information below.

TRUNKED RADIO PLAN

The U.S. Air Force, not unlike other branches of the military, is implementing several trunked radio systems in the 406-420 MHz frequency band for use at bases and facilities. There are a total of four groups of ten frequencies in the current U.S. Air Force plan.

Trunking Group 1
406.350 / 407.150 / 407.950 / 408.750 /409.550 / 415.150 / 415.950
416.750 / 417.550 / 418.350

Trunking Group 2
406.550 / 407.350 / 408.150 / 408.950 / 409.750 / 415.350 / 416.150
416.950 / 417.750 / 418.550

Trunking Group 3
406.750 / 407.550 / 408.350 / 409.150 / 409.950 / 414.750 / 415.550
416.350 / 417.150 / 417.950

Trunking Group 4
406.950 / 407.750 / 408.550 / 409.350 / 410.150 / 414.950 / 415.750
416.550 / 417.350 / 418.350

U.S. ARMY

ARMY NATIONAL GUARD

CHANNEL	FREQUENCY
Army-Air Force Common	148.000
Operations Convoys	**30.100**
Operations Convoys	32.300
Operations Convoys	34.900
Operations Convoys	36.100
Operations Convoys	40.900
Operations Convoys	46.700
Operations Convoys	46.900
Operations Tactical	30.500
Operations Tactical, 200 kHz steps	32.100-32.900
Operations Tactical, 200 kHz steps	34.100-34.900
Operations Tactical, 200 kHz steps	36.100-36.900
Operations Tactical	38.500
Operations Tactical	38.900
Operations Tactical	40.100
VHF Tactical Pool, 25 kHz steps	139.000-139.475

CORPS OF ENGINEERS

CHANNEL	FREQUENCY
Operations Repeaters (input 164.775)	163.4125R
Operations Direct	163.4125
Operations Repeaters (input 165.1875)	163.4375R
Operations Direct	163.4375
(Flood Control)	
Hydrology	169.425

Tone: 136.5 Htz

CRIMINAL INVESTIGATION DIVISION

CHANNEL	FREQUENCY
Operations Primary	413.425
Operations Secondary	413.525
Operations Tactical	412.900

EXPLOSIVE ORDNANCE DISPOSAL

CHANNEL **FREQUENCY**

Camp Pendleton / Fort Rosecrans / Fort Irwin:
49.700, 49.800, 49.810, 49.830, 49.845, 49.860, 49.875, 49.890

U.S. ARMY

FORT IRWIN

CHANNEL	FREQUENCY
Base Police TAC-1	143.350
Base Police TAC-2	143.100
Base Police	143.275
Chemical Warfare Training	33.850
Fire & Crash	141.050
Fire & Crash - Inspectors	33.550
Fire Department	142.325
Fire Department	150.575
Fire - San Bernardino County	154.070
Medical Services	40.900
Range Communications	38.900
Range Communications	48.450
Training Center	148.850
VIP Escorts	30.350 / 32.250

War Games "Blackbox", "Dragon"
30.250 / 30.350 / 31.750 / 31.900 / 32.000 / 32.100 / 32.150 / 32.500
32.950 / 33.050 / 33.150 / 33.250 / 36.500 / 36.950 / 37.000

LOS ALAMITOS ARMED FORCES CENTER

CHANNEL	FREQUENCY
Base Utility	142.375
Fire & Crash Net	143.050
Airfield	
Approach / Departure	124.650 / 343.900
Arcom Operations	30.100
Arcom Operations	139.050 / 230.900
Control Tower	123.850 / 347.500
Flight Operations	127.950 / 237.200
Ground Control	118.550 / 356.600
Guard Operations	65.050
Guard Operations	142.950 / 233.800
Pilot to Metro Service	239.800
Air-to-Ground	40.950
Air-to-Ground	149.800

MILITARY AFFILIATED RADIO SYSTEM

CHANNEL	FREQUENCY
Operations Repeaters - Southern California	148.650 R
Operations Repeaters - San Diego	143.990 R

Input 143.415 for 148.650
Input 148.010 for 143.990

U.S. MARINE CORPS

CAMP PENDLETON MARINE BASE

USAGE	FREQUENCY
Ambulance & Medics - Paramedics	**149.625 R**
Ambulance & Medics - Navy Regional	143.525
Ambulance & Medics - Paging	149.350
Amphibious Landings "Splash"	41.000
Animal Control/Game Warden MIKE	150.650 R
Disaster Control Net (Command Center): 35.800 / 48.200 / 39.700 / 47.800 / 48.200	
EOD Teams "India"	138.800
EOD Teams Tactical	149.400
Fire Department - Dispatch	**138.600 R**
Fire Department - Tactical	138.800
Fire Department - Administration	140.500
Fire Department - San Onofre	139.525
Fire Mutual Aid RED	155.085
Fire & Crash Net	140.525
Guards Prison CHARLIE	149.425
Guards Prison IRONSIDE/SNOWBELL	149.500
Guards PAPA CHARLIE	149.275
Helicopter Inbound/Gunfight/Land Zones	148.325
LCAC - Air Cushion Landing Craft (Las Pulgas area)	149.250
Military Police "Tac 1"	138.375
Military Police "Tac 2"	139.675
Military Police "Tac 3"	**138.025 R**
Motor Transport	148.025
Motor Transport	149.025
Multiple Services Trunked: 406.350, 407.150, 407.550, 407.950, 408.350, 408.950, 409.150, 409.550, 409.950	
North Range Control "Crestline"	40.450
North Range Control "Crestline"	123.200
North Range Control "Crestline"	360.300
Paging - VIP's	138.150
Public Works FOXTROT KILO	148.725 R
Provost Marshal "Control"	149.370
Provost Marshal - Tactical	149.090
Range Control (Device Control/Combat Town)	141.400
Range Control: 233.900 (Air-to-Air), 249.900, 251.700 (Air-to-Air), 265.000 (Air-to-Air), 301.900, 338.100, 350.900, 360.300, 361.100	
South Range Control "Long Rifle"	30.350
Training Center Utility	142.550
Training Center Utility	143.650
Transportation HOTEL	149.010
War Games	249.900 / 294.700
Weapon Supplies + Crypto 277.900, 283.400, 307.400, 312.700, 313.800, 385.000, 387.900	
Weapons & Ammo Dump	148.375 / 149.050

U.S. MARINE CORPS

CAMP PENDLETON MARINE BASE

USAGE	FREQUENCY
Airfield	
Air Terminal Information Service (ATIS)	267.600
Approach Control	360.100
Approach / Departure Control	127.300 / 323.000
Approach Control UHF	281.700 / 338.100
Approach Control UHF	350.500 / 362.600
Control Tower Primary	126.200 / 382.200
Control Tower Secondary	340.200
Control Tower Backup	41.950
Departure Control VHF	134.100
Departure Control UHF	300.400
Ground Control	272.400 / 360.200
Ground Utility	236.300 / 237.900
Pilot to Metro Service	344.600
Range Control "Long Rifle"	123.200 / 301.900

Air-to-Ground Frequencies:
233.900, 249.300, 249.900, 251.700, 255.200, 265.000, 265.800, 272.300, 277.100, 278.900, 280.600, 284.400, 299.300, 300.000, 301.900, 310.200, 318.900, 321.900, 325.400, 336.200, 336.900, 337.100, 340.400, 345.300, 353.200, **361.100**, 362.200, 369.700, 372.700, 382.100, 387.400

Air-to-Air Frequencies:
262.600, 269.700, 307.500, 310.100

EL TORO AIR STATION

CHANNEL	FREQUENCY
Base Police TAC 1	138.950
Crash & Fire Net	140.100
EOD & Paging	148.350
Fire Department - Structural	139.550
Medical Services	143.525
Public Works	143.500
Refuelers "Mercury"	151.865
Search & Rescue	139.500
Airfield	
Air Terminal Information Service (ATIS)	117.200 / 284.200
Clearance Delivery	120.450 / 301.300
Coast Approach	132.700 / 380.200
Coast Approach	124.650 / 343.900
Coast Approach	121.300 / 363.100
Coast Approach / Departure	124.100 / 323.100
Ground Control	120.450 / 383.800
Tower	133.100 / 271.700

U.S. MARINE CORPS

SAN DIEGO MARINE CORPS RECRUIT DEPOT

CHANNEL	FREQUENCY
Fire Dispatch - Station 15	**140.875**
Police Tac 1	140.675
Police Tac 2	140.700
Police Tac 3	148.270
Security Operations 1	140.025
Security Operations 2	140.150
Security Tactical	141.500
Security - Naval Common	138.800 / 140.850
Medical Services	143.525
Public Works	140.750
Undetermined	40.550

TUSTIN AIR STATION

CHANNEL	FREQUENCY
Air Crash Net	142.900
Security 1	142.750
Security 2	142.850
Airfield	
Air Terminal Information Service (ATIS)	384.300
Clearance Delivery	274.900
Coast Approach / Departure	132.700 / 380.200
Ground Control	380.800
Tower	132.100 / 350.100

TWENTYNINE PALMS TRAINING CENTER

CHANNEL	FREQUENCY
Base Commander	140.225
Fire & Crash Net	140.100
Fire - Structural	140.025
Military Police "Control"	138.675
Medical Services	138.975 R
PMO Tactical	138.575
Paging Services	138.600
Public Works "Alpha Base"	138.850
Range Control "Bearmat"	**49.850**
Sentry "Bag Seed"	138.525
War Games + Crypto 48.100 / 50.200 / 50.300 / 62.050 / 63.400 / 65.950	
Airfield	
Ground Control	41.700 / 362.200
Pilot to Metro Service	308.300
Tower	126.200 / 340.200

U.S. NAVY

CHULA VISTA SPACE SURVEILLANCE STATION

USAGE	FREQUENCY
Space Satellite Operations	143.500

CORONADO AMPHIBIOUS BASE

USAGE	FREQUENCY
Reminder Kilo - Roving Security	
Tactical 1	**143.550**
Tactical 2	142.550
Tactical 3	140.000
Fire - Federal Dispatch	142.625
Fire - Coronado FD	154.415
Multiple Services Trunking:	
406.350 / 406.950 / 407.150 / 407.550 / 407.950	
408.350 / 408.950 / 409.150 / 409.550 / 409.950	
Shore Patrol	148.375
SEAL Teams	407.425
SEAL Teams	413.000
SEAL Teams	413.150
Training Exercises - South	36.550
Training Exercises - North	40.800

EL CENTRO NAVAL AIR FACILITY

USAGE	FREQUENCY
Security	**148.350 R**
Fire & Crash	140.525 R
Fire & Crash	140.100
Tests	140.280
Tests	140.520
Tests	140.600
Undetermined	149.010
Undetermined	149.730
Airfield	
Air Terminal Information Service (ATIS)	271.400
Tower	119.100 / 360.200
Tower UHF Secondary	340.200
Ground Control	126.200 / 263.400
Nickel Base VF-211	361.800

Air-to-Ground Frequencies:
240.200, 240.400, 240.600, 241.400, 241.600, 241.800, 250.000, 251.400, 252.400, 259.700, 271.400, 276.400, 277.200, 283.200, 300.500, 305.000, 310.700, 324.500, 375.000, 384.200, 384.800

U.S. NAVY

FALLBROOK WEAPONS STATION

USAGE	FREQUENCY
Security TAC 1	**142.675 R**
Security TAC 2	140.500
Security TAC 3	140.520
Security DES	142.825*
PAPA - Industrial Control	138.850
Tests	140.280
Tests	140.520
Tests	140.600
Undetermined	138.750

*Repeater input to 142.675.

IMPERIAL BEACH OUTLAYING LANDING FIELD

USAGE	FREQUENCY
Security TAC 1	**138.725**
Fire Dispatch - Battalion 2	**140.450**
Fire Structural	140.975
Fire & Crash Crews	140.030
Fire & Crash Crews	140.300
Fire & Crash Tactical	140.100
Station Operations TAC 3	148.795
Multiple Services Trunking:	
406.350 / 406.950 / 407.150 / 407.550 / 407.950	
408.350 / 408.950 / 409.150 / 409.550 / 409.950	
Airfield	
Air Terminal Information Service (ATIS)	276.200
Approach Control San Diego	119.600 / 306.700
Approach / Departure San Diego	125.150 / 285.200
Control Tower	120.650 / 271.400
Departure Control San Diego	269.100
Ground Control	285.900
Ground Utility	356.800

IMPERIAL BEACH RADIO RECEIVING FACILITY

USAGE	FREQUENCY
Security	**140.975**
Fire - Federal 4	140.450
Fire Structural	140.975
Base Maintenance	139.515

U.S. NAVY

MIRAMAR AIR STATION

USAGE	FREQUENCY	TONE
Airfield Operations (Control)	140.100	none
Airfield Operations (Crash Net)	141.100	none
Airfield Operations (Supplies/Fuel)	139.575	none
Airfield Operations (Support Equipment)	150.550	none
Alcohol Rehabilitation Center (Security)	139.475	114.8
Base Police Primary	**150.105**	136.5
Base Police Secondary	**150.075**	136.5
Base Police Tactical	150.150	136.5
Explosive Ordnance Disposal (DELTA)	150.200	136.5
Fire Department F3 - Battalion 3	**140.250**	none
NDRC - Internal Security	142.525	none
Naval Medical Services	**143.525**	none
Public Works & HazMat	141.000	none
Multiple Services Trunking: 406.350, 407.150, 407.550, 407.950, 408.350, 408.950, 409.150, 409.550, 409.950		
VF 24 Maintenance	140.900	none
VF 302 Maintenance	139.550	none
Airfield		
Air Terminal Information Service (ATIS)	280.400 / 361.900	
Air-to-Air Advisory	235.400	
Approach / Departure San Diego	119.600 / 135.200	
Approach / Departure San Diego	124.350 / 323.000	
Approach / Departure San Diego	125.150 / 285.200	
Approach / Departure San Diego	132.200 / 281.800	
Approach Control	266.800	
Approach Control	269.150	
Approach Control	289.800	
Approach Control	300.400	
Approach Control	310.800	
Approach Control	313.200	
Approach Control	322.000	
Approach Control	346.000	
Approach Control	362.600	
Approach Control	374.900	
Approach Control	380.300	
Approach Control	381.400	
Approach / Departure Control UHF: 269.100, 281.800, 284.400, 284.600, 285.200, 290.400, 306.700, 318.800, 328.500, 350.800, 355.600, 360.800, 381.500, 382.000, 384.500, 385.500		
Clearance Delivery	301.300	
Control Tower	133.625 / 315.600	
Control Tower Secondary	340.200	
Departure Control (East/Northeast)	120.050 / 132.200	
Departure Control (Northwest)	119.600 / 363.100	
Ground Control	380.800	
Weather Information	362.100	

U.S. NAVY

MIRAMAR AIR STATION

USAGE	FREQUENCY
Air Squadrons	
VF1 Wolfback	267.700
VF2 Bountyhunters	253.300
VF11 Red Rippers	322.100
VF24 Renegades	299.700
VF31 Tomcatters	284.500
VF51 Screaming Eagles	263.500
VF111 Sundowners	283.300
VF124 Gunfighters	250.200
VF126 Bandits (Adversary)	275.500 / 355.600
VF211 Flying Checkmates	313.000
VF213 Black Lions	265.250
VF302 Stallions (Reserve)	339.500
VFC13 Saints (Adversary)	354.700
VFC13 Saints (Adversary)	355.700
VAW88 Cottonpickers	338.300
VAW110 Firebirds	263.700
VAW112 Golden Hawks	250.800
VAW113 Black Hawks	249.800
VAW114 Hormel Hawks	253.100
VAW116 Sun Kings	344.300
VAW117 Wallbangers	253.800
Visiting Squadrons	255.000
Air-to-Ground Miscellaneous: 239.700, 250.200, 263.500, 284.500, 323.100, 344.400, 355.700, 362.600, 376.800, 383.700	
Fighter Airborne Intercept Control	
"Happy Hunter"	285.800
"Happy Hunter"	304.200
"Happy Hunter"	326.100
"Happy Hunter"	355.100
Top Gun School	
Air-to-Air	309.300
Airfield - Maintenance	253.800
Airfield - Maintenance	262.700
Mock Fights	263.500
Mock Fights	273.100
Mock Fights	301.100
Mock Fights	308.100
Mock Fights	354.900
Testing	310.800
Tactical	309.300
Tactical	332.350
Tactical	350.900
Tactical	384.500
Visiting Squadrons	255.000

U.S NAVY

NORTH ISLAND NAVAL AIR STATION

USAGE	FREQUENCY
LIMA - Security TAC 1	**140.375**
MIKE - Security TAC 2	140.575
Fire & Crash Crews	140.050
Fire Dispatch	**142.625**
Deep Water Survival School	142.800
Multiple Services Trunked: 406.350, 407.150, 407.550, 407.950, 408.350, 408.950, 409.150, 409.550, 409.950	
NIS Repeater (F-1)	140.075 R
NIS Direct (F-2)	140.075
NIS Tactical (F-3)	140.650
Public Works	138.775 / 407.575
Survival Evasion Resistance Escape Training	140.760
Special Weapons	140.125
Transportation - NOVEMBER	143.575
Transportation - Supplies	149.200
Utility Operations - rare use	27.575
Airfield	
Air Operations	355.500
Air Terminal Information Service (ATIS)	283.000
Air-to-Air Communications: 250.400, 253.000, 253.900, 256.900, 275.700, 280.200, 281.300, 285.100, 291.300, 307.600, 311.700, 315.500, 325.100, 326.900, 342.700, 352.500	
Air-to-Ground Communications: 310.000, 342.900, 346.700, 347.900, 376.800, 384.400, 390.300	
Approach / Departure San Diego 290.400 / 350.800 / 355.600 / 382.000 / 385.500	
Approach Control	313.200 / 318.800
Approach Control	**389.900**
Clearance Delivery	128.400 / 356.800
Control Tower	135.100 / 336.400
Control Tower - Secondary	340.200
Departure Control San Diego	125.150 / 285.200
Flight International (Air-to-ground)	135.975
Ground Control	118.000 / 352.400
Pilot to Metro Service	344.600
Fleet Air Control & Surveillance Facility (FACSFAC)	
< Please also see page 240 >	
Air Refueling, Primary	276.500
Air Refueling, Secondary	319.600
Fighter Discrete (Air-to-Air Tac) 263.300, 268.500, 270.600, 273.100, 301.100, 304.900, 308.100, 314.700, 315.300, 326.500, 344.100, 349.800, 354.200, 354.900, 359.500, 376.800	
Fleet Early Tactical Warning Net	**277.800**
Flight Check In	120.850 / 289.900
Flight Check In	118.650 / 285.200
Flight Check In	314.700

U.S. NAVY

SAN CLEMENTE AUXILIARY LANDING FIELD

CHANNEL	FREQUENCY
Airfield	
Air Terminal Information Service	268.800
Approach Control	385.100
Approach / Departure	265.100
Approach / Departure	305.300
Control Tower	126.750 / 278.800
Ground Control VHF/UHF	119.550 / 352.400
San Clemente Radio USB	6.723

SAN DIEGO FEDERAL FIRE DEPARTMENT

CHANNEL	FREQUENCY
Fire Battalion 1	**142.625**
Fire Battalion 2	**140.450**
Fire Battalion 3	**140.250**
Fire Battalion 4	**140.875**
Fire Battalion 5	**140.200**
Fire Command	140.225
Fire Emergency Medical Services	143.525
Ballast Point Fire	140.675
Imperial Beach Fire & Crash	140.300
Imperial Beach Fire & Crash	140.100
Imperial Beach Structural	140.975
North Island Fire & Crash	140.050
Point Loma Fire	140.825

Federal Fire Battalions & Stations

Battalion 1	Station 1 - North Island NAS
	Station 2 - North Island NAS
	Station 3 - Coronado Amphibious Base
Battalion 2	Station 4 - Imperial Beach OLF
	Station 5 - Imperial Beach OLF
	Station 6 - 32nd Street Naval Station
	Station 7 - 32nd Street Naval Station
Battalion 3	Station 8 - Miramar NAS
	Station 9 - Miramar NAS
Battalion 5	Stations 10 & 11 - San Clemente NALF
Battalion 4	Station 12 - San Diego NOSC
	Station 13 - San Diego Submarine Base
	Station 14 - Naval Training Center
	Station 15 - Marine Corps Recruit Depot
	Station 16 - Balboa Regional Medical Center

U.S. NAVY

SAN DIEGO NAVAL ADMINISTRATIVE

CHANNEL	FREQUENCY
Naval Administrative Command	138.550
Naval Data Center - Security	138.750

SAN DIEGO NAVAL OCEAN SYSTEMS CENTER

CHANNEL	FREQUENCY
Security TAC 1	140.825
Security TAC 2	140.350
Secure Voice DES	140.275
Federal Fire Dispatch (Engine 12)	**140.875**
Operations F-1	141.950
Operations F-2	142.500
Multiple Services Trunked: 406.350, 407.150, 407.550, 407.950, 408.350, 408.950, 409.150, 409.550, 409.950	
Public Works	138.775
Pacific Fleet Common	277.800
Pacific Fleet Common	275.700
Pacific Fleet Common	255.300
Range Communications	40.530
Range Communications - Supplies	40.790
Range Communications - Watch Towers	41.090
Range Communications - Lighting	142.000
Range Communications	142.075
Range Communications - Shared with San Clemente	149.070
Range Communications - Shared with San Clemente	149.130
Range Communications - Shared with San Clemente	149.430
Range Communications - Shared with San Clemente	149.490
Range Communications UHF	**272.800**
Range Communications UHF	**277.100**
Range Communications UHF	385.500
"Starburst" Anti-Submarine Warfare Net	264.000
"Starburst" Anti-Submarine Warfare Net	267.400
"Starburst" Anti-Submarine Warfare Net	307.400
"Starburst" Anti-Submarine Warfare Net	348.100
"Starburst" Anti-Submarine Warfare Net	352.200
"Starburst" Anti-Submarine Warfare Net- Helicopters	281.300
Testing	277.300
Testing	277.700
Testing	336.200
Testing	337.300
Testing	387.400
Undetermined: 256.500, 262.900, 340.400, 355.500, 390.300, 406.500	

U.S. NAVY

SAN DIEGO NAVAL RADIO FACILITY

CHANNEL	FREQUENCY
Command Early Warning Net	**328.200**
Security Operations	**148.750**
Operations UHF	342.600
Operations UHF	350.000
Operations UHF	355.400
Santiago Peak Relay	143.600 R

SAN DIEGO NAVAL REGIONAL MEDICAL CENTER

CHANNEL	FREQUENCY
Ambulances & Medical Transport	143.525
Naval Hospital - Operations	143.625
Naval Hospital - Security	143.725 R
Security TAC 1	142.600
Security TAC 2	142.620
Paging Services	149.350

SAN DIEGO NAVAL REGIONAL RADIO NETWORK

The military in San Diego County is moving to a trunked radio system in the 403-421 MHz range. The Motorola Spectra radios were first installed at North Island NAS and 32nd Street Naval Yard in early 1993. Other facilities that are making the transition to 400 MHz trunking include MCRD, NTC, Miramar NAS, Imperial Beach OLF, San Diego NOSC, the San Diego Submarine Base, and Camp Pendleton.

Repeater sites are at Balboa Hospital, Camp Pendleton, Chollas Heights, Imperial Beach, Point Loma, and Mt. Woodson. Some of the benefits of the system are greater spectrum efficiency, increased privacy, less interference, telephone interconnect capability, and complete interoperabililty with other military services.

Signals are strong across the county and the data channels for each group rotate once every 24 hours. The frequency 407.150 MHz is where most phone patches are heard.

Group 1

406.350 / 406.950 / 407.150 / 407.550 / 407.950
408.350 / 408.950 / 409.150 / 409.550 / 409.950

Group 2

404.250 / 404.2625 / 405.500 / 405.5125 / 407.000 / 407.0125 / 412.475
412.4875 / 413.000 / 413.050 / 416.5125 / 416.625 / 420.025 / 420.050

U.S. NAVY

SAN DIEGO NAVAL STATION - 32ND STREET

CHANNEL	FREQUENCY
Fire - Federal Dispatch F-2	**140.450**
Security TAC 1	**148.350**
Security TAC 2	148.325
Security TAC 3	148.750
Environmental Control	140.625
Multiple Services Trunked: 406.350, 407.150, 407.550, 407.950, 408.350, 408.950, 409.150, 409.550, 409.950	
Naval Medical Services	143.525
NIS Repeater (F-1)	140.075 R
NIS Direct (F-2)	140.075
NIS Tactical (F-3)	140.650
Pierside & Shipboard Portables	36.530
Pierside & Shipboard Portables	36.570
Pierside & Shipboard Portables	36.590
Pierside & Shipboard Portables	36.630
Pierside & Shipboard Portables	36.870
Pierside & Shipboard Portables	40.350
Pierside & Shipboard Portables	40.550
Port Operations - Ships	142.600
Port Operations - Ships	142.725
Port Operations - Ships	142.825
Port Operations - Tugs	139.650
Port Operations - Tugs	139.725
Port Operations - Tugs	139.925
Port Operations - Tugs Primary	139.975
Ship Docking Coordination	413.000
Shore Patrol - Primary	**148.375**
Shore Patrol - Tactical	142.675
Weapons Systems Testing	409.100
Weapons Systems Testing	409.125

SAN DIEGO NAVAL SUBMARINE BASE

CHANNEL	FREQUENCY
Security F-1 "KILO"	**408.000**
Security F-2 "KILO"	**409.450**
Security F-3 "KILO"	**410.025**
Security VHF	148.300
Security VHF	149.050
Fire Dispatch	**140.875**
Multiple Services Trunked: 406.350, 407.150, 407.550, 407.950, 408.350, 408.950, 409.150, 409.550, 409.950	

U.S. NAVY

SAN DIEGO NAVAL SUBMARINE BASE

USAGE	FREQUENCY
Ballast Point - Security	**149.050**
Security Backup	140.575
Federal Fire Department (Engine 13)	**140.675**
Multiple Services Trunked: 406.350, 407.150, 407.550, 407.950, 408.350, 408.950, 409.150, 409.550, 409.950	
Port Services	143.650
Submarine Training Facility SUBTRAFAC	142.600
Submarine Sonar Testing	140.475

SAN DIEGO NAVAL SUPPLY CENTER

CHANNEL	FREQUENCY
Security Police	**148.325**
Security Police	140.325
Multiple Services Trunked: 406.350, 407.150, 407.550, 407.950, 408.350, 408.950, 409.150, 409.550, 409.950	
General Operations	148.400
Transportation	149.375

SAN DIEGO NAVAL TRAINING CENTER

CHANNEL	FREQUENCY
Security 1	**138.875**
Security 2	**138.775**
Fire Dispatch	**140.875**
Fire Tactical	140.125
Fire Tactical	140.225
NIS Repeater (F-1)	140.075 R
NIS Direct (F-2)	140.075
NIS Tactical (F-3)	140.650

SAN DIEGO NAVAL PUBLIC WORKS CENTER

CHANNEL	FREQUENCY
Paging	141.950
Public Works	140.150
Transportation	138.525
Transportation	148.290
Undetermined	138.950

U.S. NAVY

NAVAL INVESTIGATIVE SERVICE (NIS)

CHANNEL	FREQUENCY
(F-1) Operations Repeaters	**140.075** R
(F-2) Operations Direct	140.075
(F-3) Operations Tactical	140.650
(F-4) Operations Tactical	140.675
(F-5) Operations Tactical	139.525

Repeater Input: 140.775 MHz
Tone: 173.8 Htz

NAVY BLUE ANGELS SQUADRON

CHANNEL	FREQUENCY
Demonstrations 1	263.350
Demonstrations 2	**275.350**
Demonstrations 3	307.700
Demonstrations 4	302.100
Demonstrations 5	345.900
Maintenance Alpha	142.000
Maintenance Baker	143.000
Maintenance Charlie	141.560
Maintenance Delta	142.025
Air-to-Air Tactical	241.400
Air-to-Air Tactical	245.900
Air-to-Air Tactical	250.800
Air-to-Air Tactical	360.400
Air-to-Air Tactical	384.400
Air-to-Air Tactical	391.900
Air-to-Air Tactical	395.500

MILITARY AFFILIATED RADIO SYSTEM

CHANNEL	FREQUENCY
Operations Repeaters*	**148.950 R**
Operations & Teletype Direct	148.410
VHF-Low Band (rare use)	40.100

Repeater Input: 150.125

SOUTHERN CALIFORNIA OFFSHORE OPERATIONS AREA
FLEET AREA CONTROL AND SURVEILLANCE FACILITY

Beaver ATIS	282.000
Beaver VHF	118.650
Beaver VHF	120.850
Beaver Check-in	289.900
Beaver RTB	266.900
Beaver RTB	272.600
Beaver RTB	285.700
Beaver Discrete	273.100
Beaver Discrete	301.100
Beaver Discrete	308.100
Beaver Discrete	315.300
Beaver Discrete	344.100
Beaver Discrete	354.900
Beaver Discrete	376.800
Early Warning Net	328.200
Fleet Common	255.300
Fleet Common	275.700
Fleet Common	277.800
Fleet Common	280.200
Fleet Multichannel	359.400
Happy Hunter AIC	304.200
Hunter Discrete	255.600
Hunter Discrete	284.900
Hunter Discrete	285.800
Hunter Discrete	318.700
Hunter Discrete	355.100
Plead VHF	127.550
Plead Check-in	337.000
Plead Check-in	338.000
Plead Discrete	263.300
Plead Discrete	263.800
Plead Discrete	268.800
Plead Discrete	273.300
Plead Discrete	326.800
Plead Discrete	341.300
Starburst ASW	264.000
Starburst ASW	267.400
Starburst ASW	274.600
Starburst ASW	348.100
Starburst ASW	349.800
Starburst ASW	357.900
Starburst ASW	358.000

Air-to-Air	233.900
Air-to-Air	234.400
Air-to-Air	238.050
Air-to-Air	253.050
Air-to-Air	255.020
Air-to-Air	259.200
Air-to-Air	265.000
Air-to-Air	270.050
Air-to-Air	279.100
Air-to-Air	302.600
Air-to-Air	303.625
Air-to-Air	304.500
Air-to-Air	305.000
Air-to-Air	305.250
Air-to-Air	309.300
Air-to-Air	310.650
Air-to-Air	311.150
Air-to-Air	311.400
Air-to-Air	315.300
Air-to-Air	333.450
Air-to-Air	342.900
Air-to-Air	345.000
Air-to-Air	348.100
Air-to-Air	354.900
Air-to-Air	364.800
Air-to-Air	380.500
Air-to-Air	382.100
Air-to-Air	384.500
Air-to-Air	393.800

RELATED FREQUENCIES

Chocolate Mtn. Aerial Gunnery Range
272.900, 310.650
"Desert Control" Nevada Test Range
263.600, 274.000
Flight International (North Island)
135.975
"Marshall" San Clemente NALF CCA/FCLP's
305.300

AIC = Airborne Intercept Control
BEAVER = W-291 Warning Area Control
DISCRETE = Air-to-Air Tactical
HAPPY HUNTER = Airborne Intercept Control
PLEAD = Pacific Missile Firing Range Control
RTB = Return to Base
STARBURST = Anti-Submarine Warfare Control

NONGOVERNMENT

AIRCRAFT

AERONAUTICAL RADIO SERVICE

LOCATION	FREQUENCY DATA
Avalon, Santa Catalina Island	129.450 / 129.750
Bakersfield	129.200 / 129.225 / 129.325 / 129.375
"	130.250 / 131.550 / 131.625 / 131.750
Blythe	131.975
Borrego Springs	131.750
Burbank	129.100 / 129.200 / 129.225 / 129.325
"	129.350 / 129.500 / 129.625 / 129.850
"	129.950 / 130.100 / 130.150 / 130.300
"	130.500 / 130.550 / 130.700 / 130.900
"	131.200 / 131.550 / 131.750 / 131.900
Camarillo	129.100
Carlsbad	128.975 / 130.000 / 130.450 / 130.750
Costa Mesa	131.550
El Segundo	131.700
Fullerton	130.550
Gardena	132.000
Helendale	131.750
Imperial	131.750
La Jolla	129.050
Long Beach	129.200 / 129.225 / 129.275 / 129.325
"	130.425
"	130.600 / 130.900 / 131.050 / 131.075
"	131.400
"	131.550 / 131.600 / 132.000
Los Angeles	128.825 / 128.875 / 128.925 / 128.950
"	128.975 / 129.025 / 129.100 / 129.150
"	129.175 / 129.225 / 129.250 / 129.300
"	129.325 / 129.400 / 129.500 / 129.550
"	129.600 / 129.675 / 129.700 / 129.775
"	129.875 / 129.900 / 129.950 / 130.000
"	130.025 / 130.075 / 130.100 / 130.175
"	130.225 / 130.250 / 130.325 / 130.375
"	130.475 / 130.500 / 130.525 / 130.600
"	130.625 / 130.700 / 130.725 / 130.775
"	130.825 / 130.850 / 130.950 / 130.975
"	131.025 / 131.000 / 131.150 / 131.175
"	131.200 / 131.250 / 131.300 / 131.325
"	131.375 / 131.400 / 131.425 / 131.525
"	131.550 / 131.600 / 131.625 / 131.650
"	131.675 / 131.725 / 131.750 / 131.800
"	131.825 / 131.850 / 131.875 / 131.900
"	131.925 / 131.950 / 131.975 / 132.000
"	460.800 / 460.850 / 460.875 / 464.775
"	465.800 / 465.850 / 465.875 / 469.775
"	856-860.8875 T
Mount Laguna	128.900
Mount Palomar	129.550
Newport Beach	129.350 / 131.550

AIRCRAFT

AERONAUTICAL RADIO SERVICE

LOCATION	FREQUENCY DATA
Ontario	121.900 / 129.200 / 129.225 / 129.325
"	129.350 / 129.500 / 129.600 / 129.725
"	130.100 / 130.150 / 130.175 / 130.250
"	130.300 / 130.425 / 130.450 / 130.575
"	130.700 / 130.825 / 130.900 / 131.100
"	131.200 / 131.350 / 131.550 / 131.600
"	131.750 / 131.800 / 131.925 / 132.000
Oxnard	128.950
Palmdale	130.500
Palm Springs	128.900 / 129.200 / 129.225 / 129.300
"	129.325 / 129.725 / 129.975 / 130.100
"	130.450 / 131.200 / 131.550
San Bernardino	130.650 / 131.900
San Diego	129.150 / 129.200 / 129.225 / 129.250
"	129.300 / 129.325 / 129.500 / 129.600
"	129.925 / 129.975 / 130.100 / 130.175
"	130.225 / 130.350 / 130.700 / 130.800
"	130.900 / 130.950 / 131.100 / 131.125
"	131.200 / 131.450 / 131.550 / 131.625
"	131.750 / 131.775 / 131.925
"	861-865.600 T
Santa Ana	128.825 / 129.200 / 129.225 / 129.300
"	129.325 / 129.600 / 130.100 / 130.150
"	130.550 / 130.575 / 130.700 / 130.900
"	131.000 / 131.100 / 131.500
Santa Barbara	129.200 / 129.225 / 129.325 / 129.425
"	129.500 / 129.750 / 129.950 / 130.425
"	130.450 / 130.750 / 131.400 / 131.550
"	131.750 / 131.950
Santa Maria	128.950 / 130.800 / 130.850 / 131.050
"	131.750
Santiago Peak	130.000 / 130.750
Torrance	129.000
Van Nuys	128.950 / 130.350 / 130.575 / 130.875
"	131.850
Vandenberg Air Force Base	129.750

AERONAUTICAL TELEPHONE

LOCATION	BASE FREQUENCY	MOBILE FREQUENCY
Los Angeles	454.575	459.575
San Diego	454.825	459.825

AIRCRAFT

AIRFONE

LOCATION	FREQUENCY DATA
Los Angeles	944.604 / 944.610 / 944.616 / 944.622
"	944.628 / 944.634 / 944.640 / 944.646
"	944.652 / 944.658 / 944.664 / 944.670
"	944.676 / 944.682 / 944.688 / 944.694
"	944.700 / 944.706 / 944.712 / 944.718
"	944.724 / 944.730 / 944.736 / 944.742
"	944.748 / 944.754 / 944.760 / 944.766
"	944.772 / 944.778 / 942.784 / 994.794

AIRLINE GROUND COMMUNICATIONS

NAME	LOCATION	FREQUENCY	CALL SIGN
Air California	Burbank	460.650	KNFK-768
Air California	Los Angeles	460.700	WNHR-522
Air California	San Diego	460.700	WNHQ-302
Airflite Inc.	Long Beach	151.775	KNIY-373
Airflite Inc.	Long Beach	460.675	WNDW-239
Air New Zealand	Los Angeles	151.955	KKP-274
Air Pacific Intl.	Montrose	507.8125	KGE-586
Air Resorts Airlines	Carlsbad	463.9375	KB-42209
Air Resorts Airlines	Carlsbad	463.9625	KB-42209
Air Resorts Airlines	Carlsbad	464.0125	KB-42209
Alaska Airlines	Burbank	460.825	KNAD-351
Alaska Airlines	Long Beach	460.825	KNCT-998
Alaska Airlines	Ontario / Santa Ana	460.800	WNSP-454
American	Burbank / Ontario / Santa Ana	460.650	WNHR-523
American	Los Angeles / San Diego	460.700	WNHR-522
American	Los Angeles	460.775	KJD-675
American	Los Angeles	462.850	WXB-251
American	San Diego	460.775	KYH-456
America West	Burbank	464.600	WNDA-881
America West	Los Angeles / San Diego	464.600	WNDF-483
America West	Los Angeles	461-463.8125	KD-50104
China Airlines	Los Angeles	462.650	KAB-0769
China Airlines	Los Angeles	457.5375	KD-30709
China Airlines	Los Angeles	457.5625	KD-30709
Continental	Burbank	460.700	WNIG-209
Continental	Los Angeles	460.700	KNBC-928
Continental	Los Angeles	464.475	WNRD-668
Continental	Los Angeles	464.875	KNBC-928
Continental	Ontario / Santa Ana	460.850	WNMU-804
Continental	San Diego	460.800 / 460.825	WNMS-987
Delta	Burbank	460.750	WNAX-714
Delta	Los Angeles	151.805	KBR-339
Delta	Los Angeles	460.750	WRZ-759
Delta	Los Angeles	460.825	KJE-633
Delta	Los Angeles	460.875	KLO-779
Delta	Los Angeles	462.725	KAE-2197
Delta	Los Angeles	464.575	KYC-309
Delta	Southern Cal.	154.570	KB-25301
Delta	Southern Cal.	154.600	KB-25301
Delta	San Diego	460.750	WRZ-736

AIRCRAFT

AIRLINE GROUND COMMUNICATIONS

NAME	LOCATION	FREQUENCY	CALL SIGN
Eastern Airlines	Los Angeles	460.650	WNGU-375
Foreign Airlines	Los Angeles	460.800	N/A
Frontier	Los Angeles	460.700	KNBC-928
Frontier	San Diego	460.650	KNCV-527
Frontier-Pacific	Long Beach	507.7375	WIF-355
Flying Tigers	Los Angeles	460.825	WNJW-893
Jimsair Aviation	San Diego	460.650	KNIX-521
Jimsair Aviation	San Diego	460.800	U/L
M.G.M. Grand Air	Los Angeles	460.700	WNJN-612
Mar Aviation	Brawley	460.775	WXI-371
Northwest Airlines	Los Angeles	460.700	KQP-839
Northwest Airlines	Nationwide	460.8875	WNRT-570
Northwest Airlines	Ontario	460.725	WNWZ-916
Northwest Airlines	San Diego	460.850	WNQR-563
Pacific East	Los Angeles	460.825	KNDA791
Pan American	Los Angeles	464.675	KQZ-410
Philippines Air	Los Angeles	464.875	KNJU-630
Republic	Los Angeles	460.700	KQP-839
Republic	San Diego	460.700	KA-92653
Ross Aviation	Brawley	460.700	KYM-858
Singapore Air	Los Angeles	461.500	WQT-434
Transamerica Air	Los Angeles	464.875	WSO-377
Trans World	Burbank	460.675	WNKK-946
Trans World	Los Angeles	460.675	KTL-459
Trans World	Los Angeles	464.525	KAR-252
Trans World	Nationwide	151.775	N/A
Trans World	San Diego	460.675	KNGU-265
USAir	Burbank	464.875	WNGQ-260
USAir	Los Angeles	461.6125	WNMY-268
USAir	Los Angeles	461.7625	WNMY-268
USAir	Los Angeles	461.9625	WNMY-268
USAir	Los Angeles	462.7875	WNMY-268
USAir	Los Angeles	462.9125	WNMY-268
USAir	Los Angeles	463.9875	WNMY-268
USAir	Los Angeles	466.6125	KB-85121
USAir	Los Angeles	464.475	WNMY-268
USAir	San Diego	460.800	WNQK-533
USAir	San Diego	460.825	WXG-711
USAir	San Diego	461.6125	KB-73162
USAir	San Diego	464.200	WRY-663
USAir	San Diego	464.575	KB-77238
USAir	San Diego	464.875	KVP-418
USAir	Southern Cal.	151.625	KB-64288
USAir	Southern Cal.	154.570	KK-3116
United	Burbank	460.725	KRF-694
United	Los Angeles	460.725	KML-555
United	Los Angeles	460.875	KLO-779
United	San Diego	460.725	KRV-899
World Airways	Los Angeles	460.850	KNJN-861

AIRCRAFT

AIRLINE FLIGHT OPERATIONS

NAME	LOCATION	FREQUENCY	CALL SIGN
AeroMexico	Tijuana, Mexico	131.000	N/A
Air California	Statewide	129.350	N/A
Air California	Statewide	130.175	N/A
Air Canada	Los Angeles (LAX)	130.475	N/A
Air France	Los Angeles	129.020	N/A
Airflite Inc.	Long Beach	122.875	KFR-4
Air Jamaica	Los Angeles (LAX)	130.100	N/A
Air New Zealand	Los Angeles (LAX)	129.150	N/A
Air Resorts	Nationwide	130.750	N/A
AirSouth	Fullerton	122.875	WXU-9
AirSouth	Fullerton	122.950	KVL-8
Air West	Nationwide	129.600	N/A
Air West	Nationwide	131.000	N/A
Alaska Airlines	Los Angeles (LAX)	131.200	N/A
Alaska Airlines	Nationwide	129.200	N/A
American Airlines	Nationwide	129.050	N/A
American Airlines	Nationwide	129.200	N/A
American Airlines	Nationwide	129.225	N/A
American Airlines	Nationwide	129.325	N/A
American Airlines	Nationwide	130.850	N/A
American Airlines	Nationwide	131.050	N/A
American Airlines	Nationwide	131.125	N/A
America West	Nationwide	129.325	N/A
America West	Nationwide	130.175	N/A
Aspen Airlines	Nationwide	132.000	N/A
Avjet Corp.	Burbank	122.950	KMM
Cessna Aircraft	Los Angeles	122.825	WSV-6
Cinema Air	Carlsbad	129.000	N/A
Condor Helicopters	Los Angeles	129.750	N/A
Continental Airlines	Los Angeles	130.500	N/A
Continental Airlines	San Diego	129.925	N/A
Delta Airlines	Nationwide	129.550	N/A
Delta Airlines	Nationwide	130.100	N/A
Delta Airlines	Nationwide	130.525	N/A
Delta Airlines	Nationwide	131.450	N/A
Eastern Airlines	Nationwide	129.200	N/A
Eastern Airlines	Nationwide	130.450	N/A
Eastern Airlines	Nationwide	130.825	N/A
Eastern Airlines	Nationwide	130.950	N/A
Evergreen Airways (CIA)	Nationwide	131.600	N/A
Federal Express Air	San Diego	131.925	N/A
Flying Tigers Airlines	Nationwide	131.825	N/A
Inland Empire	Nationwide	131.350	N/A
Japan Airlines	Los Angeles (LAX)	129.900	N/A
Jim's Air	San Diego	129.975	N/A
KLM Airlines	Los Angeles (LAX)	130.320	N/A
Korean Airlines	Los Angeles (LAX)	129.900	N/A
Martin Aviation	Burbank	122.950	KRD-9
Mexicana Airlines	Tijuana, Mexico	130.200	N/A
Mexicana Airlines	Los Angeles	128.975	N/A
Mexicana Airlines	Los Angeles	130.100	N/A
Mountain Air Service	Big Bear City	122.875	WJD-6
Northwest Airlines	Nationwide	130.100	N/A
Northwest Airlines	Nationwide	130.700	N/A
Northwest Airlines	Nationwide	130.800	N/A
Northwest Airlines	Nationwide	130.850	N/A

AIRCRAFT

AIRLINE FLIGHT OPERATIONS

NAME	LOCATION	FREQUENCY	CALL SIGN
Pan American	Nationwide	129.200	N/A
Pan American	Nationwide	129.700	N/A
Scandinavian Airlines	Nationwide	131.850	
Skywest Airlines	San Diego	131.750	N/A
Southwest Airlines	Nationwide	129.250	N/A
Southwest Airlines	Nationwide	130.175	N/A
Stateswest Airlines	Nationwide	129.200	N/A
Sun World International	San Diego	130.225	N/A
Texas International	Nationwide	129.925	N/A
Transtar	Nationwide	129.725	N/A
Trans World Airlines	Nationwide	130.225	N/A
Trans World Airlines	Nationwide	130.725	N/A
Trans World Airlines	Nationwide	131.800	N/A
United Airlines	Nationwide	129.400	N/A
United Airlines	Nationwide	129.500	N/A
United Airlines	Nationwide	130.425	N/A
USAir	Nationwide	130.075	N/A
USAir	San Diego	130.900	N/A
World Airways	Nationwide	131.525	N/A

AIRPORTS

NAME	FREQUENCY DATA
Agua Caliente Springs	M 122.900
Agua Dulce Airpark	M 122.900; V 114.500
Apple Valley	U 122.800; FSS 122.300; V 116.400
Avalon, Santa Catalina	U 122.700; A/DC 127.400; V 111.400
Baker	M 122.900; V 113.200
Bakersfield Municipal	U 122.800; V 115.400
Banning	U 122.800; V 112.200
Bermuda Dunes (Palm Springs)	A/DC 118.850 / 128.150; U 122.800; V 116.200
Big Bear City	U 122.700; V 112.200
Blythe	A/DC 128.150; FSS 122.100 / 122.400; U 122.800; V 117.400
Borrego Springs	U 122.800; V 114.000
Brackett Field (La Verne)	A 124.400; A/DC 125.500; CT 118.200;
"	GC 125.000; U 122.950; V 110.400
Brawley Municipal	A/DC 128.600 / 291.700; M 122.900; V 115.900
Brown Field (Chula Vista)	A 132.350; AC 119.600 / 306.700; CD 124.400;
"	CT 126.500 / 128.250 / 288.100; DC 125.150 / 285.200;
"	GC 124.400; U 122.950; V 109.800
Burbank-Glendale-Pasadena	A 134.500 / 135.125; A/DC 120.400 / 124.600
"	134.200 / 135.050; CD 118.000; CT 118.700; FSS 122.350;
"	GC 123.900; ILS 109.500; U 122.950; V 113.100

ABBREVIATIONS

A = A.T.I.S. (Air Terminal Information Service)
AC = Approach Control
A/DC = Approach / Departure Control
CD = Clearance Delivery
DC = Departure Control
FSS = Flight Service Station
GC = Ground Control
ILS = Instrument Landing System
M = Multicom
S = Security (airports)
U = Unicom
V = Vortac
W = Weather Advisory

AIRCRAFT

AIRPORTS

NAME	FREQUENCY DATA
Burmuda Dunes	A/DC 118.850; U 122.800; 116.200
Buttonwillow	M 122.900; V 115.400
Cable (Upland)	A/DC 125.500; U 123.000; V 110.400
Calexico Intl.	U 122.800; V 115.900
California City Municipal	U 122.700; V 116.400
Calipatria Municipal	M 122.900; V 115.900
Camarillo	A 119.200; A/DC 124.700 / 135.500; CT 128.200;
"	CD 118.100; GC 121.800; U 122.800; V 115.800
Chino	A 121.150; A/DC 135.400; CT 118.500;
"	GC 121.600; U 122.950; V 112.200 / 112.400
Chiriaco Summit	M 122.900; V 116.200
Coalinga Municipal	U 122.800; V 114.3
Compton	M 122.900; V 113.600
Corona Municipal	A/DC 135.400; U 122.800; W 132.175; V 112.200
Dagget (Barstow)	FSS 122.200; U 123.000; V 113.200
Death Valley Air Strips	M 122.900; V 114.700
Delano Municipal	A/DC 118.900 / 127.100; U 122.800; V 115.400; W 119.550
Desert Center	U 122.900; V 114.700
El Monte	A 118.750; A/DC 125.500; CT 121.200; GC 125.900;
"	U 122.950; V 110.400 / 112.200
Fallbrook Community	U 122.800
Famosa	M 122.900; V 115.400
Flabob (Rubidoux)	U 122.800; V 112.200
Flying T Ranch (Private)	U 122.700
Fresno	A 121.350; A/DC 118.500 / 119.600 / 132.350;
"	CD 124.350; GC 121.700; U 122.950; V 112.900
Fullerton Municipal	A 125.050; A/DC 124.650; CT 119.100; FSS 122.450
"	GC 121.800; U 122.950; V 115.700
General Fox (Lancaster)	A/DC 126.100 / 132.500; CT 120.300; GC 121.700;
"	W 133.875; V 114.500
Gen. Rodrigues (Tijuana)	A/DC 119.500 / 120.300 / 128.300; CD 122.100;
"	CT 118.100; V 116.400
Gen. Toboada (Calexico)	CT 118.200; V 115.000
Gillespie Field (El Cajon)	A 125.450; A/DC 124.350 / 381.500; CD 125.100;
"	CT 120.700 / 123.800 / 257.800; GC 121.700;
"	ILS 110.500; U 123.050; V 117.800
Hawthorne Municipal	A 118.400; AC 124.500 / 124.900 / 128.500;
"	CT 121.100; DC 124.300 / 125.200;
"	GC 125.000; FSS 122.200 / 122.350 / 122.500;
"	ILS 110.900; V 113.600
Helendale (Sun Valley Ranch)	M 122.900
Hemet-Ryan	A/DC 134.000; U 123.000; V 112.200
Hesperia Air Lodge	U 123.000; V 112.200
Hi Desert (Joshua Tree)	U 122.800
Hoffman	U 122.800
Holtville	U 123.000; V 115.900
Imperial County	A/DC 128.600 / 291.700; U 122.700;
"	FSS 122.100 / 122.500; V 115.900
Joshua Tree (High Desert)	U 122.800
Landers (Giant Rock)	M 122.900

AIRCRAFT

AIRPORTS

NAME	FREQUENCY DATA
Lindbergh Field (San Diego)	A 134.800; AC 124.350 / 323.000; CD 125.900;
"	CT 118.300 / 270.900; DC 119.600 / 128.000
"	306.700 (East) / 363.100 (West); FSS 122.200
"	122.400 / 255.400; GC 123.900; ILS 110.900;
"	S 156.150 / 155.895; U 122.950; V 117.800
Lompoc	A/DC 118.000; U 122.700; V 113.800
Long Beach (Daugherty Fld)	A 127.750; AC 124.650; CD 118.150; CT (Left) 119.400
"	120.500; DC 127.200; GC 133.000; U 122.950;
"	V 115.700
Los Angeles Intl. (LAX)	A 133.800 / 135.650; AC 124.500 / 124.900
"	AC 128.500; CD 121.400; CT 133.900 (North)
"	120.950 (South) / 119.800 (Helicopters West)
"	120.350 (Helicopters East); DC 124.300 / 125.200;
"	GC 121.650 (North) / 121.750 (South); ILS 108.500
"	109.900 / 111.100 / 111.100; U 122.950; V 113.600
Lost Hills - Kern Co.	M 122.900; V 117.100
Mac Gillivray (Paso Robles)	M 122.900; V 114.300
McClellan-Palomar	A 120.150; A/DC 127.300; CT 118.600 / 392.000;
"	GC 121.800; U 122.950; V 115.300
Meadows Field (Bakersfield)	A 118.600; A/DC 126.450 / 118.800 / 118.900;
"	CT 118.1; GC 121.700; U 122.950; V 115.400;
Meadowlark	U 122.800
Mojave	A/DC 132.500 / 133.650; CT 127.600;
"	GC 123.900; U 122.800; V 116.400
Montgomery Field	A 126.900; AC 124.350; CD 127.450; CT 119.200
"	CT 125.700 / 269.400; DC 119.600; FSS 122.200
"	FSS 122.400; GC 121.900; ILS 111.700;
"	U 122.950; V 117.800
Murrieta (Bear Creek)	M 122.900
Murrieta (French Valley)	U 122.800; V 115.300
Needles	U 123.000; V 115.200
Oceano	U 122.700
Oceanside Municipal	A/DC 127.300; U 123.000; V 115.300
Ocotillo Wells	M 122.900
Ontario Intl.	A 124.250; A/DC 119.650 / 125.500 / 127.250
"	134.000 / 135.400; CD 118.100; CT 120.600;
"	GC 121.900; ILS 109.700; U 122.950; V 110.400
Orange County, John Wayne	A 126.000; AC 121.300 / 124.650 / 132.700;
"	CD 118.000 / 121.850; CT 119.900 / 126.800
"	CT 128.350; DC 128.100; FSS 122.450;
"	GC 120.800 / 121.850; ILS 108.300; U 122.950;
"	V 109.400 / 115.700
Oxnard	A 118.050; A/DC 124.700 / 135.500; CT 120.000
"	GC 121.900; U 122.950; V 115.800
Palmdale Prod. Test Flight Site	A/DC 126.100; CT 123.700; GC 121.900;
"	U 122.950; V 114.900
Palmdale	A/DC 126.100; CT 119.200; GC 121.900; V 114.500
Palm Springs Regional	A 118.250; A/DC 118.850 / 126.700 / 128.150;
"	CD 128.350; CT 119.700; GC 121.900;
"	FSS 122.100; U 122.950; V 115.500 / 116.200

AIRCRAFT

AIRPORTS

NAME	FREQUENCY DATA
Paso Robles (Mac Gillivray)	U 122.900; FSS 122.200 / 122.400; V 114.300
Paso Robles Municipal	A/DC 128.700; U 123.000; FSS 122.400; V 114.300
Perris Valley	M 122.900; V 112.200
Pixley	M 122.900
Ramona	A/DC 132.200; U 122.700; V 114.000
Redlands Municipal	U 122.700; V 112.200
Rialto (Miro Field)	A/DC 127.250; U 122.800; V 112.200
Rio Bravo (Bakersfield)	U 122.900; V 115.400
Riverside Municipal	A 128.800; A/DC 135.400; CT 121.000; FSS 122.050 / 122.200
"	GC 121.700; U 122.950; V 112.200 / 112.400
Riverside (Rubidoux)	U 122.800; V 112.200
Rosamond Skypark	M 122.900; V 114.500
Salton Sea (Salton City)	M 122.900
San Fernando	U 122.800
San Francisco Intl.	A 108.900 / 113.700 / 118.850; AC 132.550 / 134.500 / 135.650
"	CD 118.200; CT 120.500; DC 120.900 / 135.100; GC 121.800;
"	GC 124.250 / 128.650; ILS 108.900 / 109.500 / 111.700;
"	U 122.950; V 115.800; W 118.850
San Jose Intl.	A 126.950; AC 120.100 / 120.300 / 134.500
"	AC 135.400 / 135.400; CD 118.000; CT 124.000;
"	DC 121.300; GC 121.700; ILS 110.900 / 111.100;
"	U 122.950; V 114.100; W 133.325
San Luis Obispo	A 119.800; A/DC 119.050; CT 124.000; U 122.950;
"	ILS 109.700; V 112.400
Santa Barbara Municipal	A 127.800; A/DC 119.050 / 120.550 / 125.400;
"	CD 132.900; CT 119.700; GC 121.700; ILS 110.900;
"	U 122.950; V 114.900 / 116.500
Santa Catalina Island	A/DC 127.400; U 122.700; V 111.400
Santa Maria	A 121.250; A/DC 118.000 / 119.050; CT 118.300;
"	GC 121.900; ILS 108.900; U 122.950; V 112.400
Santa Monica Municipal	A 119.150; AC 124.500 / 124.900 / 128.500; CT 120.100;
"	DC 124.300 / 125.200; GC 121.900; U 122.940; V 110.800
Santa Paula	U 122.900
Santa Ynez	A/DC 118.000 / 119.050; U 122.800; V 114.900
Shafter-Minter Field	A/DC 126.450 / 127.100; M 122.900; V 115.400
Shephard (Private)	U 122.800
Taft-Kern	U 122.800; V 117.500
Tehachapi	U 123.000; V 108.400
Taft-Kern Co.	U 122.800; V 117.500
Thermal	A/DC 118.850 / 128.150; U 123.000; V 116.200
Torrance Municipal	A 125.600; A/DC 124.300 (West) / 127.2 (East);
"	CT 135.600 (West) / 124.000 (East); GC 120.900;
"	ILS 111.900; U 122.950; V 113.600
Twentynine Palms	A/DC 128.150; U 122.950; V 114.200
Van Nuys	A 118.450; A/DC 120.400 / 124.600 / 134.200 / 135.050;
"	CD 126.600; CT 119.000 (Helicopters) / 119.300 / 120.200;
"	GC 121.700; ILS 111.300; U 122.950; V 113.100
Valley Vista (Private)	U 122.800
Wasco-Kern Co.	U 122.800; V 115.400
Whiteman (Los Angeles)	CT 135.000; GC 125.000; U 122.950; V 113.100
Yucca Valley	U 123.000; V 112.200
Yuma Intl. (Arizona)	A 118.800 / 273.200; CT 119.300; DC 124.700.

AIRCRAFT

AIR SHOWS

NAME	FREQUENCY DATA
All Air Shows	122.750 / 122.850 / 122.900 / 122.925 / 123.100
"	151.625 / 154.540 / 154.570 / 464.500 / 464.575
Blue Angels (Navy)	141.560 / 142.000 / 143.000 / 251.600 / 263.350
"	275.350 / 302.100 / 307.700 / 345.900 / 360.400
Confederate Air Force	151.625
Thunderbirds (Air Force)	236.550 / 236.600 / 273.500 / 283.500 / 295.700
"	322.300 / 322.600 / 382.900 / 394.000 / 413.100

AIR TO AIR COMMUNICATIONS

CHANNEL	FREQUENCY DATA
Commercial Aircraft	122.750 / 123.450
Helicopters	122.750 / 122.850 / 122.900 / 122.925 / 123.025
Pacific Ocean Basin	128.950
Private Aircraft	122.700 (Coastal Area) / 122.750 / 122.850
"	122.900 / 122.925

AIR TRAFFIC CONTROL CENTERS

LOS ANGELES CENTER

LOCATION	FREQUENCY DATA
Arrival / Departure U.S.	119.000 / 126.000 / 126.400 / 127.400 / 128.050
"	132.150 / 133.400 / 134.400 / 134.550 / 135.450
Bakersfield	127.100 / 133.050 / 135.300
Baldwin Hills	118.550 / 125.650 / 132.850
Barstow	126.350 / 132.300 / 132.500 / 133.550 / 134.650
"	135.575
Blythe	128.150 / 135.450
Boron	132.625
Cedar City, Utah	124.200 / 127.350 / 135.250 / 135.550
Julian	125.650 / 125.850 / 127.400 / 128.150 / 128.600
"	135.450 / 135.750
Keeler	127.900 / 132.625
Mount Potosi, Nevada	127.900 / 132.625
Nelson, Nevada	124.200 / 124.850 / 134.650
Ontario	125.850
Riverside	126.350
Saddle Peak	125.800 / 132.600 / 134.750
San Luis Obispo	119.050 / 124.150
Santa Barbara	119.050 / 132.150 / 134.425 / 135.500
Seligman, Arizona	124.850 / 133.200 / 134.950
Tonopha, Nevada	127.900
Twentynine Palms	128.150 / 133.200
Yuma, Arizona	121.350

AIRCRAFT

BALLOONS / BLIMPS / GLIDERS / SKYWRITERS

MOST COMMON FREQUENCIES:

122.250 / 122.750 / 122.775 / 122.850 / 122.900 / 122.925 / 123.125
123.300 / 123.400 / 123.500 / 151.625 / 154.540 / 464.500 / 464.550

CIVIL AIR PATROL

CHANNEL	FREQUENCY DATA
(F-1) Operations Repeaters*	**148.150 R**
(F-2) Operations Direct	148.150
FM Portables Radios	26.620
Pacific Region Net USB	4.585
Search and Rescue	122.850 / 122.900 / 122.925 / 123.100

*Repeater Input 143.900

EMERGENCY COMMUNICATIONS

CHANNEL	FREQUENCY
Commercial & Private Aircraft (ELT's)	121.500
Military Aircraft (ELT's)	243.000

FLIGHT SCHOOLS

NAME	LOCATION	FREQUENCY	CALL SIGN
Airline Training Center	San Diego	123.500	WHG-9
El Cajon Flying Service	El Cajon	123.300	KNQ-9
Executive Air Service	San Diego	122.875	WQG-7
National Air College	San Diego	123.300	WQN-5
National University	San Diego	123.300	WNV-4
Renkens Executive Flights	Carlsbad	121.950	KNX-6

MISCELLANEOUS

NAME	LOCATION	FREQUENCY	CALL SIGN
Aero Corp.	Burbank	151.955	WNCK-441
Aerotest Inc.	Mojave	123.200	KGB-8
Airlines Crew Service	Mt. Lukens	464.825	WZW-598
Air Mobile Systems	El Toro	151.8775	WNKU-775
Air Mobile Systems	El Toro	151.9075	WNKU-775
Airtime Comm.	Chatsworth	462.825	WNED-680
Atmospheric Space Corp.	San Diego	464.500	N/A
Beech Aircraft Corp.	Mojave	123.450	KUI-8
California Aerial Surveys	Statewide	151.625	KN-8425
Farm Air Service	Glamis	461.050	KDL-988
Flightline Airport Srv.	Mt. Lukens	508.1125	WIJ-214
Flight Research Corp.	Mojave	123.450	WZE-5
Garrett Corp.	Long Beach	122.950	WDY-4

AIRCRAFT

MISCELLANEOUS

NAME	LOCATION	FREQUENCY	CALL SIGN
General Dynamics	Monument Peak	123.225	KMA-8
General Dynamics	Monument Peak	123.375	KMA-8
General Dynamics	Monument Peak	123.450	KMA-8
General Electric Radio	Mojave	123.225	WZO-6
Goodyear Blimp			
Company Channel	Nationwide	132.000	N/A
Primary Operations	Nationwide	151.625	N/A
Low Power Ops	Nationwide	465.9125	N/A
Low Power Ops	Nationwide	465.9375	N/A
Low Power Ops	Nationwide	465.9625	N/A
Hughes Helicopters	Carlsbad	123.200	WIT-7
Hughes Helicopters	Carlsbad	123.475	WIT-7
Litton Aero Products	Chatsworth	461.900	WNBB-970
Litton Aero Products	Chatsworth	462.1375	WNBB-970
Lockheed Corp.	Burbank	123.250	KFS-4
Lockheed Corp.	Burbank	123.325	KFS-4
Lockheed Corp.	Burbank	123.425	KFS-4
Lockheed Corp.	Burbank	123.900	KB-8290
Lockheed Corp.	Los Angeles	123.250	KMA-6
Lockheed Corp.	Los Angeles	123.325	KMA-6
Lockheed Corp.	Los Angeles	123.425	KMA-6
Lockheed Corp. Terminal	San Diego	464.5125	WNGQ-571
Lockheed Corp. Terminal	San Diego	464.575	WNGQ-571
Mercy Air Ambulance	Fontana	131.675	N/A
McDonnell Douglas	Long Beach	123.275	KMF-9
McDonnell Douglas	Long Beach	123.425	KMF-9
McDonnell Douglas	Long Beach	123.550	KMF-9
McDonnell Douglas	Long Beach	124.400	KA2XAW
McDonnell Douglas	Long Beach	125.950	KA2XAW
McDonnell Douglas	Monument Peak	123.200	WZG-6
McDonnell Douglas	Monument Peak	123.275	WZG-6
McDonnell Douglas	Monument Peak	123.450	WZG-6
McDonnell Douglas	Monument Peak	123.550	WZG-6
Northrop Radio Service	Long Beach	123.350	KVB-5
Northrop Radio Service	Long Beach	123.525	KVB-5
Oasis Petroleum	Long Beach	122.950	WCT-5
Pacific Aerosystems	San Diego	123.175	KKH-6
Rockwell Intl.	Los Angeles	123.350	KNV-8
Rockwell Intl.	Los Angeles	123.475	KNV-8
Rockwell Intl.	Los Angeles	123.525	KNV-8
Skyfox Inc.	Mojave	123.375	WBI-4
Tractor Flight Systems	Mojave	123.200	KLN-7
Val Air Corp.	Brawley	464.525	KZR-897
Volar Corporation	El Cajon	123.300	WZU-7

AMATEUR RADIO SERVICE

HIGH FREQUENCY BANDS

Band	Frequency
160 Meters	1800-2000 kHz
80 Meters	3500-4000 kHz
40 Meters	7000-7300 kHz
30 Meters	10100-10150 kHz
20 Meters	14000-14350 kHz
17 Meters	18068-18168 kHz
12 Meters	24890-24990 kHz
10 Meters	28000-29700 kHz

TEN METER FM BAND

FREQUENCY	LOCATION	REMARKS	TONE
29.500-29.600	Simplex - Nationwide		
29.620 -	Los Angeles	Open	107.2 Htz
29.640 -	Santiago Peak	Open	107.2 Htz
29.660 -	Arcadia	Open	107.2 Htz
29.680 -	Johnstone Peak	Open	107.2 Htz

SIX METER BAND

FREQUENCY	LOCATION	REMARKS	TONE
50.100-50.600	CW, Single-Sideband		
51.500-51.600	Simplex - Nationwide		
51.700 -	Catalina Island	Open	No Tone
51.740 -	Fullerton (WARA)	Open	No Tone
51.760 -	Sky Forest	Open	No Tone
51.780 -	Chino Hills	Open	No Tone
51.800 -	Oat Mountain	Open	67.0 Htz
51.840 -	Camarillo	Open	No Tone
51.860 -	Pasadena	Open	103.5 Htz
51.900 -	Morgan Hill	Closed (FUN)	107.2 Htz
51.920 -	Baldwin Hills	Open	No Tone
52.500 -	Mount Wilson	Open	No Tone
52.525 & 52.540	Simplex - Nationwide		
52.560 -	Mount Frasier	Open	No Tone
52.600 -	Cuyamaca Peak	Open	107.2 Htz
52.620 -	Santiago Peak	Open	103.5 Htz
52.640 -	Huntington Beach	Open	173.8 Htz
52.660 -	Palmdale	Open	No Tone
52.700 -	**TEST PAIR**		
52.740 -	Mount Lukens	Open	127.3 Htz
52.780 -	Mount Santa Rosa	Open	107.2 Htz
52.840 -	Mount Wilson	Open	71.9 Htz
52.860 -	Contractor's Point	Open	
52.880 -	Oak Mountain, Lompoc	Open	82.5 Htz
52.900 -	Fullerton	Open	82.5 Htz
52.920 -	Santa Barbara	Open	No Tone
52.980 -	Red Mountain, Ventura	Open	No Tone

AMATEUR RADIO SERVICE

TWO METER BAND

FREQUENCY	LOCATION	REMARKS	TONE
144.000-144.100	CW, Weak Signal		
144.100-144.300	Single-Sideband		
144.300-144.425	Simplex - Nationwide		
144.415	Southern California DX'ers		
144.425-144.475	OSCAR 13 - Mode J		
144.505 +	Verdugo Peak	Private	
145.120 -	Lompoc	Open	100.0 Htz
145.120 -	San Diego	Open	107.2 Htz
145.130 -	Castro Peak	Private	
145.140 -	Santa Maria	Open	
145.140 -	Santiago Peak	Private	
145.160 -	Easter Hill	Private	
145.160 -	Las Vegas	Private	
145.160 -	Santiago Peak	Private	
145.160 -	Signal Hill	Private	
145.180 -	Big Bear Lake	Open	100.0 Htz
145.180 -	El Cajon	Closed	
145.180 -	Pasadena	Open	131.8 Htz
145.180 -	Santa Barbara	Open	131.8 Htz
145.200 -	Carpinteria	Open	131.8 Htz
145.200 -	Littlerock	Open	
145.200 -	Monterey Park	Open	103.5 Htz
145.220 -	Santiago Peak	Private (CLARA)	
145.220 -	Victorville	Open	114.8 Htz
145.240 -	Heaps Peak	Open	
145.250 -	Yuma, AZ	Open (ZIA)	162.2 Htz
145.260 -	Fountain Valley	Open	136.5 Htz
145.260 -	Malibu	Open	
145.260 -	Snow Summit	Open	103.5 Htz
145.280 -	Covina	Open	141.3 Htz
145.280 -	Garden Grove	Open	103.5 Htz
145.280 -	Santa Monica	Closed	
145.280 -	Table Mountain	Open	
145.280 -	Vista	Open	
145.300 -	Mt. Disappointment	Open	
145.300 -	Ramona	Open (Low level)	
145.320 -	Redondo Beach	Open	114.8 Htz
145.320 -	San Diego	Open	107.2 Htz
145.320 -	Thousand Oaks	Open	
145.340 -	Panorama Heights	Closed	
145.340 -	Santa Rosa	Closed	
145.350 -	Arroya Grande	Open	141.3 Htz
145.360 -	Packet - Southern California		
145.380 -	Palos Verdes	Open	100.0 Htz
145.380 -	Point Loma	Open	107.2 Htz
145.400 -	Fullerton (WARA)	Open	
145.400 -	Ojai	Open	
145.410 -	Tijuana, Mexico	Open	
145.420 -	Lompoc	Open	
145.420 -	Newport Beach	Open	136.5 Htz
145.420 -	Pine Cove, Hemet	Open	88.5 Htz
145.440 -	Palomar Mountain	Open (linked to 147.210)	107.2 Htz

AMATEUR RADIO SERVICE

TWO METER BAND

FREQUENCY	LOCATION	REMARKS	TONE
145.440 -	Sunset Ridge	Open	136.5 Htz
145.460 -	Buzzard Peak	Open	No Tone
145.460 -	Tijuana, Mexico	Open	No Tone
145.480 -	Cucamonga	Open	77.0 Htz
145.480 -	Mount Lukens	Open	100.0 Htz
145.480 -	Palm Springs	Open	91.5 Htz
145.480 -	Skyline Hills	Open	127.3 Htz
145.500-145.600	Simplex - Nationwide		
145.600-145.700	Packet - Southern California		
145.700-145.800	Simplex - Nationwide		
145.800-146.000	OSCAR 13		
146.025 +	Glendale	Open	136.5 Htz
146.025 +	Indio	Open	107.2 Htz
146.025 +	San Clemente	Open	110.9 Htz
146.085 +	Bird Springs Pass	Open	No Tone
146.085 +	Duarte	Open	110.9 Htz
146.085 +	La Jolla	Closed	100.0 Htz
146.115 +	Hollywood Hills	Closed	
146.115 +	San Diego	Private (FUN)	
146.115 +	Victorville	Open	156.7 Htz
146.120 +	Cerro Bola, Baja Calif.	Open	114.8 Htz
146.160 +	Fountain Valley	Open	136.5 Htz
146.160 +	Laguna Niguel	Open	110.9 Htz
146.160 +	San Diego	Open	91.5 Htz
146.160 +	Tujunga	Open	
146.160 +	Whittier	Open	
146.175 +	Fallbrook	Open	107.2 Htz
146.175 +	Whittier	Open	
146.235 +	Palos Verdes	Private	
146.265 +	Lyons Peak	Open	107.2 Htz
146.355 +	Monterey Park	Open	71.9 Htz
146.400-146.600	Simplex channels		
(Be careful not to interfere with repeaters using nonstandard pairs.)			
146.565	T-Hunting - Southern California		
146.610 -	San Diego	Open	107.2 Htz
146.610 -	Sierra Peak	Open	103.5 Htz
146.620 -	Cambria	Open	
146.640 -	Diamond Bar	Open	
146.640 -	Mount Otay	Open	107.2 Htz
146.640 -	Ridgecrest	Open	
146.640 -	Simi Valley	Open	127.3 Htz
146.655 -	Moreno Valley	Open	103.5 Htz
146.655 -	San Marcos	Private (FUN)	
146.655 -	Whittier	Open	
146.660 -	Cerro Bola, Tijuana	Open	114.8 Htz
146.670 -	Arroya Grande	Open	
146.670 -	Lancaster	Open	
146.670 -	Jacumba	Open	
146.670 -	San Bernardino	Open	123.0 Htz
146.670 -	San Diego	Private	
146.670 -	Thousand Oaks	Open	127.3 Htz
146.700 -	Mount Palomar	RTTY, Portable (PARC)	
146.700 -	Newberry Springs	Open	
146.700 -	Sun City	Open	

AMATEUR RADIO SERVICE

TWO METER BAND

FREQUENCY	LOCATION	REMARKS	TONE
146.700 -	Tecate Peak	Open	186.2 Htz
146.730 -	Bridgeport	Open	
146.730 -	Hauser Peak	Open	100.0 Htz
146.730 -	Mount Palomar	Open (PARC)	107.2 Htz
146.730 -	Rio Honda	Open	
146.740 -	Yuma, Arizona	Open	
146.745 -	Packet - Southern California		
146.760 -	Box Springs, Riverside	Closed	91.5 Htz
146.760 -	Lake Elsinore	Open	136.5 Htz
146.760 -	San Diego	Testing	107.2 Htz
146.790 -	Anaheim	Open	103.5 Htz
146.790 -	Palmdale	Open	103.5 Htz
146.790 -	Santee	Open	107.2 Htz
146.790 -	Wildomar	Open	100.0 Htz
146.790 -	Yucca Valley	Open	136.5 Htz
146.800 -	San Luis Obispo	Open	
146.805 -	Saddle Peak	Private	
146.805 -	Signal Hill	Private	
146.805 -	Wildomar	Open	100.0 Htz
146.820 -	Johnstone / Onyx Peaks	Open	No Tone
146.850 -	Crestline	Open	No Tone
146.850 -	Thousand Oaks	Open	94.8 Htz
146.880 -	Escondido (EARS)	Open (linked to 224.140)	107.2 Htz
146.880 -	Mount Potosi, NV	Open	
146.880 -	Paso Robles	Open	
146.880 -	Riverside	Open	146.2 Htz
146.895 -	Orange (O.C. Races)	Open	136.5 Htz
146.895 -	Solvang	Open	
146.910 -	Mount Lee	Open	123.0 Htz
146.910 -	Mount Otay	Open (SBRG)	107.2 Htz
146.925 -	Mount Otay	Open (SBRG)	100.0 Htz
146.925 -	Santiago Peak	Open	114.8 Htz
146.940 -	Disneyland	Open	131.8 Htz
146.940 -	Pomona	Open	103.5 Htz
146.940 -	Rancho Mirage	Open	
146.940 -	Rumerosa, Mexico	Open	
146.940 -	Silver Peak, Bishop	Open	
146.940 -	Tijuana, Mexico	Open	
146.940 -	Victorville	Open	91.5 Htz
146.970 -	Barstow	Open	
146.970 -	Fullerton	Open	136.5 Htz
146.970 -	San Pedro Martir, Baja Calif.	Open (linked to 146.980)	114.8 Htz
146.970 -	Santa Clarita	Open	
146.970 -	Vista	Open	
146.980 -	Cerro Bola, Tijuana	Open (linked to 146.970)	114.8 Htz
146.985 -	Keller Peak	Open	
146.985 -	Running Springs	Open	146.2 Htz
146.985 -	South Mountain, Ventura	Open	103.5 Htz
147.000 +	Hollywood Hills	Private	
147.000 +	Saddle Peak	Private	
147.000 +	Santa Barbara	Open	131.8 Htz
147.015 +	Mount Wilson	Private	
147.015 +	Pine Cove, Hemet	Private	
147.015 +	Santiago Peak	Private	

AMATEUR RADIO SERVICE

TWO METER BAND

FREQUENCY	LOCATION	REMARKS	TONE
147.030 +	Barstow	Open	
147.030 +	Bel Air	Private	
147.030 +	Desert Center	Open	
147.030 +	Mount Palomar	Open (ECRA)	103.5 Htz
147.060 +	Chula Vista	Open	127.3 Htz
147.060 +	Palos Verdes	Private	
147.060 +	Silver Peak, Bishop	Open	103.5 Htz
147.060 +	Twentynine Palms	Open	136.5 Htz
147.075 +	Mount Palomar	Open (PARC)	107.2 Htz
147.075 +	Sunset Ridge	Closed	
147.090 +	Blythe	Open	
147.090 +	Catalina Island	Open	
147.120 +	Brawley	Open	103.5 Htz
147.120 +	Lompoc	Open	
147.120 +	Verdugo Peak	Open	136.5 Htz
147.120 +	Victorville	Open	
147.130 +	Mount Palomar	Open (PARC)	107.2 Htz
147.150 +	Mount Laguna	Open	
147.150 +	Pasadena (JPL)	Open	103.5 Htz
147.150 +	Phelan	Private	
147.180 +	Barstow	Open	151.4 Htz
147.180 +	Coronado	Open	110.9 Htz
147.180 +	Palos Verdes	Private	
147.180 +	Rumerosa, Mexico	Open	114.8 Htz
147.180 +	San Bernardino	Open	67.0 Htz
147.195 +	Cuyamaca Peak (SD Races)	Open	114.8 Htz
147.195 +	Lyons Peak (SD Races)	Closed	
147.195 +	Mount Woodson (SD Races)	Open	203.5 Htz
147.195 +	Palos Verdes	Open	100.0 Htz
147.210 +	Monument Peak	Open	107.2 Htz
147.210 +	Mount Otay	Open (linked to 145.440)	107.2 Htz
147.210 +	Sunset Ridge	Open	156.7 Htz
147.225 +	Castro Peak	Private	
147.240 +	Monument Peak (ECRA)	Open (Linked to AZ)	103.5 Htz
147.270 +	Cambria	Open	
147.270 +	Mount Disappointment	Open	
147.300 +	Boulder Mountain, Poway	Open	103.5 Htz
147.300 +	Hesperia	Open	
147.300 +	Mount Lukens	Private	
147.300 +	Upland	Open	123.0 Htz
147.330 +	Big Bear	Open	
147.330 +	Cerro Bola, Tijuana	Open	
147.360 +	Mt. Disappointment	Private (DARN)	
147.360 +	Mt. Wilson	Private (DARN)	
147.360 +	Palos Verdes	Private (DARN)	
147.360 +	Santiago Peak	Private (DARN)	
147.360 +	Verdugo Peak	Private (DARN)	
147.420	San Diego (input 146.475)	Private	
147.435	Los Angeles (input 146.400)	Open	103.5 Htz
147.450	Contractors Point (input 146.415)	Closed	
147.460	Big Bear City (input 146.505)	Open	
147.465	Huntington Beach (input 146.505)	Open	103.5 Htz

AMATEUR RADIO SERVICE

TWO METER BAND

FREQUENCY	LOCATION	REMARKS	TONE
147.645 -	Laguna Beach (SOARA)	Open	
147.705 -	Pasadena	Open (Earthquake Net)	131.8 Htz
147.705 -	San Diego	Private	
147.705 -	Victorville	Open	136.5 Htz
147.735 -	Loma Linda	Open	118.8 Htz
147.735 -	Oat Mountain	Open	
147.765 -	Frazier Park	Open	
147.765 -	Monrovia	Open	131.8 Htz
147.765 -	San Diego	Private	
147.765 -	Ventura	Open	
147.855 -	Anaheim	Open	110.9 Htz
147.855 -	San Diego	Open	
147.885 -	Ludlow	Open	
147.885 -	Newbury Park	Open	127.3 Htz
147.885 -	San Diego (Sharp Hospital)	Open	107.2 Htz
147.915 -	Anaheim	Open	136.5 Htz
147.915 -	Lancaster	Private	
147.915 -	Mt. David, Banning	Open	123.0 Htz
147.915 -	Spring Valley	Open	107.2 Htz
147.930 -	Simi Valley	Open	127.3 Htz
147.945 -	Canoga Park	Open	
147.945 -	Crestline	Open	123.0 Htz
147.945 -	Santa Barbara	Open	131.8 Htz
147.945 -	San Diego	Open	107.2 Htz
147.975 -	Fullerton	Open	
147.975 -	Ventura	Open	110.9 Htz
147.990 -	Mt. Otay	Private (DRONK)	
147.990 -	Mt. Woodson	Private (DRONK)	
147.990 -	Point Loma	Private (DRONK)	
147.990 -	Toro Peak	Private (DRONK)	
147.990 -	Tujunga	Private (DRONK)	

1.25 METER BAND

FREQUENCY	LOCATION	REMARKS	TONE
223.400-223.525	Simplex - Southern California		
223.550-223.600	Packet - Southern California		
223.620 -	Mt. Laguna	Private	
223.620 -	Whittier Hills	Open (RHARC)	136.5 Htz
223.640 -	Johnstone Peak	Open	100.0 Htz
223.640 -	Seal Beach	Open	114.8 Htz
223.660 -	Buzzard Peak	Open	
223.680 -	Running Springs	Closed (HROC)	
223.680 -	Sunset Ridge	Closed (HROC)	
223.700 -	Canoga Park	Open	156.7 Htz
223.700 -	Lakewood	Closed	
223.700 -	Rosamond	Open	100.0 Htz
223.720 -	Mt. Lukens	Closed	

AMATEUR RADIO SERVICE

1.25 METER BAND

FREQUENCY	LOCATION	REMARKS	TONE
223.740 -	Contractors Point	Open	114.8 Htz
223.740 -	Palos Verdes	Open	100.0 Htz
223.740 -	Riverside	Open	100.0 Htz
223.760 -	Chula Vista	Open	103.5 Htz
223.760 -	Huntington Beach	Open	94.8 Htz
223.780 -	Mt. Palomar	Open (WALA)	100.0 Htz
223.780 -	Palos Verdes	Open (WALA)	100.0 Htz
223.800 -	Lyons Peak	Closed / SD RACES	
223.800 -	Mt. Woodson	Closed / SD RACES	
223.800 -	Mission Hills	Closed / SD RACES	
223.800 -	Signal Hill	Open	156.7 Htz
223.820 -	Tijuana, Mexico	Open	No Tone
223.820 -	Whittier Hills	Open	103.5 Htz
223.840 -	Indio Hill	Open	156.7 Htz
223.840 -	Palos Verdes	Open	103.5 Htz
223.840 -	Victorville	Open	156.7 Htz
223.860 -	La Verne	Open	114.8 Htz
223.860 -	Mt. Lukens	Closed	
223.880 -	Hollywood Hills	Closed	
223.880 -	La Mesa	Open	107.2 Htz
223.880 -	Santa Rosa	Open	203.5 Htz
223.900 -	San Clemente	Open	156.7 Htz
223.900 -	Oat Mountain	Open	94.8 Htz
223.920 -	Palmdale	Open	100.0 Htz
223.920 -	Palos Verdes	Closed	
223.920 -	Santa Cruz Island	Open (SBARC)	131.8 Htz
223.940 -	Lake Elsinore	Open	136.5 Htz
223.940 -	Rasnow Peak	Open	141.3 Htz
223.960 -	Coronado	Open	110.9 Htz
223.960 -	Mission Viejo	Open	156.7 Htz
223.960 -	Mojave Desert	Open	
223.960 -	Santa Barbara	Open	
223.980 -	Johnstone Peak	Open	
223.980 -	Mt. Palomar	Open	
224.000 -	Brush Peak, Santa Barbara	Open (CONDOR)	156.7 Htz
224.000 -	Santiago Peak	Closed	
224.020 -	Big Bear	Open	
224.020 -	Diamond Bar	Closed	
224.020 -	Mt. Helix, La Mesa	Open	100.0 Htz
224.020 -	Palos Verdes	Open	131.8 Htz
224.040 -	Pasadena	Open (JPL)	156.7 Htz
224.040 -	Tepuesquet Peak	Open	
224.060 -	Mt. Laguna	Open (SANDRA)	107.2 Htz
224.060 -	Sunset Ridge	Closed	
224.080 -	La Cumbre Peak	Open (SBARC)	131.8 Htz
224.080 -	Pasadena	Open (JPL)	156.7 Htz
224.080 -	Rattlesnake Mountain, SD	Open (ECARC)	107.2 Htz
224.100 -	Hollywood Hills	Closed	
224.100 -	Toro Peak	Private	
224.120 -	Santa Margarita Peak	Portable	107.2 Htz
224.120 -	Santa Ynez Peak	Open (SBARC)	131.8 Htz

AMATEUR RADIO SERVICE

1.25 METER BAND

FREQUENCY	LOCATION	REMARKS	TONE
224.120 -	Whittier Hills	Open	151.4 Htz
224.140 -	Hollywood Hills	Private	
224.140 -	Frazier Park	Closed	
224.140 -	Rainbow Ridge	Open	
224.140 -	San Marcos	Open (linked to 146.880)	107.2 Htz
224.160 -	Cuyamaca Peak	Closed	
224.160 -	Oat Mountain	Closed	
224.160 -	Santa Ynez Peak	Open	131.8 Htz
224.180 -	Fullerton (linked to 145.400)	Open (WARA)	
224.180 -	Onyx Peak	Open (CONDOR)	156.7 Htz
224.180 -	Palmdale	Open	100.0 Htz
224.180 -	Ventura	Open	156.7 Htz
224.200 -	Crestline	Closed	
224.200 -	Lancaster	Open	
224.200 -	Mt. Otay	Open (SANDRA)	107.2 Htz
224.200 -	Oat Mountain	Open	100.0 Htz
224.200 -	Verdugo Peak	Open	123.0 Htz
224.220 -	Anza Borrego	Open	100.0 Htz
224.220 -	Santiago Peak	Open	151.4 Htz
224.240 -	Bermuda Dunes	Open	100.0 Htz
224.240 -	Saddle Peak	Closed	
224.260 -	Contractors Point	Closed	
224.260 -	Rim Forest	Open	110.9 Htz
224.260 -	San Diego	Open	107.2 Htz
224.280 -	Alta Loma	Open	110.9 Htz
224.280 -	Montecito Heights	Open	100.0 Htz
224.280 -	San Diego	Closed (SBRG)	
224.300 -	Mt. Disappointment	Open (LA RACES)	100.0 Htz
224.320 -	Mt. Wilson	Closed	
224.340 -	Castro Peak	Open	162.2 Htz
224.340 -	Keller Peak	Open	156.7 Htz
224.360 -	Loop Canyon	Open	107.2 Htz
224.360 -	Upland	Open	94.8 Htz
224.360 -	Vista	Open	131.8 Htz
224.380 -	Palos Verdes	Open	192.8 Htz
224.380 -	Riverside	Open	
224.400 -	Mt. Otay	Private	
224.400 -	Oat Mountain	Private	
224.400 -	Palos Verdes	Private	
224.420 -	Catalina Island	Open	123.0 Htz
224.420 -	Observatory Ridge	Open	
224.440 -	Monterey Hills	Open	100.0 Htz
224.440 -	San Marcos (Gopher Canyon)	Open	156.7 Htz
224.460 -	Point Loma	Open	
224.460 -	Riverside	Open (Riverside RACES)	110.9 Htz
224.460 -	Torrance	Open	131.8 Htz
224.480 -	Contractors Point	Open	110.9 Htz
224.480 -	Santa Rosa Peak	Open	
224.500 -	Anaheim	Open	156.7 Htz
224.500 -	La Habra	Open	162.2 Htz
224.500 -	Lompoc	Open	141.3 Htz

AMATEUR RADIO SERVICE

1.25 METER BAND

FREQUENCY	LOCATION	REMARKS	TONE
224.520 -	Hemet	Open	
224.520 -	Loop Canyon	Open	156.7 Htz
224.520 -	San Diego	Open	107.2 Htz
224.540 -	Camarillo	Open	156.7 Htz
224.540 -	San Diego	Open	107.2 Htz
224.540 -	Sierra Peak	Open	88.5 Htz
224.560 -	Laguna Niguel	Open	67.0 Htz
224.560 -	Running Springs	Open	110.9 Htz
224.560 -	Tujunga	Open	67.0 Htz
224.560 -	Santa Barbara	Open	131.8 Htz
224.580 -	Oat Mountain	Open	156.7 Htz
224.580 -	San Luis Obispo	Open	151.4 Htz
224.580 -	San Marcos	Private	
224.600 -	Palos Verdes	Open	100.0 Htz
224.600 -	San Diego	Open	107.2 Htz
224.620 -	Mt. Crest, El Cajon	Open	127.3 Htz
224.620 -	Palos Verdes	Open	100.0 Htz
224.640 -	Laguna Beach	Open (SOARA)	107.2 Htz
224.640 -	Lake Isabella	Open (CONDOR)	156.7 Htz
224.640 -	Santiago Peak	Open (HROC)	192.8 Htz
224.640 -	Yucca Valley	Open	
224.660 -	Costa Mesa	Open	156.7 Htz
224.660 -	Santa Maria	Open	
224.660 -	Warner Hot Springs	Open	107.2 Htz
224.680 -	Corona	Open	141.3 Htz
224.680 -	Hollywood Hills	Open (BARC)	114.8 Htz
224.680 -	Mt. Otay	Closed	
224.680 -	Paso Robles	Closed	
224.700 -	Chino	Open	107.2 Htz
224.700 -	Laguna Beach Hills	Open	
224.700 -	Thousand Oaks	Open	156.7 Htz
224.720 -	Frazier Park	Open (CONDOR)	156.7 Htz
224.740 -	Hollywood Hills	Open	203.5 Htz
224.740 -	Shandon	Open	
224.760 -	Blue Ridge	Open (EARN)	156.7 Htz
224.760 -	Cactus City (Indio)	Open (EARN)	131.8 Htz
224.760 -	Frost Peak	Open (EARN)	100.0 Htz
224.760 -	Laughlin, NV	Open (EARN)	156.7 Htz
224.760 -	Rincon (Ventura)	Open (EARN)	131.8 Htz
224.760 -	Santiago Peak	Open (EARN)	156.7 Htz
224.760 -	Silver Peak	Open (EARN)	156.7 Htz
224.760 -	Twentynine Palms	Open (EARN)	100.0 Htz
224.780 -	Mt. Lukens	Open	100.0 Htz
224.780 -	Palos Verdes	Open	127.3 Htz
224.780 -	Phelan	Closed	
224.800 -	Santa Ynez Peak	Closed	
224.800 -	Santiago Peak	Private	
224.820 -	Tranquillion Mtn., Lompoc	Open (CONDOR)	141.3 Htz
224.820 -	Santiago Peak	Open (CONDOR)	156.7 Htz
224.840 -	Altadena	Closed	
224.840 -	Hollywood Hills	Closed	
224.840 -	Silver Lake	Closed	
224.840 -	Yorba Linda	Closed	
224.860 -	Crestline	Open	

AMATEUR RADIO SERVICE

1.25 METER BAND

FREQUENCY	LOCATION	REMARKS	TONE
224.860 -	Santa Barbara	Open	156.7 Htz
224.860 -	Valencia	Open	
224.880 -	Santiago Peak	Closed	
224.880 -	Thousand Oaks	Open	141.3 Htz
224.900 -	Mt. Palomar	Open (220 Club)	107.2 Htz
224.900 -	Thousand Oaks	Open (CONDOR)	156.7 Htz
224.920 -	Clairemont, San Diego	Open (ECRA)	107.2 Htz
224.920 -	Hauser Peak	Open	94.8 Htz
224.920 -	Palm Desert	Open	156.7 Htz
224.940 -	Mt. Wilson	Open	
224.960 -	Black Mountain	Closed (COUVRA)	
224.960 -	Hollywood Hills	Closed	
224.980 -	Mt. Otay	Private	
224.980 -	Riverside	Private	
224.980 -	Santiago Peak	Private	

75 CENTIMETER BAND

FREQUENCY	LOCATION	REMARKS	TONE
432.100	Nationwide	SSB Calling	
440.000	Simplex Autopatches		
441.000	Simplex Control Channel (nonvoice)		
445.000	Simplex Nationwide		
445.925 -	Juniper Hills	Open	192.8 Htz
445.925 -	Pomona	Open	103.5 Htz
445.925 -	San Marcos	Open	107.2 Htz
445.925 -	Whittier	Open	100.0 Htz
446.000	Simplex Nationwide		
446.375 -	Santiago Peak	Private	
446.400 -	Hauser Peak	Open	100.0 Htz
446.400 -	Johnstone Peak	Open	103.5 Htz
446.400 -	Santa Barbara	Open	131.8 Htz
446.400 -	Mount Palomar	Closed	156.7 Htz
446.500	Simplex Nationwide		
446.625 -	Ensenada, Baja California	Open	100.0 Htz
447.000 -	Apple Valley	Open	100.0 Htz
447.000 -	Camarillo	Open	141.3 Htz
447.000 -	Mt. Otay	Open	107.2 Htz
447.000 -	Palos Verdes	Open	100.0 Htz
447.025 -	Mt. Otay	Open (SBRG)	107.2 Htz
447.025 -	Mt. Wilson	Open	100.0 Htz
447.325 -	Covina	Open	
447.325 -	Fontana	Open (RACES)	103.5 Htz
447.325 -	Palm Springs	Open	
447.325 -	San Diego	Open	
447.325 -	Sulpher Mountain	Open	
447.325 -	Table Mountain	Open	94.8 Htz
447.325 -	West L.A. (V.A. Hospital)	Open	103.5 Htz
447.575 -	Canoga Park	Open (Hughes ARC)	100.0 Htz
447.575 -	Indio	Open	100.0 Htz
447.575 -	Santa Anita Ridge	Open	131.8 Htz
447.575 -	Ventura	Open	100.0 Htz

AMATEUR RADIO SERVICE

75 CENTIMETER BAND

FREQUENCY	LOCATION	REMARKS	TONE
447.800 -	Mt. Palomar	Closed (ECRA)	88.5 Htz
447.825 -	Cerro Bola, Tijuana	Open	136.5 Htz
448.350 -	Chatsworth	Open	100.0 Htz
448.350 -	Crestline	Open	107.2 Htz
448.350 -	Laguna Niguel	Open	100.0 Htz
448.350 -	Santa Clarita	Open	67.0 Htz
448.350 -	Ventura	Open	141.3 Htz
448.460 -	Tijuana, Mexico	Open	
448.800 -	San Marcos Peak	Closed (remote base)	127.3 Htz
448.725 -	Mt. Palomar	Private (CACTUS)	100.0 Htz
449.425 -	Big Bear	Open	123.0 Htz
449.425 -	Oat Mountain	Open	67.0 Htz
449.425 -	Ontario	Open	107.2 Htz
449.425 -	Mt. Palomar	Open (PARC)	107.2 Htz
449.425 -	Thousand Oaks	Open	
449.500 -	Mt. Otay	Open (SANDRA)	107.2 Htz
449.775 -	Signal Hill	Open	156.7 Htz
449.825 -	Mt. Laguna	Open (SANDRA)	107.2 Htz
449.925 -	Tijuana	Open	100.0 Htz
449.975 -	Bullhead City, AZ	Open	156.7 Htz
449.975 -	Camp Pendleton	Private	100.0 Htz

33 CENTIMETER BAND

FREQUENCY	LOCATION	REMARKS	TONE
927.350 -	Pinon Hills	Open	
927.450 -	Mt. Palomar	Open	107.2 Htz
927.900 -	Lompoc	Open	
927.950 -	Orange Hills	Open	103.5 Htz

23 CENTIMETER BAND

FREQUENCY	LOCATION	REMARKS	TONE
1282.025 -	Santiago Peak	Open (O.C RACES)	88.5 Htz
1282.075 -	La Crescenta	Open	131.8 Htz
1282.100 -	Sulpher Mountain	Open (SMRA)	
1282.125 -	Los Angeles	Open	114.8 Htz
1282.150 -	Santiago Peak	Open (BARC)	
1282.175 -	Oat Mountain	Open (SARA)	123.0 Htz
1282.200 -	Signal Hill	Open	
1282.225 -	Huntington Beach	Open	103.5 Htz
1282.250 -	Mt. Wilson	Open	
1282.300 -	Mt. Otay	Open (SANDRA)	
1282.325 -	Mt. Lukens	Open	100.0 Htz
1282.350 -	Saddle Peak	Open	136.5 Htz
1282.375 -	Mt. Lukens	Open	136.5 Htz
1282.400 -	Palos Verdes	Open	103.5 Htz
1282.425 -	San Marcos	Open	107.2 Htz
1282.450 -	Mt. Wilson	Open	103.5 Htz

AMATEUR RADIO SERVICE

23 CENTIMETER BAND

FREQUENCY	LOCATION	REMARKS	TONE
1282.500 -	Mt. Wilson	Open	
1282.525 -	Signal Hill	Open (O.C. RACES)	88.5 Htz
1282.550 -	Chatsworth	Open	
1282.575 -	San Marcos	Open	
1282.600 -	Santa Barbara	Open	131.8 Htz
1282.650 -	Hollywood Hills	Open	103.5 Htz
1282.700 -	Pleasant Peak	Open	
1282.750 -	Arcadia	Open	131.8 Htz
1282.800 -	Pleasant Peak	Open	
1282.850 -	Santa Ana	Open	
1282.875 -	West Covina	Open	
1282.900 -	Verdugo Peak	Open	
1282.925 -	Sky Forest	Open	
1283.000 -	**Test Pair**		
1283.025 -	Contractors Point	Closed	
1283.050 -	Whittier	Closed	
1283.075 -	Mt. Lukens	Closed	
1283.100 -	Anaheim	Closed	
1283.125 -	Palos Verdes	Closed	
1283.150 -	Huntington Beach	Closed	
1283.175 -	Oat Mountain	Private	
1283.200 -	San Bernardino	Closed	
1283.225 -	Montecito Heights	Closed	
1283.250 -	Mt. Disappointment	Closed (DARN)	
1283.300 -	Santa Anita Ridge	Closed	
1283.325 -	Palos Verdes	Private	
1283.350 -	Irvine	Open	
1283.375 -	Pacific Palisades	Private	
1283.400 -	Huntington Beach	Private	
1283.425 -	Diamond Bar	Private	
1283.475 -	Mt. Lukens	Private	
1283.500 -	Palos Verdes	Closed	
1283.550 -	Palos Verdes	Closed	
1283.575 -	Santiago Peak	Open	114.8 Htz
1283.600 -	Anaheim	Closed	
1283.625 -	Carson	Private	
1283.650 -	Palos Verdes	Private	
1283.700 -	Santiago Peak	Closed	
1283.725 -	Hawthorne	Private	
1283.750 -	Palos Verdes	Closed (DARN)	
1283.800 -	Santiago Peak	Closed	
1283.850 -	Palos Verdes	Closed (DARN)	
1283.900 -	Sunset Ridge	Closed	
1283.950 -	Verdugo Peak	Closed	
1284.000 -	Costa Mesa	Open	
1284.025 -	Cerritos	Closed	
1284.050 -	Sunset Ridge	Closed	
1284.075 -	Mt. Lukens	Private	
1284.100 -	Mt. Lukens	Private	

AMATEUR RADIO SERVICE

23 CENTIMETER BAND

FREQUENCY	LOCATION	REMARKS	TONE
1284.125 -	Santiago Peak	Private	(Tones not shown)
1284.150 -	Oat Mountain	Private	"
1284.175 -	Santiago Peak	Private	
1284.200 -	Saddle Peak	Closed	"
1284.225 -	Mt. Wilson	Private	
1284.250 -	Castro Peak	Closed	"
1284.300 -	Mt. San Miguel	Private	
1284.350 -	Santa Monica	Closed	
1284.400 -	Downey	Closed	"
1284.450 -	Oat Mountain	Closed	
1284.500 -	Palos Verdes	Closed (DARN)	"
1284.650 -	Rattlesnake Mountain	Private	
1284.700 -	Mt. Woodson	Private	
1284.900	Morgan Hill	Private (FUN)	D162
1285.000 -	Verdugo Peak	Closed	(Tones not shown)
1285.050 -	Santiago Peak	Closed	"
1285.100 -	Mt. Lukens	Closed	
1285.150 -	Palos Verdes	Closed	"
1285.200 -	Palm Springs	Closed	
1285.250 -	Tujunga	Closed	"
1285.350 -	Sylmar	Closed	
1285.400 -	San Diego	Closed (ECRA)	"
1285.450 -	Snow Peak	Closed	
1285.475 -	Santiago Peak	Private	
1285.500 -	Santiago Peak	Private	"
1285.600 -	Mt. Lukens	Private	
1285.750 -	Running Springs	Closed	
1285.800 -	Mt. Otay	Closed (ECRA)	"
1285.900	Morgan Hill	Private (FUN)	
1285.950 -	Fullerton	Private	
1286.000 -	Running Springs	Private	
1286.050 -	La Palma	Closed	"
1286.150 -	Frazier Mountain	Closed	
1286.200 -	Hauser Peak	Closed	
1286.250 -	Keller Peak	Open	
1286.300 -	Palos Verdes	Open	100.0 Htz
1286.700 -	Mt. Lukens	Closed	(Tones not shown)
1286.750 -	Santiago Peak	Closed	"
1286.900 -	Sunset Ridge	Closed	
1286.950 -	Pleasant Peak	Closed	
1287.500 -	Mt. Lukens	Private	"
1288.000 -	Juniper Hills	Closed	
1294.125-1295.000	Simplex Southern California		
1294.500	Simplex Nationwide		
1294.900	Single-Sideband Calling Frequency - So. Calif.		
1296.100	Single-Sideband Calling Frequency - Nationwide		

SPACE SHUTTLE REBROADCASTS

145.460 Buzzard Peak, 147.150/224.040 La Canada (JPL), 447.975 Camp Pendleton, 448.675 Cactus Net, 447.800 Mount Palomar (ECRA), 1282.400 Palos Verdes, 1285.400 San Diego (ECRA).

Note: Space Shuttle rebroadcasts are played at the discretion of the repeater owner or club.

BROADCASTERS

A.M. BROADCAST BAND

FREQUENCY (kHz)	CALL SIGN	LOCATION	REMARKS
540	KPVP	Carmel Valley, CA	
540	XETIN	Tijuana, MX	Spanish
550	KCWR	Bakersfield, CA	
550	KOY	Phoenix, AZ	
560	KBLU	Yuma, AZ	
560	KSFO	San Francisco, CA	
570	KISN	Salt Lake City, UT	
570	KLAC	Los Angeles, CA	AM Stereo
580	KMJ	Fresno, CA	
590	KRSO	San Bernardino, CA	
600	KKLQ	San Diego, CA	Top 40
610	KAVS	Lancaster, CA	
610	KFRC	San Francisco, CA	
620	KIGS	Hanford, CA	
620	KTAR	Phoenix, AZ	
630	KOH	Reno, NV	
640	KFI	Los Angeles, CA	News, Talk
650	KRDX	Rancho Cordova, CA	
660	KGDP	Orcutt, CA	
670	KWNK	Simi Valley, CA	
680	KNBR	San Francisco, CA	News, Sports
690	XETRA	Tijuana, MX	News, Sports
700	KFAM	Salt Lake City, UT	
700	KOJY	Soledad, CA	
710	KMPC	Los Angeles, CA	News, Sports
720	KDWN	Las Vegas, NV	News, Sports
740 (Daytime)	KBRT	Avalon, CA	Family Radio
740 (Nightime)	KCBS	San Francisco, CA	
760	KFMB	San Diego, CA	All format
770	KPLA	Riverbank, CA	
780	KROW	Reno, NV	
790	KABC	Los Angeles, CA	News, Sports
800	XEMMM	Tijuana, Mexico	Baja Radio
810	KGO	San Francisco, CA	News, Sports
820	XEYX	Mexicali, BCN	
830	KOZ	Unknown	"Radio Oz"
840	KVEG	Las Vegas, Nevada	News, Sports
850	KCTQ	Thousand Oaks, CA	
850	KOA	Denver, CO	Powerful
860	XEMO	Tijuana, MX	Spanish
870	KIEV	Glendale, CA	"the talk of town"
870	KROL	Las Vegas, NV	
880	KUTE	Desert Hot Springs	
890	WLS	Chicago, IL	
900	KGRB	West Covina	Big Band
910	KECR	El Cajon, CA	Family Radio
920	KDES-AM	Palm Springs, CA	
920	KORK	Las Vegas, NV	
920	XESDA	Ensenada, MX	
930	KRTH	Los Angeles, CA	
940	KFRE	Fresno, CA	AM Stereo

BROADCASTERS

A.M. BROADCAST BAND

FREQUENCY (kHz)	CALL SIGN	LOCATION	REMARKS
940	KNST	Tucson, AZ	
940	XEWV	Mexicali, BCN	
950	XEKAM	Tijuana, BCN	
960	KNEZ	Lompoc, CA	
970	KCLB	Coachella, CA	
970	KNUU	Las Vegas, NV	
980	KFWB	Los Angeles, CA	News, Traffic
990	XECL	Mexicali, BCN	
1000	KCEO	Vista, CA	News, Talk
1010	XEDX	Ensenada, BCN	
1020	KTNQ	Los Angeles, CA	
1030	KJDJ	San Luis Obispo, CA	
1040	KURS	San Diego, CA	Spanish
1050	XED	Mexicali, BCN	Spanish
1060	KKVV	Las Vegas, NV	
1070	KNX	Los Angeles, CA	News, Traffic
1080	KSCO	Santa Cruz, CA	
1090	XEPRAS	Rosarito, BCN	"Radio Express"
1100	KZPM	Bakersfield, CA	
1110	KRLA	Los Angeles, CA	Contemporary
1120	XEMX	Mexicali, BCN	Spanish
1130	KSDO	San Diego, CA	News, Talk
1140	KCMJ	Palm Springs, CA	Country
1140	KMJJ	Las Vegas, NV	"Casino Radio"
1150	KIIS	Los Angeles, CA	
1160	KSL	Salt Lake City, UT	News, Talk
1170	KCBQ-AM	San Diego, CA	50's, 60's
1180	KERI	Wasco, CA	
1190	KORG	Anaheim, CA	
1190	XEMBC	Mexicali, BCN	
1210	KPRZ	San Marcos, CA	"K-Praise"
1220	KBET	Canyon Country, CA	
1230	KGFJ	Los Angeles, CA	
1230	KLAV	Las Vegas, NV	
1230	KXO	El Centro, CA	
1240	KPPC	Pasadena, CA	
1240	KSMA	Santa Maria, CA	
1240	KSON-AM	San Diego, CA	Country
1250	KDHI	Twentynine Palms, CA	
1250	KTMS	Santa Barbara, CA	
1260	KOIT	San Fernando, CA	Big Band
1270	KNWZ	Palm Desert, CA	News Radio
1270	XEAZ	Tijuana, BCN	
1280	KFRN	Long Beach, CA	Family Radio
1290	KMEN	San Bernardino, CA	
1300	KAZA	Gilroy, CA	
1310	XEC	Tijuana, BCN	
1320	KGMG-AM	Oceanside, CA	
1320	KHSJ	Hemet, CA	
1330	KWKW	Los Angeles, CA	Spanish
1340	KWXY-FM	Cathedral City	"Beautiful Music"
1340	XEAA	Mexicali, BCN	
1350	KBAD	Bakersfield, CA	

BROADCASTERS

A.M. BROADCAST BAND

FREQUENCY (kHz)	CALL SIGN	LOCATION	REMARKS
1360	KPOP	San Diego, CA	Nostalgic
1370	KWRM	Corona, CA	
1380	XEHF	Tecate, BCN	
1390	KGER	Long Beach, CA	Family Radio
1400	KBLX	Berkeley, CA	
1410	KCAL	Redlands, CA	Spanish
1410	KFMS	Las Vegas, Nevada	Country
1420	XEXX	Tijuana, Mexico	"Radio Latina"
1430	KNTA	Santa Clara, CA	
1440	KVHL	Santa Maria, CA	
1450	KSPA	Escondido, CA	Nostalgia
1450	KSPI	Palm Springs, CA	News, Talk
1450	KVEN	Ventura, CA	News, Talk
1460	KTYM	Inglewood, CA	
1470	KUTY	Palmdale, CA	
1470	XERCN	Tijuana, MX	
1480	KWIZ	Santa Ana, CA	
1490	KMET	Banning, CA	
1490	KICO	Calexico, CA	
1500	KRCK	Burbank, CA	
1510	KNSE	Ontario, CA	Country
1520	KTRO	Port Hueneme, CA	
1530	KFBK	Sacramento, CA	News, Sports
1540	KSKQ-AM	Los Angeles, CA	Spanish
1550	KXED	Los Angeles, CA	Spanish
1560	KNZR	Bakersfield, CA	
1570	KPRO	Riverside, CA	Inspirational
1580	KBLA	Santa Monica, CA	Korean
1590	KCIN	Victorville, CA	
1590	XEHC	Ensenada, CA	
1600	KMNY	Pomona, CA	
1600	KTAP	Santa Maria, CA	

F.M. BROADCAST BAND

FREQUENCY	CALL SIGN	LOCATION	REMARKS
88.100	KLON	Long Beach, CA	Jazz
88.100	KSGN	Banning, CA	Low Power
88.100	KUCR	Riverside, CA	State Univ.
88.300	KSDS	San Diego, CA	Jazz
88.300	KUCR	Riverside, CA	State Univ.
88.500	KCSN	Northridge, CA	State Univ.
88.500	KSBR	Los Angeles, CA	
88.700	KSPC	Claremont, CA.	College Radio
88.700	KUBO	Tijuana, CA	Spanish
88.900	KSGN	Victorville, CA	
88.900	KXRR	Temecula, CA	
88.900	KXLU	Los Angeles, CA	
89.100	KUOR-FM	Redlands, CA	State Univ.

BROADCASTERS

F.M. BROADCAST BAND

FREQUENCY	CALL SIGN	LOCATION	REMARKS
89.300	KPCC	Pasadena, CA	
89.500	KPBS	San Diego, CA	News, Talk
89.700	KNHS	Torrance, CA	School Dist.
89.700	KSGN	Riverside, CA	
89.900	KCRW	Santa Monica, CA	Comm. College
90.100	KBPK	Buena Park, CA	School Dist.
90.100	KSUL	Long Beach, CA	School Dist.
90.300	XHITZ	Tijuana, Mexico	Top 40
90.500	KSAK	Walnut, CA	
90.700	KPFK-FM	Los Angeles, CA	
91.100	KCPB	Thousand Oaks, CA	
91.100	XTRA	Tijuana, Mexico	Rock "91-X"
91.500	KUSC	Los Angeles, CA	
91.900	KVCR	San Bernardino, CA	
92.100	KOWF	San Marcos, CA	Country
92.300	KKBT	Los Angeles, CA	Classical
92.500	XHRM	Tijuana, Mexico	"Power Station"
92.700	KCMJ-FM	Indio, CA	
92.700	KQLH	Riverside, CA	"Q-Lite Hits"
93.100	KNX-FM	Hollywood, CA	"Arrow 93"
93.300	KECR	El Cajon, CA	Family Radio
93.300	KZLA	Los Angeles, CA	
93.500	KFOX/KRZE	Redondo Beach, CA	Korean
93.500	KFSG	Barstow, CA	Low Power
93.500	KNTF	Ontario, CA	
93.700	KCLB	Coachella, CA	"the Rock"
93.900	KZLA-FM	Los Angeles, CA	Country
94.100	KFSD	San Diego, CA	Classical
94.300	KDUC	Barstow, CA	
94.300	KIKF	Garden Grove, CA	Country
94.500	KWST	Brawley, CA	Country
94.700	KTWVFM	Los Angeles, CA	"The Wave"
94.900	KBZS	San Diego, CA	"K-BEST 95"
95.100	KBBY	Ventura, CA	
95.100	KFIG	Los Angeles, CA	Mix
95.500	KLOS	Los Angeles, CA	Rock & Roll
95.700	XHKY	Tijuana, Mexico	Fiesta Mexicana
95.500	KLOS	Los Angeles, CA	
95.700	KQYN	Twentynine Palms, CA	
95.700	XHYK	Tijuana, Mexico	Spanish
95.900	KKOS	Carlsbad, CA	Contemporary
96.100	KSIQ	Brawley, CA	
96.300	KFSG	Los Angeles, CA	Christian
96.500	KYXY	San Diego, CA	Soft & Contemporary
96.700	KBET	Palm Desert, CA	
96.700	KWIZ-FM	Santa Ana, CA	
97.100	KLSX	Los Angeles, CA	Classic Rock
97.300	KSON-FM	San Diego, CA	Country
97.500	KHST	Riverside, CA	Country
97.700	KAVS	Mojave, CA	
97.700	KQVO	Calexico, CA	Spanish
97.900	KSKQ-FM	Long Beach, CA	Spanish

BROADCASTERS

F.M. BROADCAST BAND

FREQUENCY	CALL SIGN	LOCATION	REMARKS
98.100	KIFM	San Diego, CA	Lite
98.100	KRXV	Yermo, CA	
98.300	KDAR	West Covina, CA	Big Band
98.500	KWXY-FM	Cathedral City	"Beautiful Music"
98.700	KBOB	Los Angeles, CA	"Star 98"
98.900	XHQF	Tijuana, Mexico	Spanish
99.100	KGGI	Riverside, CA	
99.500	KKLA	Los Angeles, CA	
99.500	KXVR	Mountain Pass, CA	
99.900	KOLA	San Bernardino, CA	"Stereo 99.9"
100.100	KBNN	Ramona, CA	Contemporary
100.300	KJMB-FM	Blythe,3;. CA	
100.300	KXEZ	Los Angeles, CA	Easy Listening
100.500	KSPI-FM	Palm Springs, CA	
100.700	KFMB-FM	San Diego, CA	"B-100"
100.700	KHAY	Ventura, CA	Country
101.100	KRTH-FM	Los Angeles, CA	"K-Earth 101"
101.500	KGB	San Diego, CA	Classic Rock
101.700	KTOT	Big Bear Lake, CA	
101.900	KLITE	Santa Monica, CA	
102.100	KIOZ	Oceanside, CA	"Rock 102"
102.300	KZXY	Apple Valley, CA	
102.300	KJLH	Los Angeles, CA	Rhythm 102.3
102.500	XHUN	Tijuana, Mexico	Spanish
102.700	KIIS-FM	Los Angeles, CA	
102.900	KCLX	San Diego, CA	Classic Rock
103.100	KEZN-FM / KOCM	Santa Monica, CA	Jazz
103.100	KVVQFM	Victorville, CA	
103.500	KOST	Los Angeles, CA	
103.700	KJQY	San Diego, CA	"Sunny FM"
103.900	KBON	Lake Arrowhead, CA	Oldies
104.100	XE?	Tijuana, Mexico	Spanish
104.300	KBIG	Los Angeles, CA	"KBATE"
104.500	XLTN	Tijuana, Mexico	Radio Latina
104.700	KCAQ	Oxnard, CA	
104.700	KDES-FM	Palm Springs, CA	
105.100	KKGOFM	Los Angeles, CA	Classical
105.300	KCBQFM	San Diego, CA	"Eagle 105"
105.500	KNAC	Long Beach, CA	Hard Rock
105.500	KZAL	Desert Center, CA	
105.700	KHYE	Los Angeles, CA	"K-HI"
105.900	KPWR	Los Angeles, CA	"Power 106"
106.100	KPLM	Palm Springs, CA	
106.300	KYMS	Santa Ana, CA	
106.500	KKLQ	San Diego, CA	"Q-106"
106.700	KROQFM	Pasadena, CA	
107.100	KBAX	Fallbrook, CA	Christian
107.100	KMAX	Arcadia/Burbank, CA	
107.300	XHFG	Tijuana, Mexico	Spanish
107.500	KLVE	Los Angeles, CA	
107.500	KXO	El Centro, CA	Spanish
107.900	KWVE	San Clemente, CA	"KWAVE"

BROADCASTERS

TELEVISION BROADCAST BAND

CHANNEL	LOCATION	FREQUENCY	CALL SIGN
2	Los Angeles, CA	59.750	KCBS (CBS)
3	Santa Barbara, CA	65.750	KEYT (IND)
4	Los Angeles, CA	71.750	KNBC (NBC)
5	Los Angeles, CA	81.750	KTLA (IND)
6	Tijuana, Mexico	87.570	XETRA (FOX)
7	El Centro, CA	179.750	KLXO (IND)
7	Los Angeles, CA	179.750	KABC (ABC)
8	San Diego, CA	185.750	KFMB (CBS)
9	Los Angeles, CA	191.750	KCAL (IND)
10	San Diego, CA	197.750	KGTV (ABC)
11	Los Angeles, CA	203.750	KTTV (FOX)
12	Tijuana, Mexico	209.750	XEWT (IND)
13	Los Angeles, CA	215.750	KCOP (IND)
15	San Diego, CA	481.750	KPBS (PBS)
19	Mt. Palomar, CA	505.750	KBNT (IND)
22	Los Angeles, CA	523.750	KWHY (IND)
29	Los Angeles, CA	565.750	KBAK (IND)
30	San Bernardino, CA	571.750	KAGL (IND)
31	Twentynine Palms, CA	577.750	KVMD (IND)
33	Tijuana, Mexico	589.750	XE?
34	Los Angeles, CA	595.750	KMEX (IND)
39	San Diego, CA	625.750	KNSD (NBC)
40	Santa Ana, CA	631.750	KTBN (IND)
46	Los Angeles, CA	667.750	KIHS (IND)
50	Los Angeles, CA	691.750	KOCE (IND)
51	San Diego, CA	697.750	KUSI (IND)
52	Los Angeles, CA	703.750	KVEA (IND)
56	Los Angeles, CA	727.750	KDOC (IND)
59	Big Bear Lake, CA	745.750	KBBL (IND)
62	Riverside, CA	763.750	KSLD (IND)
63	Oxnard, CA	769.750	KADY (IND)
64	Barstow, CA	775.750	KVVT (IND)
69	San Diego, CA	805.750	KTTY (IND)

TRAVELER'S INFORMATION SERVICE (T.I.S.)

530 kHz

Anaheim WNFG-219; Burbank-Glendale-Pasadena, CA WNCM-441; Coronado, CA WNAG-556; Escondido, CA (Caltrans-HW78) WNRS-427; Los Angeles, CA WYZ-301; Pasadena, CA WNNL-841; San Bernardino, CA KNCN-250; Santa Ana-Tustin (Caltrans-HW55) KNEC-996; San Ysidro (Caltrans I-5, I-805) WNSB-415; Simi Valley, CA WNHI-810; Vista, CA (Caltrans-HW78) WMRS-426.

1.610 kHz

Barstow, CA KMC-490; Chula Vista, CA WNUB-568; Descanso, CA (Caltrans I-8) WNDQ-665; Idyllwild, CA WNKI-578; Kernville, CA (Ranger Station) KMA-523; Lancaster, CA WNKC-810; Los Altos, CA KNIP-553; San Diego, CA (Balboa Park) WNPF-405.

BUSINESS

ALARM COMPANIES

NAME	LOCATION	FREQUENCY	CALL SIGN
A.P.I. Alarm Systems	Inglewood	464.775 R	KYR-655
A.P.I. Alarm Systems	Los Angeles	460.925 R	KGJ-973
A.P.I. Alarm Systems	Los Angeles	460.975 R	KFL-394
A.P.I. Alarm Systems	Los Angeles	461.000 R	KNAE-701
A.P.I, Alarm Systems	Los Angeles	855.7625R	KNEP-581
A.P.I. Alarm Systems	Pasadena	463.700 R	KNER-202
A.P.I. Alarm Systems	Pasadena	464.600 R	WSW-449
B.W. Alarms	San Diego	856-860.975 T	WNIW-528
Downing Systems	El Toro	935.650-935.750 T	WNXF-209
Guard Systems Alarms	Monterey Park	935 MHz	WNY-288
Knight Alarm & Fire	Escondido	463.375 R	KNEG-244
Morse Alarm Co.	San Diego	856-860.825 T	SMR
San Diego Alarm Co.			
Primary	San Diego	460.975 R	KJI-908
Secondary	San Diego	460.9125	KJI-908
Secondary	San Diego	461.0125	KJI-908
Sharp Security Systems	Los Angeles	461.100 R	WNPE-511

AMUSEMENT & SPECIAL EVENTS

NAME	LOCATION	FREQUENCY	CALL SIGN
America's Cup Sailboat Races			
Productions	San Diego	470.625	N/A
Productions	San Diego	470.825	N/A
Race Officials	San Diego	470.425	N/A
Security	San Diego	471.375 R	N/A
Anaheim Stadium			
Chartered Buses	Anaheim	464.500	N/A
Concessions - Food	Anaheim	151.625	N/A
Concessions - Food	Anaheim	173.325	N/A
Concessions - Merchandise	Anaheim	153.050	N/A
General Operations F1	Anaheim	461.650 R	N/A
General Operations F2	Anaheim	461.275 R	N/A
General Operations	Anaheim	461.5125R	N/A
General Operations	Anaheim	464.425 R	N/A
Medical Services	Anaheim	462.0625	N/A
Security - Primary	Anaheim	453.950 R	KML-963
Security - Anaheim PD	Anaheim	460.500 R	KUN-399
Stadium Ushers	Anaheim	461.925	N/A
Budweiser Thunder Boat Races			
Security (F-1)	San Diego	151.625	N/A
Security (F-2)	San Diego	154.600	N/A
Rescue & Fire Crews	San Diego	156.300	N/A
Operations	San Diego	464.550	N/A
California Institute of the Arts: 464.3375 / 464.3625 / 464.3875 / 464.4625 / 464.6625 / 464.7125			
Carlsbad Race Track			
Security / Operations	Carlsbad	173.2375 / 173.3625	N/A
Del Mar Fair			
(Itinerate frequencies are used by concessionaires)			
General Operations	Del Mar	464.525 R	WNXE306
Maintenance	Del Mar	151.865	N/A
Security (F-1)	Del Mar	464.750 R	N/A
Security (F-2)	Del Mar	461.225 R	KNGC-205

BUSINESS

AMUSEMENT & SPECIAL EVENTS

NAME	LOCATION	FREQUENCY	CALL SIGN
Del Mar Race Track			
Security (F-1)	Del Mar	151.715	WST-774
Maintenance (F-2)	Del Mar	151.865	KVL-545
Disneyland			
Security Primary	Anaheim	**461.850 R**	WNHY-774
BLACK - Security	Anaheim	461.375 R	WNHY-774
BLUE - Security	Anaheim	464.4625	WNHY-774
BROWN - Maintenance	Anaheim	464.5375	WNHY-774
GREEN - Operations / Railroad	Anaheim	464.4875	WNHY-774
ORANGE - Maintenance	Anahiem	464.4125	WNHY-774
PURPLE - Buses	Anaheim	464.575 R	KAD-2116
SILVER - Concessions	Anaheim	464.0375	WNHY-774
RED - Operations / Rides	Anaheim	464.6375	WNHY-774
WHITE - Trains	Anaheim	464.7625	WNHY-774
YELLOW - Special Events	Anaheim	464.5125	WNHY-774
SP1 - Ride Security	Anaheim	154.570	N/A
SP2 - Submarine Ride	Anaheim	154.600	N/A
Disneyland Hotel - Security	Anaheim	151.745	KNGV-380
Disneyland Hotel	Anaheim	461.7375	KB-92700
Disneyland Hotel	Anaheim	463.6375	KD-47934
Disneyland Hotel	Anaheim	464.9875	KD-47934
Ham Repeater - Matterhorn	Anaheim	146.940 R	WA6BIZ
McDonald's Restaurant	Anaheim	154.600	N/A
Paging Services	Anaheim	154.625	KNGV-380
SMR Trunking System	Anaheim	861-865.5125 T	WNRN-887
Wireless Floats: 194.0, 194.6, 195.2, 195.6, 196.0, 204.0, 205.0, 205.6, 206.0, 206.6, 207.2, 207.6, 208.0			
Gene Autry Western Museum	Los Angeles	467.9125	KD-34514
Grand Central Market (LA):	461.0375 / 461.1125 / 463.5875		
Greyhound Exposition Center	Anaheim	461.325 R	WNQE-823
"	461.425 / 461.750 / 463.450 / 464.450 / 464.600 / 464.725 / 464.825		
Hollywood Park Racetrack			
Operations / Security	Inglewood	154.515	KK-8699
Operations / Security	Inglewood	154.570	KK-8699
Operations / Security	Inglewood	154.600	KK-8699
"	463.4625 / 463.9625 / 464.1125 / 464.1875 / 464.4625 / 464.5125 / 464.5625		
Irvine Amphitheater	Irvine	173.225	N/A
Irvine Amphitheater	Irvine	173.275	N/A
Irvine Amphitheater	Irvine	173.325	N/A
John Paul Getty Center	Santa Monica	464.800 R	WNKE-688
John Paul Getty Museum			
Security / Operations	Malibu	464.675 R	KNDW-808
Security / Operations	Malibu	464.725 R	KNDW-808
Security / Operations	Malibu	464.825 R	KNDW-808
Security / Operations	Malibu	464.950 R	KNDW-808
Knott's Berry Farm			
BLACK - Paging	Buena Park	158.460	KML-654
BLUE - Entertainment	Buena Park	151.685	KML-654
GREEN - Security	Buena Park	151.655	KML-654
PURPLE - Security	Buena Park	151.625	KML-654
RED - Parking	Buena Park	151.745	KML-654
WHITE - Ride Maintenance 1	Buena Park	154.570	KML-654
YELLOW - Ride Maintenance 2	Buena Park	154.600	KML-654
LA Dodger Stadium (Baseball)			
Dugout-Bullpen	Los Angeles	151.625	KT-3730
"	461.4625 / 463.2125 / 464.2875 / 464.4625		

BUSINESS

AMUSEMENT & SPECIAL EVENTS

NAME	LOCATION	FREQUENCY	CALL SIGN
LA Dodger Stadium			
General Stadium Ops	Los Angeles	151.715	KT-3730
Parking Control	Los Angeles	151.745	KT-3730
Stadium Ushers	Los Angeles	154.570	KT-37
Los Angeles Equestrian Center	Los Angeles	464.575 R	N/A
Los Angeles Equestrian Center	Los Angeles	464.675 R	N/A
Los Angeles Equestrian Center	Los Angeles	464.875 R	N/A
Movieland Wax Museum			
General Operations	Buena Park	151.805	WRF-786
Security	Buena Park	151.625	WRF-786
Museum of Contemporary Art	Los Angeles	463.425 R	WNVC-230
Natural History Museum	Los Angeles	464.5125	N/A
Newport Art Museum	Newport Beach	464.3125	
Norton Simon Museum	Pasadena	467.925	KB-70457
Orange County Fairgrounds F1	Orange	464.500	N/A
Orange County Fairgrounds F2	Orange	469.500	N/A
Orange County Raceway	Anaheim	151.625	N/A
Pntages Theater	Hollywood	154.570	N/A
Raging Waters Park	San Dimas	463.9875	KB-71408
Raging Waters Park	San Dimas	464.1375	KB-71408
Raging Waters Park	San Dimas	464.9875	KB-71408
Six Flags Magic Mountain			
General Operations	Valencia	154.515	KTU-627
General Operations	Valencia	151.895	KTU-627
Parking	Valencia	154.570	KTU-627
Ride Maintenance	Valencia	151.715	KTU-627
Security	Valencia	151.665	KTU-627
Security	Valencia	154.540	KTU-625
Security Tactical	Valencia	151.955	KTU-627
S.D. Convention Center			
Security	San Diego	453.900 R	N/A
Shows / Operations	San Diego	453.325 R	N/A
Operations	San Diego	461.175	N/A
Engineering	San Diego	467.0125	N/A
Meeting Services Intl.	San Diego	460.6625R	N/A
Staff Services	San Diego	461.0125	N/A
Staff Services	San Diego	462.0125	N/A
Staff Services	San Diego	463.0125	N/A
Staff Services	San Diego	464.0125	N/A
Audio/Visual Primary	San Diego	463.550	N/A
S.D. Jack Murphy Stadium			
24 Hour Security (F-1)	San Diego	453.800 R	KA-87252
24 Hour Security (F-2)	San Diego	458.800	KA-87852
Concessions	San Diego	461.2875	KB-68829
Concessions	San Diego	461.3625	KB-68829
Concessions	San Diego	461.4375	KB-68829
Concessions	San Diego	461.5125	KB-68829
Event Security (F-1)	San Diego	151.925	N/A
Event Security (F-2)	San Diego	151.955	N/A
Maintenance	San Diego	453.200 R	WNCR-798
Paramedics	San Diego	154.570	N/A
Parking Control	San Diego	154.600	N/A
Parking - Lot A	San Diego	461.6375	KB-32921
Parking - Lot B	San Diego	462.8875	KB-32921
Parking - Lot C	San Diego	464.5125	KB-32921
Parking - Lot D	San Diego	464.5625	KB-32921

Padres Baseball Club: 464.100 / 464.150 / 464.200 / 464.250 / 464.3875 / 464.425 / 464.550

BUSINESS

AMUSEMENT & SPECIAL EVENTS

NAME	LOCATION	FREQUENCY	CALL SIGN
San Diego Sports Arena			
Concessions	San Diego	469.4125	N/A
Parking	San Diego	154.600	N/A
Pro Staff Security	San Diego	467.3875	N/A
San Diego Zoo			
General Operations	San Diego	464.425 / 464.575	KNCR-311
Security	San Diego	453.200 / 453.800	KTV-721
Maintenance	San Diego	154.570	KD-37270
Phone Interconnect	San Diego	461.0125R	KD-44530
San Jacinto Winter Park			
Security & Maintenance 463.475 / 464.425 / 464.525 / 464.575 / 464.975			
Sea World			
Security Primary	San Diego	466.0875	KA-90452
Security Secondary	San Diego	467.0875	KA-90452
Security Secondary	San Diego	464.2125	KA-90452
Agriculture	San Diego	464.4375	KA-90452
Educational Services	San Diego	464.5875	KB-41356
Landscapers (primary)	San Diego	466.6625	KB-41356
Landscapers (secondary)	San Diego	461.6625	KB-41356
Merchandise	San Diego	462.050 R	WNWW-453
Paging Services	San Diego	462.850 R	KNHK-346
Technical Services	San Diego	464.5125	KA-90452
Food Services	San Diego	464.4625	KA-90452
Park Operations	San Diego	462.0625	KA-90452
Productions Department	San Diego	462.9875	KB-42045
New System: 936.200 / 936.225 / 936.2375 / 936.250 / 936.6625 / 936.675 / 936.725 / 937.175			
Universal Studios			
General Operations	Universal City	151.685	WNIN-912
Film Productions	Universal City	152.870	KMK-699
Film Productions - Paging	Universal City	152.930	KMK-699
Film Productions - Paging	Universal City	153.050	KMK-699
Film Productions - Primary	Universal City	152.990	KMK-699
Film Productions - Tactical	Universal City	173.225	KMK-699
Film Productions - Tactical	Universal City	173.375	KMK-699
Ride Maintenance	Universal City	151.895	KNAN-960
Security - Parking Lot	Universal City	464.975 R	WNLY-912
Security - Plaza "Control"	Universal City	464.375 R	WNLY-912
Security - Secondary	Universal City	464.575 R	WNLY-912
Security - Tactical	Universal City	464.825 R	WNLY-912
Studio Tour / Amphitheater	Universal City	152.900	KMK-699
Shows / Special Events	Universal City	152.960	KMK-699
Shows / Special Events	Universal City	451.7625R	N/A
Wild Animal Park			
Security (F-1)	San Diego	453.200 R	KTV-721
Transportation (F-2)	San Diego	453.800 R	KQR-954
Animal Services (F-3)	San Diego	453.350 R	WNFJ-501
(900 MHz System in 1996)			

CABLE TELEVISION

NAME	LOCATION	FREQUENCY	CALL SIGN
A.D.C. Cable	Lawrence Welk Village	461.350 R	N/A
American Cablevision	San Diego	860.500 R	KNEM-219
Cabletronics	Vista	151.895	WPCC-504
Calvidio Cable TV	Alpine	463.450 R	N/A

BUSINESS

CABLE TELEVISION

NAME	LOCATION	FREQUENCY	CALL SIGN
Cox Cable	San Diego	856.600 R	WNLI-362
Country Cable - Portables	San Diego	469.7625	KB-78381
Crown Cable	Azusa	464.525 R	WYF-511
Daniel's Cablevision	Carlsbad	856-860.925 T	SMR
Dimension Cable	Orange County	35.720	N/A
Dimension Cable	Orange County	35.960	N/A
Dimension Cable	San Marcos	860.550 R	N/A
Dimension Cable	San Marcos	863.300 R	N/A
Pacific Cablevision	San Marcos	861-865.750 T	SMR
Southwestern Cable	San Diego	151.805	N/A
Southwestern Cable	San Diego	800 MHz	SMR
Telecable Services	Borrego Springs	151.655	KNDJ-932
Televents of Coronado	Coronado	151.925	N/A
Yuma Cablevision	El Centro	151.745	KBL-655

DISPOSAL SERVICE

NAME	LOCATION	FREQUENCY	CALL SIGN
Edco Disposal Service	Poway	856-860.925 T	WNDK-887
Edco Disposal Service	Poway	856-860.975 T	WNDK-887
Edco Disposal Service	Poway	862.500	WNGK-263
Fallbrook Refuse District	Fallbrook	856-860.925 T	WNIQ-595
Fallbrook Refuse District	Fallbrook	856-860.975 T	WNIQ-595
Mashburn Sanitation	Escondido	463.200 R	N/A
Orange Disposal Service	Orange	854.5625R	WNNE-240
Sanitainer Inc.	San Diego	858.575	KNHH-527
Western Waste	Gardena	854.6875R	WNPW-338
Western Waste	San Diego	856-860.825 T	WNDF-744

HOTELS & INNS

NAME	LOCATION	FREQUENCY	CALL SIGN
Airport Century Inn	L.A.X.	151.715	WNAC-246
Airport Century Inn	L.A.X.	464.675	WNBV-778
Airport Park Hotel	L.A.X.	464.925	WNDV-684
Anaheim Hilton	Anaheim	464.375 R	N/A
Ambassador Hotel	Los Angeles	462.925	WZT-728
American Canyon Hotel	Palm Springs	464.925	WRZ-482
American Canyon Hotel	Palm Springs	464.950	WNFE-753
Atlas Hotels - Operations	San Diego	151.685	KNCV-260
Atlas Hotels - Security	San Diego	464.525	KNFZ-851
Bahia Hotel	San Diego	151.805	KVW-499
Barnabeys Hotel	Los Angeles	471.8125	WIJ-311
Barnabeys Hotel	Manhattan Beach	461.200	WNGA-942
Bayview Plaza Hotel	Santa Monica	464.825	WNFN-663
Bel Age Hotel	Los Angeles	461.1375	N/A
Bel Age Hotel	Los Angeles	463.2375	N/A
Bel Air Sands Hotel	Los Angeles	462.900	WNCD-381
Best Western Inn	Anaheim	154.570	KB-74395
Beverly Hills Hotel	Beverly Hills	151.955	KNCW-425

Beverly Hilton Inn
Operations / Security: 461.075 / 461.975 / 463.425 / 464.175 / 464.3125

BUSINESS

HOTELS & INNS

NAME	LOCATION	FREQUENCY	CALL SIGN
Beverly Plaza Hotel	Los Angeles	464.325 R	KNJF-758
Beverly Plaza Hotel	Los Angeles	465.000 R	KNJF-758
Beverly Plaza Hotel	Los Angeles	464.325 / 465.000	KNJF-758
Beverly Wilshire	Beverly Hills	464.325 R	KSJ-600
Beverly Wilshire	Beverly Hills	464.750 R	KSJ-600
Biltmore Hotel	Los Angeles	464.575	KNHJ-269
Biltmore Hotel	Los Angeles	464.825	WNIG-954
Biltmore Hotel	Los Angeles	464.875	KNIJ-891
Buena Park Hotel	Corona	463.750	KNJW-503
Buena Park Hotel	Corona	464.150	KNGU-914
Bunker Hill Hotel	Los Angeles	462.175	KNEU-414
Burbank Hilton	Burbank	463.850	KNAU-232
Burbank Hilton	Burbank	464.675	KNAU-232
Burbank Hilton	Burbank	464.775	KNAU-232
California Inn	Los Angeles	151.775	WRU-250
Candy Cane Inn	Anaheim	464.875	WHNO-328
Catamaran Hotel	San Diego	151.805	KNFR-934
Century Plaza Hotel	Los Angeles	154.515	KNBU-489
Century Plaza Hotel	Los Angeles	462.800	KIC-587
Century Plaza Hotel	Los Angeles	463.250	WNAH-372
Checkers Hotel	Los Angeles	463.975	WNPM-222
Clarion Hotel	Ontario	464.375	WNPF-431
Cockatoo Inn	L.A.X.	464.675	KB-61692
Cockatoo Inn	L.A.X.	464.925	KNDB-412
Compri Hotel	San Diego	151.955	WNKT-238
Country Side Inn	Newport Beach	151.685	WNCG-260
Courtesy Inn	L.A.X.	151.745	WYF-751
Dana Inn	San Diego	151.745	KNJL-788
Days Inn of America	L.A.X.	464.875	WNDQ-317
Disneyland Hotel (Anaheim)			
Intercom & Paging:	151.745 / 461.7375 / 463.6375 / 464.9875		
Doubletree Hotel (Corona)			
Security	464.175 / 464.225 / 464.275		
Doubletree Hotel	San Diego	856-860.850 T	WNIS-712
Doubletree Hotel	Temecula	464.525 R	N/A
Dragon Gate Inn	Los Angeles	158.460	KNAX-977
Embassy Hotels	San Diego	856-860.950 T	WNIX-351
Embassy Suites	Anaheim	464.375	WNJD-847
Embassy Suites	Covina	463.275	KNBN-262
Embassy Suites	El Segundo	464.325	WNBY-994
Embassy Suites	Irvine	461.1625	KA-88414
Embassy Suites	Oxnard	151.685	WNFR-951
Embassy Suites	Palm Desert	463.975	WNDN-592
Emerald Hotel	Anaheim	464.775	KNIG-640
Emerald Hotel	Anaheim	464.925	WNKF-874
Erawan Garden Hotel	Indio	463.550	KNJF-739
Erawan Garden Hotel	Palm Springs	461.425	WNHB-707
Four Seasons Hotel	Los Angeles	463.275	WNHE-588
Four Seasons Hotel	Los Angeles	465.000	WNLY-378
Four Seasons Hotel	Newport Beach	464.325 R	WNFH-271
Four Seasons Hotel	Newport Beach	464.525	WNFH-272
Frontier Hotel	Los Angeles	151.805	WNFC-421
Gene Autry Hotel	Palm Springs	154.515	KRQ-532
Granada Royale Hotel	Palos Verdes	464.400	WNAB-526
Grand Hotel	Anaheim	464.3875	KJ-4145
Great Pacific Hotels	San Diego	151.685	KNFH-219

BUSINESS

HOTELS & INNS

NAME	LOCATION	FREQUENCY	CALL SIGN
Hacienda Hotel	El Segundo	464.400 R	KNIS-592
Half Moon Inn (San Diego)			
Intercom & Paging	151.835 / 461.6375 / 463.8125 / 464.2375 / 464.5125 / 464.5625		
Hampton Inn	Anaheim	464.7375	KB-76492
Hanalei Hotel	San Diego	154.625	WSM-228
Harbortown Marina Hotel	Ventura	151.895	KNIE-541
Hilton Hotel (Anaheim)			
Maintenance & Security	464.325 / 464.375 / 464.525		
Hilton Hotel (Beverly Hills)			
Security Contract	461.075 / 463.425 / 464.175		
Hilton Hotel	Burbank	463.850	KNGY-340
Hilton Hotel	Burbank	464.675	KNGY-340
Hilton Hotel	Hollywood	464.500	KNIU-410
Hilton Hotel	Irvine	464.375	WXR-427
Hilton Hotel	Irvine	464.675	WNAR-530
Hilton Hotel (Airport Hilton)	L.A.X.	464.325	KNGH-942
Hilton Hotel (Airport Hilton)	L.A.X.	464.925	KNGH-942
Hilton Hotel	Los Angeles	151.685	KNEB-213
Hilton Hotel	Los Angeles	464.875	KNBY-852
Hilton Hotel	Newport Beach	464.375	WXR-427
Hilton Hotel	Palm Springs	151.685	WXS-758
Hilton Hotel	Sherman Oaks	462.925	KNIM-300
Hilton Inn			
Security (F-1)	San Diego	154.570	KA-88447
Security (F-2)	San Diego	154.600	KA-88447
Holiday Inn	Anaheim	151.715	KB-65792
Holiday Inn	Buena Park	463.250 R	KNFA-257
Holiday Inn	Burbank	463.525 R	KNBU-422
Holiday Inn	Fullerton	151.805	KB-45385
Holiday Inn	Hollywood	154.540	KB-33452
Holiday Inn	Irvine	463.300 R	WNEB-516
Holiday Inn	Los Angeles	154.515	KVH-803
Holiday Inn	Los Angeles	154.600	KVH-803
Holiday Inn	Los Angeles Cty	463.650 R	KNAZ-361
Holiday Inn	Ontario	462.1125	KB-45894
Holiday Inn	Pasadena	464.925 R	KNJV-350
Holiday Inn, Embarcadero (San Diego)			
Security Ops	154.515 / 461.300 / 461.575/ 464.325		
Holiday Inn, Montgomery Field	San Diego	464.975 R	WNJS-762
Holiday Inn	Santa Monica	464.475 R	KNBC-248
Holiday Inn	Torrance	464.425 R	KNHR-782
Hollywood Roosevelt	Hollywood	464.175 R	WNFM-200
Hotel Bel Air (Los Angeles)			
Security:	463.725 / 464.375 / 464.525		
Hotel Del Coronado	Coronado	462.825 R	KNJF-726
462.850 / 463.3125 / 463.4875 / 463.6375 / 464.0625 / 464.475 / 464.6375 / 464.975			
Hotel Ibis (paging)	Anaheim	158.460	WNAZ-499
Hotel Intercontinental (San Diego)			
Security Primary 464.425 R			
Security & Maintenance	463.6375 / 463.6625 / 464.1375 / 464.4125		
Hotel Nikko	Los Angeles	462.850 R	WNGR-278
Hotel San Diego			
(F-1) Primary	San Diego	464.325	N/A
(F-2) Secondary	San Diego	469.325	N/A
Howard Johnson's	Anaheim	151.865	KNCF-365
Huntley House Hotel	Santa Monica	151.865	KNFY-219

BUSINESS

HOTELS & INNS

NAME	LOCATION	FREQUENCY	CALL SIGN
Huntley House Hotel	Santa Monica	151.865	KNFY-219
Hyatt at Los Angeles	Los Angeles	151.805	KWR-476
Hyatt House on Sunset	Hollywood	151.655	KNCW-239
Hyatt Regency	Long Beach	464.575	KNFM-723
Hyatt Regency	Long Beach	464.925	KNFM-723
Hyatt Regency	Los Angeles	154.540	KWI-534
Imperial Motel	Los Angeles	151.775	WSX-317
Inland Empire Hotel	Running Springs	464.200	WNAT-592
Inn at the Park (Anaheim)			
Security	151.745 / 464.375 / 464.4375		
Intl. Airport Hotel	L.A.X.	151.745	KNBX-346
Intl. Airport Hotel	L.A.X.	151.805	KNBX-346
Island Resorts	Avalon, San Pedro	464.850	KNFW-289
J.C. Resorts Inn & Golf.	Laguna Beach	151.865	WNBA-963
Jolly Rodger Inn	Anaheim	151.805	WZV-925
King's Inn	San Diego	151.685	KNDV-266
Lake Arrowhead Lodge	Lake Arrowhead	461.025	KNFV-835
La Quinta Hotel	La Quinta	463.475	KNBT-962
La Quinta Hotel	La Quinta	464.425	WNKX-365
Le Baron Hotel	San Diego	154.515	N/A
La Costa Hotel & Spa	Carlsbad	154.600	KA-68941
La Costa Hotel & Spa	Carlsbad	463.500	WNAX-498
La Jolla Coast Motel	La Jolla	464.300	WSU-854
La Jolla Village Inn	San Diego	461.375	KMDL-980
La Valencia Hotel	San Diego	154.515	WZE-942
Le Meridien Hotel	Coronado	461.1375	KB-88907
Little American Westgate	San Diego	151.715	KXK-907
Little American Westgate	San Diego	157.740	KMX-907
Los Angeles Airport	Los Angeles	464.325	KNGH-942
Los Angeles Airport	Los Angeles	464.925	KNGH-942
Manchester House Motel	L.A.X.	151.745	WYF-751
Marriott Desert Springs Resort	Palm Springs	464.675	WNGH-452
Marriott Hotel (Anaheim)			
461.2375 / 461.5375 / 462.575 / 463.300 / 464.250 / 464.375 / 464.425 / 464.475 / 464.675 / 464.9875			
Marriott Hotel	Buena Park	464.0875	KB-86186
Marriott Hotel	El Segundo	464.9375	KB-88378
Marriott Hotel	Indio	462.125	KNCA-661
Marriott Hotel (Irvine)			
Maintenance & Security:	461.300 / 463.575 / 465.000		
Marriott Hotel	La Jolla	464.325	WNBA-445
Marriott Hotel (LAX Airport)			
Security & Intercom	461.0375 / 462.9125 / 464.3375 / 464.450 / 464.825		
Marriott Hotel	Los Angeles	464.375	WRS-568
Marriott Hotel	Marina Del Rey	464.375	KKQ-394
Marriott Hotel	Marina Del Rey	464.950	KKQ-394
Marriott Hotel	Newport Beach	464.250	KNHU-888
Marriott Hotel	Newport Beach	464.925	WXR-427
Marriott Hotel (Palm Springs)			
Security Service	464.575 / 464.925 / 464.975		
Marriott Hotel	Rancho Mirage	464.400	WNBQ-543
Marriott Hotel	Santa Ana	464.9375	KB-88378
Marriott Hotel	Torrance	461.550	WNAC-283
Marriott Hotel	Torrance	464.625	WNAC-283
Marriott Hotel	Woodland Hills	464.525	WNBA-441
Marriott Hotel	Woodland Hills	464.575 R	WNBA-441
Marriott Marina-Security	San Diego	464.375 R	N/A
Marriott Marina-Housekeeping	San Diego	464.925 R	N/A

BUSINESS

HOTELS & INNS

NAME	LOCATION	FREQUENCY	CALL SIGN
Marriott Marina - Houskeeping	San Diego	464.925 R	N/A
Maxim De Paris Hotel	Palm Springs	463.500 R	WNFR-689
Maxim De Paris Hotel	Palm Springs	463.525 R	WNFR-689
Mayfair Hotel	Los Angeles	154.600	KB-54651
Meridien Hotel	Newport Beach	462.875	KNJJ-632
Miramar Sheraton Hotel	Santa Monica	461.325	KNDV-273
Mission Inn	Riverside	463.500	WNCR-924
Mission Inn	Riverside	464.325	KNEK-657
Mountain High Inn	Los Angeles	154.540	KNFT-494
New Otani Hotel	Los Angeles	464.525	WSX-595
Nightlite Inns	San Diego	463.975	KNES-767
Omni Hotel	San Diego	157.740	WNJU-214
Omni Hotel	San Diego	464.675	WNJU-214
Pacifica Hotel	Culver City	151.805	KWC-788
Pacifica Hotel	Culver City	464.525	KB-60631
Pacific Shore Hotel	Santa Monica	461.175	KNAY-503
Pacific Shore Hotel	Santa Monica	461.250	KNAY-503
Pacific Southwest Motels	Chula Vista	461.550	KNAA-329
Palm Desert Resort	Palm Desert	461.025	WNFT-504
Penny Sleeper Inn	Anaheim	154.570	KB-45275
P.G.A. West Resort	La Quinta	464.675	WNKG-220
Plaza Intl. Hotel	Mt. Otay	462.150	WNCB-951
Plaza Intl. Hotel	San Diego	463.625	WNAD-701
Premier Hotel	Universal City	154.515	WNIP-679
Premier Hotel	Universal City	464.525	WNIP-679
Quality Inn	Anaheim	151.685	KB-28351
Quality Inn	Los Angeles	31.040	WNDC-565
Quality Inn	Los Angeles	464.975	KB-73993
Radisson Hotel	Palm Springs	464.875	WNFR-382
Radisson Hotel	Palm Springs	464.925	WNFR-382
Radisson Plaza Hotel	Irvine	464.825	N/A
Radisson Plaza Hotel	Irvine	464.925	N/A
Radisson Plaza Hotel (Manhattan Beach)			
Security Service:	151.805 / 461.825 R / 463.975 R		
Ramada Inn - Security F1	Anaheim	464.475 R	N/A
Ramada Inn - Security F2	Anaheim	464.875	N/A
Ramada Inn	Anaheim	466.2625	KB-85396
Ramada Inn	Beverly Hills	461.225	KNFZ-416
Ramada Inn (Culver City)			
Maintenance & Security	151.925 / 463.525 / 463.975		
Ramada Inn	L.A.X.	464.875	WRT-389
Ramada Inn	Los Angeles	35.940	KCV-558
Ramada Inn	Los Angeles	464.475	WNGP-51
Ramada Renaissance Hotel	Long Beach	463.850	WNGR-939
Ramada Resort Hotel	Palm Springs	461.750	WTS-734
Rancho Bernardo Inn	San Diego	151.685	KNBX-647
Rancho Santa Fe Inn	Rancho Santa Fe	464.9875	KB-34588
Red Lion Inn	Riverside	463.8125	KB-21033
Red Lobster Inn	San Diego	461.3875	KB-83897
Registry Hotel	Irvine	464.825	KNIS-570
Resident Inn	La Jolla	154.515	WNJU-700
Ritz Carlton Hotel	Laguna Niguel	464.100	KNIP-422
Ritz Carlton Hotel	Orange	464.625	KNJA-91
Rodeway Inn	San Diego	157.740	KNJL-495
Rosewood Hotel	Los Angeles	464.575	KB-42226
Royal Quality Inn	San Diego	151.715	KNFH-918

BUSINESS

HOTELS & INNS

NAME	LOCATION	FREQUENCY	CALL SIGN
Sands Hotel	San Diego	463.975	WYE-334
Sea Lodge	La Jolla	151.865	KNDH-248
Seven Seas Lodge	San Diego	461.6875	KNHF-968
Seven Seas Lodge	San Diego	461.900	KNHF-968
Shell Beach Motel	La Jolla	464.800	WNHM-907
Sheraton Grande Hotel	Los Angeles	462.175	KNEU-414
Sheraton Hotel	Newport Beach	464.850	WNHH-717
Sheraton Hotel	Pasadena	464.925	WNAC-251
Sheraton Hotel	Pasadena	464.975	WNAC-251
Sheraton Hotel	Redondo Beach	463.300	WNHS-438
Sheraton Hotel	San Fernando	154.540	KNCF-251
Sheraton Hotel	Universal City	464.775	WXE-660
Sheraton Inn (San Diego)			
Housekeeping & Office:	151.805 / 151.835 / 461.3125		
Sheraton Plaza (Los Angeles)			
Security Service	462.175 / 464.325 / 464.775		
Sheraton Plaza	Palm Springs	151.925	WQC-836
Sheraton Plaza La Reina	Los Angeles	464.775	WQL-417
Sheraton Town House	Los Angeles County	462.925	KNFY-600
Skyways Airport Hotel	L.A.X.	158.460	KNHZ-891
Spa Hotel	Palm Springs	464.475	KHR-304
Sportmens Lodge	North Hollywood	151.775	KNGZ-999
Stouffer Hotels (Los Angeles)			
Security Service:	462.0625 / 464.175 / 464.975 / 465.000		
Summer House Inn (San Diego)			
Security & Housekeeping	464.375 R / 461.6375 / 461.6875 / 464.2875 / 464.5125 / 464.5625 / 464.6875		
Tivoli Hotel	L.A.X.	151.835	KFI-965
Town & Country Hotel	San Diego	465.000	WXT-341
Travelodge Airport Inn	L.A.X.	151.805	KNBY-643
Travelodge at the Park	Anaheim	464.575	WNDC-421
Travelodge	San Diego	151.685	KNFS-739
Turf & Sand Hotel	Laguna Beach	151.775	KB-40907
United Suites	Montrose	461.175	WNBB-881
United Suites	Newport Beach	463.400	WNBB-880
U.S. Grant Hotel (San Diego)			
Security & Room Service:	461.850 / 464.175 / 464.975 / 465.000		
Vacation Bay Hotel	Laguna Beach	151.745	KNBR-315
Vacation Village Resort (San Diego)			
Paging & Security:	151.655 / 157.740 / 461.375		
Vagabond Hotel	San Diego	464.425	WNHE-462
Walkers Caravan Hotel	L.A.X.	151.685	KNEZ-491
Western Intl. Hotel	Los Angeles	464.975	WYZ-846
Westin Bonaventure (Los Angeles)			
Security Service	461.225 / 463.275 / 464.975 / 464.475		
Westin Hotel	Costa Mesa	464.975	KNBM-350
Westin Hotel	Costa Mesa	464.975 R	N/A
Westin Hotel	Los Angeles	464.325	KBH-670
Westlake Plaza Hotel	Westlake Village	464.325	KNHG-666
Westlake Plaza Hotel	Westlake Village	464.375	KNHG-666
Westwood Marquis Hotel	Los Angeles	464.325	WNML-258
Winegardner & Hammons	Santa Ana	464.875	WNFB-248
Wrather Hotel (Anaheim)			
Security& Housekeeping	462.575 / 464.2375 / 464.2875 / 464.3625 464.575		
Wrather Hotel (Long Beach)			
Maintenance & Security	464.375 / 464.775 / 462.800		
Wyndham Hotel	Palm Springs	462.8875	KB-16583
Wyndham Hotel	Palm Springs	463.2125	KB-16583

BUSINESS

INDUSTRIAL RADIO SYSTEMS

NAME	LOCATION	FREQUENCY	CALL SIGN
Allied Chemical Co.	El Segundo	151.835	KXB-878
Alpha Beta Co.	La Habra	464.775 R	KVI-472
American Honda Motors	Gardena	461.350 R	WNCP-201
Anheuser Busch Plant	Van Nuys	153.140	KRH-770
Anheuser Busch Plant	Van Nuys	158.415	KRH-770
Anheuser Busch Plant	Van Nuys	461.9125	KB-8305
Anheuser Busch Plant	Van Nuys	462.2875R	KD-27192
Anheuser Busch Plant	Van Nuys	464.525 R	WNIK-582
Anheuser Busch Plant	Van Nuys	464.775 R	WNIK-582
Anheuser Busch	West Hollywood	463.925 R	WNSD-343
Arco Communications	Carson	852.8125 / 853.9375	WNMD-522
Arco Communications	Long Beach	48.720 / 49.140	KBK-933
Arco Communications	Long Beach	463.725 R	KUL-564
Bee Petroleum Corp.	Glendale	852.2375 R	WXE-953
B.F. Goodwrench Co.	Long Beach	462.525 R	KFC-686
Burroughs Corp.	San Diego	153.185	KNHZ-320
California Institute of Technology	Pasadena	508.1875	N/A
California Institute of Technology	Pasadena	508.2125	N/A
Campbell Industries	San Diego	153.095	KYO-631
Chevron Industries - FD	Culver City	158.310	KBQ-896
Chevron Industries	Culver City	451.550 R	KJC-248
Chevron Industries	Culver City	451.675 R	WNFJ-366
Clean Coastal Waters	Avalon	159.980	KCD-769
Clean Coastal Waters	Avalon	454.000	KDG-703
Crystal Silica	Oceanside	153.020	KNBW-903
G.A. Technologies	San Diego	153.230	KNEV-860
G.A. Technologies	San Diego	153.320	KNEV-860
General Atomic (San Diego):			
464.325, 464.375, 464.425, 464.475, 464.525, 464.575, 464.675, 464.825, 464.975			
General Dynamics			
Paging	San Diego	35.980	KRB-959
Operations	San Diego	462.175 R	KML-679
Police / Fire	San Diego	462.250 R	KTS-415
"	San Diego	462.325 R	KTS-415
"	San Diego	462.400 R	KTS-415
Gillette Communications	San Diego	154.6625	WNCR-209
Gold Fields Mining	Glamis	451.900	WNCD-576
Gold Fields Mining	Glamis	452.000	WNCG-613
Gold Fields Mining	Glamis	464.625	KNBQ-543
Gulf General Atomic			
Security (F-1)	San Diego	153.230	KLM-232
Security (F-2)	San Diego	153.320	KL-3338
Gypsum Communications	Plaster City	151.550	N/A
Gypsum Communications	Plaster City	151.595	N/A
Hawthorne Engine Systems	Escondido	860.400	WNPJ-446
Hawthorne Machinery	Escondido	857.400	WNPJ-446
Hewlett Packard	San Diego	153.350	KLA-943
Hughes Aircraft Co.			
Plant Operations	Carlsbad	35.080	KNIU-869
Plant Operations	El Segundo	451.625 R	KXW-919
Pant Operations	El Segundo	462.450 R	KNB-71
Plant Operations	El Segundo & SD	462.500 R	KNFA-929
Plant Operations	San Diego	463.725	KNEV-210
I.B.M. Research & Development	La Habra	462.825 R	KLR-622
I.T. Hazardous Waste	San Diego	463.450 / 463.750	N/A

BUSINESS

INDUSTRIAL RADIO SYSTEMS

NAME	LOCATION	FREQUENCY	CALL SIGN
I.T.T. Barton Instruments	Industry	462.525 R	KNGA-350
Irwin Industries	Long Beach	508.0875 R	KNQ-710
Ivac Corp.	San Diego	153.380	KNAT-848
Jet Propulsion Lab	Pasadena	463.5625R	KD-39296
Kelco Company	San Diego	153.170	KXU-409
Lockheed Corp. FD	Burbank/Palmdale	158.295	KJJ-234
Lockheed Corp.	Burbank	153.200	KMB-548
Lockheed Corp.	Burbank	462.475 R	KMK-328
Lockheed Corp.	Burbank	936.925 / 937.400	WNIS-343
Lockheed Engineering	San Diego	464.5125	WNGQ-571
Lockheed Engineering	San Diego	464.525	WNGQ-571
Long Beach Container Terminal: 451.600 / 464.475 / 464.575 / 464.675 / 464.825			
Ma Com Govt. Systems	San Diego	158.310	WRT-262
Marilu Petroleum	San Diego	153.335	WNHT-645
McDonnell Douglas Corp.	Culver City	462.200 R	KNDY-653
McDonnell Douglas Corp.	Culver City	508.8875 R	WIE-890
Miller Brewing Co.	Irwindale	451.525 R	WNFR-290
N.C.R. Corp.	San Diego	153.260	KNJP-444
National Science Foundation (Glendale)			
471.0375 / 471.2125 / 471.7875 / 471.9625 / 471.9875 / 472.2875 / 472.3375 / 472.5125 / 472.7625			
National Steel & Ship Bldg.			
Itinerate	San Diego	31.040	KJB-338
Engineering	San Diego	151.715	KNGC-816
Parts / Repairs	San Diego	151.805	N/A
Fire / Police	San Diego	462.250 R	KAR-768
General Operations	San Diego	462.350 R	KAR-768
Paging	San Diego	462.450 R	KAR-768
Workers Portables	San Diego	151.625	KNBR-684
Workers Portables	San Diego	151.655	KNBR-684
Workers Portables	San Diego	154.570	KNBR-684
Northern Telecom	San Diego	153.170	KNDP-422
Northrop Radio Service	El Segundo	462.350 R	KNFP-943
Pepper Industries	San Diego	159.480	WNIY-420
Pharmaseal Laboratories	Irwindale	153.170 / 158.460	WNFF-762
Pharmaseal Laboratories	Irwindale	464.425 R	WRL-980
Phoenix Research Corp.	La Mesa	153.275	WNKR-609
Powerline Oil Co.	Glendale	508.8125R	KFJ-401
R.M.I. Shipyards	National City	464.325	WRF-585
Rockwell Intl. - FD	Downey	153.260	KMG-948
Rockwell Intl.	Downey	461.400 R	WRV-379
Rockwell Intl.	Downey	462.325 R	KNAP-891
Rockwell Intl.	Downey	464.400 R	WNNH-513
Rockwell Intl.	Downey	464.875 R	KSU-676
Rockwell Intl. - FD	Palmdale	464.875	N/A
Rohr Industries			
Fire / Paging	Chula Vista	153.320	KVA-773
Undetermined	Chula Vista	153.365	KLC-591
Undetermined	Chula Vista	158.295	KNAJ-695
Undetermined	Chula Vista	158.310	WRZ-703
Security	Chula Vista	462.375 R	WXU-811
Undetermined	Chula Vista	461.9875	KA-8528
Undetermined	Chula Vista	462.1625	KA-68528
Internal Portables	Chula Vista	72.300	KM-7161
Internal Portables	Chula Vista	72.340	KM-7161
Internal Portables	Chula Vista	72.360	KM-7161

BUSINESS

INDUSTRIAL RADIO SYSTEMS

NAME	LOCATION	FREQUENCY	CALL SIGN
Rohr Industries			
Internal Portables	Chula Vista	72.380	KM-7161
Internal Portables	Chula Vista	72.440	KM-7161
Internal Portables	Chula Vista	72.480	KM-7161
Internal Portables	Chula Vista	72.520	KM-7161
Internal Portables	Chula Vista	72.560	KM-7161
Internal Portables	Chula Vista	72.600	KM-7161
Shell Communications			
Security	Carson	462.250 R	KMH-403
General Ops.	Carson	153.170	KMD-577
General Ops.	Carson	154.625	KZR-914
Operations Trunked	Carson	937.450-937.500 T	WNJD-722
Sinclair Paints	Commerce	508.8125 R	WII-355
Solar Turbines			
Operations	San Diego	153.200	WSU-678
Operations Primary	San Diego	462.200	KZP-543
Undetermined	San Diego	462.300	KZP-543
Sony Corp.	San Diego	451.425	WNKR-612
Sony Corp.	San Diego	154.600	N/A
Southwest Marine	San Diego	451.225	KNBB-601
Sparks Inc.	San Diego	153.230	KSW-859
Sparks Inc.	San Diego	153.260	KNHA-375
T.R.W. Corp.	San Diego	153.350	WNAN-250
Teledyne Industries	San Diego	153.140	KRY-794
Texaco Communications	Culver City	48.520	WBR-672
Todd Pacific Shipyards (San Pedro)			
Workers Portables:	151.625 / 151.655 / 151.685 / 151.715 / 151.745 / 151.805 / 151.835		
"	151.865 / 151.895 / 151.925 / 151.955 / 154.570 / 154.600 / 158.310		
Unisys Corporation	San Diego	158.310	KGH-939
Xerox Corporation (El Segundo)			
	151.775 / 151.865 / 153.140 / 153.260 / 153.335 / 158.460 / 464.825		

MISCELLANEOUS

NAME	LOCATION	FREQUENCY	CALL SIGN
Allied Clinical Labs	San Diego	463.750	WNCZ-485
Association for Retarded Citizens	El Monte	44.500	KNFU-885
Audio Visual at Hilton Towers	Anaheim	464.975	N/A
Avis Rent A Car	Santa Ana	464.975	N/A
Baldridge Racing Service	Lakeside	151.655	WSR-505
Baptist Gardens	Long Beach	154.540	KJI-835
Bell Gardens Bicycle Club	Bell Gardens	463.825 R	WNKY-291
Bluffs Homeowner Assoc.	Newport Beach	151.865	KB-83689
Bonneville Satellite Corp.	San Diego	858.650	KNHK-815
Boy Scouts of America	Long Beach	471.1875 R	WII-282
Bridgestone Tire	Carson	464.425	N/A
Broadway Dept. Stores	Los Angeles	151.625	WNGR-248
Bullock's Dept. Stores	Los Angeles	151.625	KB-68976
Bullock's Dept. Stores	Southern Ca.	151.865	KNBU-810
Cabazon Bingo Hall	Indio	461.025	WNDZ-717
Cabrillo Crane Co.	San Diego	158.400	N/A

BUSINESS

MISCELLANEOUS

NAME	LOCATION	FREQUENCY	CALL SIGN
California Republican Party	Burbank	151.625	KB-72919
California Republican Party	Burbank	151.955	KB-72919
Callaway Vineyard	Temecula	461.0375	KB-86257
Camp Jubilee	Pinion Hills	151.715	WNED-337
Campus Crusade for Christ	San Bernardino	151.745	WNBE-993
Campus Crusade for Christ	San Bernardino	151.775	KWY-785
Carl's Junior Restaurants			
Food Orders VHF	Nationwide	151-174 MHz	N/A
Food Orders UHF	Nationwide	450-470 MHz	N/A
Catalina Canyon	Avalon	464.925 R	WNQV-622
Century Plaza Tower	Los Angeles	464.975 R	N/A
Church of Scientology - Security	San Jacinto	464.775	N/A
Citizen's Agency For Joint Intel.	Southern California	461.4625R	N/A
Citizen's Emergency (LA)			
461.7875 / 461.8375 / 463.800 / 463.825 / 464.125 / 467.1875 / 467.8375 / 468.800 / 468.825 / 469.125			
Commerce Casino (LA)			
464.325 / 464.375 / 464.425 / 464.575 / 464.650 / 464.875 / 464.925			
Coronado Cays Security	Coronado	464.575	WNDA-570
Coronado Shores Council	Coronado	464.675	WNDA-570
Crenshaw Christian Center	Los Angeles	151.865	KNCN-745
Crow Winthrop	Irvine	464.475	N/A
Crow Winthrop	Irvine	464.775	N/A
Crystal Cathedral	Garden Grove	464.575	N/A
Daley Corp.	San Diego	152.870	KMK-843
Daley Corp.	San Diego	865.400	WNLE-793
De Anza Harbor			
Operations Primary	San Diego	464.975 R	KIS-364
Portables	San Diego	464.925	KB-43889
Del Mar Turf Club	Del Mar	151.745	KNBT-608
Downey Family YMCA	Downey	151.685	KMH-400
DRONK Radio Network	San Diego	151.715	WNBE-554
El Cajon Speedway			
Pit Operations	El Cajon	154.570 / 154.600	N/A
Emerald Bay Homeowners Assoc.	Laguna Beach	464.875 R	N/A
Fallbrook Golf Club	Fallbrook	151.895	N/A
Fairbanks Ranch			
Security Primary	Rancho Santa Fe	464.775 R	WNRZ-780
Security Secondary	Rancho Santa Fe	464.375 R	WNRZ-780
Security Tactical	Rancho Santa Fe	464.7375 / 464.8875	WNRZ-780
Ferris Cellular Comm.	San Marcos	856-860.800 T	WNHB-437
Filtrol Corp.	Los Angeles	464.375 R	N/A
Filtrol Corp.	Los Angeles	464.425 R	N/A
Filtrol Corp.	Los Angeles	464.575 R	N/A
Filtrol Corp.	Los Angeles	464.675 R	N/A
Firestone Tire	Rancho Cucamonga	464.375 R	N/A
Forest Lawn Cemetary	Los Angeles	464.875 R	WNDV-651
Friendly Hills Pool & Spa	Whittier	151.865	KNGK-293
Goodwill Industries	San Diego	856-860.825 T	WNDJ-691
Grossmont Gardens	La Mesa	154.515	WNAP-762
Hamilton Cove	Avalon	151.895	WNKC-899
Hertz Rent A Car	Anaheim	464.575 R	N/A
Hillcrest North Apartments	San Diego	154.540	KOI-243
Hillside Memorial Park	Los Angeles	151.835	KEQ-531
Huntington Library	Los Angeles	151.715	KLT-746
International Labor Union 586	San Diego	864.550 R	WNHB-957

BUSINESS

MISCELLANEOUS

NAME	LOCATION	FREQUENCY	CALL SIGN
Jetco Homes (Security)	Victorville	461.025 R	WNGZ-639
Jun's Electronics	Culver City	151.625	KA-82757
La Jolla Seville	La Jolla	151.655	WNAT-612
Lakeside Golf Course	North Hollywood	151.745	WYU-749
Landmark Square	Long Beach	463.275	WNXI-723
Lawrence Welk Village	Escondido	151.805	WNPH-572
Lawrence Welk Village	Escondido	154.515	WNBB-714
Lucky Stores	Buena Park	464.775	KUC-813
Lusardi Construction			
Supervisors	Escondido	856.500 T	WNAS-463
Supervisors	Escondido	857.600 T	WNAS-463
Supervisors	Escondido	858.600 T	WNAS-463
Supervisors	Escondido	859.550 T	WNAS-463
Supervisors	Escondido	860.650 T	WNAS-463
Mar Con Concrete	San Marcos	860.375 R	N/A
Maritime Museum Assoc.	San Diego	151.715	KNAM-425
McDonald's Restaurants			
Food Orders Primary	Nationwide	35.020	N/A
Food Orders Primary	Nationwide	154.600	N/A
Metropolitan Waste	Montebello	861-865.9875 T	WGH-871
Mountain High Ski Area (Wrightwood)			
Sky Patrol & Maintenance:	464.375 / 464.425 / 464.475 / 464.575 / 464.675 / 464.975		
Mount Baldy Ski Patrol	Mt. Baldy	462.9875R	WNSA-298
National Missing Persons Bureau	Los Angeles	151.6125	N/A
National Missing Persons Bureau	Los Angeles	151.6375	N/A
Nelson & Sloan Concrete	Escondido	861-865.650 T	WNJY-888
Nelson & Sloan Concrete	Escondido	861.350	WNGT-858
National Ordnance Co.	San Diego	856-860.825 T	WNDK-893
Normandie Casino	Gardena	464.725	KD-44231
North River Ranch	Santa Ana	158.835	N/A
Nutralite Products	Buena Park	158.460	KCF-819
Oakwood Garden Apartments	San Diego	151.655	KNHM-758
Oakwood Garden Apartments	San Diego	151.715	KNIE-786
Ocean Hills Country Club	Oceanside	464.700	WNLP-939
Ocean Hills Country Club	Oceanside	464.875	WNLP-939
Orange County Rescue Mission	Santa Ana	461.275 R	N/A
Orange County Swap Meet	Orange	462.550	N/A
Oscar Meyer Foods Plant	Los Angeles	151.835	KXW-235
Pacific Boat Racing	La Habra	151.625	KZ-8342
Pacific Coast Paging	San Marcos	152.480	WNBB-919
Palomar Transit Mix	Escondido	856-860.925 T	WNDS-765
Palos Verdes Golf Club	Palos Verdes	154.570	KA-81131
Pauma Valley Country Club	Pauma Valley	154.515	WNCR-485
Pauma Valley Farm	Pauma Valley	857.550	KNGK-469
Penasquitos Village	San Diego	154.540	KWW-795
Pinery Tree Farms	Escondido	452.075 R	WQJ-305
Poway Valley REACT	El Cajon	151.925 / 151.955	WPBE-756
Price Club Stores			
Security Primary	Nationwide	154.600	KD-24138
Security Alternate	Nationwide	154.570	KD-24138
Pro Golf Association	North Hollywood	151.625	KB-24788
Queen Mary - Security	Long Beach	464.375	N/A
Queen Mary - Operations	Long Beach	464.775	N/A
R.E. Hazard Const.	San Diego	856.350	WNGX-933
R.E. Hazard Const.	San Diego	857.500	KNJH-374

BUSINESS

MISCELLANEOUS

NAME	LOCATION	FREQUENCY	CALL SIGN
R.M. Pyles Boys Camp	Valencia	151.715	KLH-551
Racquetball World	Fullerton	151.805	KB-63026
Rancho Santa Fe Golf Club	Rancho Santa Fe	464.475 R	N/A
Regency Card Club	Bell	462.625 / 462.725	WNWW-233
Riker Laboratories	Northridge	151.805	KB-54431
Rolling Hills Community	Rolling Hills	151.775	KNCP-945
Royal Vista Terrace	Durarte	154.540	WNXV-945
Sail America Foundation	San Diego	151.7875	WNNA-377
Sail America Foundation	San Diego	151.9375	WNNA-377
Salt Creek Golf Course	Laguna Niguel	151.715	KB-60671
Salvation Army	San Diego	464.900 R	WXB-341
San Diego Ready Mix	Poway	857.575 R	KNHK-816
San Diego Ready Mix	Poway	859.575 R	KNHK-816
Scottish Rite Center	San Diego	464.950	N/A
Shadowridge Country Club	Vista	154.540	KB-46108
Shrine Civic Center	Los Angeles	151.865	WQU-246
Silicon Systems Inc.	Irvine	464.425 R	N/A
Silicon Systems Inc.	Irvine	464.525 R	N/A
Silicon Systems Inc.	Irvine	464.575 R	N/A
Singing Hills Country Club	El Cajon	462.0375 / 462.8375	KD-20984
Sorrento Ready Mix	San Diego	857.350	WNIL-637
Sports Car Club			
Remote Control: 452.325 / 452.425 / 452.725 / 452.775 / 452.800 / 452.825 / 452.900 / 452.925			
Star Builders Supply	San Diego	858.650	KNHA-572
Stardust Country Club	San Diego	463.275	KQE-605
Stoneridge Country Club	Poway	154.515	KB-77040
Sunrise Ballooning	Simi Valley	151.805	KNAJ-643
Sycuan Bingo Palace	Lakeside	461.050	N/A
The Canine Spa	Cathedral City	151.745	WYS-558
"Tight 10" Offroad Racing	San Diego	151.805	KD-44595
"Tight 10" Offroad Racing	San Diego	157.470	KD-44595
Trowbridge Recreation	Hacienda Heights	462.075 R	KNAZ-817
U.F.O. Video Network	Los Angeles	461.000	N/A
Valet Air Park	L.A.X.	151.835	KNHY-651
Village Oaks	Pasadena	461.100	WNHQ-418
Vista Lock & Key	Vista	858.550	WNFZ-556
Volunteer Assistance Team			
Primary	San Diego	154.515	KZ-8600
Alternate	San Diego	154.600	KZ-8600
Warner Springs Ranch			
Operations	Warner Springs	154.540	KNEP-778
Portables	Warner Springs	464.4875	KB-32398
Portables	Warner Springs	464.5625	KB-32398
Portables	Warner Springs	464.7875	KB-32398
Watts Labor Comm. Action	Los Angeles	151.835	WZY-876
Western Aviation Investigation	Fullerton	464.100 R	N/A
Wilshire Christian Manor	Los Angeles	461.0375	KB-32866
World Trade Center	Los Angeles	151.895	KZK-947
Y.M.C.A. of Los Angeles	Los Angeles	151.775	KA-96520
Y.M.C.A. of Malibu	Malibu	151.955	WNFJ-415
Y.M.C.A. of Pasadena	Pasadena	151.865	KNGK-616
Youth Action For Christ	La Crescenta	461.075	KNJK-539

BUSINESS

MOTION PICTURE, MUSIC & TELEVISION

NAME	LOCATION	FREQUENCY	CALL SIGN
A & M Records	Los Angeles	472.0375	N/A
Burbank Studios			
Fire / Security	Burbank	152.990	N/A
General Ops / Security	Burbank	152.960	N/A
Paging	Burbank	158.460	N/A
Productions	Burbank	152.870	N/A
Productions	Burbank	173.225	N/A
Productions	Burbank	173.325	N/A
Security	Burbank	153.050	N/A
CBS Studio Center	Los Angeles	173.225	N/A
Columbia Pictures			
Security F1	Burbank	464.525 R	WNQU-566
Security / Operations	Burbank	463.600 R	WNQU-566
Productions	Los Angeles	461.0625	N/A
Productions	Los Angeles	461.4625	N/A
Dexter Film Service	Los Angeles	173.225	KMF-614
Fox Television Production			
General Operations	Los Angeles	856-860.3375 T	SMR
General Operations	Los Angeles	856-860.6375 T	SMR
Ops / Productions: 461.1125 / 461.1375 / 461.1625 / 461.2125 / 461.2375 / 461.2625 / 461.2875 / 461.3125, 466.1125 / 466.1375 / 466.1625 / 466.1875			
Security	Los Angeles	153.050	KNJG-435
Security / Transportation	Los Angeles	152.930	KME-515
Productions	Los Angeles	173.325	KNJG-435
Maintenance	Los Angeles	173.375	KNJG-435
Remotes	Los Angeles	455.5875	KJT-464
Hanna Barbera Productions	Hollywood	152.870	WNLX-866
Hanna Barbera Productions	Rancho Palos Verdes	151.865	KD-23453
Kenny Loggins, Inc.	Beverly Hills	173.225	KB-22741
Kenny Loggins, Inc.	Beverly Hills	173.275	KB-22741
Lorimar Productions	Culver City	469.8375	KB-74372
Lorimar Productions	Culver City	471.8375	N/A
Lorimar Productions	Los Angeles	935 MHz	WNMX-540
Motion Picture Coordination	Los Angeles	461.000 / 461.150	N/A
Music Center			
General Ops.	Los Angeles	154.570	KB-36279
General Ops.	Los Angeles	154.600	KB-36279
Security	Los Angeles	462.175 R	WNDC-958
Productions:	461.600 / 463.375 / 463.475 / 464.075 / 464.425 / 464.925		
Modern Film Effects	Los Angeles	508.4125	N/A
Panavision	Los Angeles	861-865.3875	SMR
Paramount Pictures			
Security	Los Angeles	173.225 / 173.275	KJG-259
Operations	Los Angeles	154.540 / 154.600	KO-5802
Productions:	151.625 / 151.925 / 151.955 / 152.300 / 152.360 / 152.420 / 152.930		
"	152.960 / 152.990 / 157.560 / 157.620 / 157.680 / 173.325 / 173.375		
Rainbow West Films	Los Angeles	173.225	N/A
Rainbow West Films	Los Angeles	173.275	N/A
Rainbow West Films	Los Angeles	173.325	N/A
Rainbow West Films	Los Angeles	173.375	N/A
Sound Tronics			
Operations Primary	Burbank	462.075 R	WNIG-657
Productions:	152.870 / 152.900 / 152.930 / 152.960 / 152.990		
"	153.050 / 173.225 / 173.275 / 173.325 / 173.375		

MOTION PICTURE, MUSIC & TELEVISION

NAME	LOCATION	FREQUENCY	CALL SIGN
Spelling-Goldberg Productions	Los Angeles	467.800	KB-65783
Spelling-Goldberg Productions	Los Angeles	467.825	KB-65783
Stephen J. Cannell Productions	Pasadena	461.275 R	WNCE-261
Stephen J. Cannell Productions			
Renegade Series	Southern Calif.	152.930	KB-60987
Teasley Enterprises	Burbank	173.375	KLE-687
Technicolor	North Hollywood	153.020	KUV-252
Warner Brothers Studios:			
General Operations	Burbank	508.3875 R	WIH-897
Productions:	461.0625 / 461.2625 / 461.3125 / 461.5625 / 461.7625 / 461.9125		
"	462.0375 / 463.2875 / 463.4375 / 463.6875 / 463.8625 / 464.1125		
	464.3875 / 464.4625 / 464.5125 / 464.6625 / 464.8625 / 508.3875		
Walt Disney Imagination	Glendale	464.325 R	KNAP-467
Walt Disney Pictures:	461.0375 / 461.1375 / 461.2375 / 461.3375 / 461.4375 / 461.5375		
"	461.6375 / 461.7375 / 461.8375 / 461.9375 / 462.0375 / 462.1375		
	463.3375 / 463.4375 / 464.5375 / 464.7625 / 464.8875 / 464.9125		
	464.9375 / 464.9875		
Walt Disney Productions	Burbank	464.5375	KB-60748
Walt Disney Trunked Systems:	861-865.3375, 861-865.5875, 861-865.7375		
Walt Disney World Productions	Anaheim	464.775 R	WNSP-303

SECURITY SERVICE

NAME	LOCATION	FREQUENCY	CALL SIGN
1st Security	Whittier	464.600 / 464.650	N/A
A1A Security (Chatsworth)			
463.675 / 464.000 / 464.100 / 464.125 / 464.225			
ABA Recovery			
Operations (F-1)	Chula Vista	461.550 R	KNCQ-304
Operations (F-2)	Chula Vista	461.650 R	KNCQ-304
A B L E Security	Indio	463.550 R	WNFM-295
ABM Security			
Security Guards	San Diego	461.125	N/A
Security Guards	San Diego	463.775	N/A
ADT - Sterling Security			
Alarms	LA & San Diego	460.950 R	N/A
Security Guards	San Diego	461.175	N/A
Security Guards	San Diego	464.725	WNIG-206
Security Primary	San Diego	850 MHz	SMR
AFI Security	Los Angeles	464.800 / 464.850	WNFQ-612
AFI Security	Pasadena	461.350 R	WNKN-892
API Security	Culver City	463.700 / 464.600	N/A
API Security (LA)			
460.925 / 460.975 / 461.000 / 463.700 / 464.775 / 464.600 / 507.8125 / 855.7625			
APS - Security Guards	Southern Calif.	464.550	WNSE-854
ARM Security	Pasadena	461.200 R	WNWZ-231
Able Patrol & Guard	San Diego	464.075 R	N/A
Acme Security	Crestline	463.525 R	WNWV-667
Action Security	Orange	463.200 R	N/A
Action Security	San Diego	462.000 R	N/A

BUSINESS

SECURITY SERVICE

NAME	LOCATION	FREQUENCY	CALL SIGN
Admiral Security	Riverside	461.550 R	WNKY-472
Alliance Security	San Diego	463.250 / 800 MHz	N/A
All Safe Security	Los Angeles	508.2625R	WII-733
Allup Security Corp.	San Diego	463.750 R	KNHU-716
Appollo Security	Victorville	461.275 / 464.550	N/A
Argus Diversified	Los Angeles	464.925 R	WNJR-279
Associated Security	LA & Riverside	463.750 R	WNSP-688
Assurance Patrol	San Diego	461.575 R	WXG-535
A Team Security	Malibu	461.550 R	WNNT-932
A Team Security	Malibu	461.650 R	WNNT-932
B.P.S. Guard Service	Beverly Hills	508.2125R	N/A
B.P.S. Guard Service	Mt. Wilson	464.725 R	WNCB-791
B.P.S. Guard Service - HT's	Southern Calif.	461.3625	KA-43530
Baker Protection	Orange	460.900 R	KML-284
Baker Protection	San Diego	865.800 R	WNCE-516
Baker Protection (portables)			
466.1625 / 466.2625 / 466.7875 / 467.0875 / 468.5875			
Balboa Patrol Service (need)			
Beach Security	San Diego	860.600 R	KNIQ-699
Beverly Security			
Security Guards	San Diego	461.450	N/A
Security Guards	San Diego	464.050 R	N/A
Best Security	Los Angeles	463.650 R	WNIL-225
Bridwell Patrol	San Diego	461.375	N/A
Bogosian Security	Campo	464.325 R	N/A
Boyde & Associates	Los Angeles	463.700 / 800 MHz	N/A
Burmuda Dunes Security	Indio	462.025 R	WNWY-479
Burns International	Orange	461.825 R	N/A
Burns International	Southern Calif.	464.600 R	N/A
C& C Security	Ventura	461.375 R	WNQR-966
C.H.S. Coast Protective	Los Angeles	461.225 R	WNPD-768
C & M Fire Investigations	Los Angeles	471.6625R	WII-201
Camacho Security	Thousand Oaks	461.625 R	WNMT-258
Carlsbad Security			
Operations Repeater	Escondido	463.450 R	KNGM-817
Operations Tactical	Escondido	461.5875	KB-70876
Portables	Carlsbad	173.2375	N/A
Portables	Carlsbad	173.2625	N/A
Portables	Carlsbad	173.3625	N/A
Carlson Security	Escondido	464.800 R	KNAW-394
Cavalli Security	San Diego	464.075 R	N/A
Central Bureau of Invest.	Los Angeles	464.950 R	WNND-645
Central Security Patrol	LA / Riverside	471.9625R	KCF-553
Centurion Protection	Los Angeles	463.350 R	KNGN-614
Centurion Protection	Los Angeles	464.700 R	KNJF-722
Circle T Security	Escondido	461.225 R	WNDL-368
Circle T Security	San Diego	464.350 R	WNDL-368
City of Industry Security	City of Industry	151.955	WRX-887
Code Four Protection	Los Angeles	463.5625	WNWP-471
Colombia Security	San Diego	461.725 R	KNIX-493
Consair Certified Security	Buena Park	461.825 R	WNYK-521
Control Security	Corona	461.975 / 463.675	KNCA-228
Control Zone Protection	Corona	464.750 R	WNJD-521
Cornerstone Security	Long Beach	461.575 R	WNNG-546
Coronado Cays Security	Coronado	464.575 R	KNBK-537

BUSINESS

SECURITY SERVICE

NAME	LOCATION	FREQUENCY	CALL SIGN
Crime Prevention Specialists	Los Angeles	463.500 R	KNAZ-926
Curtin Security	Los Angeles	800 MHz	SMR
Deans Security	San Pedro	464.400 R	KNGK-891
De Anza	San Diego	462.100 R	N/A
De Anza	Winterhaven	464.175 R	KNCV-666
Double Eagle Patrol	San Diego	461.900 R	KNCK-594
Double Eagle Patrol	San Diego	463.500 R	KNCK-594
Duber Industrial	Los Angeles	464.775 R	KNHT-534
Duber Industrial	San Diego	856-860.825 T	SMR
Eagle Security	San Diego	856-860.900 T	SMR
Elite Patrol & Guard	Orange	461.625 R	KNHM-351
Emerald Bay Security	Orange	464.875 R	KNGH-949
Evan's Security	Long Beach	464.775 R	KNDG-546
Excel Security	Los Angeles	464.950 R	KNFV-932
Eye in the Sky Security	Orange	461.475 R	WNHK-472
First Security	Los Angeles	464.450 R	WNSA-724
Forest Security	Los Angeles	463.900 R	WNUW-956
G.H.G. Security	Pasadena	463.750 R	KNEM-840
G.M.I. Security			
Operations Primary	San Diego	461.675 R	WYS-277
Portables	San Diego	461.6375	KB-38903
Portables	San Diego	464.2375	KB-38903
Portables	San Diego	464.2625	KB-38903
Portables	San Diego	464.5125	KB-38903
Portables	San Diego	464.5625	KB-38903
General Security	Palos Verdes	472.2875R	WIG-920
General Security	Torrance	861-865.9875 T	N/A
Gibralter Security	Los Angeles	472.2875R	WIJ-905
Gold Star Security	Los Angeles	151.835	WNVI-325
Golden West K-9	Chatsworth	461.050 R	WNJX-577
Golden West K-9	Chatsworth	461.150 R	WNRo-464
Golden West K-9	La Crescenta	471.2375R	WIH-991
Golden West K-9	Pasadena	464.075 R	WNJU-231
Guardian Angels	151.625 / 154.540 / 154.570 / 154.600 / 170.220		
Hayes Security	Los Angeles	462.150 R	WNQU-721
Hayes Security	Los Angeles	463.575 R	WNQU-721
Hayes Security	Southern Calif.	464.625 R	WNSS-713
High Desert Industrial Security	Victorville	151.865	WNPW-511
Hollywoodland Patrol	Hollywood	31.200	KNJM-652
Holmquist Security	San Diego	464.900 R	N/A
Hoover Protection	Los Angeles	464.150 R	KNHW-344
Inter-Con Security	Los Angeles	463.725 R	KNFI-322
International Security	Los Angeles	861-865.5125 T	KNJF-225
International Protection	Running Springs	464.025 R	WNRD-748
J.M.K. Security (need)			
Jack Kalman & Associates	San Diego	461.550 R	KNGD-656
Jack Kalman & Associates	San Diego	463.725 R	KNIS-538
K-9 Sentry Dogs	Orange / Riverside	461.150 R	WNWR-707
Knight Security	Escondido	463.375 R	N/A
Knight Security	Orange	461.950 R	N/A
Kojack Security	Orange	464.150 R	N/A
L.& A. Protection	Orange	462.025 R	WNIR-255
Mariman Security - Portables	Los Angeles	151.625	KA-70215
Master Security	Escondido	464.800 R	KNCU-695
Master Security	Los Angeles	861-865.2125 T	N/A

BUSINESS

SECURITY SERVICE

NAME	LOCATION	FREQUENCY	CALL SIGN
Mercury Security	Los Angeles	464.425 R	WNJ-585
Metro Coast Investigations	Riverside	461.500 R	WNSQ-490
Mission Viejo Security	Mission Viejo	508.2125 R	KSR-477
Multiple Assignment on 467.7875			
(Hughes Research Labs, Newport & Orange Security, Personal Protection Srv, Strong Arm Protection)			
Multiple Assignment on 467.8375			
(Protective Service Corp., Smith Security, Strong Arm Protection, Target Stores Security)			
National Guardian	Los Angeles	863.1125 / 864.1125	WNAF-902
Nightwatch	Escondido	464.250 R	WNIM-881
Nims Security	San Diego	464.150	WNDC-402
Northeastern Security	San Diego	856-860.950 T	SMR
Odona Central	Los Angeles	461.875 R	KNFj-990
Omni Mega	San Diego	461.925 R	WNAH-495
Orange County Patrol	Orange	471.5875	WIH-245
Pacific Security	San Diego	461.550 R	KNIL-438
Pala Security	Pala	154.570	KB-79112
Pedus Security	Los Angeles	463.975 R	KKT-647
Pedus Security	San Diego	461.475 R	KNCC-741
Pedus Security - Portables	Southern Calif.	151.625	KA-96276
Pedus Security - Portables	Southern Calif.	154.600	KA-96276
Peninsula Security	463.2152 / 466.0625 / 466.2125		
Pinkerton's			
Operations North SD	Escondido	461.825 R	KNJK-909
Operations South SD	San Diego	464.200 R	KNHF-483
Plant Portables	San Diego	151.625 / 151.655	KNEF-251
"	461.475 / 461.800 / 461.950 / 935 MHz		
Pinkus Security	Los Angeles	463.550 R	WNME-563
Point Loma Security	San Diego	461.875 R	KNDD-420
Professional Security	Escondido	861-865.650 T	WNPF-473
Pro-Tect Professional	San Diego	464.200 R	KNJU-796
Pro-Tect Professional	Southern Calif.	151.775	KA-94273
Pro-Tect Professional	Southern Calif.	151.925	KA-94273
Pro-Tect Professional	Southern Calif.	151.955	KA-94273
Pro-Tect Professional	Southern Calif.	154.515	KA-94273
Protectors Security	San Diego	462.125 R	KNCY-442
Rampart Security	Victorville	461.600 R	N/A
Rancho Santa Fe Security			
Portables	Encinitas	151.745	KB-64484
Portables	Encinitas	154.600	KB-64484
Primary	Encinitas	850 MHz	N/A
Ranger Security	Orange	461.950	N/A
SPS Security	Los Angeles	861-865.0875 T	N/A
Saddleback Valley Security	Mission Viejo	462.025 R	WNIU-379
Santec Security	San Diego	464.600 R	N/A
Scola Security	Oat Mountain	463.900 R	WNVM-473
Security Bureau	Palm Springs	461.975 R	WNWY-699
Security Dispatch	Pasadena	464.275 R	WNPT-729
Security Enforcement	Los Angeles	461.125 / 462.025	WNWF-374
Security Response	Los Angeles	508.3375 R	WIK-279
Sentinel Security	San Clemente	463.600 R	KNCH-343
Sepac Security	Long Beach	464.050 R	WNJZ-648
Sepac Security	Los Angeles	935.650-935.750 T	WNMY-508
Shield Security	El Toro	461.475 R	WNWB-695
Shield Security	Los Angeles	853.4875R	WNSX-408

BUSINESS

SECURITY SERVICE

NAME	LOCATION	FREQUENCY	CALL SIGN
Shield Security	Pasadena	461.275 R	WNUM-913
Shield Security	Rolling Hills	464.400 R	WNVS-405
Shield Security	San Diego	462.150 R	WNGP-304
Shield Security	San Diego	461.200 R	WNGP-304
Shield Security	San Diego	464.9125R	WNGP-304
Sierra Security	Los Angeles	464.675 / 464.875	WNJR-882
Silverlake Security	San Bernardino	463.800 R	N/A
Silverton Security	Los Angeles	508.1875R	WII-742
Spartan Security	San Diego	464.900	KNEK-52
Spectaguard, Inc.	Long Beach	461.850 R	WNPI-507
Stanley Smith Security	Los Angeles	935 MHz	SMR
Stanley Smith Security	Orange County	464.750 R	N/A
Stanley Smith Security	Southern Calif.	151.835	KB-29522
Stanley Smith Security	Southern Calif.	464.550	KB-90584
Strategic Security	San Diego	151.925	N/A
Strong Arm Protection	Escondido	859.625 R	KNGD-890
Sun Security	San Diego	464.050 R	WSK-338
Tandem Security	Los Angeles	461.725	KNFY-862
Thompson Security	Riverside	461.500 R	WNEC-789
Thorton Security	Banning	461.875 R	WNXK-851
Tom Reddin Security	Crestline	461.800 R	WNXN-583
Tom Reddin Security	Los Angeles	461.475 R	WNMI-722
Tom Reddin Security	Los Angeles	461.950 R	WNMI-722
Tom Reddin Security	Los Angeles	507.2875R	KNAO-278
Tom Reddin Security - HT's	Southern Calif.	463.7625	KB-72420
Topp's For Crowd Control	Los Angeles Area	151.625	KA-78324
Tri Star Security	Orange	461.025	WNDF-341
Trol Security	R. Palos Verdes	151.565	N/A
Trojan Security	Los Angeles	507.8125R	WIG-912
Turner Investigations	Los Angeles	464.650 R	WNGJ-765
USPS Security	Los Angeles	463.700 R	WNIC-618
United International Security	Los Angeles / Riverside	461.100 R	KNAR-299
United International - Portables	Southern Calif.	463.225	KA-50475
United Security Industries	Los Angeles	472.3625R	WIG-710
United Security Industries	Southern Calif.	461.0125	KB-63938
United Security Industries	Southern Calif.	466.0125	KB-63938
Universal Protection	Los Angeles	854.3875R	WNWB-482
Universal Protection	Northridge	461.900 R	WNHA-358
Universal Protection	Santa Ana	461.150 R	WNKC-896
Portables:	463.9625 / 463.9875 / 464.0375 / 464.3375		
U.S. Guards Co.	Pasadena	464.075 R	WYB-544
U.S. Guards Co.	Southern Calif.	464.5125	WYB-544
VIP Security	Northridge	464.200 R	WNMI-316
Virgill Patrol	Chatsworth	938.6875-938.750 T	WNQJ-797
Vista Patrol	North SD County	151.715	N/A
Wackenhut Corp. (CIA)	Nationwide	467.700	N/A
Wackenhut Corp. (CIA)	Nationwide	467.800	N/A
Waddingtons Security	San Diego	154.515	N/A
Wells Fargo Guard Service	Palmdale	464.025 R	WNSE-690
Wells Fargo Guard Service	Southern Calif.	151.805 / 154.600	N/A
West Coast Detectives	Los Angeles	464.450	KON-611
Westec Security	Escondido	859.475 R	KNDN-331
Western Security	Los Angeles	461.625 R	WNWT-805
Western Security	Los Angeles	464.775 R	WNWT-805

BUSINESS

SHOPPING CENTERS

NAME	LOCATION	FREQUENCY	CALL SIGN
Anaheim Plaza	Anaheim	463.775 R	N/A
Barstow Mall	Barstow	469.325 R	N/A
Beverly Mall	Beverly Hills	464.425 R	N/A
Brea Mall	Brea	464.375 R	N/A
Buena Park Mall	Buena Park	464.375 R	N/A
Buena Park Mall	Buena Park	464.525 R	N/A
Buena Park Mall	Buena Park	464.575 R	N/A
Buena Park Mall	Buena Park	464.925 R	N/A
Carson Mall	Carson	464.675 R	N/A
Century City Mall	Century City	461.0125	N/A
City Shopping Center	Orange	463.625 R	N/A
Clairemont Square	San Diego	461.550	N/A
Del Norte Plaza	Escondido	154.600	N/A
Eagle Rock Plaza	Eagle Rock	464.525 R	N/A
El Camino Real	Carlsbad	464.675 R	WPAP-270
Escondido Mall	Escondido	151.865	KNGT-436
Esplanda Mini Mall	Oxnard	151.895	N/A
Esplanda Mini Mall	Oxnard	463.225 R	N/A
Fallbrook Mall	Canoga Park	464.825 R	N/A
Fashion Island - Security F1	Newport Beach	461.825 R	N/A
Fashion Island - Security F2	Newport Beach	461.900 R	N/A
Fashion Island - Security F3	Newport Beach	464.325 R	N/A
Fashion Valley F1	San Diego	464.925 R	N/A
Fashion Valley F2	San Diego	464.825 R	WST-605
Fox Hills Mall	Culver City	461.5875	N/A
Gateway Plaza	La Mirada	461.125 R	KNIA-521
Glasshouse Square	San Diego	154.570	N/A
Glendale Galleria Mall	Glendale	464.425 R	N/A
Glendale Galleria Mall	Glendale	464.475 R	N/A
Glendale Galleria Mall	Glendale	482.050 R	N/A
Glendale Mall (F-1)	Glendale	464.425 R	N/A
Glendale Mall (F-2)	Glendale	464.475 R	N/A
Grand Central Market	Los Angeles	461.0375	N/A
Grand Central Market	Los Angeles	461.1125	N/A
Grossmont Center	La Mesa	464.475 R	N/A
Hawthorne Plaza	Hawthorne	461.375 R	N/A
Hemet Valley Mall	Hemet	154.600	N/A
Horton Plaza (F-1)	San Diego	464.825 R	WNDL-272
Horton Plaza (F-2)	San Diego	464.375 R	WNDL-272
Indio Fashion Mall	Indio	464.3875	N/A
Inland Center Mall	San Bernardino	464.525 R	N/A
Janss Mall	Santa Monica	464.9125R	WNVQ-544
Kings Mall	Newport Beach	151.865	N/A
Laguna Hills Mall - Security	Laguna Hills	461.700 R	N/A
Laguna Hills Mall - Maintenance	Laguna Hills	461.7625	N/A
Laguna Hills Mall	Laguna Hills	464.975 R	N/A
La Brea Mall	La Brea	464.375 R	N/A
La Jolla Village Square	San Diego	464.475 R	N/A
Lakewood Center Mall - Primary	Lakewood	464.325 R	N/A
Lakewood Center Mall - Tactical	Lakewood	464.3125	N/A
Laurel Plaza - Phone Patches	North Hollywood	464.675 R	N/A
Long Beach Plaza	Long Beach	464.775 R	N/A
Los Cerritos Mall	Cerritos	464.525 R	N/A
Main Place - Security	Santa Ana	464.325 R	N/A
Marina Pacifica Mall	Long Beach	154.515	N/A

BUSINESS

SHOPPING CENTERS

NAME	LOCATION	FREQUENCY	CALL SIGN
Mira Mesa Shopping Center	San Diego	461.8625	KD-43643
Mira Mesa Shopping Center	San Diego	464.9125	KD-43643
Mission Valley	San Diego	463.4875	N/A
Mission Viejo Mall	Mission Viejo	464.975 R	N/A
North Coast Village	San Diego	151.805	N/A
North County Fair - Security	Escondido	464.675 R	WNDU-893
North County Fair - Maintenance	Escondido	461.1625	N/A
Oceanside Swap Meet	Oceanside	155.800	N/A
Old Town Mall	Torrance	464.5125	N/A
Orange Mall	Orange	464.975 R	N/A
Orange Swap Meet	Orange	464.550	N/A
Palm Springs Mall	Palm Springs	464.825 R	N/A
Parkway Plaza	El Cajon	464.875 R	N/A
Pasadena Center	Pasadena	464.525 R	N/A
Pasadena Center	Pasadena	508.4125R	N/A
Pasadena Gateway Plaza	Pasadena	464.525 R	N/A
Plaza Bonita Center	National City	464.825 R	KNBJ-308
Plaza Camino Real	Carlsbad	464.475 R	WZQ-503
Plaza Camino Real	Carlsbad	464.675 R	WPAP-270
Promenade Mall	Woodland Hills	154.540	N/A
Puente Mall	Puente Hills	464.975 R	N/A
Redlands Mall	Redlands	464.475 R	N/A
Redlands Mall	Redlands	464.975 R	N/A
Riverside Plaza	Riverside	463.2625	N/A
Riverside Plaza	Riverside	463.2875	N/A
Riverside Shopping Center	Riverside	151.835	N/a
Saint James Plaza	Imperial Beach	154.515	WNBE-239
Santa Anita Fashion Park	Arcadia	464.525 R	N/A
Seaport Village	San Diego	464.4625	KA-81139
South Coast Plaza - Security	Costa Mesa	464.425 R	N/A
South Coast Plaza - Security	Costa Mesa	464.475 R	N/A
South Coast Plaza - Phone Patches	Costa Mesa	464.675 R	N/A
South Coast Plaza - Security	Costa Mesa	464.825 R	N/A
Stonewood Shopping Center	Los Angeles	464.6875R	KB-93055
The City Mall	Anaheim	463.625 R	N/A
Topanga Plaza	Canoga Park	154.540	N/A
Tyler Mall - Security F1	Riverside	464.375 R	N/A
Tyler Mall - Security F2	Riverside	464.775 R	N/A
Tyler Mall	Riverside	464.4125	N/A
Tyler Mall	Riverside	464.4625	N/A
Tyler Mall	Riverside	464.5375	N/A
University Town Center	San Diego	464.875 R	N/A
Victorville Mall	Victorville	464.925 R	N/A
Webster's Shopping Center	Altadena	464.150 R	N/A
Webster's Shopping Center	Altadena	464.200 R	N/A
West Covina Fashion Mall	West Covina	461.375 R	N/A
Westland Shopping Center	Los Angeles	464.925 R	N/A
Westminster Mall - Security	Westminster	464.375 R	N/A
Whitwood Center	Whittier	464.975 R	N/A

GENERAL MOBILE RADIO SERVICE

CHANNEL ALLOCATIONS

CHANNEL	FREQUENCY
CH-1 - Repeater Operations	462.550
CH-2 - Repeater Operations	462.575
CH-3 - Repeater Operations	462.600
CH-4 - Repeater Operations	462.625
CH-5 - Repeater Operations	462.650
CH-6 - Repeater Operations	**462.675 ***
CH-7 - Repeater Operations	462.700
CH-8 - Repeater Operations	462.725
Low Power Simplex	462.5625
Low Power Simplex	462.5875
Low Power Simplex	462.6125
Low Power Simplex	462.6375
Low Power Simplex	462.6625
Low Power Simplex	462.6875
Low Power Simplex	462.7125

* This frequency is used nationally by R.E.A.C.T. (Radio Emergency Associated Communications Team). Listed below are R.E.A.C.T. organizations in the Southern California area.

TEAM NAME	LOCATION
Antelope Valley REACT	Lancaster
Coachella Valley REACT	Palm Desert
Comupac REACT	Montclair
Corona-Norco REACT	Corona
Hidden Valley REACT (disbanded)	Escondido
Inland Valley REACT	Lake Elsinore
Las Vegas REACT	Las Vegas, NV
LEMSAR REACT	Lake Elsinore
Long Beach REACT	Long Beach
Mid Valley REACT	Fresno
Mount Crest REACT	N/A
Orange County REACT	Anaheim
Poway Valley REACT	Poway
Rio Hondo REACT	Montebello
Riverside REACT	Riverside
Roadrunner REACT	Palm Springs
Saddleback REACT	Trabuco Canyon
San Gabriel Valley REACT	San Gabriel
Santa Monica REACT	Santa Monica
Seacoast REACT	Costa Mesa
South Gate REACT	South Gate
Southwestern REACT	San Diego
Tri-City REACT	Lakewood

LAND TRANSPORTATION SERVICE

ARMORED CAR TRANSPORT

NAME	LOCATION	FREQUENCY	CALL SIGN
Armored Transport Intl	Riverside	159.930	KVV-685
Armored Transport Intl	San Diego	452.775 R	KZC-582
Brinks Inc.	Chula Vista	44.200	KNAU-498
Brinks Inc.	Southern Calif.	467.0125 / 467.8125	KD-28581
Loomis Armored	Los Angeles	508.6375 R	WIG-462
Loomis Armored	San Diego	856-860.825 T	WNDK-892
Wells Fargo Armored	Los Angeles	463.725 R	WNMX-313

BUS LINES

NAME	LOCATION	FREQUENCY	CALL SIGN
Airport Stage Lines (Esondido) 856.500 / 857.600 / 858.600 / 859.550 / 860.650 (trunked)			
Angeltown Charter Lines	Burbank	508.6375R	KNP-517
Antelope Valley Buses	Lancaster	43.800	KNAB-693
Bayshore Transit	National City	452.800 R	WNLG-261
Brock Bus Lines	Montrose	854.3375R	WXK-350
California Paratransit	San Diego	856.375 R	KTL-687
California Stage Lines	San Diego	856-860.775 T	WNGM-922
Century Transit Co.	Burbank	508.9875R	WCI-255
Community Bus Lines	La Canada	463.525 R	WNAD-484
Community Transit	San Diego	463.275 R	WSA-428
Community Transit	San Diego	464.275 R	KNIY-314
Community Transit	San Diego	800 MHz	SMR
Coronado Livery	Coronado	151.745	KNAE-701
East County Transit	Chula Vista	463.425 R	KNJB-929
Goodall's Chartered	San Diego	861-865.650 T	WNUT-734
Greyhound Lines	Los Angeles	856-860.9375 T	WNFW-400
Greyhound LInes	Los Angeles	936.6625	WNNL-394
KO-AM Transportation	San Diego	464.600 R	KNIU-642
Laidlaw Transit	La Jolla	464.850 R	KNFV-795
Laidlaw Transit	La Jolla	858.550 R	KNEM-220
Laidlaw Transit	Los Angeles	853.1625R	WDM-813
Laidlaw Transit	Los Angeles	853.2125R	WDM-813
Laidlaw Transit	Los Angeles	853.2375R	WDM-813
Laidlaw Transit	Los Angeles	938.6375-938.675 T	WNIS-851
Mark IV Charter Lines	Los Angeles	853.7125R	WXR-591
Mark IV Charter Lines	Los Angeles	935.750	WNIT-933
Mark IV Charter Lines	Los Angeles	937.425	WNJP-421
Mark IV Charter Lines	San Diego	861.450 R	WNGW-502
Mexicoach	San Diego	860.400 R	KNER-631
Mexicoach	San Diego	861-865.650 T	WNJY-867
Moonlight Transit Corp.	Chatsworth	31.160	WNLV-639
North County Transit District			
Bus Operations	North SD County	856-860.850 T	SMR
Bus Operations	North SD County	860.350 R	KYF-381
Bus Operations	North SD County	860.275 R	KYF-381
Lifeline Paratransit	North SD County	462.725 R	KAD-7496
Remote Link	Escondido	452.6625	KB-66036
Remote Link	Escondido	452.7125	KB-66036
Road Maintenance	Escondido	452.7375	KB-66036
Yard Maintenance	Escondido	452.8125	KB-66036

LAND TRANSPORTATION SERVICE

BUS LINES

NAME	LOCATION	FREQUENCY	CALL SIGN
Omni-Trans			
Bus Operations	Montclair	452.625 R	WNLI-811
Bus Operations	Montclair	452.650 R	WNFD-469
Bus Operations	Montclair	43.760	WNMC-417
Bus Operations	Montclair	44.460	WNMJ-385
Bus Operations	Montclair	44.540	WNMJ-386
Bus to Bus	Southern Calif.	44.500 / 44.520	KB-31811
Orange County Transit			
Buses Trunked	Orange County	856-860.4875 T	KNCM-802
Dial A Ride	Orange County	508.4625	WIJ-588
Dial A Ride	Orange County	508.5125	WIJ-588
Dial A Ride	Orange County	508.6625	WIJ-588
Dial A Ride	Orange County	508.7125	WIJ-588
Paul's Line	San Diego	151.865	WNMA-331
Paul's Line	San Diego	462.100	WNMV-622
Peerless Shuttle	San Diego	856-860.825 T	SMR
Peerless Shuttle	San Diego	856-860.950 T	SMR
Riverside Transit Agency	Riverside	452.450 R	KNAX-337
Riverside Transit Agency	Riverside	452.650 R	KNAX-337
Robert's Holiday Line	Los Angeles	851.1875R	KGE-571
San Diego Airporter	San Diego	856-860.750 T	WNIX-760
San Diego Express	San Diego	857.500	KNJU-861
San Diego Mini Tours	San Diego	858.650	KNIV-795
San Diego Mini Tours	San Diego	856-860.875 T	SMR
San Diego Transit District			
Trunked Operations	San Diego	856-860.325 T	KTL-687
Sundance Stage Lines			
Operations SD North	Mt. Palomar	856-860.975 T	SMR
Operations SD Central	Mt. Woodson	856-860.925 T	SMR
Operations SD Downtown	Mt. Otay	856-860.750 T	SMR
Sunline Transit Agency	Indio	461.325 R	WNWI-654
Sunline Transit Agency	Palm Springs	453.200 R	WXB-756
Super Shuttle Intl.			
Operations Primary	Los Angeles	800 MHz	SMR
Operations Los Angeles	Los Angeles	508.0125 R	KVX-374
Operations Pasadena	Pasadena	856.4125 R	KNHA-498
Operations San Diego	Mt. Palomar	856-860.725 T	SMR
Taylor Bus Service			
Operations Primary	San Diego	155.235	WSW-736
Operations Mobiles	San Diego	155.220	N/A

RAILROADS

AMTRAK CORPORATION

CHANNEL	FREQUENCY
Road - Los Angeles to San Diego	160.650
Road - Los Angeles to San Luis Obispo	161.550
Administrative & Police*	**161.055**
Train end-to-end	160.455
Yard Operations - Los Angeles Depot*	160.455

*Tone: 146.2 Htz

LAND TRANSPORTATION SERVICE

RAILROADS

ATCHISON TOPEKA & SANTA FE RAILWAY

CHANNEL	FREQUENCY
Road Channels	
Lambert to San Bernardino	161.190
Los Angeles to San Bernardino	160.560
Los Angeles to San Diego	160.650
Barstow Yard	
Calling East	**160.560**
Calling East	160.275
Calling East	160.530
Calling East	160.770
Calling West	**161.370**
Calling West	161.280
Calling West	161.340
Carmen	160.440 R
Carmen	452.900 R
Maintenance	160.980
Mini hump	160.305
Hobart Yard	
Carmen	161.070 R
Ramps	161.190
PBX Telephones	
P2 - Los Angeles Metro	160.245 R
P2 - National City	160.245 R
P3 - Sierra Peak	160.260 R
P4 - Crestline	160.335 R
P4 - Oceanside	160.335 R
P5 - Mojave to Four Corners	160.425 R
Maintenance of Way	161.010
Railroad Police	161.205 / 161.220
Railroad Police - San Diego	160.650
Yard Channel (in use 20 yrs)	161.370

PACIFIC SOUTHWEST RAILWAY MUSEUM

CHANNEL	FREQUENCY
Operations Campo	**160.455**
Operations Secondary	160.530

SAN DIEGO & IMPERIAL VALLEY RAILWAY

CHANNEL	FREQUENCY
(F-1) Operations Primary	**161.505 R**
(F-2) Operations Campo	160.455
(F-3) San Diego Trolley - South Line	160.665 R
(F-4) San Diego Trolley - East Line	160.380 R
(F-5) ATSF Railway	160.650
(F-6) Switching Primary	161.505
(F-7) Operations Campo	160.530

LAND TRANSPORTATION SERVICE

RAILROADS

SAN DIEGO TROLLEY

CHANNEL	FREQUENCY
(F-1) Operations South Line	**160.665 R**
(F-2) Operations East Line	**160.380 R**
(F-3) Yard Personnel	**160.710 R**
(F-4) Central & Special Units	**160.350 R**
(F-5) Security Personnel	**161.565 R**
Mt. Otay Link	463.5875

Tone: 131.8 Htz

Trolley Identification

100's	Administration	370's	Special Enforcement Detail
200's	Transportation Supervisors	400's	Maintenance / Shops
300's	Security-South	500's	Code Compliance Inspectors
320's	Security-Metro	600's	Special Units
330's	Security-East	700's	Board of Directors
350's	Yard Security	800's	Yard Operations
360's	Undercover Security		

SOUTHERN CALIFORNIA METROLINK

CHANNEL	FREQUENCY
(F-1) Road - Mission Tower to Claremont	161.415
(F-2) Yard Channel (Taylor Yard)	160.815
(F-3) Road - Mission Tower to Burbank Junction	160.545
(F-4) Maintenance of Way	160.995
(F-5) Southern Pacific Road - North of Burbank Junction	161.550
(F-6) Southern Pacific Road - South of Burbank Junction	161.550
(F-7) Union Pacific Road	160.515
(F-8) ATSF Road - LA to San Bernardino	160.560
(F-9) ATSF Road - LA to San Diego	160.650
(F-10) ATSF Road - Lambert to San Bernardino	161.190
(F-11) Amtrack Yard Operations	160.455

SOUTHERN CALIFORNIA METRORAIL*

CHANNEL	FREQUENCY
New System 1995	939.450 T
New System 1995	939.4625 T
New System 1995	939.475 T
New System 1995	939.4875 T
New System 1995	939.500 T
Sheriff - MetroRail Blue Line	**484.1625R**

*The installation of the 900 MHz system has been delayed. Please see page 66 for the information regarding the current MetroRail radio system.

LAND TRANSPORTATION SERVICE

RAILROADS

SOUTHERN PACIFIC RAILWAY

CHANNEL	FREQUENCY
Western Region Road Channels	
WR51 - Fresno to Mojave	161.550
WR52 - Mojave to West Colton / Burbank to Palmdale	**160.320**
WR53 - Los Angeles to San Luis Obispo	161.550
WR54 - Los Angeles to West Colton	**161.550**
WR55 - West Colton to Yuma, AZ	161.550
WR56 - Yuma, AZ to Gila Bend, AZ	160.320
Intermodal Container Terminal	160.350
Intermodal Container Terminal	161.400
PBX Telephones	**160.395 R**
PBX Telephones	**160.890 R**
PBX Telephones	**160.950 R**
Maintenance	161.400
Maintenance - Portables	151.625
Railroad Police	161.205
Railroad Police	161.220
Railroad Shops	160.620
Switching	161.430
West Colton Yard	
Carmen	452.900 R
Hump	160.740 R
Yard	160.380
Yard Channel	161.100
Yard Channel	161.430

UNION PACIFIC

CHANNEL	FREQUENCY
(F-1) Road Operations	160.410
(F-2) Road & Switching	**160.740**
(F-3) Road Operations - Los Angeles	**160.515**
(F-4) Hump Yard	160.710
(F-5) Yard Operations-Los Angeles	160.680
(F-6) Harbor Channel	160.980
Carmen - Los Angeles	452.950 R
Maintenance of Way	161.025
PBX Telephone	160.290 R
Railroad Police	161.205
Railroad Police	161.220
Yard Operations	160.620
Yard Operations	161.115
Yard Operations	161.175
Yard Operations	161.340

LAND TRANSPORTATION SERVICE

TAXI CABS

NAME	LOCATION	FREQUENCY	CALL SIGN
ABC Cab Co.			
Primary	San Diego	151.655	KNDD-611
Secondary	Chula Vista	152.375	KNDD-611
American Cab Co.			
Primary	San Diego	800 MHz	N/A
Backup	San Diego	152.345	WQS-964
American Yellow Cab Co.	Los Angeles	452.350	WNPU-396
Amigo Cab Co.	San Diego	452.150	KNCU-918
Amigo Cab Co.	San Diego	462.000	WNHQ-237
Antelope Valley Taxi	Lancaster	152.330	WNMN-472
Beverly Hills Professional	Beverly Hills	452.100	KNAJ-220
Big Bear Cab Co.	Big Bear	152.390	WYZ-520
Bill's Cab Co.	Solana Beach	464.250 R	WXU-508
Border Cab Co.	Calexico	152.390	WQW-999
Calexico Taxi	Calexico	152.270	WSH-702
California Cab Co.	Calexico	152.360	KNCR-264
City Cab Co.	Barstow	152.270	KMJ-486
Coast Cab Co.			
Beaches (F-1)	San Diego	152.315	WYK-605
Other Areas (F-2)	San Diego	152.285	WYK-605
Coast Yellow Cab Co.	Santa Ana	152.450	KMD-726
Coronado Yellow Cab	Coronado	151.715	KNAE-713
Culver City Yellow Cab	Culver City	452.350 R	KNHA-453
Desert Cab Co.	San Diego	856-860.675 T	WNFJ-886
Diamond Cab Co.	San Diego	152.360	KDK-288
Encinitas Yellow Cab	Escondido	461.450 R	KNFF-368
Escondido Yellow Cab	Escondido	152.300	KME-662
Hunter Cab Co.	Palm Desert	152.345	WNKN-464
La Jolla Cab Co.	La Jolla	152.375	KMG-513
Los Angeles Checker Cab	Los Angeles	472.9875R	WCL-626
Los Angeles City Cab	Los Angeles	452.050 R	KKV-753
Metro Cab	Escondido	800 MHz	N/A
North City Cab	Escondido	856-860.875 T	SMR
Oceanside Yellow Cab	Oceanside	152.450 / 157.710	KMA-404
Orange Cab Co.			
Primary	San Diego	452.225	KJJ-780
Secondary	San Diego	452.275	KJJ-780
Orange Coast Yellow Cab	Santa Ana	152.375	KMA-665
Orange County Yellow Cab	Orange	452.150	KMA-699
Orange County Yellow Cab	Orange	452.200	KMA-699
Orange County Yellow Cab	Orange	452.275	KMA-699
R.& L. Taxi			
Base / Mobile	Brawley	152.270 / 157.530	KMA-640
Radio Cab	San Diego	152.390	KFA-808
Ray's Taxi	Chula Vista	452.100	KNBZ-663
Red Cab Co.	San Diego	800 MHz	N/A
Red Top Cab Co.	San Ysidro	152.405	KVM-801
Red Top Cab Co.	San Ysidro	452.200	KNBZ-663
Red Top Cab Co.	Van Nuys	452.150	KNBA-816
Riverside Taxicab Co.	Riverside	152.450	KMB-400
San Diego Co-Op	San Diego	152.300	WRM-207
San Diego Co-Op	San Diego	152.405	WRM-207

LAND TRANSPORTATION SERVICE

TAXI CABS

NAME	LOCATION	FREQUENCY	CALL SIGN
San Diego Yellow Cab			
North Base	San Diego	152.330	KMB-259
East Base	San Diego	152.420	KMB-259
South Base	San Diego	152.450	KMB-259
Dial-A-Ride	San Diego	152.270	KMB-259
Santa Monica Cab	Culver City	452.400	KMA-572
Silver Cab Co. (Primary)	San Diego	152.300	N/A
Silver Cab Co. (Backup)	San Diego	152.405	N/A
Sky Cab Co.	Los Angeles	452.275	KNBU-587
U.S.A. Cab Co.	San Diego	152.345	KQG-837
United Cab Co. (F-1)	San Diego	452.450 R	WNDE-655
United Cab Co. (F-2)	San Diego	452.200	WNDE-655
United Independent Taxi	Los Angeles	452.500	KIL-251
Yellow Cab of Culver City	Culver City	452.350	KNHA-453
Yellow Cab of Orange	Anaheim	452.200	KMA-699
Yellow Cab of Orange	Anaheim	452.150	KMA-699
Yellow Cab of Orange	Anaheim	452.275	KMA-699
Yellow Cab of Orange	Anaheim	452.500	KMA-699
Yellow Coast Cab	Santa Ana	507.3375R	WBT-807

TOWING COMPANIES

NAME	LOCATION	FREQUENCY	CALL SIGN
A-1	Carlsbad	800 MHz	N/A
AAA of Southern Ca.			
Operations Primary	Southern Calif.	150.920	KAU-987
Operations Secondary	Southern Calif.	150.965	KAU-987
Metro	San Diego	452.575	KAU-987
North SD County	San Marcos	**452.525**	KAU-9870
Simplex	Southern Ca.	452.5875	KEF-892
(The AAA is also on 800 MHz SMR's)			
ABC Towing	Southern Ca.	856-860.5625 T	WNHW-768
Abe's / Kiley's	San Marcos	463.575 R	WNAT-639
Abre Towing	San Diego	461.675 R	WXU-733
Advanced Towing	Los Angeles	856-860.7625 T	WNQB-795
Alcity Towing	San Diego	856-860.675	N/A
Allied Gardens			
North SD County	Mt. Palomar	856-860.975 T	WNGX-922
San Diego	Mt. Otay	856-860.750 T	WNGX-922
San Diego	Mt. Woodson	856-860.925 T	WNGX-922
Al's Towing	Escondido	157.500	KJP-484
Al's Towing	Ontario	508.6625R	WIJ-441
American	Orange	461.575 R	WNSF-548
American	San Diego	463.425 R	KJP-846
Andrews 24 HR	Carlsbad	464.700 R	KNFC-773
Anthony Charlton Towing	Escondido	861-865.650 T	WNRv-517
Anytime Towing			
Operations Trunked	Escondido	856.500 T	WNCS-295
Operations Trunked	Escondido	857.600 T	WNCS-295
Operations Trunked	Escondido	858.600 T	WNCS-295
Operations Trunked	Escondido	859.550 T	WNCS-295
Operations Trunked	Escondido	860.650 T	WNCS-295
Archie's Towing	Orange & LA	935.650-936.700	WNMX-592
Arnold's Garage Towing	San Diego	463.250 R	WNRD-614

LAND TRANSPORTATION SERVICE

TOWING COMPANIES

NAME	LOCATION	FREQUENCY	CALL SIGN
A to Z Metro Towing	San Diego	856.400 / 861.350	WNGT-848
Beeman's	Spring Valley	157.515	KZI-379
Bedrock Towing "486 A"	San Diego	463.300 R	WNIT-473
Best Towing	Huntington Park	508.8125 R	KNR-806
Big Tow & Texas Street	San Diego	857.475	KNGK-512
Bitar Exxon	Orange	856-860.5125 T	KNJV-498
Blue Bird Towing	Orange	856-860.7875 T	WGA-794
Bob Baker Chevrolet	San Diego	856.475 R	KNGD-886
Bob Stall Chevrolet	San Diego	856-860.825 T	WNCH-738
Borrego Springs Chevron	Borrego Springs	35.900	WNFR-829
Bryant's Towing	Inglewood	150.875	KSL-967
Butlers Bestway	Chula Vista	858.650 R	KNGR-345
Cardinal Centre Towing	Escondido	800 MHz	WNLU-600
Carl Bergman	Escondido	464.350 R	WZE-401
Carpenter's Garage	Oceanside	464.800 R	KEH-920
Casas Towing	San Diego	464.325 R	WNQY-749
Casey's Towing	Chula Vista	463.275 R	KNBI-359
Central Towing	San Diego	800 MHz	N/A
Colonial Ford	San Diego	860.600 R	KNHY-606
Courtesy Chevrolet	San Diego	860.550 R	KNGK-511
Courtesy Tow	Los Angeles	936.1375-.1875	WNJJ-476
E.& S. Towing	Escondido	861-865.650 T	WNJY-891
Economy Tow	San Marcos	856-860.675 T	N/A
El Norte Unocal	Escondido	800 MHz	N/A
Emerald Towing	Orange	472.2875R	WIJ-530
Fairmont Towing	San Diego	463.625 R	WNLA-754
Far West Towing	Orange	852.5375R	KRQ-430
Freeway Patrol Service (CHP)	Orange County	44.140 / 44.160	N/A
Fuller Ford	San Diego	856-860.950 T	WNGH-327
Geer's Towing	Orange	463.200 R	KNHM-319
Gene's Towing	Imperial Beach	150.830	KNBT-599
Graham's Towing	La Habra	157.515	WNNG-686
Great Western	San Diego	461.925 R	WNLY-684
Grossmont Shell	La Mesa	157.500	KSS-347
Hansen's Towing	Orange	463.975 R	KNDU-527
Hillcrest Towing	Pasadena	150.845	N/A
Irvine Tow	Irvine	800 MHz	SMR
J.C. Towing	San Diego	462.000 R	KNAA-360
J.D. Club Towing	Beverly Hills	151.715	WSI-884
Jim's	Oceanside	857.550 R	KNHY-588
Jone's Towing			
North SD County	Mt. Palomar	856-860.900 T	WNCZ-651
San Diego	Mt. San Miguel	856-860.825 T	WNCZ-651
Johnstone Towing	San Diego	856-860.925 T	WNIZ-607
La Jolla Towing	La Jolla	463.450 R	WCW-946
Lamb Chevrolet	National City	856.475 R	KNGK-514
Larry's	Dulzura	858.625 R	KNGL-506
Leucadia Exxon	Leucadia	859.500 R	KNHK-864
Lora Towing	Chula Vista	464.000 R	WNWK-741
McBride's	San Diego	858.525 R	KNGY-753
Mike Moor's 24 Hour	San Diego	461.900 R	KRV-888
Mission Garage	San Diego	150.845	KMK-339
Mission Valley	San Diego	858.525 R	KNGY-753
Mueller's	El Cajon	857.475 R	KNIQ-693
Police Towing (SDPD)	San Diego	856.475 R	N/A
Police Towing (SDPD)	San Diego	863.500 R	N/A

LAND TRANSPORTATION SERVICE

TOWING COMPANIES

NAME	LOCATION	FREQUENCY	CALL SIGN
Quality Towing	Escondido	461.225 R	N/A
S.& N. Ward Towing	Pasadena	157.470	N/A
S.& R. Towing	Oceanside	151.715	WNMF-917
San Dieguito Towing	Encinitas	800 MHz	N/A
Sierra Automotive	Lancaster	150.860	WNVY-888
Southside Towing	Bell Gardens	463.275 R	WSZ-741
Ralph's Towing	San Diego	860.600	KNHK-801
Rancho Towing	San Diego	461.100 R	WST-781
Rancho Del Oro Towing Operations Primary	Mt. Palomar	856-860.975 T	WNSM-817
T.E. Towing	Jacumba	461.375 R	WNLP-841
Texas Street	San Diego	857.475 R	KNGK-512
Tom Moynahan	National City	150.890	WNBE-787
Tri City Towing	Vista	461.175 R	WNNG-795
Two Bit Tow	Apple Valley	461.350 R	N/A
Two Bit Tow	Mt. Otay	856-860.950 T	WNHH-232
University Ford Chrysler	San Diego	858.650 R	KNHH-466
Vista Towing	Vista	860.475 R	KNGL-507

TRUCKING & DELIVERY

NAME	LOCATION	FREQUENCY	CALL SIGN
3 C Trucking	Long Beach	462.100 R	WGA-273
20th Century Trucking	Glendale	160.185	KNDA-812
A.A. Parking & Trucking	Calexico	464.075 R	WNMI-533
American Messenger Srv.	San Diego	452.850 R	WNGV-818
Able Transportation	San Diego	856-860.825 T	WNDJ-673
Ace Trucking	Calexico	461.375 R	WNLI-729
Action Courier	San Diego	858.400 R	KNGK-516
Airborne Freight Corp.	San Diego	861.300 R	WNCS-299
Airborne Freight Corp.	San Diego	862.300 R	WNCS-299
Auto Fast Freight	Southern Calif.	160.005	KZB-837
Barnard Trucking	Glendale	852.2375R	KSU-303
Border Turf Lines	Chula Vista	461.725 R	WNDS-533
Border Turf Lines	Chula Vista	463.350 R	WNCS-463
Border Turf Lines	Chula Vista	463.800 R	WNCS-463
Bud McDaniel Transport	San Diego	856-860.675 T	WNIM-759
Cal Pacific Transport	San Diego	856-860.775 T	WNGY-780
Commercial Transport	San Diego	856-860.925 T	WNGG-913
Di Salvo Trucking	Glendale	159.975	KNDX-499
Federal Express Corp.	Los Angeles	851.3875R	KXU-969
Federal Express Corp.	Los Angeles	856-860.8125	WNDM-578
Federal Express Corp.	Los Angeles	856-860.8375	KNEC-206
Federal Express Corp.	Los Angeles	856-860.8625	KNEC-212
Federal Express Corp.	Los Angeles	856-860.9125	WNCE-533
Federal Express Corp.	Los Angeles	939.200-939.250 T	WNIT-676
Federal Express Corp.	Los Angeles	939.650-939.750 T	WNIS-324
Federal Express Corp.	San Diego	856.550 R	KNDH-661
Federal Express Corp.	San Diego	856.625 R	KNHY-640
Federal Express Corp.	San Diego	861.550 R	WNCE-514

LAND TRANSPORTATION SERVICE

TRUCKING & DELIVERY

NAME	LOCATION	FREQUENCY	CALL SIGN
Gypsum Trucking Co.	Lynwood	159.780	KDX-851
Greyhound Express	San Diego	856.575 R	WNAS-515
Home Merchants Delivery	San Diego	856-860.725 T	WNCV-570
Jack Jones Trucking	Glendale	508.7375 R	KZR-765
Janis Trucking	San Diego	856-860.725 T	WNCV-284
Jet Delivery Service	Spring Valley	859.400 R	KNGK-515
Jim Brown Trucking	Bonsall	859.625 R	WNBF-511
Jon Hunt Trucking	San Diego	856-860.950 T	WNGT-499
LA Messenger Express	Glendale	159.765	WNGF-874
La Salle Trucking	San Diego	856-860.825 T	WNDU-695
M.A.T. Parcel Express	San Diego	856-860.925 T	WNHS-990
M.A.T. Parcel Express	San Diego	856-860.975 T	WNHS-990
Mallet & Sons Trucking	San Diego	856-860.725 T	WNDB-704
Merchants Home Delivery	San Diego	856-860.925 T	WNHA-732
Mex Cal Truckline	San Diego	856-860.750 T	WNFG-434
Mex Cal Truckline	San Diego	856-860.975 T	WNFG-434
Nelson & Sons Trucking	La Crescenta	508.7375R	WIJ-820
One Hour Delivery	Orange	44.260	WZT-980
Pony Express Courier	San Diego	856-860.950 T	WNDW-600
Premier Custom Courier	Glendale	452.700 R	WNHG-507
Pronto Messenger Service	Los Angeles	159.930	KMM-227
Purolator Courier	San Diego & Corona	452.675 R	WQT-640
Rapid Messenger Service	Glendale	159.885	WNGD-744
Reliable Delivery	Commerce	508.6375R	WIJ-517
Reyes Brothers Trucking	Corona	507.3375R	WIK-921
Robert Lynch Trucking	San Diego	856-860.750 T	WNFP-777
Robert Lynch Trucking	San Diego	856-860.925 T	WNFP-777
Robert Lynch Trucking	San Diego	856-860.975 T	WNFP-777
Rocha Trucking	Calexico	464.075 R	WNMB-735
Rudy Gaitan Trucking	San Diego	508.6375R	WIK-980
Sanberg Trucking	Corona	508.7375R	KTZ-766
San Diego Trucking	San Diego	856-860.950 T	WNID-435
Shaker Express	Poway	857.450 R	WNIY-204
Sharmax Freightlines	Lakeside	856.400 R	WNAU-444
Space Age Delivery	San Diego	856-860.750 T	WNIW-548
Space Age Delivery	San Diego	856-860.925 T	WNIW-548
Space Age Delivery	San Diego	856-860.975 T	WNIW-548
Standon Trucking	Rialto	43.960	KWE-345
Sun Valley Trucking	Poway	858.350 R	WNIY-204
Sun Valley Trucking	Poway	859.350 R	WNIY-204
Super Delivery Service	Los Angeles	508.5375R	KNM-277
Superior Fast Freight	El Toro	159.720	WNAE-453
Target Express Co.	Chino & Los Angeles	508.7875R	WIJ-306
Transcon Lines	San Diego	159.585	KCE-610
United Drayage Co.	Los Angeles	508.5875R	WCC-733
Valley Truck Lines	Calexico	464.075 R	WNMI-509
Victorville Barstow	La Crescenta	44.280	KCK-363
Viking Freight Systems	Montrose	452.800 R	WNBS-219
Viking Freight Systems	Pomona	452.875 R	WNDC-529
Vito Transportation	Montrose	452.725 R	WNFY-575
Warren Trucking Co.	Pasadena	452.700 R	WNUZ-556
Well's Trucking	Escondido	857.650 R	KNGL-508

MARITIME

HIGH FREQUENCY COMMERCIAL BANDS

CALLING FREQUENCIES

Gulf of Mexico (Initial): 4183.0, 6278.0, 8367.0, 12551.0, 16735.0, 22281.5, 25171.5 kHz
Gulf of Mexico (Alternate): 4183.5, 6278.5, 8367.5, 12551.5, 16735.5, 22282.0 kHz
Northern Pacific (Initial): 4185.0, 6279.0, 8368.5, 12552.5, 16736.5, 22282.5, 25172.5
Northern Pacific (Alternate): 4185.5, 6279.5, 8369.5, 12553.0, 16737.0, 22283.0 kHz
Southern Pacific (Initial): 4186.0, 6280.0, 8370.0, 12554.0, 16737.5, 25172.5 kHz
Southern Pacific (Alternate): 4186.5, 6280.5, 8370.5, 12554.5, 16738.5, 22284.0 kHz
Worldwide (Initial): 4184.0, 6276.0, 8368.0, 12552.0, 16736.0, 25172.0 kHz
Worldwide (Alternate): 4184.5, 6276.5, 8369.0, 12553.5, 16738.0, 22281.0 kHz

DISTRESS & SAFETY FREQUENCIES

Radiotelephone: 2182.0, 4125.0, 6215.0, 8291.0, 12290.0, 16420.0 kHz
Narrowband Direct Printing: 2174.5, 6268.0, 8376.5, 12520.0, 16695.0 kHz
Digital Selective Calling: 2187.5, 4207.5, 6312.0, 8414.5, 12577.0, 16804.5 kHz

KMI, HIGH SEAS RADIOTELEPHONE

2450.0/2003.0, 2506.0/2406.0, 4357.0/4065.0, 4402.0/4110.0, 4405.0/4113.0, 8728.0/8204.0, 8743.0/8219.0, 8782.0/8258.0, 13077.0/12230.0, 13080.0/12233.0 13083.0/12236.0, 13161.0/12314.0, 17245.0/16363.0, 17248.0/16366.0, 17311.0/16429.0, 22735.0/22039.0, 22762.0/22066.0, 22777.0/22081.0, 22801.0/22105.0

To reach a High Seas Operator, dial 1-800 SEA-CALL

MARITIME BUSINESS FREQUENCIES

2096.0, 4145.0, 4148.0, 6224.0, 6227.0, 6230.0, 8294.0, 8297.0, 12353.0, 12356.0, 12359.0, 16528.0, 16531.0, 16534.0, 22159.0, 22162.0, 22165.0, 22168.0, 22171.0

SSB CALLING / REPLY FREQUENCIES PAIRS (SHIP / COAST)

4125.0/4417.0 6215.0/6516.0 8255.0/8779.0 12290.0/13137.0 16420.0/17302.0 18795.0/19770.0 22060.0/22756.0 25097.0/26172.0

SSB WORKING FREQUENCY BANDS

4060.0-4143.0 (Ship), 4351.0-4435.0 (Coast); 6209.0-6218.0 (Ship), 6510.0-6519.0 (Coast); 8113.0-8285.0 (Ship), 8713.0-8809.0 (Coast); 12230.0-12338.0 (Ship), 13077.0-13185.0 (Coast); 16360.0-16501.0 (Ship), 17242.0-17365.0 (Coast); 19755.0-19776.0 (Ship), 18780.0-18801.0 (Coast); 22000.0-22138.0 (Ship), 22696.0-22834.0 (Coast); 25070-25079 (Ship), 26145-26154 (Coast)

Note: The above frequencies are in kilohertz (kHz).

MARITIME

VHF MARINE BAND

CHANNEL NO.	USE	FREQUENCY
1	International Use	156.000
2	International Use	156.100
3	International Use	156.150
4	International Use	156.200
5	Port Operations	156.250
6	Intership Safety	156.300
"	Search & Rescue	**156.300**
7	Commercial	156.350
8	Commercial (Intership)	156.400
9	Commercial / Recreation	**156.450**
10	Commercial	156.500
11	Commercial	156.550
12	Port Operations	156.600
13	Navigation (1-watt)	156.650
14	Port Operations	156.700
15	Environmental (Shore)	156.750
16	Calling Channel & Distress	**156.800**
17	State Control (1-watt)	156.850
18	Commercial	156.900
19	Commercial	156.950
20	Port Operations	157.000
21	Coast Guard	**157.050**
22	Coast Guard	**157.100**
23	Coast Guard - Air Ops San Diego	**157.150**
24	Marine Operator-San Pedro	**161.800 R**
25	Marine Operator-Santa Barbara	**161.850 R**
26	Marine Operator-San Pedro	**161.900 R**
27	Marine Operator-Redondo Beach	**161.950 R**
28	Marine Operator-San Diego	**162.000 R**
29	International Use	157.450
30	International Use	157.500
31	International Use	157.550
32	International Use	157.600
33	International Use	157.650
34	International Use	157.700
35	International Use	157.750
36	International Use	157.800
37	International Use	157.850
38	International Use	157.900
39	International Use	157.950
40	International Use	158.000
41	International Use	158.050
42	International Use	158.100
43	International Use	158.150
44	International Use	158.200
45	International Use	158.250
46	International Use	158.300
47	International Use	158.350
48	International Use	158.400

MARITIME

VHF MARINE BAND

CHANNEL NO.	USE	FREQUENCY
49	International Use	158.450
50	International Use	158.500
51	International Use	158.550
52	International Use	158.600
53	International Use	158.650
54	International Use	158.700
55	International Use	158.750
56	International Use	158.800
57	International Use	158.850
58	International Use	158.900
59	International Use	158.950
60	International Use	156.025
61	International Use	156.075
62	International Use	156.125
63	International Use	156.175
64	International Use	156.225
65	Port Operations	156.275
66	Port Operations	156.325
67	Commercial (Intership)	156.375
68	Non-Commercial	**156.425**
69	Non-Commercial	**156.475**
70	Non-Commercial (Intership)	156.525
71	Non-Commercial	**156.575**
72	Non-Commercial (Intership)	156.625
73	Port Operations	156.675
74	Port Operations	156.725
75	International Use	156.775
76	International Use	156.825
77	Port Operations (Intership)	156.875
78	Non-Commercial	**156.925**
79	Commercial	156.975
80	Commercial	157.025
81	Safety	157.075
82	U.S. Coast Guard	157.125
83	U.S. Coast Guard	157.175
84	Marine Operator-Dana Point	**161.825 R**
85	Marine Operator-Redondo Beach	**161.875 R**
86	Marine Operator-San Diego	**161.925 R**
87	Marine Operator-Redondo Beach	**161.975 R**
88	Commercial	157.425
89	International Use	157.475
90	International Use	157.525
91	International Use	157.575
92	International Use	157.625
93	International Use	157.675
94	International Use	157.725
95	International Use	157.775
96	International Use	157.825

MARITIME

VHF MARINE BAND

CHANNEL NO.	USE	FREQUENCY
97	International Use	157.875
98	International Use	157.925
99	International Use	157.975
NOAA Weather Reports	San Diego	**162.400**
NOAA Weather Reports	Los Angeles	**162.550**

AUTOMATED MARITIME TELECOMMUNICATIONS SYSTEM

USE	FREQUENCIES
Administration / Maintenance (kHz)	2087.4, 2183.4, 2783.4, 4117.1,
Administration / Maintenance (kHz)	6520.2, 8726.5, 13105.2, 17293.2
Coastal Stations	216.0125-217.9875 MHz
Ships / Offshore Oil Rigs	218.0125-219.9875 MHz

AMTS is a new computer-controlled radio system that offers a maritime telephone service, precise vessel tracking, and selected data transmission (FAX, RTTY). Shore stations are limited to 1 KW ERP (effective radiated output). Vessels use 10 watt transmitters. Channel spacing on the 216 MHz band is 12.5 kHz.

NEWS MEDIA

NEWSPAPERS

ESCONDIDO TIMES ADVOCATE

CHANNEL	FREQUENCY	CALL SIGN
Operations Trunked	856-860.875 T	WNRJ-229
Paper Deliveries	452.975 R	WNDZ-346

LOS ANGELES TIMES

CHANNEL	FREQUENCY	CALL SIGN
Operations Trunked*	861-865.5125 T	SMR
Operations Trunked*	861-865.8125 T	SMR
Operations Trunked*	861-865.9375 T	SMR
Operations VHF	173.275	WNQS-214
Operations VHF	173.325	KNCZ-957
Operations VHF	173.375	KNCZ-957
Paper Deliveries	452.975 R	WNML-711

*Additional information needed.

SAN DIEGO UNION TRIBUNE

CHANNEL	FREQUENCY	CALL SIGN
News Reporters North	856-860.675 T	WNBF-492
News Reporters South	856-860.950 T	WNBF-502
News Reporters Central	856-860.775 T	WNBF-496
Press Photographers	458.000	N/A
Building Security	453.000	N/A
Operations VHF	173.225	N/A

RADIO STATIONS

NAME	LOCATION	FREQUENCY	CALL SIGN
KABC-AM 790 (Traffic Reports)	Los Angeles	450.050 R	KRO-642
KABC-AM 790	Los Angeles	450.825 R	KRO-642
KAVL & KAVS AM/FM	Lancaster	153.290	KF-8410
KAVL & KAVS AM/FM	Lancaster	26.330	KMM-786
KAVR-AM/FM	Apple Valley	161.760	N/A
KBBV-AM 1050	Big Bear	(Licensed for all VHF channels)	
"	450.010 / 450.050 / 450.450 / 450.550 / 450.650 / 450.750		
KBBY	Ventura	161.640	KO-6385
KBBY	Ventura	161.670	KO-6385
KBON & KCKC AM/FM	Lake Arrowhead	450.010	KPL-687
KCBQ-AM	San Diego	161.670	KMJ-887
KCBQ-AM	San Diego	161.760	KFK-935
KCBQ-FM	San Diego	455.650	N/A
KCLB & KCHV AM/FM	Coachella	450.050 R	KJC-527

NEWS MEDIA

RADIO STATIONS

NAME	LOCATION	FREQUENCY	CALL SIGN
KCMJ-FM	Palm Springs	161.730	KB-96773
KCMJ-FM	Palm Springs	450.020	KPE-904
KCMJ-FM	Palm Springs	450.980	KPS-930
KCPB	Los Angeles	450.900	N/A
KCRW-AM/FM	Los Angeles	450.1125R	N/A
KDAY-AM 1580	Los Angeles	161.670	N/A
KEZY-FM	Anaheim	455.5125R	N/A
KFAC	Los Angeles	26.070	KMG-541
KFAC	Los Angeles	26.110	KMG-541
KFAC	Los Angeles	26.450	KMG-541
KFI-AM	Los Angeles	170.150	KEP-825
KFI-AM	Los Angeles	450.725 R	KPK-405
KFMB-AM/FM	San Diego	450.2625R	KPJ-931
KFMB-AM/FM	San Diego	450.7625R	KPH-592
KFSD	San Diego	161.670	KU-8413
KFWB-AM	Los Angeles	450.850 R	KEZ-366
KFWB-AM (Traffic Reports)	Los Angeles	450.650 R	KEZ-366
KGB	San Diego	450.925	KGM-270
KGB	San Diego	455.925	KGM-270
KGGI-FM	Riverside	450.990	KPJ-507
KGIL-AM/FM	Los Angeles	161.670	N/A
KGMG-AM/FM	Oceanside	450.010	KPJ-682
KGMG-AM/FM	Oceanside	450.925 R	N/A
KGMG-AM/FM	Vista	450.980	KPJ-683
KGMG-AM/FM	Vista	455.020	KPJ-678
KHAY-FM	Ventura	161.700	KO-3813
KHAY-FM	Ventura	450.010	KPJ-886
KHJJ-AM	Lancaster	455.925	KC-24250
KICO AM	Calexico	161.700	KLO-503
KIIS-FM	Los Angeles	161.6875	KPE-766
KISS-FM (Traffic Reports)	Los Angeles	450.950 R	KPH-442
KIIS-FM	Los Angeles	450.980	KPH-442
KKLQ-FM	San Diego	455.2125	KML-441
KKOS	Carlsbad	Cellular	N/A
KKUR-FM	Ojai	455.925	KC-24250
KKZZ-FM	Lancaster	455.925	KC-24250
KLIT-FM	Glendale	450.250	KH-7555
KLIT-FM (Traffic Reports)	Glendale	450.750 R	KMH-853
KLOS (Traffic Reports)	Los Angeles	450.050	N/A
KLOS	Los Angeles	450.825	N/A
KLSX-FM (Traffic Reports)	Los Angeles	450.450 R	KPL-212
KLSX-FM	Los Angeles	455.2875	WZZ-771
KLZZ-AM/FM	San Diego	26.110	N/A
KLZZ-AM/FM	San Diego	35.960	N/A
KLZZ-FM	San Diego	450.2125	N/A
KLZZ-FM	San Diego	450.250	N/A
KMEN-AM 1290	San Bernardino	450.990	KPJ-507
KMPC-AM (Traffic Reports)	Los Angeles	450.750 R	KMH-583
KNTF-FM	Ontario	455.800	KB-55890
KNX-AM 1070	Los Angeles	450.0875R	N/A
KNX-AM 1070 (Traffic Reports)	Los Angeles	450.700 R	N/A
KNX-AM 1070	Los Angeles	450.800 R	N/A
KOWN-AM/FM	Escondido	153.050	KGI-281
KOWN-AM/FM	Escondido	161.730	KGI-281
KOWN-AM/FM	Escondido	166.250	KMH-392

NEWS MEDIA

RADIO STATIONS

NAME	LOCATION	FREQUENCY	CALL SIGN
KPBS-FM	San Diego	455.800	KQ-4634
KPBS-FM	San Diego	455.875	KQ-4634
KPBS-FM	San Diego	455.980	KQ-4634
KPFK-FM	Los Angeles	450.020	WHE-739
KPFK-FM	Los Angeles	450.650 R	WHE-738
KPOP-AM	San Diego	450.925	KGM-270
KPSI	Los Angeles	450.925	N/A
KPWR-FM	Los Angeles	450.900	N/A
KRLA-AM	Los Angeles	450.450 R	KPL-212
KRLA-AM	Los Angeles	455.2875	WZZ-771
KRTH-AM	Los Angeles	161.6275	KPJ-782
KRTH-AM	Los Angeles	455.450	KPJ-781
KRTH-AM	Los Angeles	455.850	KPJ-781
KSDO-AM	San Diego	161.730	KUN-343
KSDO-AM	San Diego	455.2375	KQ-8519
KSON	San Diego	166.250	N/A
KSON	San Diego	450.6875R	KPH-572
KSON	San Diego	455.0875	KPH-572
KTOT-FM (See KBBV)			
KTWV-FM	Los Angeles	450.850	KEZ-366
KUCI	Los Angeles	455.800	KU-2550
KUSC	Los Angeles	455.900	N/A
KWKW-AM	Los Angeles	152.870	KMK-353
KWKW-AM	Los Angeles	161.670	KMK-353
KWKW-AM	Los Angeles	166.250	KMK-353
KWXY	Los Angeles	161.700	N/A
XTRA (91X)	San Diego	Cellular	N/A

STUDIO TRANSMITTER LINKS

Most news media services use studio transmitter links, better known as STL's, to provide communications for live shots and other remote broadcasts. Numerous frequencies are shared by the media in the 2, 3, 4, and 7 GHz band.

TELEVISION STATIONS

KCBS-TV CHANNEL 2

CHANNEL	FREQUENCY	CALL SIGN
(F-1) News Desk	**450.2875R**	KQV-755
(F-2) Secondary, Helicopter*	450.5125R	KEH-377
(F-3) CBS Network News	450.4125	KEH-377
(F-4) CBS Network News	450.6125	KEH-377
(F-5) Remotes	450.800	KEH-377
(F-6) News Desk Secondary	450.0875R	KQV-755
CBS Network News	450.450	KOS-344
News Simplex	455.010	KKN-664

***Tone:** 71.9 Htz

NEWS MEDIA

TELEVISION STATIONS

KNBC-TV CHANNEL 4

CHANNEL	FREQUENCY	CALL SIGN
News Desk Primary	161.760	N/A
Engineering	450.3875R	KEH-396
IFB / Remotes	450.5375R	KFH-896
IFB / Remotes	450.5625R	KFH-896

Tone: 123.0 Htz, 131.8 Htz

KTLA-TV CHANNEL 5

CHANNEL	FREQUENCY	CALL SIGN
(F-1) News Desk	450.350 R	KPJ-777
(F-2) News Secondary	450.250 R	KPJ-777
(F-3) Operations Direct	450.350	KB-55232
(F-4) Operations Direct	450.250	KB-55232

Tone: 162.2 Htz (all channels)

KABC-TV CHANNEL 7

CHANNEL	FREQUENCY	CALL SIGN
ABC News Desk*	450.1625R	KFA-632
News Desk & Live Shots*	450.1375R	KQA-845
Engineering - Anaheim*	450.3125R	KQA-845
Engineering - Mt. Lukens	450.550 R	KRO-643
IFB / Remotes	450.150	KA-74769
IFB / Remotes	455.150	KA-74769
IFB / Remotes	450.225	KQA-845
Traffic Watch	450.050 R	KR0-642
Undetermined	455.230	KFA-632

***Tone:** 123.0 Htz

KFMB-TV CHANNEL 8

CHANNEL	FREQUENCY	CALL SIGN
(F-1) News Desk	450.3375R	KEH-339
(F-2) Primary Direct	450.3375	KEH-339
(F-3) Secondary Operations	450.4125R	KEH-340
(F-4) Secondary Direct	450.4125	KEH-340
CBS News Desk L.A.	450.450	KRL-433
Studio Transmitter Links	948-950 MHz	N/A
Wireless Microphones	186.000	BLP-00051
Wireless Microphones	174.000	BLP-00765
Wireless Microphones	203-205 MHz	N/A

Tone: 141.3 Htz (all channels)

NEWS MEDIA

TELEVISION STATIONS

KCAL-TV CHANNEL 9

CHANNEL	FREQUENCY	CALL SIGN
(F-1) News Desk (LA)	450.2125R	KPK-889
(F-2) Engineering	455.0875	KPK-889
(F-3) Operations Direct	450.2125	KPK-889
(F-4) Engineering	450.0875	KPK-889

Tone: 91.5 Htz (all channels)

KECY-TV CHANNEL 9

CHANNEL	FREQUENCY	CALL SIGN
News Desk (El Centro)	856-860.725 T	SMR
IFB / Remotes	26.110	KQS-278
IFB / Remotes	455.980	KOS-320

KG-TV CHANNEL 10

CHANNEL	FREQUENCY	CALL SIGN
(F-1) News Desk Primary	**161.700**	KNM-789
(F-2) Car to Car	161.730	KNM-789
(F-3) News Desk Secondary	161.760	KNM-789
Backup Channel (rare use)	166.250	KNM-789
Backup Channel (rare use)	171.150	KNM-789
ABC News Desk	450.1625R	KFA-632
Engineering (Microwave Control)	450.2375R	KPE-945
Live Audio & Remotes	450.575	KPE-945
Live Audio & Remotes	450.600	KPE-945
Secondary Dispatch	450.3635R	KPE-945
Sky 10 Helicopter	455.2875	N/A
Wireless Microphones	174.000	BLP-00045
Wireless Microphones	203-205 MHz	N/A

Tones: 88.5 (VHF) and 71.9 (UHF)

KTTV CHANNEL 11

CHANNEL	FREQUENCY	CALL SIGN
(F-1) News Desk	450.5875R	N/A
(F-2) IFB / Remotes	450.950 R	N/A

Tone: 88.5 Htz

XEWT-TV CHANNEL 12

CHANNEL	FREQUENCY	CALL SIGN
"Notivisa Informa" (news & remotes)	169.750 R	N/A

NEWS MEDIA

TELEVISION STATIONS

KCOP-TV CHANNEL 13

CHANNEL	FREQUENCY	CALL SIGN
News Desk	508.4125R	N/A
Engineering	455.3125R	WZZ-706
Engineering	161.640 / 161.670	KPH-273

Productions: 152.870 / 152.990 / 153.170 / 153.290

KPBS-TV CHANNEL 15

CHANNEL	FREQUENCY	CALL SIGN
Engineering & Productions	450.1125R	KUN-650
Engineering & Productions	450.4875R	KUN-650

KSCI-TV CHANNEL 18

CHANNEL	FREQUENCY	CALL SIGN
Operations Primary	450.1125R	KZH-896

KCET-TV CHANNEL 28

CHANNEL	CALL SIGN
Remotes: 152.930 / 152.990 / 153.050 / 153.110 / 153.170 / 153.230	KV-8040

KMEX-TV CHANNEL 34

CHANNEL	FREQUENCY	CALL SIGN
(F-1) News Desk	450.1125R	KPF-248
(F-2) News Desk Direct	450.1125	KPF-248
(F-3) News Desk	450.4875R	KPF-893
Operations VHF (rare use)	151.6425 / 154.5525	KNJK-361

Tone: 186.2 Htz, 146.2 Htz on F-3.

KNSD-TV CHANNEL 39

CHANNEL	FREQUENCY	CALL SIGN
(F-1) Operations Primary	**450.4375R**	KEH-327
(F-2) Operations Direct	450.4375	KEH-327
(F-3) Operations Secondary	450.4625R	KEH-327
(F-4) Operations Direct	450.4625	KEH-327
(F-5) NBC News Desk L.A.	450.1875R	KPG-823
Inter News Agency Emergency	151.4625	WNCR-209
Wireless Microphones	203-205 MHz	N/A

Tone: 141.3 Htz (channels 1-4)

NEWS MEDIA

TELEVISION STATIONS

KUSI-TV CHANNEL 51

CHANNEL	FREQUENCY	CALL SIGN
(F-1) News Primary	450.0625R	WHE-820
(F-2) News Direct	450.0625	WHE-820
(F-3) News Secondary	450.0375	WHE-820

Tone: 127.3 Htz (all channels)

KDOC-TV CHANNEL 56

CHANNEL	FREQUENCY	CALL SIGN
(F-1) Studio & Remotes	450.1125R	KPE-717

KDAY-TV CHANNEL 63

CHANNEL	FREQUENCY	CALL SIGN
(F-1) News Operations	450.4375R	KPH-686
(F-2) News Operations	450.0375R	KPH-686

KTTY-TV CHANNEL 69

CHANNEL	FREQUENCY	CALL SIGN
Studio & Productions	450.020	KPE-837

CNN AMERICAN

CHANNEL	FREQUENCY	CALL SIGN
News Crews Los Angeles	**450.775 R**	KPH-396
News Crews Los Angeles	450.3125R	KPH-396
News Simplex / Productions	450.150	KPH-396
News Simplex / Productions	450.550	KPH-396
News Simplex - Nationwide	450.9625	N/A
News Simplex - Nationwide	450.9875	N/A

Tone: 146.2 Htz (all channels)

FOX TELEVISION

CHANNEL	FREQUENCY	CALL SIGN
Production Operations - Los Angeles	455.5875	KJT-464

NEWS MEDIA

TRAFFIC REPORTS

NAME	FREQUENCY	CALL SIGN
Airwatch Communications		
Operations LA & Orange	**450.825 R**	KPJ-853
Operations San Diego	450.7125	KPJ-774
Operations San Diego	455.7125	KPJ-774
Monica Zeck's Co. (F-1)	455.925	KGM-554
Monica Zeck's Co. (F-2)	450.925	KGM-554
KABC Radio 790	450.050 R	KRO-642
KFWB FM 98 News		
Traffic Reports (Airwatch)	450.825 R	KEZ-366
Traffic Reports	**450.850 R**	KEZ-366
KEZY Traffic Radio	450.6375	
KISS-FM News/Traffic	450.950 R	KPH-422
KLIT-FM News/Traffic	450.750 R	KMH-58
KLSX-FM News/Traffic	450.450 R	KPL-212
KMPC News/Traffic	450.750 R	KMH-583
KNX AM 1070		
Traffic Reports	450.700 R	N/A
Traffic Reports	450.800 R	N/A
Metro Traffic Control		
Los Angeles	455.5125	KPL-411
Los Angeles	455.6375	KPL-411
San Diego	450.500	KPL-453
San Diego	455.500	KPL-453
San Diego	455.050	KPK-981
REACT Liaison	462.675 R	KAE-0095

MISCELLANEOUS

NAME	FREQUENCY	CALL SIGN
Diario 29 Periodico (Mexico)	450.550 R	N/A
Goodyear Blimp Operations		
Air Coordination: 151.625, 464.500, 464.550		
Fixed Links: 465.9375, 465.9625		
El Mexicano Periodico (Mexico)	166.000 R	N/A
James A. Kay News Photo (LA)	471.1875R	WIL-712
James A. Kay News Photo (LA)	471.9375R	WIH-868
James A. Kay News Photo (LA)	472.2125R	KJV-843
News Helicopters (Air-to-Air)	**126.775**	N/A
News Helicopters (Interagency)	122.850	N/A
News Photographers (GMRS)	462.550 R	N/A
Orange Coast College - Productions	450.4875R	KVW-737
Orange Coast College - Production	450.150	KT-41173
Time Life Broadcasting (California)	161.700	KF-9831
Time Life Broadcasting (San Diego)	450.250	KRL-429
Time Life Broadcasting (San Diego)	166.250	KN-6988
"Z" Periodico (Mexico)	456.750 R	N/A

Most news services are also equipped with cellular telephones.

SCHOOLS

COMMUNITY & PRIVATE COLLEGES

CAMPUS	LOCATION	FREQUENCY	CALL SIGN
Chaffey Community			
Campus Police	Alta Loma	154.650	WNDU-902
Maintenance & Ops	Alta Loma	151.775	KQA-371
Coast Community Colleges			
Corona Operations	Corona	461.475 R	WNNW-253
Costa Mesa Campus	Costa Mesa	154.540	WNAB-203
Orange - Campus Patrol	Orange	464.175 R	WXH-390
All Facilities: 463.2375 / 463.2625 / 463.3875 / 463.8125 / 467.1125 / 468.2375 / 468.2625 / 468.3875 / 468.8125			
Golden West College			
Campus Police	Huntington Beach	151.655	KFO-381
Undetermined	Huntington Beach	463.3875R	N/A
Grossmont Community	El Cajon	155.580	KJL-596
Long Beach Community			
Campus Patrol	Long Beach	154.650 R	KNDY-988
Campus Patrol Tactical	Long Beach	155.115	WNHP-965
Los Angeles Community Colleges			
Campus Patrol	Los Angeles County	154.650 R	KNJY-515
Campus Patrol	Los Angeles County	155.685	KNJY-515
Campus Patrol	Los Angeles County	156.090	KP-2569
Maintenance	Los Angeles County	151.685	N/A
Mira Costa College	Oceanside	464.575 R	KNBR-497
Palomar Community			
Campus Patrol	San Marcos	464.475 R	WNML-216
Campus Patrol	San Marcos	464.425 R	WNML-216
General Operations	San Marcos	151.655	KB-25653
Registration Operations	San Marcos	27.125	CB 14
Pepperdine University	Malibu	151.490	N/A
Pepperdine University	Malibu	154.515	WYA-500
Pepperdine University	Malibu	154.600	WNDE-461
Pepperdine University	Malibu	471.1875R	WII-484
Saddleback Community			
Campus Police	Mission Viejo	154.650 R	KCE-631
Athletics Department	Mission Viejo	464.425 R	WNJZ-934
Student Escort Service	Mission Viejo	154.515	KXB-220
Telephone Interconnect	Mission Viejo	453.5375R	KCE-631
San Diego Community Colleges			
Security Primary	San Diego	453.450 R	KXD-597
Santa Monica Community			
Campus Patrol	Santa Monica	154.650	KNCV-305
Campus Security	Santa Monica	151.835	WNMH-879
Campus Security	Santa Monica	154.570	WNMH-879
Southwestern College			
Campus Security	Chula Vista	464.4625	KB-40070
Maintenance	Chula Vista	469.4625	KB-40070
Telephone Interconnect	Chula Vista	461.4625R	KB-40070
United States Intl. University			
Security	San Diego	151.655	KNHB-735
Security & Operations	San Diego	151.715	WNBC-401

SCHOOLS

SCHOOL DISTRICTS

NAME	LOCATION	FREQUENCY	CALL SIGN
Bassett USD - Security	La Puente	154.650	WNBE-374
Bassett USD	La Puente	151.895	KUU-955
Borrego USD	Borrego	155.280	WNGA-927
Cajon Valley Union	El Cajon	45.960	KVP-583
Cajon Valley Union	El Cajon	154.570	N/A
Cajon Valley Union	El Cajon	863.000	WNHS-768
Capistrano USD	San Juan Capistrano	46.000	KRB-453
Carlsbad - Buses	Carlsbad	858.500	WNGK-907
Carlsbad - Security	Carlsbad	154.600	KB-71643
Chino School District - Buses	Chino	155.235	KCS-319
Chino School District - PD	Chino	453.500 R	KNFB-577
Chula Vista - Buses	Chula Vista	155.265	KMA-344
Chula Vista - Security	Chula Vista	155.625	N/A
Clairemont Christian	San Diego	46.000	N/A
Coachella Valley USD	Coachella	155.280	WQE-453
Desert Sands USD	Indio	155.175	KFH-374
Escondido - Buses	Escondido	151.865	KTH-839
Escondido - Security	Escondido	151.835	KTH-839
Fallbrook - Buses	Fallbrook	155.265	WQU-361
Fountain Valley School District	Fountain Valley	151.805	KGB-492
Grossmont Union - Buses	La Mesa	861.500	WNBY-342
Hesperia USD	Hesperia	155.175	WNUZ-472
Huntington Beach School Dist.	Huntington Beach	155.265	KWI-746
Jamul Dulzura - Buses	San Diego	463.250	WQJ-718
Jamul Dulzura - Security	Jamul	154.570	KB-67512
Lake Elsinore USD	Lake Elsinore	853.8625R	WNQI-811
La Mesa-Spring Valley	El Cajon	864.000	WNHS-767
Lemon Grove - Buses	Lemon Grove	155.280	WSY-257
National City - Security	National City	464.7375 / 464.9375	KB-46267
Oceanside - Buses	Oceanside	155.295	WRD-858
Oceanside - Security	Oceanside	151.805	KA-56992
Oceanview School Dist.	Huntington Beach	151.835	WNBG-762
Oceanview School Dist.	Huntington Beach	464.875	WNBN-730
Palo Verde USD	Blythe	155.235	KUU-381
Perris School District	Perris	155.220	WQF-243
Poway - Buses	Poway	151.895	KNAV-340
Poway - Security	Poway	154.600	KB-67684
Ramona - Buses	Ramona	151.925	WZU-629
Rim of the World USD	Lake Arrowhead	37.940	KFE-221
Rim of the World USD	Lake Arrowhead	154.570 / 154.600	KD-43069
San Bernardino USD	San Bernardino	460.075 R	WFR-380
San Diego - Alarms	San Diego	39.620	N/A
San Diego - Buses	San Diego	800 MHz	WNHS-765
San Diego - Police F1	San Diego	453.300 R	WNFJ-501
San Diego - Police F2	San Diego	453.350 R	WNFJ-501
San Diego - Security	San Diego	154.570 / 154.600	KB-73495
San Dieguito	Leucadia	151.925	KZS-283
San Marcos - Buses	San Marcos	155.235	WXK-701
Southbay - Buses	Imperial Beach	155.265	WXP-422
Sweetwater Union	San Diego	857.425	WNBY-264
Valhalla - Security	El Cajon	464.500	KB-36164
Vista - Buses	Vista	862.000 R	WNHS-769
Vista - Security	Vista	155.280	KNBF-296

UTILITIES

C.F.E. LIGHT & POWER CO.

CHANNEL	FREQUENCY
Operations Tijuana	163.750 R
Operations Tijuana	163.875 R
Statewide Link	163.800 R
Operations Mexicali	164.025 R

IMPERIAL IRRIGATION DISTRICT

CHANNEL	FREQUENCY
(F-1) Imperial Valley Power	153.650 R
(F-2) Imperial Valley Water	153.680 R
(F-3) Coachella Valley Power	153.470 R
(F-4) Imperial Valley Water	153.485 R
(F-5) Future Operations	153.605 R
(F-6) Utility Operations (low power)	153.635
Operations Central	451.075 R
Operations Central	451.125 R
Operations Winterhaven	463.325 R

Tone: 146.2 (VHF Channels)

CALL SIGN IDENTIFICATION

Brawley WGE-776, WNAZ-380; Calexico WNAZ-380; Calipatria WGE-773;
El Centro KNDZ-623, KTM-36, WGE-774, WNAZ-380, WNCH-396; Holtville WNAZ-380;
Imperial KKK-505, KME-558, WGE-777, WNGG-715, Niland WNCH-369;
Westmoreland WGE-771; Winterhaven WRP-447

SAN DIEGO GAS & ELECTRIC CO.

CHANNEL	FREQUENCY
Countywide Trunking System	**856.150 T**
Countywide Trunking System	**856.300 T**
Countywide Trunking System	**856.450 T**
Countywide Trunking System	**857.150 T**
Countywide Trunking System	**857.175 T**
Countywide Trunking System	**857.300 T**
Countywide Trunking System	**857.375 T**
Countywide Trunking System	**858.150 T**
Countywide Trunking System	**858.300 T**
Countywide Trunking System	**858.375 T**
Countywide Trunking System	**859.150 T**
Countywide Trunking System	**859.300 T**
Countywide Trunking System	**859.350 T**
Countywide Trunking System	**860.150 T**
Countywide Trunking System	**860.300 T**
Countywide Trunking System	**860.425 T**
Liaison with Asplund Tree Service	856-860.800 T
Encino Power Plant - Security	451.525
Southbay Power Plant - Security	451.625
Station Security	158.235
Station Operations	158.400
Telemetry - Various locations	154.46375

UTILITIES

SAN DIEGO GAS & ELECTRIC CO.

RADIO CODES

800	Message received	808	Change channel
801	Routine Assignment	809	Call by phone
802	Urgent Assignment	810	Stand By
803	Any Assignment	811	In service
804	Nothing For You	812	Out of service
805	Cancel Last Message	813	Job progress
806	What is your Location	814	Too weak to read
807	Repeat last message	888	Emergency

CALL SIGN IDENTIFICATION

Camp Pendleton KNJU-841; Carlsbad KNJU-841; Dulzura KNJU-841; El Cajon KNJU-841; Escondido WRW-987; Jacumba KNJU-841; Mount Palomar KNJU-841; Mount Otay KNJU-841; Mount Woodson KNJU-841; San Diego KNJU-841

SAN ONOFRE POWER PLANT

CHANNEL	LOCATION	FREQUENCY	CALL SIGN
Fire - Camp Pendleton	San Onofre	138.600 R	USMC
Fire - Camp Pendleton	San Onofre	139.525 R	USMC
Fire - Camp Pendleton	San Onofre	140.500	USMC
Fire - Orange County "System 4" Fire Control	San Onofre	140.520	USMC
General Operations	San Onofre	464.675 R	KNBU-490
Paging	San Onofre	465.000 R	KNBU-491
Security Primary	San Onofre	451.025 R	N/A
Security Tactical	San Onofre	451.300 R	N/A
Unit 1 Operations	San Onofre	451.250 R	N/A
Units 2/3 Operations	San Onofre	451.200 R	N/A
U.S. Department of Energy	Camp Pendleton	167.850	USDOE
U.S. Department of Energy	Camp Pendleton	168.450	USDOE
UHF Telemetry	San Onofre	451.1125	KF-4247
UHF Telemetry	San Onofre	451.1625	KF-4247
UHF Telemetry	San Onofre	451.2125	KF-4247
UHF Telemetry	San Onofre	451.2625	KF-4247
UHF Telemetry	San Onofre	451.3125	KF-4247
VHF Portables - Red	San Onofre	151.625	KF-4247
VHF Portables - Blue	San Onofre	154.570	KF-4247
VHF Portables - Green	San Onofre	154.600	KF-4247

UTILITIES

SOUTHERN CALIFORNIA EDISON

CHANNEL	FREQUENCY
High Frequency Net USB	7.5521
Operations Blythe	153.620
Operations Southern California	37.800
Operations Southern California	158.130
Operations Southern California	158.145
Operations Southern California	158.160
Operations Southern California	158.175
Operations Southern California	158.190
Operations Southern California	158.205
Operations Southern California	158.220
Operations Southern California	158.235
Operations Southern California	158.250
Operations Southern California	158.265
Operations Trunked	856.275 T
Operations Trunked	857.275 T
Operations Trunked	858.425 T
Operations Trunked	859.425 T
Operations Trunked	860.275 T
Operations Trunked	862.350 T
Operations Trunked	863.350 T

TELEPHONE COMPANIES

NAME	LOCATION	FREQUENCY	CALL SIGN
M.C.I. Communications	San Diego	462.525	N/A
Pacific Telephone	San Diego	151.985 R	KNFL-955
Pacific Telephone	San Diego	152.840 R	KWU-207
Pacific Telephone	San Diego	859.450 R	KNIQ-673
Tel-Mexico	Guerrero Negro, B.C.	153.950	N/A
Tel-Nor	Tijuana, B.C.	170.325 R	N/A

WATER DISTRICTS

NAME	LOCATION	FREQUENCY	CALL SIGN
Anaheim Municipal Water	Anaheim	47.940	KMD-621
Anaheim Municipal Water	Anaheim	451.475 R	KMD-621
Anaheim Municipal Water	Anaheim	173.210	WNKT-503
Antelope Valley Water	Antelope Valley	451.150 R	WBQ-843
Azusa Municipal Water	Azusa	451.100 R	KMe-591
Blythe Municipal Water	Blythe	48.020	KMJ-467
Borrego Springs	Borrego Springs	153.650	N/A
California American	Imperial Beach	173.250	KMG-271
California American	Whittier	37.480 R	KME-298
California American	Whittier	48.100	KMG-635
California Water Service	Statewide	47.900	Varies
Calleguas Municipal	Thousand Oaks	47.720	WFQ-974
Carlsbad Municipal	Carlsbad	37.760	WSU-605
Casitas Municipal	Ventura	173.300	KML-588
Channel Islands Co.	Oxnard	153.575	KNFP-565

UTILITIES

WATER DISTRICTS

NAME	LOCATION	FREQUENCY	CALL SIGN
City of Industry Water Co.	Industry	451.200 R	WNRF-637
Coachella Valley Water	Coachella	48.100	KMB-551
Coachella Valley Water	Coachella	48.140	KMB-551
Coachella Valley Water	Coachella	48.200	KMB-551
Colton City Water	Colton	47.780	KMJ-356
Crestline Water	Crestline	37.820	KKK-300
Crestline	Crestline	451.100 R	KFE-240
Del Luz Heights	Fallbrook	47.740	WNAV-759
Del Luz Heights	Fallbrook	48.260	WNAV-759
Desert Water Agency	Palm Springs	153.665	WYG-898
Desert Water Agency	Palm Springs	452.100 R	Varies
Devore Water Co.	Devore	451.100 R	WNBR-551
Dominguez Water Corp.	Compton	37.660	KNI-49
Dominguez Water Corp.	Carson	48.000	KNI-49
East Valley Water District	San Bernardino	47.920	KMH-924
East Valley Water District	San Bernardino	48.220	KMH-942
Eastern Municipal Water	Riverside	48.040	KJR-533
Eastern Municipal Water	Riverside	48.400	KJR-533
Elsinore Valley	Lake Elsinore	37.600	KGU-722
Elsinore Valley	Lake Elsinore	48.280	KGU-722
El Toro Water District	Laguna Hills	451.100 R	KGS-434
Encina Joint Powers	Carlsbad	47.800	KNFR-949
Encina Joint Powers	Carlsbad	48.340	KNFR-949
Fallbrook Public Water	Fallbrook	173.250	WNLY-901
Garden Grove City Water	Garden Grove	153.710	KMJ-537
Glendale City Water	Glendale	47.980 / 48.460	KME-313
Heber Geothermal	Heber	451.100 R	WNAL-662
Helix Water	Lakeside	153.590	KME-221
Hesperia Water District	Crestline	451.175 / 451.575	KMH-942
Hesperia Water District	Crestline	451.625	KMH-942
Hi Desert Water District	Yucca Valley	452.050 R	WGQ-560
Indio Public Water	Indio	153.650	KNHZ-368
Irvine Ranch Water	Newport Beach	451.375 R	KOE-539
Joshua Basin County Water	Joshua Tree	47.820	KYO-713
Joshua Basin County Water	Joshua Tree	48.820	KYO-713
La Habra Heights Water	La Habra	47.880	WNKA-215
La Puente Valley	La Puente	47.800	KNBN-945
Lake Arrowhead	Running Springs	153.665	KBD-951
Lake Arrowhead	Running Springs	173.300	KBD-951
Lake Hemet	Hemet	48.480	KDR-270
Lake Hemet	Hemet	173.390	WBZ-987
Lakeside Irrigation	Lakeside	48.100	N/A
Lakeside Irrigation	Lakeside	153.560	KSR-706
Lakeside Irrigation	Lakeside	153.680	KSR-706
Lake Virgenes	Calabasas	451.100 R	WNKA-633
Leucadia Water	Carlsbad	47.800	KLL-409
Leucadia Water	Carlsbad	48.340	KLL-409
Lincoln Avenue Water	Altadena	37.520	KEJ-449
Loma Linda City Water	Loma Linda	173.390	WFN-512
Long Beach City Water	Long Beach	451.150 R	KME-993
Metropolitan Water District	Southern Calif.	48.540 R	Various
Metropolitan Water District	Southern Calif.	48.560 R	Various
Metropolitan Water District	Southern Calif.	48.440	Various

UTILITIES

WATER DISTRICTS

NAME	LOCATION	FREQUENCY	CALL SIGN
Murrieta County Water	Murrieta	451.125 R	WNXE-938
Olivenhain Municipal	Encinitas	47.920	KEY-477
Olivenhain Municipal	Encinitas	48.400 R	KEY-477
Orange County Water	Yorba Linda	49.160 / 49.460	WNHB-735
Otay Municipal	Dulzura	47.900	KDH-545
Otay Municipal	San Diego	48.380 R	WXY-896
Otay Municipal	Spring Valley	47.900	WGE-753
Otay Municipal	Spring Valley	48.380	WXY-896
Padre Dam Municipal	Santee	48.100 R	KFE-239
Padre Dam Municipal	Santee	153.635	KNDF-245
Padre Dam Municipal	Santee	173.300	KCJ-63
Palo Verde Irrigation	Blythe	173.210	WBK-505
Palm Desert Water	Palmdale	451.100 R	KNAY-345
Pomerado County Water	Poway	153.635	KVD-799
Poway Municipal Water	Poway	48.140	N/A
Poway Municipal Water	Poway	153.635	KVD-799
Rainbow Municipal	Fallbrook	173.300	KAO-986
Ramona Municipal	Ramona	47.880	KMK-279
Ramona Municipal	Ramona	48.480	KMK-279
Rancho Santa Fe	Rancho Santa Fe	47.780	KFM-815
Rancho Santa Fe	Rancho Santa Fe	48.280	KFM-815
Rincon Del Diablo	Escondido	462.475 R	KNFX-704
Riverside City Water	Riverside	47.720 / 48.020	KA-5977
Riverview Water	Lakeside	153.650	KVI-520
Running Springs Water	Running Springs	37.820	KLQ-527
San Diego Co. Water (F-1)	San Diego	47.760 R	various
San Diego Co. Water (F-2)	San Diego	47.820 R	various
San Dieguito Water	Encinitas	48.240 R	KDO-555
San Gabriel Valley Municipal	Glendale	507.4125R	KWT-467
Santa Ana City Water	Santa Ana	451.200 R	KMF-829
Santa Margarita Water	Mission Viejo	173.2375	KNEN-458
Southern Ca. Water	El Centro	153.425	KMC-929
Southern Ca. Water	El Centro	153.530	KMC-929
Southern Ca. Water	El Centro	153.695	KMC-929
Southern Ca. Water	El Centro	451.200 R	WFS-956
Sweetwater Authority	Chula Vista	153.655	WRB-507
Tijuana Water Dept.	Tijuana	171.100 R	N/A
Vallecitos Water District	San Marcos	153.650	KDW-858
Vallecitos Water District	San Marcos	158.175	KDW-858
Valley Center Water District	Valley Center	47.940	KOE-92
Valley of Enchantment Water	Crestline	464.350 R	N/A
Vista Irrigation District	Vista	48.060	KMG-580
Yuma County Water District	Winterhaven	173.210	WNGS-810

MISCELLANEOUS

NATIONWIDE
CITIZEN'S BAND RADIO

CHANNEL NO.	FREQUENCY	CHANNEL NO.	FREQUENCY
1	26.965	21	27.215
2	26.975	22	27.225
3	26.985	23	27.255
4	27.005	24	27.235
5	27.015	25	27.245
6	27.025	26	27.265
7	27.035	27	27.275
8	27.055	28	27.285
9	**27.065**	29	27.295
10	27.075	30	27.305
11	27.085	31	27.315
12	27.105	32	27.325
13	27.115	33	27.335
14	27.125	34	27.345
15	27.135	35	27.355
16	27.155	36	27.365
17	27.165	37	27.375
18	27.175	38	27.385
19	27.185	39	27.395
20	27.205	40	27.405

Certain agreements are generally observed on the CB channels, including AM mode on channels 1-34 and SSB on channels 35-40. Channel 9 is set aside for emergency communications; channel 19 is for traveler assistance; and channel 21 is used by many truckers.

Within the frequency range of the CB band there are several channels reserved for one-way non-voice communications: 26.995 (3A), 27.045 (7A), 27.095 (11A), 27.145 (15A), and 27.195 (19A). These channels are used for radio controlled toys, low-power paging services, car alarms, rural electric power companies, and water and gas utility companies.

FREEBAND OPERATIONS

The frequency range from 25.5-28.0 is home to thousands of unlicensed radio stations surrounding the legitimate CB Band. Over the years, a kind of unofficial band plan has developed for the usage of these frequencies.

Generally speaking, the 25.5-26.965 MHz range is used by AM and FM operators. Channelization is typical in 10 kHz increments (e.g., 26.815, 26.825, 26.835). Most conversation here is either technically oriented or similiar to that found on the legal AM channels.

SSB is the dominate mode between 27.405-28.0 MHz with LSB being used by most english speaking stations. Conversation here consists of technical topics, DX contacts, and ragchewing.

NATIONWIDE

CELLULAR TELEPHONES

Cellular systems divide each service area into cells of frequencies which are reused many times over to accommodate thousands of conversations. As the mobile telephone travels, the computer at the cell site constantly checks its signal quality and automatically switches it to a stronger radio site as the signal weakens. Cellular frequencies are 30 KHz spaced and mobiles are 45 MHz lower than the base frequency.

Base A: 869.000-879.360, 890.000-891.500 MHz
Base B: 880.650-890.000, 891.500-894.000 MHz

Mobile A: 824.000-835.650, 845.000-846.500 MHz
Mobile B: 835.650-845.000, 846.500-849.000 MHz

CORDLESS TELEPHONES (PRE-1984)

BASE: 1695 kHz, 1725 kHz, 1755 kHz.
MOBILE: 49.830, 49.845, 49.860, 49.875, 49.890.

CORDLESS TELEPHONES (POST 1984)

BASE/MOBILE: 46.610/49.670, 46.630/49.845, 46.670/49.860, 46.710/49.770, 46.730/49.875, 46.770/49.830, 46.830/49.890, 46.870/49.930, 46.930/49.990, 46.970/49.970.

CORDLESS TELEPHONES (PROPOSED)

BASE/MOBILE: 43.720/48.760, 43.740/48.840, 43.820/48.860, 43.840/48.920, 43.920/49.020 43.960/49.090, 44.120/49.100, 44.160/49.160, 44.180/49.200, 44.200/49.240 44.320/49.280, 44.360/49.360, 44.400/49.400, 44.460/49.460, 44.480/49.500

The newest cordless phone band operates between 902-928 MHz. Baby monitors and other low power devices also use the above frequencies.

GENERAL REFERENCE

AMATEUR NATIONWIDE SIMPLEX: 29.500, 223.500, 146.520, 446.000, 446.500

AMERICAN RED CROSS: 47.420 (nationwide), 47.460, 47.500, 47.660.

LAW ENFORCEMENT: 155.475, 866.0125, 866.5125, 867.0125, 867.5125, 868.0125, 868.5125

MOTION PICTURE: 152.870, 152.900, 152.930, 152.960, 152.990, 153.020, 173.225, 173.275, 173.325

NATIONAL WEATHER SERVICE: 162.400, 162.425, 162.450, 162.475, 162.500, 162.525, 162.550

REACT TEAMS: 27.065, 462.575, 462.675

SEARCH & RESCUE: 47.460, 155.160, 155.235

WALKIE-TALKIES (unlicensed): 49.830, 49.845, 49.860, 49.875, 49.890

MOBILE TELEPHONES

VHF LOW BAND: 35.260, 35.300, 35.340, 35.380, 35.420, 35.460, 35.500, 35.540, 35.620, 35.660.

VHF HIGH BAND: 152.030, 152.060, 152.090, 152.120, 152.150, 152.180, 152.210, 152.510, 152.540, 152.570, 152.600, 152.630, 152.660, 152.690, 152.720, 152.750, 152.780, 152.810.

UHF BAND: 454.025, 454.050, 454.075, 454.100, 454.125, 454.150, 454.175, 454.200, 454.225, 454.250, 454.275, 454.300, 454.325, 454.350, 454.375, 454.400, 454.425, 454.450, 454.475, 454.500, 454.525, 454.550, 454.575, 454.600, 454.625, 454.650.

NATIONWIDE

SATELLITE & SPACE COMMUNICATIONS

This is an area of communications that scanner listeners rarely explore. However, the military SATCOMM, the Space Shuttle, the Russian Space Program, and weather satellites may be easily monitored by anyone with a quality receiver and a VHF antenna. The best frequency ranges to search are 136-138 MHz, 261-262 MHz, 265-266 MHz, and 268-269 MHz.

FLEET SATCOMM

261.450 Bravo Yankee 1
261.475 Bravo Yankee 2
261.500 Bravo Yankee 3
261.525 Bravo Yankee 4
261.550 Bravo Yankee 5
261.575 Bravo Yankee 6
261.600 Bravo Yankee 7
261.625 Bravo Yankee 8
261.650 Bravo Yankee 9
261.675 Bravo Yankee 10
261.700 Bravo Yankee 11
261.725 Bravo Yankee 12
261.750 Bravo Yankee 13
261.775 Bravo Yankee 14
261.800 Bravo Yankee 15
261.825 Bravo Yankee 16
261.850 Bravo Yankee 17
261.875 Bravo Yankee 18
261.900 Bravo Yankee 19
261.925 Bravo Yankee 20
261.950 Bravo Yankee 21
261.975 Bravo Yankee 22

OTHER CHANNELS

121.750 Russian Space Program
137.150 Tiros Satellite
137.300 Tiros Satellite
137.500 Meteor Satellite
137.620 Meteor Satellite
142.415 Russian Space Program
143.625 Russian Space Program
259.700 Space Shuttle
279.000 Space Shuttle
296.800 Space Shuttle

SPECIAL EVENTS

These are the frequency ranges you want to check first at all special events including county fairs, auto races, air shows, amusement parks, ballooning, etc. Also try these frequencies for hotels, plant security, shopping centers, hospitals, school security, private security, and construction crews.

Frequency bands to search:

151.490-151.955
154.490-154.625
461.000-475.000

Business Low Power:

30.840, 33.140, 35.040, 42.980, 151.625, 151.655, 151.685, 151.715, 151.745, 151.805, 151.835, 151.865, 151.895, 151.925, 151.955, 154.540, 154.600, 464.325, 464.375, 464.425, 464.475, 464.500, 464.525, 464.550, 464.675, 464.775, 464.875, 464.925, 464.975, 465.000, 465.500, 465.550, 467.850, 467.875, 467.900, 467.925

Special Industrial Radio:

43.020, 151.505, 158.400, 451.800, 456.800

WIRELESS DEVICES

The following list is used by drive-thru restaurants such as McDonald's, intercom systems, parade floats, low power walkie-talkies, police surveillance, and wireless communicators.

VHF Band:
30.840, 33.120, 33.140, 35.020, 37.100, 39.600, 40.220, 42.980, 47.240, 49.830-49.860, 72.100, 72.300, 72.500, 73.100, 73.300, 73.600, 75.100, 75.300, 75.500, 75.700, 75.900, 150.775, 150.790, 151.625, 151.715, 151.775, 154.570, 154.600, 169.200, 169.425, 169.445, 169.550, 170.245, 171.905, 174.600, 174.800, 175.000, 177.000, 177.600, 180.000, 180.600, 181.600, 183.600, 184.000, 186.000, 186.600, 190.600, 192.600, 194.600, 195.600, 196.600, 199.600, 202.400, 204.800, 205.600, 206.400, 208.200, 209.150,

UHF Band:
457.525, 457.550, 457.5625, 457.575, 457.5875, 457.600, 460.8875, 460.9125, 460.9375, 460.9625, 460.9875, 461.1125, 465.8875, 465.9125, 465.9375, 465.9625, 465.9875, 467.7875, 467.8125, 467.8375, 467.8625, 467.8875, 467.9125, 467.9375, 467.9625, 467.9875, 470-473 MHz.

DISASTER COMMUNICATIONS

During the rash of natural disasters and disturbances that have struck the Southern California region in the last several years, the following frequencies have been noted to be some of the most interesting and informative to monitor. Remember also to tune in the local AM/FM radio stations as well.

Amateur Radio Service

52.525 Simplex
145.160 ALERT Net
145.460 Earthquake Net
146.520 Simplex
146.880 Ventura County RACES
146.895 Orange County RACES
147.195 San Diego County RACES
147.360 DARN Net
147.705 Earthquake Net
223.500 Simplex
223.760 Orange County RACES
223.800 San Diego County RACES
223.880 Southern Calif. RACES
224.300 Los Angeles County RACES
224.460 Riverside County RACES
446.475 Statewide RACES
447.325 Orange County RACES
447.475 Orange County ARES Net
1282.025 Orange County RACES
1282.525 Orange County RACES
1283.100 Orange County RACES

Large Brush Fires Tactical

166.6125 Tactical
167.950 Air-to-Ground
168.700 Command Net 1
168.100 Command Net 2
168.075 Command Net 3
169.175 Command Net 5
168.775 Cache Repeater
168.050 Tactical 1
168.200 Tactical 2
168.600 Tactical 3
169.925 Tactical 4
169.950 Tactical 5

Statewide EMS System

46.260 Riverside County
46.260 San Bernardino County
46.320 San Diego/Imperial
46.340 San Diego County

Orange County

460.025 County RED (Sig Alerts)

Newsmedia

450.1875 NBC Link
450.700 KNX News 1070
450.850 KFWB News 98
462.675 REACT Liaison

Civil Air Patrol

5236.0 KHz National Command Net
7635.0 KHz National Command Net
10586.2 KHz National Command Net
14396.5 KHz National Command Net
14902.0 KHz National Command Net
14902.0 KHz National Command Net
122.900 Air-to-Air
123.100 Search & Rescue
148.150 VHF Primary

Fire Air Tactical

151.220 CDF Air Drops
151.280 CDF Air Tactical
151.295 CDF Air Tactical
151.310 CDF Air Tactical
168.550 BLM Air Tactical
166.675 Air Tactics 1
169.150 Air Tactics 2
169.200 Air Tactics 3
170.000 Air-to-Ground

Fire Mutual Aid

154.160 State OES
154.220 State OES
154.755 State OES
154.280 Fire WHITE 1
154.265 Fire WHITE 2
154.295 Fire WHITE 3
156.075 State Calcord
487.2375 State FIRMARS
868.9875 State FIREMARS

Statewide SECURE

2326.0 KHz USB Fixed/Mobile
2419.0 KHz USB Fixed/Mobile
2422.0 KHz USB Fixed/Mobile
2802.0 KHz USB Fixed/Mobile
5195.0 KHz USB Fixed
7480.0 KHz USB Fixed
7802.2 KHz USB Fixed
7805.0 KHz USB Fixed

Los Angeles County

37.900 Civil Defense
47.240 Caltrans Sig Alerts
155.400 Hospitals HEAR
170.200 Emergency Net
453.250 LA City Local Govt.
484.4875 Glendale Water & Power

FEMA

5212.0 KHz
10494.0 KHz
138.225 Repeaters
139.825 Repeaters
142.975 Tactical
143.000 Tactical

Fire Tactical VHF Air

118.950 VHF-1 Air Tactics
122.850 VHF-2 Air Tactics
122.975 VHF-3 Air Tactics
123.050 VHF-4 Helicopters
123.075 VHF-5 Helicopters
122.925 VHF-6 Air Tactics
119.950 VHF-7 Air Tactics
119.975 VHF-8 Air Tactics
122.900 VHF Air-to-Air
127.475 VHF Tactical
127.925 VHF Tactical

Police Mutual Aid

42.340 CHP BLUE
42.280 CHP BLUE
45.080 CHP IVORY
154.920 CLEMARS
155.475 NLEMARS
158.790 CLERS
460.025 CLEMARS
460.375 State Police
484.2125 LA Small City
484.2375 LA Small City
868.5125 CLEMARS

Red Cross Network

6998.5 KHz SITOR-A
13569.5 KHz SITOR-A
13915.0 KHz SITOR-A
13950.0 KHz SITOR-A
13965.0 KHz SITOR-A
14375.0 KHz SITOR-A
47.420 Disaster Net
463.475 LA Repeaters
463.500 LA Repeaters

Others

143.990 Army MARS
148.650 Army MARS
148.950 Navy MARS
148.725 Coast Guard

Many agencies and services rely upon radio links during a disaster since the telephone system is routinely disrupted or overloaded. Cellular telephone service may be out completely.

SCANNER MODIFICATIONS

SOME BASIC PRECAUTIONS

Simple modifications may be performed to many popular scanners if you have the proper knowledge, small wire snippers and minimal ability. For example, the Radio Shack PRO-2006 may be modified to receive the 824-851 and 869-894 MHz frequency ranges. Remember to always disconnect the power to the radio and be prepared to undo the modification. Some procedures require good soldering skills. Listening to certain restored frequencies also may be against the law.

If you have any doubts on your ability to perform a modification, please consult a licensed technician. The Author and Publisher assume no liability for the consequences of performing these procedures and the warranty of the radio may be voided.

BASE / MOBILE SCANNERS

REALISTIC PRO-2004

Remove the screws from the case, slip off the cover, and locate circuit board PC-3. Pry off the cover to the silver box and you will see the matrix of diodes. Snip one lead of diode **D-513** (Cellular Restoration) and add a 1N914 diode in position **D-510** (100 additional memories).

Note: The 1N914 diode is Radio Shack part #276-1122 and sometimes diode position D-510 is not labeled.

REALISTIC PRO-2005

Again slip off the cover and locate diode **D-502** and clip the exposed lead. It is directly behind the number 3 key on the keypad (Cellular Restoration). Next add a 1N914 diode in position **D-501** (Scan Speed increase).

REALISTIC PRO-2006

Again slip off the cover. Then locate and clip diodes **D-502** (Cellular Restoration) and **D-503** (Scan Speed).

REALISTIC PRO-2021/2022

After sliding off the case, locate the diode matrix in the CPU section by prying off the cover to the silver box. Locate diode **D-44** and snip one lead to regain the cellular band. THIS MODIFICATION IS NOT ALWAYS REPORTED TO WORK.

REALISTIC PRO-2026

Remove the back cover and locate the small circuit board found in the lower righthand corner. Look carefully and you will find the **JUMPER L-201** (wire). Snip it to regain the cellular frequencies.

UNIDEN 760XLT / 950XLT
(Recent Models with BNC Connector)

Remove the four screws holding the bottom cover and locate the SANYO IC (LC3517BM-15). You will see a long row of solder pins above the IC. Find Pin 26 and cut the two traces leading to it. Next get out your soldering iron.

Solder bridge **Pins 26** & **27** together.
Solder bridge **Pins 19** & **20** together.
Solder bridge the two leads of the transistor above **Pin 27**.
Cut one lead of the **47K** chip resistor marked "427."

UNIDEN 760XLT / 950XLT
(Older Models without BNC Connector)

Remove the four screws holding the bottom cover as above and locate the SANYO IC (L3517BM-15) and clip **Pin 20**. That's all there is to it. Carefully reassemble the radio.

HANDHELD SCANNERS

Handhelds are much more difficult and tricky to work on and all procedures require more time and better soldering skills. If you have any doubts of your ability, consult a good technician.

Realistic scanners all open up the same way. Remove the knobs, antenna, and the four screws from the case. Gently pry open the case and slide the cover back up over the antenna. Disconnect the two circuit board connectors and unsolder the antenna leads. Next unsolder the leads from the board that go to the volume control. Now unfasten and remove the board by disconnecting the in-line pin connector and hexadecimal board mounts. Remove the three small screws holding the metal shield to reveal the circuit board and the diodes.

PRO-37: Clip one lead of diode D-13
PRO-39: Clip one lead of diode D-6

UNIDEN 200XLT (Early Models)

Remove the battery pack and the antenna. Remove the two screws holding the battery and the two holding the case. Gently open the case, pull out the front panel, and locate resistor **R215** which is just above the microprocessor (UC-1147). Note that the resistor is directly above the "den" letters. Clip one lead of **R215**. Carefully reinstall the front panel and make sure the dual in-line connector is properly installed when the radio is reassembled.

UNIDEN 200/205XLT (Recent Models)

Gain access to the circuitry using the same procedure as above and again locate the microprocessor (UC-1147). Locate the chip resistor above the "den" letters. This time add a **10K** resistor from the chip resistor to **Pin 8** of the adjacent IC (K1013). Carefully reassemble the radio.

RADIO CODES

IMPERIAL COUNTY REGIONAL CODES

10-1	Reception poor
10-2	Reception good
10-3	Change channels
10-4	Message acknowledged
10-5	Relay message
10-6	Busy, stand by
10-7	Out of service
10-7A	Out of service, home
10-8	In service
10-9	Repeat last message
10-10	Off duty
10-13	Advise road/weather conditions
10-15	Prisoner in custody
10-17	Pick up paperwork
10-18	Assist the officer at ______ location.
10-19	Return to station
10-20	What is your location?
10-21	Telephone
10-21A	Telephone, home
10-22	Disregard
10-23	Standby
10-26	No wants
10-27	Drivers license check
10-28	Request registration check
10-29	Check for wants
10-30	Does not conform to regulations
10-34	Assistance required at office
10-36	Confidential information
10-36A	Armed subject
10-36F	Felon subject
10-37	Identify operator
10-39	Message delivered
10-97	Arrived at scene
10-98	Leaving scene
11-7	Prowler
11-8	Person down
11-10	Take a report
11-12	Loose live stock
11-24	Abandoned vehicle
11-25	Traffic hazard
11-27	DDL request/driver detained
11-28	Registration/driver detained
11-41	Ambulance required
11-42	Ambulance not required
11-44	Coroner's case (dead body)
11-45	Attempted suicide
11-48	Furnish transportation
11-53	Security check
11-55	Hazardous or chemical spill
11-60	Investigate water leak
11-66	Signals out of order
11-71	Fire
11-79	Accident - Ambulance sent
11-80	Accident - Serious injury
11-81	Accident - Minor injury
11-82	Accident - Non-injury
11-83	Accident - No details
11-84	Traffic control
11-85	Request tow truck
11-86	Special detail
11-88	Citizen assist
11-98	Meet ______.
11-99	Officer needs help

Code 2 Urgent traffic
Code 3 Emergency
Code 4 No further assistance needed
Code 5 Stakeout
Code 6 Clear the area
Code 7 Out of service to eat
Code 33 Clear this channel for emergency traffic.

RADIO CODES

LOS ANGELES & ORANGE REGIONAL CODES

10-1	Receiving Poorly
10-2	Receiving Well
10-3	Stop Transmitting / Change Channels
10-4	Acknowledged Message
10-5	Relay Message
10-6	Busy
10-7	Out of Service - Radio Off
10-7B	Out of Service (Home)
10-8	In Service
10-9	Repeat last Message
10-10	Out of vehicle - Subject to call
10-11	Transmitting too rapidly
10-12	Visitors Present
10-13	Advise Weather Conditions
10-14	Provide Escort
10-15	En Route with Prisoner in custody
10-16	Pick up Prisoner
10-17	Pick up Paperwork
10-18	Complete Assignment ASAP (Also Riverside)
10-19	Return to Station
10-20	What is your location?
10-21	Telephone
10-21A	Advise home I will return ___.
10-21H	Telephone your home
10-22	Disregard or cancel last message
10-23	Stand by
10-24	Trouble at Station (rare use)
10-25	Do you have contact with ___.
10-27	Any return on number or subject?
10-28	Vehicle registration check (complete record
10-28C	Change in ownership
10-29	Vehicle registration and wants check
10-29F	Subject is wanted for felony
10-29FD	Subject is wanted for felony and is dangerous
10-29H	Confidential info on your subject
10-29T	Subject is wanted for traffic warrant (s)
10-29V	Property is reported stolen
10-30	Improper conduct on radio
10-31	Request unit and frequency (LASO)
10-31	Subject has no record, no wants
10-32	Subject wanted
10-32F	Subject wanted. Felon
10-33	Emergency Traffic
10-34	Request routine clearance
10-35	Confidential Information on subject
10-36	Correct Time
10-37	Identify operator
10-38	Request clearance to run subject (s)
10-39	Request to clear call
10-40	Is ___ available by telephone
10-40A	Is ___ available by radio?
10-42	Pickup officer
10-45	Servicing equipment
10-46	Stand by, going to better radio spot
10-48	Ready to take information
10-49	Proceed to ___.
10-86	Traffic Check
10-97	Arrived at scene
10-98	Finished last assignment
10-99	Emergency. Officer needs help!

Ambulance Traffic: Accident - Ambulance Enroute

Code 1	Acknowledge receipt of message
Code 1	Routine Response
Code 2	Proceed Immediately
Code 2H	High Priority Call
Code 3	Proceed with Lights & Sirens
Code 4	No Further Assistance Required
Code 4A	Situation is not secure, however there are sufficient units at the location
Code 4N	No Assistance needed. No evidence of crime
Code 5	Stake out. All units stay away.
Code 6	Out for investigation. Request SRC
Code 6G	Investigation - Gang Activity
Code 6H	Investigation - Homicide
Code 7	Out of Service to eat
Code 8	Fire Alarm (still on official list)
Code 10	Request clear frequency
Code 12	False Alarm
Code 14	Resume normal operations
Code 20	Notify news media
Code 33	Emergency
Code 37	Vehicle reported stolen
Code 77	Possible ambush, use caution
Code 77N	No answer at call back. use caution
Code 100	In position to intercept
Code 1B	Clear your Buffer (LASO)
Code 1M	Check your MDT (LASO)

Code Alex A: Establishes 53 checkpoints in Orange County
Code Alex B: Establishes major checkpoints in Orange County
Code Charlie C: Mutual Aid Call for riots in Orange County
Code Echo: Bank & Jewelry robbery Radio Directional Finding (RDF)

RADIO CODES

LOS ANGELES & ORANGE REGIONAL CODES

901	Ambulance Call
901A	Ambulance Call - Attempt Suicide
901H	Ambulance Call - Dead Body
901K	Ambulance Call - Overdose
901N	Ambulance needed
901S	Ambulance Call - Sick Person / Shooting
901T	Ambulance Call - Traffic Accident
901Y	Request Ambulance if needed
902	Person sick or injured
902A	Accident - Attempt Suicide
902H	Accident - Enroute to Hospital
902M	Accident - Medical Aid Request
902N	Accident - No injuries
902T	Traffic Collision - No details
903	Aircraft - Crash
903L	Aircraft - Low Flying
904	Fire Call
904A	Fire Alarm
904B	Brush Fire
904C	Car Fire
904F	Forest Fire
904G	Grass Fire
904I	Illegal Burn
904S	Structure Fire
904T	Trash Fire
905	Animal Call
905A	Abuse to Animals
905B	Animal Bite
905D	Dead Animal
905I	Injured Animal
905N	Noisy Animal
905S	Stray Animal
905V	Vicious Animal
906	Rescue Call
906K	Rescue Dispatched
906N	Rescue Needed
907	Disturbance Call
907A	Loud Radio or TV
907B	Ball Game in Street
907K	Paramedics Dispatched
907N	Paramedics Requested
907Y	Do You Need paramedics
908	Begging in Public
909	Traffic Congestion
909T	Traffic Hazard
910	Can You Handle?
911	Advise Person
911A	Contact informant
911B	Contact officer
911N	Do no contact informant
912	Are we clear?
913	You are clear
914C	Request Coronor
914D	Request Doctor
914F	Request Fire Department
914H	Heart attack
914N	Concerned Party Notified
914S	Suicide
915	Dumping Rubbish
916	Officer holding misdemeanor suspect
916A	Officer holding felony suspect
916B	Citizen holding suspect
916C	Citizen holding felony Suspect
917A	Abandoned Vehicle
917S	Suspicious Vehicle
918	Mental Patient
918A	Mental Patient - Escaped
918V	Mental Patient - Violent
919	Keep the peace
920	Missing Adult
920A	Found Adult
920C	Missing Child
920F	Found Child
920J	Missing Juvenile
921	Prowler
921P	Prowler - Peeping Tom
922	Illegal Peddling
922A	Pandering
923	Illegal Shooting
924	Station Detail
924B	Notify the Station
924M	Messenger Service
925	Suspicious Person
925A	Suspicious Person in Auto
926	Tow Truck needed
926A	Tow Truck dispatched
927	Suspicious Circumstances
927A	Person pulled from phone
927B	Open door or window
927C	Check Vicinity
927D	Possible Dead Body
927H	9-1-1 Hangup
927S	Person Screaming
928	Found Property
928L	Lost Property
929	Person down
930	See the man regarding complaint
930A	See the Manager
931	See the women
931A	See the security guard
932	Women or child abuse
933	Open Window
934	Undetermined
935	Undetermined

RADIO CODES

LOS ANGELES & ORANGE REGIONAL CODES

949	Gasoline Spill	975	Can Suspect Hear your Radio?
950	Check Burn Permit	980	Emergency Traffic
951	Request Fire Investigator	981	Frequency is Clear
952	Report Conditions	982	Are we being received?
953	Check Report of Smoke	982	Bomb Threat
954	Arrived at Scene	983	Explosion
955	Fire under control	995	Labor Dispute
956	Available for assignment	**996**	Explosion (LASO)
957	Fire Out on Arrival	996A	Unexploded Bomb
960X	Vehicle Hot Stop	996B	Bomb Explosion
961	Take a Report / Vehicle Stop	**996T**	Bomb Threat
962	Suspect is armed	997	Officer Needs Help!
963	Undetermined	998	Officer involved in Shooting
966	Sniper	999	Officer Needs Help! - Urgent!
967	Outlaw Motorcylist		
969	Undetermined		

LOS ANGELES COUNTY FIRE CODES

Code A	Request Arson Investigator
Code F	Fatality
Code K	Undetermined
Code L	Request Law Enforcement
Code N	Notify News Media
Code R	Emergency Response

LASO TRANSIT SERVICE BUREAU*

830A	Accident - Train vs. Vehicle	837	Remove Power for Train (overhead)
830P	Accident - Train vs. Pedestrian	838	Restore Power for Train (overhead)
831	Train Derailment	840	Check Crowd Conditions
832	Stalled Train	841	Hazmat Incident
833	Undetermined	842	No Service - Advise Passengers

*MetroRail Blue Line. Radio traffic is on 484.1625 (base) & 487.1625 (mobile).

We again would like to extend a special thanks to the contributors who assisted us with the radio codes. It is quite well-known that many codes are used that do not appear on any official list and listener input is the only method of compiling an accurate guide to these codes.

RADIO CODES

RIVERSIDE COUNTY REGIONAL CODES

10-1	Receiving Poorly	10-32	Subject has wants
10-2	Receiving Well	10-33	Emergency Traffic
10-3	Stop Transmitting	10-34	Resume Normal Traffic
10-4	Acknowledged Message	10-35	Confidential Information
10-5	Relay Message	10-36	Correct Time
10-6	Busy - Not Available for Calls	10-37	Name of Operator
10-7	Out of Service	10-42	Transport Item / Pick up ___.
10-8	In Service	10-50	Shakedown
10-9	Repeat last Message	10-57	Call Unverified
10-10	Busy - Available for Calls	10-84	File Number
10-11	Talk Slower	10-85	Request Report
10-12	Units Covering	10-86	Do You Have Traffic?
10-13	Road / Weather Conditions	10-87	Meet ___ at ___.
10-14	Escort / Transport Item or Person	10-88	Call by Telephone
10-15	Enroute to Jail	10-89	Request Helicopter
10-16	Pick up Prisoner at ___.	10-97	Arrived at Scene
10-17	Pick up Paperwork	10-98	Completed Last Assignment
10-18	Complete Assignment ASAP	11-1	Bomb Threat
10-19	Enroute to Station	11-10	Request Backup
10-20	Location	11-11	Officer Needs Assistance
10-21	Call by Telephone	11-80	Accident - Serious Injury
10-22	Disregard	11-81	Accident - Minor Injury
10-23	Stand By	11-82	Accident - Property Damage
10-28	Registration Check	11-83	Accident - No Details
10-29	Check Wants & Warrants	11-85	Need Tow Truck
10-30	Subject has no Warrants	11-99	Officer Needs Help - Fast!

Code 2	Proceed Immediately
Code 3	Proceed with Lights & Sirens
Code 4	No Further Assistance Needed
Code 6	Follow-up Call
Code 7	Out For Lunch
Code 8	Coroner's Case
Code 9	No Units Available
Code 10	Helicopter Not Available
Code 33	Clear Channel for Emergency Traffic!

RADIO CODES

SAN DIEGO COUNTY REGIONAL CODES

The following radio codes are generally used by all law enforcement agencies in the county of San Diego excluding federal and state organizations.However, a few codes are unique to specific agencies.

10-1	Receiving poorly
10-2	Receiving well
10-3	Change channels
10-4	Ackowledged Message
10-5	Relay Message
10-6	Busy
10-7	Out of service
10-8	In service
10-9	Repeat last message
10-10	Remain in service
10-16	Prisoner in custudy
10-17	Routine report
10-19	Return to station
10-20	What is your location?
10-21	Telephone
10-21H	Telephone your home
10-22	Disregard
10-23	Standby
10-26*	No wants or warrants
10-28	Vehicle registration check
10-29	Check for local wants (person or plate)
10-29	NCIC - Check for all wants
10-34	Are you clear?
10-35	Dangerous person alert
10-36	What time is it?
10-41**	Beginning tour of duty
10-42**	Ending tour of duty
10-46	Assist motorist
10-53	Road blocked
10-60	Attempt to contact
10-87	Meet the officer, remain in service
10-88	Request cover unit
10-89	Bomb threat
10-92	Radio check
10-97	Arrived at scene
10-98	Finished last assignment
11-6	Discharge of firearm
11-7	Prowler
11-8	Person down
11-10	Conduct an investigation
11-11	Checking the area, making a pickup
11-12	Injured animal
11-13	Dead animal
11-14	Animal bite
11-15	Ball game in street
11-27	Felony record, no wants
11-28	Misdemeanor record, no wants
11-29	No record, no wants, not stolen
11-30	Incomplete phone call
11-31	Person calling for help
11-40	Notify if ambulance is required
11-41	Ambulance required
11-42	Ambulance not required
11-44	Coroner's case (dead body)
11-45	Attempted suicide
11-46	Report of death
11-47	Injured person
11-48	Provide transportation
11-49	Vehicle stop
11-50	Vehicle stop (license check)
11-51	Pedestrian / Field interview
11-52	Are you O.K.? (If response is other than code word, cover will be sent)
11-53	Security check
11-55	Hazardous spill
11-60	Investigate water leak
11-66	Traffic signals out of order
11-71	Fire
11-80	Serious injury accident
11-81	Minor injury accident
11-82	Non-injury accident
11-83	No detail accident
11-84	Traffic control needed
11-85	Tow truck requested
11-86	Special assist needed
11-88	Assist citizen / disabled motorist
11-99	Officer needs help fast! cover required

* Escondido Police Department

** San Diego Police Department

RADIO CODES

SAN DIEGO COUNTY REGIONAL CODES

Code 2	Proceed immediately
Code 3	Proceed with Lights & Sirens
Code 4	No further assistance needed
Code 5	Stakeout
Code 6	Remain clear of the area
Code 7	Out of service to eat
Code 8	Out of service for restroom
Code 10	SWAT Alert
Code 11	SWAT Staging
Code 12	High risk warrant search
Code 14	Dive Team Call-out
Code 33	Clear this channel for emergency traffic
Code 37	Subject wanted M - Misdemeanor F - Felon Subject T - Traffic V - Vehicle (V-Victor)
Code 52*	Check subject for 4th ammendment waiver
Code 100	Station under Siege

Code BLUE: Bus or Taxi in trouble

<u>Emergency Traffic:</u>

Clears the channel for only priority traffic. Therefore, an officer in a dangerous or life-threatening situation can immediately contact the dispatch center. Periodic beeping tones are sometimes used to indicate this type of situation.

*El Cajon PD

<u>Cover Now!</u>

Please expedite the sending of a cover unit!

SAN DIEGO FIRE & PARAMEDIC CODES

Code 10	Critical Trauma Patient
Code 20	Acute Truama Patient
Code 30	Trauma Case
Code 40	Serious Case (IV started)
Code 50	Basic Transport (not serious)
Code N	Newsworthy Incident

SAN DIEGO PD DISPOSITION CODES

A	Arrest made, report will be submitted
K	No report required
R	Report made, will be submitted
U	Unfounded

RADIO CODES

STATE OF CALIFORNIA HIGHWAY PATROL

10-1	Receiving Poorly
10-2	Receiving Well
10-3	Change Channels
10-4	Acknowledge (O.K.)
10-5	Relay Message
10-6	Busy
10-7	Out of Service
10-8	In Service
10-9	Repeat last message
10-10	Off duty
10-11	Identify mobile call sign
10-12	Computer Delay
10-14	Provide Escort
10-15	Prisoner in custody
10-17	Relay papers, supplies, etc.
10-19	Return to station
10-20	What is your location?
10-21	Telephone
10-22	Disregard
10-23	Stand By
10-28	Vehicle Registration Check
10-29	Check for wants
10-37	What time is it?
10-39	Message or item delivered
10-97	Arrived at scene
10-98	Assignment Completed
11-10	Take a report
11-12	Loose Livestock
11-24	Abandoned Vehicle
11-25	Traffic Hazard
11-26	Disabled Vehicle
11-27	DDL Check
11-28	Registration Check
11-41	Ambulance Needed
11-44	Possible Fatality
11-48	Provide Transportation
11-66	Defective Traffic Signals
11-78	Paramedics Dispatched
11-79	Paramedics Sent
11-80	Accident - Serious Injury
11-81	Accident - Minor Injury
11-82	Accident - Property Damage
11-83	Accident - No Details
11-84	Direct Traffic
11-85	Tow Truck Needed
11-86	Bomb Threat
11-87	Bomb Found
11-98	Meet the Officer at ___.

Code 2	Urgent
Code 3	Emergency (Lights & Siren)
Code 4	No Further Assistance Needed
Code 5	Stakeout
Code 7	Out of Service
Code 20	Notify Press
Code 33	Emergency Traffic

The letter "X" used with any code specifies that a female is involved.
The letter "A" may be used to designate an abandoned vehicle.

STATE OF CALIFORNIA CALTRANS

10-10	Off duty
10-12	Can you write?
10-14	Do you copy so far?
10-16	Contact ___
10-17	Unable to contact ___
10-18	Use different channel
10-19	Return to office
10-23	Standby
10-30	Emergency Traffic
10-32	Dead animal
10-40	Normal radio traffic
10-43	Message delivered
10-44	Do you have traffic?
10-45	No traffic for this unit
10-48	Traffic accident
10-49	Stalled vehicle
10-50	Emergency traffic - All units stand by

RADIO CODES

STATE OF CALIFORNIA

11350 - Possession of a controlled narcotic
11351 - Possession of a controlled narcotic for sale
11357 - Possession of marijuana
11358 - Cultivation of marijuana
11359 - Possession of a controlled narcotic
11550 - Under the influence of a narcotic

PENAL CODES

146a - Impersonating an officer
148 - Resisting arrest or interfering
182 - Conspiracy
187 - Murder
203 - Mayhem
207 - Kidnapping
207A - Attempted Kidnapping
211 - Armed Robbery
220 - Assault with intent to commit mayhem, rape, or robbery.
241 - Assault against a peace officer
242 - Battery
245 - Assault with a deadly weapon
246 - Shooting at inhabited dwelling
261 - Rape
261A - Attempted Rape
270 - Failure to provide
273 - Child abuse
273A - Child abuse - neglect
278 - Child abduction
285 - Incest
286 - Sodomy
288 - Oral Copulation
314 - Indecent Exposure
330 - Gambling
374 - Illegal dumping
415 - Disturbing the peace
416 - Refusing to dispurse
417 - Displaying a deadly weapon in a threatening or rude manner
451 - Arson
459 - Burglary
470 - Forgery
480 - Felony hit and run
481 - Hit and run
484 - Petty theft
487 - Grand theft
488 - Petty theft
496 - Possession of stolen property
499b - Joyriding
502 - Drunk driving (C.V.C. 23152)
503 - Auto theft (C.V.C. 10851)
504 - Tampering with a vehicle (SDPD)
505 - Reckless driving (C.V.C. 23103)
518 - Extortion (SDPD)
537 - Defrauding an innkeeper
586 - Illegal parking (SDPD)
587b - Trespassing on trains
594 - Vandalism
595 - Runaway Vehicle
597 - Cruelty to animals
602 - Trespassing
602.5 - Breaking and entering
647a - Engaging in lewd act in public
647b - Prostitution
647c - Begging
647d - Soliciting for lewd act
647e - Loitering
647f - Drunk in public
647g - Prowling
647i - Illegal lodging
653m - Making annoying telephone calls
12020 - Possession of an illegal weapon
12025 - Carrying a concealed weapon (CCC)

VEHICLE CODES

4000A - Unregistered vehicle
10851 - Stolen vehicle
10852 - Tampering with vehicle
10853 - Vandalizing vehicle
12500 - Driving without a license
14601 - Driving with a suspended license
15500 - Acquisition of vehicle by juvenile
20001 - Hit and run
20002 - Hit and run, property damage
20003 - Hit and run, death or injury
21650 - Driving on wrong side of street
22348a - Driving faster than allowed speed
22350 - Driving unsafe speed
22450 - Running stop sign
22500 - Illegal parking
23103 - Reckless driving
23110 - Throwing objects
23152 - Driving under the influence
23153 - Drunk driving and causing bodily injury
23222a - Open alcohol container in vehicle
23222b - Possession of marijuana in vehicle

WELFARE AND INSTITUTION CODE

5149.5 - Near Mental Case (not a real code)
5150 - Mental Case

RADIO CODES

LAW ENFORCEMENT ABBREVIATIONS & LINGO

"77" - Residence
AB - Airborne
ABLE - Airborne Law Enforcement (SDPD)
AC - Aircraft Crash (LASO)
ACC - Area Communications Coordinator (LAPD)
AD - Armed and Dangerous
ADW - Assault with a Deadly Weapon
AF - Arrest Felony (SDSO)
AID - Accident Investigation Detail
AJ - Arrest Juvenile (SDSO)
AKA - Also Known As
AM - Arrest misdemeanor (SDSO)
AMA - American Medical Association
AMH - Anaheim Memorial Hospital (OCSO)
AR - Alarm Reset (SDSO)
ARJIS - Automated Regional Justice Information - System
ARO - Animal Regulations Office
ASAP - As Soon As Possible
ASP - Alarm System Passage (number)
ASTREA - Aerial Support To Regional Enforcement Agencies (SDSO Helicopter)
ASU - Anti-Smuggling Unit (I.N.S.)
BC - Bar Check
BFA - Black Female Adult
BFJ - Black Female Juvenile
BMA - Black Male Adult
BFJ - Black Female Juvenile
BO - Out of Order / Business Office
BOL - Be On the Lookout
BP - Border Patrol
BT - Bomb Threat
Cars - Car to car channel
CB - Cardiff Beach (SDSO)
CB - Citizen Band
CC - Cancelled after dispatch (SDSO)
CC - Civil Case
CC - Communications Center
CCP - Center City Parkway (Escondido PD)
CCW - Carrying Concealed Weapon (CCC)
CD - Civil Defense
CDF - California Department of Forestry
CDF - Colinas Detention Facility (SDSO)
CDL - California Drivers License
CDM - Camino Del Mar (San Diego)
Chipy - California Highway Patrol
CHP - California Highway Patrol
CI - Citation Issued (SDSO)
CI - Confidential Informant
CJ - County Jail
CLEMARS- California Law Enforcement Mutual Aid Radio System (frequency 154.920 MHz)
CLETS - California Law Enforcement Teletype System
CMH - County Mental Hospital
CMS - County Medical Services
CP - Command Post / Complaining Party
CO - Commanding Officer
COV - City of Vista (SDSO)
CP - Command Post
CPD - Carlsbad PD / City Property Damage
CPPS - California Plant Protection Security
CRO - Community Relations Officer
CSO - Community Service Officer
CSP - California State Police
CSU - Crime Suppression Unit
CVA - Cardiovascular (stroke) call
CVC - California Vehicle Code
CYA - Cover Your Ass (rare use)
DB - Dead Body
Deuce - Drunk Driver (C.V.C. 23152)
DL - Drivers License
DMV - Department of Motor Vehicles
DOA - Dead on Arrival
DOB - Date of Birth
DOJ - Department of Justice
DR - Daily Report (case number)
DUI - Driving Under the Influence (C.V.C. 23152)
DWI - Driving While Intoxicated
EBS - Emergency Broadcast System
ECPD - El Cajon Police Department
ECR - El Camino Real (OCSO)
EHS - Escondido High School (EPD)
EPD - Escondido Police Department
EOC - Emergency Operations Center
EOS - End of Shift / Out of order
EOW - End of Watch
ET - Emergency Traffic
ETA - Estimated Time of Arrival
ETOH - Intoxicated
FA - False Alarm (Sheriff disposition)
FD - Fire Department
FI - Field Interview / Follow up
FM - Found Merchandise
FPD - Fire Protection District
FSP - Free Service Patrol (CHP)
FTA - Failure To Appear
FTP - Failure To Pay (fines)
FTO - Field Training Officer
FYI - For Your Information
GA - Same as GOA (Sheriff disposition)
GOA - Gone On Arrival
GSW - Gun Shot Wound
GTA - Grand Theft Auto (C.V.C. 10851)
HA - Hell's Angels
HCFA - Heartland Communications Facility Auth.
H & R - Hit and Run
H & S - Health and Safety (Penal Code Violations)
HBD - Has Been Drinking
HFA - Hispanic Female Adult
HFJ - Hispanic Female Juvenile
HMA - Hispanic Male Adult
HMJ - Hispanic Female Juvenile
Hot Stop - Vehicle stop on a suspected felon
HTF - Heartland Training Facility (El Cajon)
HUA - Head Up Ass (rare use)
IC - Internal Command
IO - Information Only
ISO - Imperial County Sheriff's Office
LGA - Lemon Grove Avenue (SDSO)

RADIO CODES

LAW ENFORCEMENT ABBREVIATIONS & LINGO

INS - Immigration Naturalization Service
Landline - Telephone
LAPD - Los Angeles Police Department
LPD - La Mesa Police Department
LTF - License To Follow
LSF - Lomas Santa Fe (SDSO)
MA - Mission Avenue (SDSO)
MARS - Military Affiliated Radio System
MDT - Mobile Data Terminal
MICU - Medical Intensive Care Unit
M Mary - Misdemeanor (Escondido PD)
MO - Modus Operandi (Method of Operation)
MP - Military Police
MRS - Mobile Radio Service
MTA - Metropolitan Transit Authority (LA County)
NLEMARS- National Law Enforcement Mutual Aid Radio System
NCTD - North County Transit District (SD County)
NCIC - National Crime Information Computer
NCS - North County Substation (Vista 30)
NFD - No Further Detail
NFI - No Further Information
NCPD - National City Police Department
NTF - Narcotics Task Force
OCSO - Orange County Sheriff's Office
OES - Office of Emergency Services
OP - Observation Post
OPD - Oceanside PD / Ontario PD
OSI - Office of Special Investigations
PAB - Police Administration Building
PAR - Palomar Airport Road (Carlsbad PD)
PC - Penal Code / Probable Cause
PD - Police Department
PMC - Palomar Memorial Convalescent
PMH - Palomar Memorial Hospital (Escondido)
PR - Person/Party Reporting
PRO - Public Relations Officer
PVH - Paradise Valley Hospital (SDPD)
QA - Same as QOA (SDSO)
QOA - Quiet On Arrival (SDSO)
RACES - Radio Amateur Civilian Emergency Srv.
RADAR - Radio Repair (SDSO)
RB - Rancho Bernardo (SDPD)
RD - Reporting District
REACT - Radio Emergency Associated Communications Team
RF - Report Filed (SDSO)
RJ - Runaway Juvenile
RO - Registered Owner
ROV - Registration of Voters
RP - Reporting Party/Reporting Person
RPVE - Rancho Palos Verdes Estates
RSF - Rancho Santa Fe (SDSO)
RSO - Riverside Sheriff's Office
RTF - Return To File (paper work)
SAFE - Service Auth. For Freeway Emergencies
SBPD - San Bernardino Police Department
SBSO - San Bernardino County Sheriff's Office
SBRPCA - South Bay Regional Pub. Comm. Authority
SCE - Southern California Edison
SDG&E - San Diego Gas and Electric
SDPD - San Diego Police Department
SDSO - San Diego Sheriff's Office
SEB - Special Enforcement Bureau (LAPD)
SED - Special Enforcement Detail (SDSO)
SMPD - Santa Monica PD
SO - Sheriff's Office
SOL - Stopped on Location
SRC - Sheriff's Radio Communications (LASO)
Station "A" - San Diego Police Department
Station "C" - Carlsbad or Chula Vista PD
Station "D" - Coronado Police Department
Station "D" - San Diego Community College PD
Station "F" - San Diego Fire Department
Station "G" - San Diego County District Attorney
Station "H" - Escondido Police Department
Station "K" - El Cajon Police Department
Station "L" - San Diego Marshal's Office
Station "M" - San Diego Sheriff's Office
Station "O" - Oceanside Police Department
Station "Q" - National City Police Department
Station "R" - San Diego County Search & Rescue (155.160 MHz)
Station "X" - San Diego County Emergency Medical Net, Emergency Operations Center
Station 1 - La Mesa Police Department
Station 20 - SD Sheriff's Substation - Encinitas
Station 30 - SD Sheriff's Substation - Vista
Station 32 - SD Sheriff's Substation - Fallbrook
Station 35- SD Sheriff's Substation - San Marcos
Station 40 - SD Sheriff's Substation - Poway
Station 44 - SD Sheriff's Substation - Ramona
Station 50 - SD Sheriff's Substation - El Cajon
Station 52 - SD Sheriff's Substation - Alpine
Station 70 - SD Sheriff's Substation - Imperial Beach
Station 80 - SD Sheriff's Substation - Valley Center
SWAT - Special Weapons And Tactics
TA - Traffic Accident
TC - Traffic Collision
TCH - Tri-City Hospital (Oceanside)
TKS - Thanks
TOC - Time of Call
TRO - Temporary Restraining Order
UC - Undercover Agent / University of California
UF - Unfounded (SDSO)
UL/UTL - Unable To Locate (SDSO)
USD - Unified School District
VDF - Vista Detention Facility (SDSO)
VFD - Volunteer Fire Department
VID - Vista Irrigation District (SDSO)
VIN - Vehicle Identification Number
VSO - Ventura Sheriff's Office
VPS - Vista Patrol Station (SDSO)
WC - Watch Commander
WC - Will Cooperate (SDSO)
WFA - White Female Adult (also CFA)
WFJ - White Female Juvenile (also CFJ)
WMA - White Male Adult (also CMA)
WMJ - White Male Juvenile (also CMJ)
X - Female (CHP)

RADIO CODES

MILITARY TIME

CIVILIAN TIME	MILITARY TIME	CIVILIAN TIME	MILITARY TIME
12:00 Midnight	00:00/24:00	12:00 Noon	12:00
1:00 AM	01:00	1:00 PM	13:00
2:00 AM	02:00	2:00 PM	14:00
3:00 AM	03:00	3:00 PM	15:00
4:00 AM	04:00	4:00 PM	16:00
5:00 AM	05:00	5:00 PM	17:00
6:00 AM	06:00	6:00 PM	18:00
7:00 AM	07:00	7:00 PM	19:00
8:00 AM	08:00	8:00 PM	20:00
9:00 AM	09:00	9:00 PM	21:00
10:00 AM	10:00	10:00 PM	22:00
11:00 AM	11:00	11:00 PM	23:00

PHONETIC ALPHABET

The Phonetic Alphabet is used by all law enforcement agencies for the purpose of clear communications. The first set of phonetics shown are used by city, county, and state law enforcement agencies. The second set of phonetics shown after the slashes are used by federal agencies and amateur radio operators. For example, if an officer of the California Highway Patrol or sheriff's office stopped a vehicle with the license number: RDJ 559, he would communicate with the dispatcher: ROBERT - DAVID - JOHN - 5 - 5 - 9. In another example, if an officer of the Federal Bureau of Investigation stopped a vehicle with the license number: KFJR444, he would communicate with the dispatcher: KILO - FOXTROT - JULIET - ROMEO - 4 - 4 - 4.

A - Adam / Alpha
B - Boy / Bravo
C - Charles / Charlie
D - David / Delta
E - Edward / Echo
F - Frank / Foxtrot
G - George / Golf
H - Henry / Hotel
I - Ida / India
J - John / Juliet
K - King / Kilo
L - Lincoln / Lima
M - Mary / Mike
N - Nora / November
O - Ocean / Oscar
P - Paul / Papa
Q - Queen / Quebec
R - Robert / Romeo
S - Sam / Sierra
T - Tom / Tango
U - Union / Uniform
V - Victor / Victor
W - William / Whiskey
X - Xray / Xray
Y - Yellow / Yankee
Z - Zebra / Zulu

CROSS-REFERENCE

FREQUENCY	NAME
530 kHz	Traveler's Information Service (T.I.S.)
540-1710 kHz	AM Broadcast Band - See Page 267
1610 kHz	Traveler's Information Service (T.I.S.)
2326 kHz	SECURE - Interstate Coordination (F-8)
2419 kHz	SECURE - Night Operations (F-4)
2422 kHz	SECURE - Night Operations (F-5)
2804 kHz	SECURE - Night Operations (F-7)
2812 kHz	SECURE - Night Operations (F-6)
5140 kHz	SECURE - Day & Night Fixed (F-3)
5195 kHz	SECURE - Interstate Coordination (F-9)
7480 kHz	SECURE - Primary Day Operations (F-1)
7802 kHz	SECURE - Primary Day Operations (F-2)
7805 kHz	SECURE - Interstate Coordination (F-10)
9023 kHz	Sierra Pete
25.040	Government Chemical Cleanups (Nationwide)
25.080	Government Chemical Cleanups (Nationwide)
25.870	KBBV (remotes)
26.070	KBBV / KFAC (remotes)
26.110	KBBV / KFAC / KLZZ (remotes)
26.110	KECY-TV Channel 9 (remotes)
26.160	KBBV (remotes)
26.260	KBBV (remotes)
26.310	KBBV (remotes)
26.330	KAVL / KAVS / KBBV (remotes)
26.350	KBBV (remotes)
26.410	KBBV (remotes)
26.450	KBBV / KFAC (remotes)
26.620	Civil Air Patrol (VHF Portables)
26.965	Citizens Band Channel 1
26.975	Citizens Band Channel 2
26.985	Citizens Band Channel 3
26.995	Citizens Band Channel 3A (non-voice)
27.005	Citizens Band Channel 4
27.015	Citizens Band Channel 5
27.025	Citizens Band Channel 6
27.035	Citizens Band Channel 7
27.045	Citizens Band Channel 7A (non-voice)
27.055	Citizens Band Channel 8
27.065	Citizens Band Channel 9
27.075	Citizens Band Channel 10
27.085	Citizens Band Channel 11
27.095	Citizens Band Channel 11A (non-voice)
27.105	Citizens Band Channel 12
27.115	Citizens Band Channel 13
27.125	Citizens Band Channel 14
27.135	Citizens Band Channel 15
27.145	Citizens Band Channel 15A (non-voice)
27.155	Citizens Band Channel 16
27.165	Citizens Band Channel 17
27.175	Citizens Band Channel 18
27.185	Citizens Band Channel 19
27.195	Citizens Band Channel 19A (non-voice)
27.205	Citizens Band Channel 20
27.215	Citizens Band Channel 21
27.225	Citizens Band Channel 22
27.235	Citizens Band Channel 24
27.245	Citizens Band Channel 25
27.255	Citizens Band Channel 23
27.265	Citizens Band Channel 26
27.275	Citizens Band Channel 27
27.285	Citizens Band Channel 28
27.295	Citizens Band Channel 29
27.305	Citizens Band Channel 30
27.315	Citizens Band Channel 31
27.325	Citizens Band Channel 32
27.335	Citizens Band Channel 33
27.345	Citizens Band Channel 34
27.355	Citizens Band Channel 35 (Sideband)
27.365	Citizens Band Channel 36 (Sideband)
27.375	Citizens Band Channel 37 (Sideband)
27.385	Citizens Band Channel 38 (Sideband)
27.395	Citizens Band Channel 39 (Sideband)
27.405	Citizens Band Channel 40 (Sideband)
27.575	North Island NAS (Base Utility "India")
27.585	Military (Low Power)
28-29.7	Ten Meter Ham Band - See Page 254
30.100	California Army National Guard (Convoys)
30.130	San Diego Naval Data Center
30.150	North Island NAS (experimental)
30.350	Camp Pendleton - South Range Control

FREQUENCY	NAME
30.450	Point Loma-San Clemente (Coordination)
30.500	California Army National Guard
30.500	North Island NAS "Sideflare"
30.760	Baldwin Park USD (Paging)
30.760	Torrey Enterprises (San Diego)
30.860	State Fish & Game Department
30.920	Arcadia USD
31.000	Pioneer Memorial Hospital (Brawley)
31.040	National Steel & Shipbuilding (NASCO)
31.140	CDF - Riverside County
31.180	CDF - Riverside County
31.240	USAir (Maintenance-Lindbergh Field)
31.260	State Fish & Game Department
31.850 R	State Highway Patrol (Tijuana)
32.050	Camp Pendleton (tactical activity)
32.050	Point Loma-San Clemente (Coordination)
32.070	San Clemente NALF
32.100	California Army National Guard (Tactical)
32.300	California Army National Guard
32.450	Camp Pendleton (Tactical)
32.460	Seal Beach (Security)
32.700	California Army National Guard
32.800	Camp Pendleton (tactical)
32.900	California Army National Guard
33.020	Los Angeles Street Department
33.040	Medic 1 Ambulance (San Dimas)
33.040	Santa Fe Memorial Hospital (LA)
33.060	Los Angeles Street Department
33.080	Antelope Valley USD (Buses)
"	California Search & Rescue
"	Hughes Lake/Elizabeth USD
"	Rescue 1 Ambulance (Santa Fe Springs)
33.100	Rescue 3 Ambulance (Barstow)
"	Saugus USD
"	Service Search & Rescue (Glendale, Rialto)
"	William Hart UHSD (Chatsworth)
33.160	Newhall Ambulance (Chatsworth)
33.400	San Marino High Schools
33.420	San Bernardino FPD's (Fireground)
33.440	San Bernardino FPD's (Mountain Blue)
33.640	San Bernardino FPD's (Nountain Yellow)
33.660	California State OES #2 (Input/154.220 out)
33.740	La Habra Heights VFD
33.760	San Bernardino FPD's
33.920	Riverside County (Data System)
33.980	California State OES #1 (Input/154.160 out)
34.100	California Army National Guard
34.230	U.S. Fish & Wildlife Service (F3-Car-to-Car)
34.300	California Army National Guard
34.410	U.S. Department of Agriculture (Riverside)
34.500	California Army National Guard
34.700	California Army National Guard
34.810 R	U.S. Fish & Wildlife Service (F1-R, F2-S)
34.830	U.S. Fish & Wildlife Service (F3)
34.900	California Army National Guard
35.000	Children's Hospital (Orange)
35.020	McDonald's Restaurants (Food Orders)
35.080	Paramount USD
35.200	Paging Companies (Several)
35.220	Radio Relay Corp. (Paging)
35.240	Page America Communications
35.300	Hightech Communications (Paging)
35.320	Paging Companies (Several)
35.340	Mobilecomm of California (Paging)
35.380	Pacific Telephone (Paging)
35.460	Pacific Telephone (Paging)
35.560	M.C.I. Airsignal of California (Paging)
35.580	Gencom Incorporated (Paging)
35.600	I.C.S. Communications (Paging)
35.640	Bay General Community Hospital (Chula Vista)
"	Marian Hospital (Santa Maria)
"	Saddleback Community Hospital (Laguna Hills)
35.680	Foothill Presbyterian Hospital (Glendora)
"	Mission Community Hospital (Mission Viejo)
"	Pacifica Hospital (Huntington Beach)
35.700	San Diego Hilton Hotel (Paging)
35.720	Dimension Cable (Orange County)
35.720	Sweetwater UHSD (Paging)
35.860	Atkins Nursery (Fallbrook)
35.880	Coronado Towing (Coronado)

FREQUENCY	NAME
35.880	Tony Cacho Farms (San Diego)
35.900	Borrego Chevron Towing
35.940	Miller Irrigation (Fallbrook)
35.960	Dimension Cable (Orange County)
35.960	Escondido USD (Paging)
35.980	Lutz Garage Towing (Alpine)
36.100	California Army National Guard
36.150	North Island NAS (experimental)
36.300	California Army National Guard
36.370	U.S. Dept. of Agriculture
36.400	El Toro Marine Base
36.500	California Army National Guard
36.530	U.S. Navy - Shipboard Portable Radios A
36.550	Camp Pendleton (tactical)
36.550	North Island NAS (Training)
36.570	U.S. Navy - Shipboard Portable Radios B
36.590	U.S. Navy - Shipboard Portable Radios C
36.600	U.S. Marine Corps (Amphibious Vehicles)
36.630	U.S. Navy - Shipboard Portable Radios D
36.670	U.S. Dept. of Agriculture
36.700	California Army National Guard
36.710	Army Hazardous Materials Escort
36.870	U.S. Navy - Shipboard Portable Radios E
36.890	U.S. Army Hazardous Materials Escort
36.900	California Army National Guard
37.020	L.A. County Beaches & Harbors
37.020	State of Calif. - O.E.S. Packet Network
37.040	OCSO - Detentions & Jail Paqing
37.040	SD County Marshal/Sheriff Office (Pagers)
37.060	L.A. County Beaches & Harbors
37.080	L.A. County Beaches & Harbors
37.100	South East Area Animal Control (Downey)
37.120	L.A. County SCAQMD
37.140	Los Angeles City Electronic Department
37.160	L.A. County SCAQMD
37.180	Allison Water Management Agency (Orange)
37.180	Monrovia Public Works
37.200	L.A. County Beaches & Harbors
37.220	L.A. County SCAQMD
37.220	L.A. County Beaches & Harbors
37.240	L.A. County Beaches & Harbors
37.240	Riverside County (packet/voice)
37.260	South East Mosquito Abatement District
37.260 R	State of California (Food & Agriculture)
37.280	L.A. County Beaches & Harbors
37.280	Riverside County (packet/voice)
37.300	L.A. County SCAQMD
37.320	L.A. County Beaches & Harbors
37.320	Santa Ana (Bookmobile)
37.340	L.A. County Beaches & Harbors
37.340	Riverside County (Local Government)
37.360	L.A. County Beaches & Harbors
37.380	L.A. County Beaches & Harbors
37.480	Union Oil Company (Brawley)
37.520	Lincoln Avenue Water District
37.580	Anaheim City Utilities
37.760	Carlsbad Municipal Water District
37.800	Southern California Edison Company
37.820	Union Oil Company (Brawley)
37.900	Los Angeles Civil Defense
37.940	Rim of the World USD (San Bernardino)
37.980	Pacifica Hospital (Huntington Beach)
37.980 R	SD County Civil Defense (F-1)
38.100	Marine Corps (Amphibious Vehicles)
38.450	California Army National Guard
38.470	California Army National Guard
38.900	California Army National Guard
39.020	State of California (Prisons)
39.025	Mexican Coast Guard
39.040	State of California (Youth Authority)
39.060	Pomona PD (Mobiles)
39.100 R	CHP - Los Angeles
39.100	State of California (Youth Authority)
39.120	State of California (Youth Authority)
39.140	CHP - Statewide Intersystem
39.180	Blythe Public Works
39.180	Temple City Public Works
39.200	CHP - Southern California
39.200	State of California (Prisons)
39.220	CHP - Southern California

FREQUENCY	NAME
39.240	LA County SCAQMD
39.260	CHP - Southern California
39.280	LA County SCAQMD
39.300	CHP - Southern California
39.320	LA County EBS Data Net
39.340	CHP - Southern California
39.360	Arizona State - Commission of Agriculture
39.360	CHP - Southern California
39.380	LA County SCAQMD
39.400	CHP - Southern California
39.420	CHP - Southern California
39.460 R	CLEMARS Gold #6 (s) / Gold #7 (MR)
39.480	LA County EBS Voice Net
39.500	Baldwin Park Public Works
"	Buellton CSD
"	Rancho Cucamonga Public Works
39.580	Azusa Public Works
39.580	Manhattan Beach Public Works
39.600	CHP - Southern California
39.680	CHP - Southern California
39.720	CHP - Southern California
39.800	CHP - Southern California
39.880	CHP - Southern California
39.900	South East Mosquito Abatement District
39.920	CHP - Southern California
39.940	Arizona State - Commission of Agriculture
39.980	Los Angeles County EBS/SCAQMD (Alerts)
40.100	California Army National Guard
40.150 R	San Diego Naval OSC
40.170	U.S. Army - O.S.I. (F-1)
40.190	U.S. Army - O.S.I. (F-2)
40.220	U.S. Army - O.S.I. (F-3)
40.260	U.S. National Highway Traffic Safety
40.350	U.S. Navy - Shipboard Portable Radios F
40.410 R	Point Loma-San Clemente (Coordination)
40.450	Camp Pendleton - North Range Control
40.450	PEMEX, Mexican Petroleum Ops
40.530	Navy Range Communications (Coastal)
40.550	San Diego MCRD
40.790	SD Naval OSC - Base Supplies
40.800	North Island NAS (Training)
40.820	Marine Corps (Air Crash Crews)
40.900	California Army National Guard - Convoys
40.940	California Army National Guard - Convoys
40.950	U.S. Army (Air to Ground)
40.970	U.S. National Highway Traffic Safety
41.090	SD Naval OSC - Range Control
41.210	Coast Guard - All Stations
41.230	Coast Guard - All Stations
41.270	U.S. BLM (Logistics 1)
41.430	Department of Health & Human Services
41.470	Department of Health & Human Services
41.500	Army National Guard (Air Operations)
41.550	U.S. BLM (Logistics 2)
41.630	U.S. EPA (nationwide)
41.650	Army National Guard
41.830	U.S. Health & Human Services (Pagers)
41.880	North Island NAS (India)
41.900	California Army National Guard
41.950	U.S. Marine Corps
42.020	CHP - Turquoise (mobiles)
42.080	CHP - Silver Base (Dispatch)
"	(Fresno #65)
42.120	CHP - Gold Base (Dispatch)
"	(El Cajon #108)
"	(Santa Fe Springs #83)
42.140	Unknown user (call sign KMG-570)
42.160	CHP - Purple (mobiles)
42.180	CHP - Blue (mobiles)
42.200	CHP - Gold (mobiles)
42.240	CHP - Green (mobiles)
42.280	CHP - Red/Silver (mobiles)
42.300	CHP - Yellow (mobiles)
42.340	CHP - Statewide Common Blue
"	(Air Operations/Pursuits)
"	(LA Comm Center #5)
42.360	CHP - Maroon 2 Base (Dispatch)
"	(San Luis Obispo #28)
"	(Verdugo Hills #98)
42.400	CHP - Purple Base (Dispatch)

FREQUENCY	NAME
42.400	(Santa Ana #86/Ventura #31)
42.420	CHP - Tan Base (Dispatch)
"	(Antelope Valley #89)
"	(Fort Tejon #102/Newhall #78)
"	(Oceanside #92/Rancho California #115)
42.440	CHP - Pink/Red Base (Dispatch)
"	(Banning #97/Blythe #6/Indio #76)
"	(Malibu #109/West LA #79)
42.460	CHP - Black Base (Dispatch)
"	(Central LA #15)
42.480	CHP - Gray Base (Dispatch)
"	(San Juan Capistrano #113)
"	(Westminister #55)
42.500	CHP - Brown Base (Dispatch)
"	(West Valley Office #56)
42.520	CHP - Yellow Base (Dispatch)
"	(East LA #82)
42.540	CHP - Green Base (Dispatch)
"	(Ontario #123/Riverside #74)
"	(Santa Barbara #29)
42.560	CHP - White Base (Dispatch)
"	(Barstow #73/Mojave #59)
"	(Moronga Basin #106/Needles #114)
"	(South LA #77/Victorville #12)
42.600	CHP - Copper/Turquoise Base (Dispatch)
"	(Arrowhead #101/San Bernardino #75)
42.620	CHP - Aqua Base
42.640	CHP - Maroon 1 & 2 (mobiles)
42.660	CHP - Orange (mobiles)
42.680	CHP - Gray (mobiles)
42.700	CHP - Black (mobiles)
42.720	CHP - White (mobiles)
42.740	CHP - Copper (mobiles)
42.760	CHP - Pink (mobiles)
42.820	CHP - Brown (mobiles)
42.840	CHP - Aqua/Tan (mobiles)
42.880	CHP - Orange Base (Dispatch)
"	(Baldwin Park #81)
"	(San Diego Central #87)
42.920	CHP - Maroon 1 Base (Dispatch)
"	(El Centro #85/Winterhaven #116)
42.960	Allen's Professional Ambulance
"	Girl Scouts of America
"	Oceanview Farms (Oceanside)
42.980	Business Itinerate (many users)
43.000	Scheafer's Ambulance (Business)
"	Paramount USD
"	Rancho La Costa
43.080	Pinery Tree Farms (Escondido)
"	Ramona Farms
43.200	Cook Telecommunications (Pagers)
43.220	Gencom Incorporated (Pagers)
43.240	Gencom Incorporated (Pagers)
43.300	Comven, Inc. (Pagers)
43.580	Gencom Incorporated (Pagers)
43.640	Saint Joseph Hospital (Orange)
"	White Memorial Med Center (LA)
43.680	Eisenhower Med Center (Rancho Mirage)
"	Loma Linda University Med Center
43.780	Southern California RTD - Staff
43.960	Purolator Armored Car Service
44.040	California Cartage Co. (Chula Vista)
44.200	Brinks Armored Transport
44.340	G.I. Trucking (San Diego)
44.420	California Cartage Co. (Chula Vista)
44.500	Downey Transit District
44.500	El Monte Transit District
44.600	Red Cross Dial A Ride (San Diego)
44.680 R	State Parks & Recreation (F1-R, F2-S)
44.720 R	State Parks & Recreation (F1-R, F2-S)
44.760	Orange County Sanitation District
44.800	State Parks & Recreation (F-3)
44.880	State Parks & Recreation (F-4)
45.020	CHP - Ivory (Air-to-Ground)
45.080	Costa Mesa City PD
"	San Dimas Public Works
45.120	City of Del Mar (Local Government)
"	Paramount Public Works
45.140	Orange Co. Harbor Dept. (F-1)
45.160	City of Imperial (Local Government)

FREQUENCY	NAME
45.160	City of Oceanside (Local Government)
"	Orange Co. Harbor Dept. (reported)
"	San Dimas Public Works
45.180	Garden Grove PD AVL
45.200	Pico Rivera Public Works
45.220 R	Long Beach Community College
45.240	Santa Fe Springs Local Government
"	Torrance Local Government
45.280	City of Lemon Grove (Local Government)
"	City of Lynwood (Local Government)
"	City of Paramount (Local Government)
"	City of Riverside (Local Government)
45.300	Orange Co. Harbor Dept (F-4)
45.320	Cypress City Public Works
"	Cudahy Public Works
45.340	Palos Verdes Estates PD
45.360	Baldwin Park (Public Works)
"	El Monte Parks & Recreation (F-1)
"	City of San Clememte (Local Gov.)
45.380	El Monte Parks & Recreation (F-2)
"	SD Animal Regulations Office (F-3)
45.400	Orange County Harbor Dept. (F-2)
"	Orange County Lifeguards
45.420	Arcadia PD
45.440	Anaheim City Public Works
45.480	Huntington Beach City Utilities
45.520	Brea City Public Works
"	Fountain Valley City Public Works
45.560	U.C.S.D. - Engineering Department
45.580	Vernon PD (Vehicle Location System)
45.600	City of Carlsbad (Local Government)
"	City of Redlands (Local Government)
"	City of San Fernando (Local Govern.)
45.620	Los Angeles Harbor Department
45.640	City of La Habra (Public Works)
"	City of Laguna Beach (Public Works)
"	City of Redlands (Public Works)
45.660 R	Palos Verdes Estates PD
45.700	SD Animal Regulations Office (F-1)
45.740	Los Angeles Harbor Department
45.780 R	SD Sheriff's Office (Relay of F-4)
45.860	CLEMARS (Input to 39.460 MHz)
45.880	CHP - Future Intersystem
45.900	SD Animal Regulations Office (F-2)
45.920	Orange County Lifeguards (Stations 51-56)
45.940 R	SD Sheriff's Office (Relay of F-4)
45.960	Cajon Valley USD (El Cajon)
"	Lifefleet Ambulance "Red" (Laguna Hills)
"	Professional Ambulance (Glendale)
46.000	Capistrano USD
"	Claremont Christian School
"	Howard/Lifecare Ambulance (Redlands)
"	Professional Ambulance (Glendale)
46.040	Goodhew Ambulance (LA & Orange)
"	Howard/Lifecare Ambulance (Redlands)
"	Lifefleet Ambulance (Laguna Hills)
46.060	San Diego Co. EMS (simplex)
46.140 R	California Interregional EMS
46.200	Alhambra FD
46.260 R	San Bernardino Co. EMS
46.260 R	San Diego Co. EMS (simplex)
46.260 R	Riverside Co. EMS
"	San Diego Co. EMS (simplex)
46.320 R	California State - OES Region VI
"	Inyo Co. EMS
"	Imperial Co. EMS
"	Riverside Co. EMS
"	San Bernardino Co. EMS
"	San Diego Co. EMS
46.340 R	San Diego Co. EMS (primary)
46.360	Avalon FD
46.380	Imperial Co. EMS (primary)
"	San Diego Co. EMS (simplex)
46.420	San Diego Co. EMS (simplex)
46.440	San Diego Co. EMS (simplex)
46.460	SD Rancho Fire "Low Band"
46.480	San Diego Co. EMS (simplex)
46.500	Alhambra FD
46.520	City of Orange Public Works
46.540	Avalon Harbor Department

FREQUENCY	NAME
46.540	Home Gardens Water District (Corona)
"	Norwalk Public Works
46.560	Orange Co. EMS
46.580	Banning Local Government
"	Monterey Park Public Works
"	Palos Verdes Estates Public Works
46.610	Cordless Telephone (Base)
46.630	Cordless Telephone (Base)
46.670	Cordless Telephone (Base)
46.700	California Army National Guard
46.710	Cordless Telephone (Base)
46.730	Cordless Telephone (Base)
46.750	El Centro NAS (Air Drops)
46.770	Cordless Telephone (Base)
46.770 R	U.S. Dept. of Agriculture (Police/Prevention)
46.810	U.S. BLM (Logistics 4)
46.830	Cordless Telephone (Base)
46.870	Cordless Telephone (Base)
46.900	California Army National Guard
46.930	Cordless Telephone (Base)
46.970	Cordless Telephone (Base)
46.970 R	U.S. Dept. of Agriculture (Police/Prevention)
46.990	U.S. BLM (Logistics 5)
47.040	Caltrans (Black)
47.060	Caltrans (Brown)
47.080	Caltrans (Orange)
47.100	Caltrans (Yellow)
47.240	Caltrans Sig Alerts (Los Angeles)
47.420	American Red Cross (Disaster Net)
47.460	American Red Cross
"	Desert Rescue Squad (Barstow)
"	Lifefleet Ambulance "Yellow" (Laguna Hills)
"	Santa Monica-Malibu Ambulance
47.500	American Red Cross (Los Angeles)
"	Courtesy Ambulance
47.540	American Red Cross
"	Wilson Ambulance (Palmdale)
47.580	American Red Cross
"	Schaefer's Ambulance (Southern Calif.)
47.620	American Red Cross
"	Bower's Ambulance (Long Beach)
"	Crippen Ambulance (Covina)
"	High Desert Mountain Rangers
"	Medivac Ambulance Service (San Diego)
47.660	American Red Cross
"	Medivac Ambulance (El Monte)
47.700	Valley Water Department (Azusa)
47.740	De Luz Heights Water District
47.780	Colton City Water Department
"	Rancho Santa Fe Water District
47.800	La Puente Water District
"	Leucadia County Water District
47.820	San Diego County Water Authority
47.880	La Habra Heights Water District
"	Ramona Municipal Water District
47.900	Bellflower Water District
"	Otay Municipal Water District
47.920	Olivenhain Municipal Water District
47.940	Anaheim Public Works
47.940	Valley Center Water District
47.980	Glendale Electronic Department
48.060	Torrance Water District
48.060	Vista Irrigation District (VID)
48.100	Padre Dam Municipal Water District
48.260	De Luz Heights Water District
48.340	Leucadia County Water District
48.380	Otay Municipal Water District
48.440	Metropolitan Water District
48.460	Glendale Power Department
48.480	Ramona Municipal Water District
48.500	San Diego County Water Authority
48.520	San Diego County Water Authority
48.540 R	Metropolitan Water District
48.560 R	Metropolitan Water District
49.160	Orange Co. Water District
49.250	Petroleos Mexicanos (Mexicali/Tijuana)
49.460	Orange Co. Water District
49.520	East Brothers Grove Service
49.600	Petroleos Mexicanos (Mexicali/Tijuana)
49.670	Cordless Telephones (Handsets)

FREQUENCY	NAME
49.700	Army EOD Teams (Fort Rosecrans)
49.770	Cordless Telephones (Handsets)
49.800	Army EOD Teams (Fort Rosecrans)
49.830	Cordless Telephones (Handsets)
49.845	Cordless Telephones (Handsets)
49.860	Cordless Telephones (Handsets)
49.875	Cordless Telephones (Handsets)
49.890	Cordless Telephones (Handsets)
49.900	Army National Guard (EOD Teams)
49.930	Cordless Telephones (Mobile)
49.970	Cordless Telephones (Mobile)
49.990	Cordless Telephones (Mobile)
50.0-54.0	Six Meter Ham Band - See Page 254
59.750	Television Channel Two (Los Angeles)
65.750	Television Channel Three (Santa Barbara)
71.750	Television Channel Four (Los Angeles)
72.020	U.C.S.D. Medical Center (Pagers)
72.040	Paging Services
72.060	Paging Services
72.140	Metrocall USA, Inc. (Pagers)
72.200	Metrocall USA, Inc. (Pagers)
72.220	Paging Services
72.260	Kaiser Hospital (Pagers)
72.260	State of California (Air Resource Board)
72.280	Tri-City Hospital District (Pagers)
72.300	Rohr Industries (Chula Vista)
72.320	Gencom Incorporated (Pagers)
72.340	Rohr Industries (Chula Vista)
72.360	I.C.S. Communications (Pagers)
72.360	Rohr Industries (Chula Vista)
72.380	Expo Stucco Inc. (Land-mobile)
72.380	Rohr Industries (Chula Vista)
72.420	U.C.S.D. Medical Center (Pagers)
72.440	Rohr Industries (Chula Vista)
72.440	State of California (Air Resources Board)
72.460	Gencom Incorporated (Pagers)
72.480	Rohr Industries (Chula Vista)
72.480	State of California (Air Resources Board)
72.520	Rohr Industries (Chula Vista)
72.520	State of California (Air Resources Board)
72.560	Rohr Industries (Chula Vista)
72.560	State of California (Air Resources Board)
72.580	Gencom Incorporated (Pagers)
72.580	Hendrix Electronics (Pagers)
72.600	Rohr Industries (Chula Vista)
72.600	State of California (Air Resources Board)
72.620	Hendrix Electronics (Pagers)
72.640	Brake Delivery Service (Mt. Otay)
"	Mobilecomm of California (Pagers)
"	Scripps Memorial Hospital (Pagers)
72.680	Page America Communications (Pagers)
72.700	Grossmont District Hospital (Pagers)
72.800	Doctors Service Bureau (San Diego)
72.820	Hendrix Electronics (Pagers)
72.840	Department of General Services (Mt. Laguna)
72.920	Gencom Incorporated (Pagers)
72.940	AAlert Paging Company of San Diego
73.100	Imperial County Local Government
73.240	State of California
73.260	State of California
73.420	State of California
74.100	State of California
74.140	State of California
74.500	CHP-El Centro Transmitter Link
74.500	State of California
75.460	Doctors Service Bureau (San Diego)
75.460	Gypsum Communications Co. (Plaster City)
75.500	AAlert Paging Company of San Diego
75.540	Doctors Service Bureau (San Diego)
75.580	State of California
75.640	Doctors Service Bureau (San Diego)
75.660	Expo Stucco (Escondido)
75.680	Brake Delivery Service (San Diego)
75.700	State of California
75.720	Metrocall USA (Pagers)
75.740	Brake Delivery Service (Escondido)
75.780	CHP - Mt. Woodson Transmitter Link
"	Hendrix Electronics (Pagers)
75.820	State of California
75.840	Comven Inc. (Pagers)

FREQUENCY	NAME
75.860	Metrocall USA (Pagers)
75.900	State of California
75.940	Hendrix Electronics (Pagers)
81.750	Television Channel Five (Los Angeles)
87.750	Television Channel Six (Tijuana)
88.1-107.9	FM Broadcast Band - See Page 269
108.0-136.0	VHF Aircraft Band - See Page 242
108.0-118.0	VHF Aircraft Band - Vortacs
118.650	North Island NAS - Beaver
118.950	Forest Fire Control - Air Tactics
119.600	Brown Field - Approach Control
119.950	Forestry - Helicopters
119.975	Forestry - Helicopters
120.425	F.B.I. Air Operations 1
120.450	F.B.I. Air Operations 2
120.475	F.B.I. Air Operations 3
120.650	Imperial Beach OLF - Control Tower
120.850	North Island NAS - Beaver
121.500	Emergency Channel
121.850	North Island Naval Air Station (Ground Ctrl)
122.000	Flight Service Station (Julian)
122.200	Flight Service Station (San Diego)
122.400	Flight Service Station (San Diego)
122.700	Air to Air (Coastal Area)
122.750	Air to Air Channel (primary)
122.800	Unicom (Many small airports)
122.850	Air-to-Air, Forestry Air Tactics
122.900	Air to Air Channel / firefighting
"	Multicom - Advisory Channel (airports)
"	Search & Rescue (CAP)
122.925	Air to Air Channel / Firefighting
122.925	Search & Rescue (CAP)
122.950	Unicom (many small/large airports)
123.000	Unicom (many small airports)
123.025	Helicopters (Air to Air)
123.050	Helicopters (Unicom)
123.075	Helicopters (Unicom)
123.100	Search & Rescue (Civil Air Patrol)
123.175	Pacific Aerosystems (San Diego)
123.200	Camp Pendleton - North Range Control
"	Hughes Helicopters
"	McDonnell Douglas Aircraft
123.225	General Dynamics - Testing
123.275	McDonnell Douglas Aircraft
123.300	El Cajon Flying Service (Flight School)
"	National Air College (Flight School)
"	National University (Flight School)
"	Volar Corporation (El Cajon)
123.375	General Dynamics - Testing
123.400	Camp Pendleton - South Range Control
123.425	McDonnell Douglas Aircraft
123.450	McDonnell Douglas Aircraft
123.450	Riverside County (Air-to-Air)
123.475	Hughes Helicopters (Carlsbad)
123.500	Airline Training Center (Flight School)
123.500	Safari Avaition
123.550	McDonnell Douglas Aircraft
123.650	Flight Service Station (San Diego)
124.350	San Diego Approach/Departure Control
124.400	Brown Field - Ground Control
125.150	Brown Field / Imperial Beach OLF Approach
125.850	L. A. Air Traffic Control Center (Julian)
126.200	Camp Pendleton Marine Corps Base - Tower
"	El Centro Naval Air Facility - Tower
126.200	Military Towers common VHF
"	Miramar Naval Air Station (Departure)
126.500	Brown Field - Control Tower
126.750	San Clemente Naval Auxiliary - Tower
127.300	Flight Service Station (Mexico)
127.300	San Diego Approach Control
127.400	L.A. Air Traffic Control Center (Julian)
127.550	North Island NAS
127.600	Flight Service Stations (Mexico)
128.150	L.A. Air Traffic Control Center (Julian)
128.250	Brown Field - Control Tower
128.400	North Island Naval Air Station - Clearance Del.
128.600	L.A. Air Traffic Control Center (Julian)
128.90-131.925	Aeronautical Radio Stations, 25 kHz spacing
131.000	Airmexico (Tijuana)
131.675	Mercy Air Ambulance

FREQUENCY	NAME
132.000	Goodyear Blimp (Operations)
132.350	Brown Field - ATIS
134.100	Departure Control (Camp Pendleton)
135.100	North Island Naval Air Station - Tower
135.200	San Diego Approach Control
135.450	L.A. Air Traffic Control Center (Julian)
135.750	L.A. Air Traffic Control Center (Julian)
138.025 R	Camp Pendleton-Military Police (F-1, F-4)
138.050	Camp Pendleton - MP's reported
138.075	Office of Special Investigations (USAF)
138.150	Camp Pendleton-Public Works (Pagers)
138.175	Office of Special Investigations (USAF)
138.225	U.S. Federal Emergency Management Agency
138.325	March A.F. Base - Munitions
138.375	Camp Pendleton-Military Police Tactical
138.525	La Jolla Naval Ship Research
138.525	San Diego Naval Public Works Center (GOLF)
138.550	San Diego Naval Administrative 1
138.600 R	Camp Pendleton-Fire Department (F-3, F-4)
138.700	U.S. Coast Guard - Cutters (rare use)
138.725	Imperial Beach Naval Field (Security DES)
138.750	Fallbrook Naval Weapons Station (Guards)
138.750	San Diego Data Center (Security)
138.775	San Diego Naval OSC (Public Works)
138.775	San Diego Naval Training Center (Security 2)
138.800	Camp Pendleton-EOD Teams
138.850	Fallbrook Naval Weapons Station (PAPA)
138.875	San Diego Naval Training Center (Security 1)
138.950	San Diego Naval Public Works Center
138.975	U.S. Marine Corps - Twentynine Palms FD
139.000	California Army National Guard (ML1)
139.025	California Army National Guard (F-1)
139.050	California Army National Guard (F-2)
139.075	California Army National Guard (F-3)
139.100	California Army National Guard (F-4)
139.125	California Army National Guard (F-5)
139.150	California Army National Guard (F-6)
139.175	California Army National Guard (F-7)
139.200	California Army National Guard (F-8)
139.225	California Army National Guard (F-9)
139.250	California Army National Guard (F-10)
139.275	California Army National Guard (F-11)
139.300	California Army National Guard (F-12)
139.325	California Army National Guard (F-13)
139.350	California Army National Guard (F-14)
139.375	California Army National Guard (F-15)
139.400	California Army National Guard (F-16)
139.425	California Army National Guard (F-17)
139.450	California Army National Guard (F-18)
139.475	Miramar Naval Air Station - Alcohol Rehab Ctr.
139.500	U.S. Marine Corps - El Toro
139.510	Imperial Beach Naval Radio Receiving
139.525	Imperial Beach Naval Radio Receiving
139.550	Miramar Naval Air Station
139.550	U.S. Marine Corps - El Toro
139.550	North Island NAS (Supplies)
139.560	Warner Springs Naval SEER
139.575	Miramar NAS Airfield (Supplies/Fuel)
139.600	El Centro NAS
139.625	San Diego Navy - Unidentified
139.650	San Diego 32nd St. (Port Services 1)
139.675	Camp Pendleton-Military Police Tactical
139.700	San Diego 32nd St. (Port Services)
139.725	San Diego 32nd St. (Port Services 2)
139.800	El Centro Naval Air Field (Input-140.525)
139.825	U.S. Federal Emergency Management Agency
139.925	San Diego 32nd St. (Port Services 3)
139.975	Salton Sea Naval Test Base
139.975	San Diego 32nd St. (Port Services 4)
140.000	Camp Pendleton-Military Police (Input F-1)
140.000	Coronado Amphibious Base (F-3)
140.025	San Diego MCRD (Security 1)
140.025	San Diego Naval Administrative (F-2)
140.030	Imperial Beach OLF (Fire & Crash)
140.050	North Island NAS (Fire)
140.075 R	Naval Investigative Service (F-1, F-2)
140.100	El Centro NAS (Fire & Crash)
140.100	Imperial Beach OLF (Fire & Crash)
140.100	Miramar Naval NAS (Airfield Control)
140.125	North Island NAS (Special Weapons)

FREQUENCY	NAME
140.125	Twentynine Palms USMC EMS
140.150	San Diego MCRD (Security 2)
140.150	San Diego Naval Public Works (India)
140.150	San Diego Naval Training Center (Fire)
140.175	North Island NAS (maintenance)
140.200	San Clemente NALF (Fire & Crash)
140.225	San Diego Naval Training Center (Fire)
140.250	Miramar NAS (Fire-Structural)
140.275	San Clemente NALF
140.275	Point Loma - San Clemente (Coordination)
140.280	El Centro NAS (Tests)
140.280	Fallbrook Naval Weapons Station
140.300	Imperial Beach OLF (Fire & Crash)
140.325	San Diego Naval Supply Center (Fire)
140.350	San Diego Naval OSC (Delta)
140.375	North Island NAS (Security Tac 1)
140.400	U.S. Navy (San Pedro Fuel Dock)
140.425	U.S. Marine Corps. (El Toro)
140.450	San Diego 32nd St. / Imperial B. (Fire)
140.475	San Diego Naval Submarine Facility
140.500	Camp Pendleton-Fire Tactical
140.500	Fallbrook Naval Weapons Station (common)
140.520	El Centro NAS (Tests)
140.520	Fallbrook Naval Weapons Station (Tests)
140.525	Camp Pendleton-Fire & Crash Crews
140.525	El Centro NAS (Airfield)
140.575	North Island NAS (Security 2)
140.600	El Centro NAS (Tests)
140.600	Fallbrook Naval Weapons Station (Tests)
140.620	Imperial Beach OLF
140.625	San Diego 32nd St. (Environmental)
140.650	Naval Investigative Service (F-3)
140.675	San Diego MCRD (Tac 1)
140.700	San Diego MCRD (Tac 2)
140.710	Camp Pendleton Tactical
140.725	U.S. Navy San Diego (undetermined)
140.750	San Diego Marine Corps Recruits (Pub. Works)
140.760	North Island Naval Air Station (Tugs)
140.760	Warner Springs Naval SERE
140.800	U.S. Navy (undetermined)
140.825	San Diego Naval OSC (Security / Fire)
140.850	Camp Pendleton - Navy Security Common)
140.850	Marine Corps Recruit Depot - Security
140.875	San Diego Marine Corps Recruits (Fire)
140.895	Miramar NAS (Station Utility)
140.895	North Island NAS (Station Utility)
140.900	El Centro Naval Facility (Weapons)
140.900	Fallbrook Naval Weapons Station (Tests)
140.900	Miramar NAS - Maintenance
140.925	Camp Pendleton-Fire Department (Input F-4)
140.950	Camp Pendleton (undetermined)
140.975	Imperial Beach Naval Radio Receiving
141.000	Miramar Naval Air Station (Public Works)
141.100	Miramar Naval Air Station (Fire & Crash)
141.175	Salton Sea Naval Test Base
141.500	Marine Corps Recruit Depot (security)
141.500	Salton Sea Naval Test Base
141.950	San Diego Naval Public Works Center (Paging)
141.875	Federal Emergency Management Agency
141.975	Long Beach Naval Station - Shore Patrol
142.000	San Diego Naval OSC (Range Comm.)
142.025	Camp Pendleton (undetermined)
142.050	San Diego Naval OSC (Range Comm.)
142.050	San Clemente NALF
142.075	San Diego Naval OSC (Range Comm.)
142.100	San Diego Naval OSC (Range Comm.)
142.100	San Clemente NALF
142.225	Los Angeles A.F. Station - Security
142.250	El Centro Naval Air Facility MAC (AM)
142.350	Federal Emergency Management Agency
142.375	Federal Emergency Management Agency
142.425	Federal Emergency Management Agency
142.500	Fallbrook Naval Weapons Station (F-1)
142.515	Miramar Naval Air Station
142.525	Miramar Naval Air Station (Internal Security)
142.550	Coronado Amphibious Base
142.550	Camp Pendleton - Training Utility
142.575	Camp Pendleton (Undetermined)
142.600	San Clemente NALF (Tests)
142.600	San Diego Naval Station (Port Services 5)

FREQUENCY	NAME
142.620	San Diego Naval Medical Center (ROMEO)
142.625	U.S. Navy Blue Angels (Ground Support)
142.675	Fallbrook Naval Weapons Station (F-2)
142.675	SD Navy Shore Patrol - Tactical
142.700	San Diego Naval Station (Emergency)
142.725	San Diego Naval Station (Port Services 6)
142.750	San Diego Naval Electronic Engineering Lab
142.775	Fallbrook Naval Weapons Station (F-2) Input
142.800	El Centro Naval Air Facility (Range Comm.)
142.800	Camp Pendleton (undetermined)
142.800	North Island NAS (Diving School/Ramps)
142.825	Fallbrook Naval Weapons Station Secure DES
142.825	San Diego Naval Station (Port Services 7)
142.850	San Diego Navy-Point Loma (FOXTROT)
142.975 R	Federal Emergency Management Agency
143.000 R	Federal Emergency Management Agency
143.000	Navy Blue Angels AM
143.275	Mexican Mobile Telephone (Tijuana)
143.325	Mexican Mobile Telephone (Tijuana)
143.375	Mexican Mobile Telephone (Tijuana)
143.400	Mexican Mobile Telephone (Tijuana)
143.415	Army M.A.R.S. (F-1) Input
143.450 R	Air Force M.A.R.S. (F-1) Primary
143.475	Mexican Mobile Telephone (Tijuana)
143.500	Chula Vista Space Surveillance Station
143.525	San Diego Naval Medical Services
143.550	Coronado Naval Amphibious Base (Security TAC-1)
143.575	Mexican Mobile Telephone (Mexicali)
143.575	North Island NAS (Transportation)
143.600	San Diego Naval Radio Transmitter Facility
143.625	San Diego Naval Station (Emergency)
143.650	Camp Pendleton - Training Utility
143.650	San Diego Naval Submarine Facility
143.700	Mexican Mobile Telephone (Mexicali)
143.715	Miramar NAS (Station Utility)
143.715	North Island NAS (Station Utility)
143.750	Civil Air Patrol - San Diego
143.800	Mexican Mobile Telephone (Mexicali)
143.875 R	U.S. Coast Guard - Auxilliaries
143.950	Air Force M.A.R.S. (F-2) Tactical
143.990	Army M.A.R.S. (F-3)
144.0-148.0	Two Meter Amateur Band - See Page 255
148.000	California Army National Guard
148.010	Army M.A.R.S. (F-3) Input
148.025	Camp Pendleton - Motor Transport
148.085	Edwards A.F. Base - Runway Maintenance
148.125	Mexican Red Cross (Interagency)
148.150	Civil Air Patrol (California Wing)
148.175 R	Baja Judical Public Works (Tijuana & Mexicali)
148.175	March A.F. Base - Maintenance C
148.225	Baja Judical Public Works (Tijuana)
148.245	March A.F. Base - Munitions
148.270	Marine Corps Recruit Depot (Security)
148.275	Edwards A.F. Base - Range Control
148.290	El Centro Naval Air Facility
148.290	North Island NAS (General Ops)
148.290	San Diego Naval Public Works Center
148.300	Camp Pendleton (Base-mobile)
148.300	Ensenada Traffic Police
148.300	SD Naval Submarine Base (Security)
148.325	San Diego Naval Supply Center (Security)
148.330	Imperial Beach Naval Facility (FL-1, MO-1)
148.350	El Centro Naval Air Facility (Security)
148.350	Salton Sea Naval Test Base (Security)
148.350	San Diego 32nd St. Naval Station TAC-1
148.375	San Diego Navy Shore Patrol - Primary
148.400	San Diego Naval Supply Center (Operations)
148.410	U.S. Navy M.A.R.S. (F-2)
148.425	Camp Pendleton (undetermined)
148.450	Air National Guard - Van Nuys AF Fire & Crash
148.4875	Edwards A.F. Base - EOD Teams
148.500	Air National Guard - Van Nuys AF Disaster Net
148.500	Ensenada Traffic Police
148.515	March A.F. Base - Maintenance B
148.550	Los Angeles A.F. Station - Commander's Net
148.650 R	Army M.A.R.S. (F-1) & (F-2)
148.725	Camp Pendleton-Public Works (FOXTROT KILO)
148.725 R	U.S. Coast Guard Auxiliaries (F1-L.A.)
148.750	San Diego 32nd St Naval Station TAC-2
148.795	Imperial Beach Naval OLF (Operations)

FREQUENCY	NAME
148.825 R	U.S. Coast Guard Auxilliaries (F2)
148.850	San Clemente NALF (paging)
148.875 R	U.S. Coast Guard Auxiliaries (F3)
148.910	Edwards A.F. Base - Police 1
148.950 R	U.S. Navy MARS (F-1)
149.000	El Centro Naval Air Facility (Input-148.350)
149.010	Camp Pendleton-Transportation HOTEL
149.010	El Centro Naval Air Facility (Station Utility)
149.025	Camp Pendleton-Motor Transport
149.025	North Island NAS (Station Utilty)
149.050	Camp Pendleton - Weapons "Kilo 7"
149.050	San Diego Naval Submarine Facility
149.070	San Diego Naval OSC (Range Comm.)
149.070	San Clemente NALF (Range Comm.)
149.100	Camp Pendleton - Runway/Avionics
149.125	Miramar NAS (portables)
149.130	San Diego Naval OSC (Range Comm.)
149.130	San Clemente NALF (Range Comm.)
149.150	California Air National Guard - Northern
149.175	March A.F. Base - Maintenance A
149.200	Los Angeles A.F. Station - Security
149.200	North Island NAS (Supplies)
149.265	Air National Guard - Van Nuys AF Station
149.275	Camp Pendleton-Area Guard (PAPA CHARLIE)
149.275	Tijuana-Mexicali Toll Road
149.300	Camp Pendleton-Common
149.350	Camp Pendleton-Medical Pagers
149.350	San Diego Naval Medical Center Pagers
149.370	Camp Pendleton-Provost Marshal
149.375	San Diego Naval Supply Center (Transport)
149.400	Camp Pendleton-Base Ops
149.425	Camp Pendleton-Prison Guards (CHARLIE)
149.430	San Diego Naval OSC (Range Comm.)
149.430	San Clemente NALF (Range Comm.)
149.475	March A.F. Base - Base Taxi
149.490	San Clemente NALF (Range Comm.)
149.500	Camp Pendleton-Area Guard (IRONSIDE)
149.545	California Air National Guard - Southern
149.550	U.S. Navy (Balboa Hospital)
149.5625	Edwards A.F. Base - Engineering
149.600	Air National Guard - Van Nuys AF Security
149.625	Camp Pendleton-Paramedics
149.650	Edwards A.F. Base - Transportation
149.675	California Air National Guard
149.730	El Centro NAS (Station Utility)
149.775	Camp Pendleton (Input to 150.650)
149.900	Edwards A.F. Base - Fuel & Lubrication
149.925	Civil Air Patrol (Command & Tactical)
149.925	Edwards A.F. Base - Police 2
150.050	March A.F. Base - Maintenance E
150.075	Miramar NAS Base Police 2
150.075	San Clemente NALF
150.090	Camp Pendleton (undetermined)
150.105	Miramar NAS Base Police 1
150.125	U.S. Navy Mars (Input-148.410)
150.150	Miramar NAS (Security)
150.175	March A.F. Base - Flight Line Refueling
150.200	Miramar Naval Air Station (E.O.D.)
150.225	March A.F. Base - Medical Services
150.345	March A.F. Base - Fire & Crash Net
150.375	Camp Pendleton USMC
150.390	U.S. Navy Communications Station SD
150.450 R	Tijuana Public Works (La Mesa Sector)
150.550	Miramar Naval Air Station (Support Equipment)
150.650	Camp Pendleton-Game Warden (MIKE)
150.750	San Clemente NALF
150.830	Gene's Union Towing (Imperial Beach)
150.845	Hillcrest Towing (Glendale)
150.845	Mike Moor's 24 HR Towing (San Diego)
150.890	Tom Moynahan Towing (National City)
150.905	Auto Club of Southern Ca. (CH-1)
150.920	Auto Club of Southern Ca. (CH-4)
150.935	Auto Club of Southern Ca. (CH-2)
150.950	Auto Club of Southern Ca. (CH-5)
150.965	Auto Club of Southern Ca. (CH-3)
150.965	Auto Club of Southern Ca. (San Diego)
151.010 R	Caltrans Maintenance / Traffic Control
151.025 R	Riverside County Animal Control
151.025	Ventura County Local Government
151.055 R	Riverside County Roads

FREQUENCY	NAME
151.070 R	Beverly Hills Street Dept.
151.085 R	Orange Co. Road Department
151.100	Burbank Street Department
151.115	Compton City Code Enforcement
151.115	SD Airport Security (Brown & Montgomery Fld)
151.115	San Diego City Electronic Department
151.115	San Diego City Street Department
151.145 R	Avalon - Santa Catalina Conservatory
151.160	C.D.F. Tactical Net
151.160 R	Orange Co. Environmental Management
151.160	Orange Co. Parks & Recreation
151.175 R	C.D.F. Dispatch (Perris East & Secondary)
151.190 R	C.D.F. Dispatch (Monte Vista-San Diego)
151.205	San Diego City Electronic Department
151.205	State of California - Water Resources Division
151.220	C.D.F. Tactical (Red Net)
151.235	C.D.F. Tactical Net
151.235	Orange Co. Probations Dept. (Security)
151.235	State of California - Water Resources Division
151.250	C.D.F. Tactical Net
151.265	C.D.F. Fire Command 2 (Statewide)
151.280	C.D.F. Air Operations BLUE
151.295	C.D.F. Air Operations GREEN
151.300 R	Ensenada Toll Road (Emergencies)
151.310	C.D.F. Air Operations YELLOW
151.325 R	C.D.F. Dispatch San Bernardino East
151.340	C.D.F. Administrative Net
151.355 R	C.D.F. Fire Command 1 (Statewide)
151.370	C.D.F. Tactical Net
151.385 R	C.D.F. Dispatch Perris West
151.400	C.D.F. Tactical Riverside
"	C.D.F. Tactical San Bernardino East
"	Tijuana Fire Dispatch
151.415 R	California Fish & Game Department
151.430 R	California Fish & Game Department
151.445 R	C.D.F. Dispatch Riverside West
151.460	Beverly Hills Parks & Recreation
151.460	Tijuana Public Works (Beaches Sector)
151.475	C.D.F. Tactical Net
151.475 R	Tijuana Public Works Department
151.600	Jai-Alai Games Palace (Tijuana)
151.625	- BUSINESS ITINERATE -
"	Budweiser Thunder Boat Races (Security)
"	Del Mar Fair (General Ops)
"	Event Medical Services (LA)
"	Foothill Presbyterian Hospital (Glendora)
"	Goodyear Blimp
"	Guardian Angels (Security Ops)
"	National Steel & Ship Building (San Diego)
151.6425	KNSD-TV Channel 39 (News Emergency Net)
151.655	A.B.C. Cab Co. (Chula Vista)
"	Azusa USD
"	Baldridge Racing Service (Lakeside)
"	Broatman Medical Center (Culver City)
"	Caliente Race Track (Tijuana)
"	Calstate San Bernardino
"	California Plant Protection (San Marcos/Vista)
"	Claremont City High Schoools
"	Golfland Amusement Parks
"	Mars Cab Co. (Oceanside)
"	Masters Automotive (Oceanside)
"	National Steel & Shipbuilding (NASCO)
"	Oakwood Gardens Apartments (Coronado)
"	Oakwood Garden Apartments (San Diego)
"	Palomar College Operations (San Marcos)
"	Pinkerton Security (F1 Portables)
"	Placentia City Schools (Security)
"	San Diego State University (Plant Operations)
"	San Diego Institute of Pathology
"	San Diego Maritime Museum (Security)
"	Sharp Memorial Hospital (San Diego)
"	Sierra Land & Livestock (Pauma Valley)
"	Sundance Stage Lines (San Diego)
"	Telecable Service Corp. (Borrego Springs)
"	U.S.I.U. (Maintenance)
"	Vacation Village Hotel (Operations)
"	Mars Cab Co. (Oceanside)
"	Pinkerton Security (F2 Portables)
151.685	Alhambra USD - Security
"	Alvarado Community Hospital

FREQUENCY	NAME
151.685	Atlas Hotels (Security)
"	Basset USD - Security
"	Bonita Union Towing
"	Butterfield Resort (Rancho)
"	Calabasas USD
"	California First Bank
"	Circle A Cab
"	Cuyamaca Resort (Julian)
"	Great Pacific Hotels
"	Hawthorne USD (Security)
"	Huntington Beach School Security
"	Rancho Bernardo Inn (Paging)
"	Sorrento Towers (San Diego)
"	Travelodge Hotels
"	Travelodge Tower (Harbor Island)
151.6975	Henley Group (Dulzura)
151.715	Baker Protection (Oceanside)
"	Burroughs Corp. (Rancho Bernardo)
"	Coronado Cab Co.
"	Del Mar Thoroughbreds (Race Track Security)
"	Dronk Radio Network (Operations)
"	El Monte USD
"	Glendora USD
"	Independent Taxi Cab Service
"	Lake San Marcos Security Patrol
"	Lakeside Union School District
"	Little America Westgate Hotel
"	Maritime Museum Association (Operations)
"	Mira Mesa Shopping Center (Security)
"	National Steel & Shipbuilding
"	Oakwood Garden Apartments
"	Royal Quality Inn
"	San Juan Capistrano Schools
"	Servomation Supply Inc. (SD Stadium)
"	Sol's Cab Co. (Oceanside)
"	St. John Hospital (Santa Monica)
"	U.S. International University
151.745	Aztec Center (S.D.S.U.)
"	Bay Vista Methodist Heights Apartments
"	Coronado Livery Buses
"	Culver City USD
"	Dana Inn & Marina
"	Del Mar Turf Club (Operations)
"	Pap Smear Testing Inc.
"	Rancho Santa Fe Security
"	San Diego State University (Plant Operations)
"	San Diego Fire Equipment Rental
"	Salk Institute of Biological Studies
"	Tri-City Hospital (Paging)
151.775	All State Insurance
"	Los Angeles USD (Security)
"	Placentia City Schools (Security)
"	Pleasant Valley Hospital (Santa Paula)
"	Premier Foods Plant Security
"	Robert F. Kennedy Hospital (Hawthorne)
151.805	American Television (Black Mtn.)
"	Bahia Hotel (Operations)
"	Bressi Ranch (Carlsbad)
"	Crawford Detective Agency (National City)
"	Crown Air (Montgomery Field)
"	Fountain Valley School District
"	Fujitsu Microelectrics (Security)
"	Horton Farms (Jamul)
"	Lawrence Welk Country Club (Security)
"	Oceanside School District (Security)
"	Pepper Industries
"	Santa Ana Schools (Security)
"	Sheraton Inn Airport
"	Villa Nueva Apartments (San Ysidro)
151.825 R	State Highway Patrol (Mexico)
151.835	C.V.R.V. Park (Chula Vista)
"	Coronado Shores Apartments (Security)
"	Escondido Unified Schools (Security)
"	Goodall's Charter Bus (Primary)
"	Grossmont Center (La Mesa)
"	Half Moon Inn (Security)
"	Hydro Products
"	Irvine USD - Security
"	Magatex Corp.
"	Manhattan USD

FREQUENCY	NAME
151.835	Oak Communication Systems (Carlsbad)
"	Santa Monica Comm. College
"	Sheraton Harbor Island Hotel
"	St. Mary Hospital (Apple Valley)
151.865	Bay Vista Methodist Heights Apartments
"	Del Mar Thoroughbreds (Operations)
"	Escondido Mall Security (Intercom/Security)
"	Escondido Unified Schools (Buses)
"	Event Medical Service (LA)
"	Kings Inn (Operations)
"	Long Beach City Building Security
"	Los Angeles Community Colleges
"	Placentia City Schools (Security)
"	Ray Wilson Corp.
"	Sea Lodge Inn (La Jolla)
"	Simcal Properties (Imperial Beach)
151.895	Anaheim Memorial Hospital
"	Basset USD - Security
"	Casa Colina Hospital (Pomona)
"	Checker Cab Co.
"	Firestone Energy Systems (El Cajon)
"	Greenwood Memorial Park
"	Islandia Associates
"	Los Alamitos General Hospital
"	Palmdale School District
"	Poway Unified School District (Buses)
"	Queen of Angels Hospital (LA)
"	Ray Wilson Corp.
"	San Diego Zoo (Maintenance/Utility)
"	Torrance USD
"	Universal Studios (Ride Maintenance)
"	University of San Diego
"	West Covina Disposal
151.900	Rosarito Beach Hotel (Mexico)
151.925	Alsa Medical Center (Hollywood)
"	American Television (Coronado)
"	Burbank Community Hospital
"	Calvideo Inc. (Alpine)
"	Equity Investigations
"	Golden Eagle Farms (Ramona)
"	Jack Murphy Stadium (Security Primary)
"	Long Beach Community Hospital
"	Palomar Memorial Hospital (Paging)
"	Ramona Unified School District (Buses)
"	San Dieguito Union High School District
"	Strategic Security Services
151.955	Brotman Medical Center (Culver City)
"	Casa Colina Hospital
"	City of God Hospital
"	Daniel Freeman Hospital (Inglewood)
"	Event Medical Services (LA)
"	Golden Eagle Farm (Ramona)
"	Hollywood Presbyterian Hospital
"	Jack Murphy Stadium (Security Tactical)
"	La Costa Country Club (Catering)
"	Midway Hospital (LA)
"	Redondo Beach High School
151.985	Pacific Telephone (Maintenance Primary)
152.0075	Hospital pagers
152.030	Gencom Incorporated (Pagers)
152.060	Gencom Incorporated (Pagers)
152.090	Gencom Incorporated (Pagers)
152.120	Gencom Incorporated (Pagers)
152.150	Gencom Incorporated (Pagers)
152.180	Gencom Incorporated (Pagers)
152.210	Gencom Incorporated (Pagers)
152.240	Gencom Incorporated (Pagers)
152.270	Yellow Cab of San Diego
152.300	Escondido Yellow Cab
152.300	Silver Cab Co. (San Diego)
152.315	Coast Cab Co. (F-1)
152.330	Yellow Cab of San Diego
152.345	U.S.A. Cab Co. (Dulzura)
152.360	Diamond Cab Co.
152.375	ABC Cab Co. (Backup)
152.375	La Jolla Cab Co.
152.390	Radio Cab Co.
152.390	(Several taxi cabs here)
152.405	Red Top Cab Co. (San Ysidro)
152.405	Silver Cab Co. (San Diego)

FREQUENCY	NAME
152.420	Yellow Cab of San Diego
152.435	American Cab Co. (Backup System)
152.450	Oceanside Yellow Cab Co. (Powerful)
152.450	Yellow Cab of San Diego
152.475	Federal Water Resources (Tijuana)
152.480	Paging Services (LA, San Diego, San Marcos)
152.510	Pacific Mobile Telephones (LA & San Diego)
152.540	Pacific Mobile Telephones (LA & Mt. Palomar)
152.570	Pacific Mobile Telephones (LA & San Diego)
152.600	Pacific Mobile Telephones (LA & Mt. Palomar)
152.630	Pacific Mobile Telephones (LA & San Diego)
152.660	Mobilecomm of California (San Diego)
152.690	Paging Services (So. California)
152.720	Ibex Communications (Paging)
152.750	Mobilecomm of California (San Diego)
152.780	Pacific Mobile Telephones (LA & San Diego)
152.810	Pacific Mobile Telephones (LA & San Diego)
152.840	Paging Services (So. California)
152.870	Daley Construction (Mt. Woodson)
152.900	Calavo Growers of California (Escondido)
152.900	Universal Studios (Amphitheater & Tour)
152.960	Jack Murphy Stadium
152.960	Mellano & Company (San Luis Rey)
152.960	Universal Studios (Shows & Special Events)
152.990	Jack Murphy Stadium
152.990	Universal Studios (Film Productions)
153.020	Crystal Silica Company (Oceanside)
153.035	Padre Transit Mix Inc. (Poway)
153.050	KOWN Radio Station - Remotes
153.050	Rohr Industries (Chula Vista)
153.065	McDonald Douglas Radio Service
153.095	Campbell Industries
153.140	Teledyne Industries
153.155	Nelco Oil Refining (Dulzura)
153.155	Whittaker Corporation (San Diego)
153.170	Kelco Company (San Diego)
153.170	Northern Telecom Inc. (San Diego)
153.185	Behring Diagnostic Inc. (San Diego)
153.185	Burroughs Corporation (San Diego)
153.200	Solar Turbines Incorporated (San Diego)
153.230	Checkeredboard Communications
153.230	Gulf General Atomic F2 - Security (San Diego)
153.230	Long Beach City Port Operations
153.230	Sparks Inc. (San Diego)
153.260	N.C.R. Corporation (Rancho Bernardo)
153.260	Sparks Inc. (San Diego)
153.290	KAVL / KAVS AM/FM (remotes)
153.290	Northern Telecom Inc. (San Diego)
153.320	Gulf General Atomic F1 - Security (San Diego)
153.320	Rohr Industries (Security-Chula Vista)
153.350	Hewlett Packard (Rancho Bernardo)
153.350	TRW Corporation (San Diego)
153.365	Rohr Industries (Chula Vista)
153.380	Ivac Corporation (San Diego)
153.440	Utility Companies
153.485	Imperial Irrigation District
153.500	Utility Companies
153.515	Utility Companies
153.560	Riverview Water District (Lakeside)
153.575	Utility Companies
153.590	Beverly Hills Public Works
"	La Canada-Flintridge Irrigation District
153.635	Covina Water District
"	Imperial Irrigation District
"	Padre Dam Municipal Water
"	Poway Water Department
153.650	Pasadena Water Department
"	Vallecitos Water District
153.665	Sweetwater Authority (Chula Vista)
153.665	Sweetwater Authority (Spring Valley)
153.680	Downey Water Department
"	Imperial Irrigation District
"	Lakeside Water District
"	Southern California Edison
153.710	Garden Grove Water Dept.
153.740	State of California Disaster Office (F-3)
153.740	State Universities (Various Locations)
153.755 R	State of California Conservation Camps
153.755 R	State of California Disaster Office (F-1)
153.755 R	State of California Fire Marshal (F-1)

FREQUENCY	NAME
153.770	Alhambra FD Tactical
"	Long Beach FD (F-2)
"	Riverside County Fire - Support Net 8
153.785	City of Placentia - Public Works
"	City of Tustin - Local Government
"	City of San Diego
"	Ventura County Local Government
153.800	Los Angeles Dept of General Services
"	LAX & Ontario Airport Ops
153.815	City of Fallbrook (Local Government)
"	Coachella Valley Parks & Recreation
"	La Verne PD & Local Government
"	Ventura County Local Government
153.830	Arcadia/El Monte/Monrovia/Sierra Madre FD
"	LACoFD Fireground (Low Power)
"	State of California Fire Intersystem (F-6)
153.845 R	La Mesa PD Dispatch (F-1)
"	Ventura County Local Government
153.860	Barstow Local Government
"	Covina PD (Mutual Aid)
"	El Segundo (Mutual Aid)
"	Glendora PD (Mutual Aid)
"	Hermosa Beach Local Government
"	Inglewood Local Government
"	Lakewood Local Government
"	La Puente Local Government
"	Pomona PD (Mutual Aid)
"	Redondo Beach PD (Mutual Aid)
153.875	Simi Valley PD
153.890	Arcadia FD
153.890	El Centro FD - Fireground
153.890	SD Fire Channel - Yellow (Tactical South)
153.905	Agoura Hills Local Government
153.920	Lancaster City Services
"	Los Angeles Parks & Recreation
"	Orange County - Animal Control (17)
153.935	Burbank Airport Police & Fire
"	Lake Crowley Park (Los Angeles)
"	Pomona Inquiry (F-2)
"	San Diego City Local Government 1
"	Ventura Local Government
153.950	Long Beach / Signal Hill FD (F-1)
153.965	Ventura County Local Government
153.980	L.A. County Beaches & Harbors (Maintenance)
"	Orange Co. Airport Operations
153.995	Alhambra/Monterey Park/San Gabriel FD
"	SD Countywide Mutual Aid - BLUE
"	S.D.S.O. Backup Dispatch & Honor Camps
154.010	Imperial County Fire - North Dispatch
"	Ventura County Fire - BLUE
154.025 R	Oceanside PD
"	Ontario / Montclair / Upland FD
154.040	Bell PD (F-3) & Local Government
"	El Centro Local Government
"	Garden Grove City Public Works
"	Huntington Beach PD & Public Works
154.055	Fillmore PD
154.055 R	San Diego PD Inquiry (F-8)
154.070	Covina / La Verne / Pomona FD
"	San Bernardino County FD (Desert Yellow)
"	Yucca Valley FD
154.085	El Monte FD
"	Salton Sea Community Services
"	Westminster Sanitation District
154.100 R	Lancaster Parks & Recreation
"	Riverside Sheriff's Office (Temecula)
154.115	City of Brawley (Local Government)
"	City of Coronado (Local Government)
"	Fontana Local Government
"	Monrovia PD & Local Government
"	Stanton Local Government
154.130	Coachella Valley FD
"	Inglewood/San Marino/Vernon FD
"	Riverside County VES COMM
154.145	Beaumont/Hemet/Murrieta/Norco FD
"	San Diego FD Response (F-2)
"	Ventura County Fire - GREEN
154.160 R	**FIREMARS - O.E.S. Tactical**
154.175	Inland Volunteer Fire Dispatch
"	Inglewood/San Marino FD

FREQUENCY	NAME
154.190	San Bernardino County FD (Central Cities)
154.205	Barstow-Victorville (Desert Gray)
"	Downey Fire Dispatch
154.220 R	FIREMARS - O.E.S. Tactical
154.235	Corona & Norco FD
154.235	North SD County Fire Dispatch
154.235	Rancho Fire Dispatch (North SD)
154.235	Ventura County Fire
154.250	Downey/Compton/Lynwood/Santa Fe Spr. FD
154.250	SD Fire Channel BROWN (Countywide Tactical)
154.265	State of California - Fire Mutual Aid WHITE 2
154.280	State of California - Fire Mutual Aid WHITE 1
154.295	State of California - Fire Mutual Aid WHITE 3
154.310	Cathedral City / Palm Springs FD
"	San Diego Fire Dispatch
"	West Covina FD
154.325	Escondido Fire Dispatch VHF GREEN
154.325	San Bernardino County Mutual Aid RED
154.340	LACoFD Fireground Battalions 6,10,13,14
154.340	Blythe FD
154.355	SD County Fire Tactical GRAY (North)
154.355	South Bay Regional Pub. Comm. (Dispatch)
154.370	Arcadia/Monrovia/Sierra Madre/El Monte FD
"	Imperial County Fire - Central Dispatch
"	Ventura County Fire - RED
154.385	Carlsbad Fire Dispatch
154.385	San Bernardino FD
154.400	Indio / Murrieta FD
154.400	LACoFD Firegound & Air Operations
154.415	Beaumont / Cathedral City FD
"	Pomona FD (fireground)
"	South Bay Fire Dispatch - San Diego
154.430	Imperial County Fire - East Dispatch
"	LACoFD Fireground Battalions 3,5,9,11,12
"	San Diego Co. Fire Tactical GOLD
154.445	State Youth Authority - El Centro Youth School
154.445	Riverside County Fire - Support Net 6
154.46375	S.D.G.& E. (telemetry)
154.475 R	Directorate of National Security (Mexico)
154.500	Police Surveillance & Bugs GOLD
154.515	American Red Cross (portables)
"	Anaheim Memorial Hospital
"	Borrego Plumbing
"	Clairemont Mission Bay Hospital
"	Covina USD - Security
"	Courtesy Ambulance (San Bernardino)
"	Georgia Pacific Corp.
"	Grossmont Gardens Apartments
"	Grossmont High School
"	Hoag Memorial Hospital (Newport Beach)
"	Holiday Inn at the Embarcadero
"	Kaiser Foundation (Anaheim)
"	La Valencia Hotel (San Diego)
"	Lawrence Welk Village (Operations)
"	Le Baron Hotel
"	Loma Palisades
"	Monte Vista High School
"	Mountain View Schools
"	Pauma Valley Country Club
"	Penasquitos Village (Security)
"	Pro-Tect Professional Security
"	Riverside County VESCOMM
"	Saddleback Comm. College (F-2)
"	Saint James Plaza (Imperial Beach)
"	San Dimas USD
"	San Miguel Hospital
"	Sea World (Parking Control)
"	Stoneridge County Club (SD)
"	U.C.S.D. (CPPS Student Escort)
"	U.C.S.D. Hospital (Security)
"	Volunteer Assistance Team
"	Waddingtons Security (SD)
"	Walnut USD
"	West Los Angeles College
154.540	American Red Cross (portables)
"	Antelope Valley USD
"	Burbank USD
"	Casamar Inc.
"	Covina USD - Security
"	Charter Hospital (Hawaiian Gardens)
154.540	Diamond Bar High Schools (Security)
"	Dominguez Hills State University
"	Hillcrest North Condos
"	Hilton Inn F1 - Security
"	Hotel Del Coronado
"	Irvine USD
"	Long Beach USD
"	Maytag Aircraft Corp. (Coronado)
"	Pomona Community College
"	San Diego Humane Society (Building Sec.)
"	Scripps Clinic & Research (Borrego Springs)
"	Sea World (Hubs Research Institute)
"	Shadowridge Country Club (Vista)
"	Southland Surveying
"	Tustin City Schools (Security)
"	Universal Studios (Plaza Security)
"	Valley Disposal Service (El Cajon)
"	Volunteer Assistance Team
"	Warner Springs Ranch
"	West Los Angeles College
"	Whittier College PD
154.570	Covina USD (Security)
"	Delta Airlines
"	El Cajon Speedway (Pits & Security)
"	Guardian Angels (Security Ops)
"	Helix High School (La Mesa)
"	Hilton Inn Security (San Diego)
"	Hydro Products (San Diego)
"	Jamul La Flores School District
"	Linear Corporation (Carlsbad)
"	Orange Schools Security
"	Pala Reservation Security
"	San Diego Sports Arena - Security
"	San Diego USD - Security
"	San Juan Capistrano Schools (Security)
"	Santa Monica Comm. College
"	Southwest Exhibitors (La Jolla)
"	University of San Diego High School
"	West Covina USD
154.600	American Red Cross (portables)
"	Beverly Medical Center (Beverly Hills)
"	Budweiser Thunder Boat Races - Security
"	Carlsbad High School - Security
"	Chamber Building (San Diego)
"	El Cajon Speedway (Pits)
"	Gompers Secondary School (San Diego)
"	Hilton Inn F2 - Security (San Diego)
"	Horizon Christian Fellowship
"	Hotel San Diego
"	Irvine, City of
"	La Costa Hotel & Spa (Security)
"	La Habra Schools
"	Lincoln High School (San Diego)
"	Mann Middle School (San Diego)
"	Mesa Vista Hospital (Security)
"	Old Globe Theater (San Diego)
"	Poway USD - Security
"	Rancho Santa Fe Security
"	S.D. Jack Murphy Stadium (Parking)
"	S.D. Sports Arena (Parking)
"	San Diego Adventist Academy
"	Hawthorne City Schools (Security)
"	San Diego High Schools (security)
"	San Juan Capistrano Schools (Security)
154.625	American Red Cross (Claremont)
"	Bank of America (San Diego)
"	Buck Knives (El Cajon)
"	Children's Hospital (San Diego)
"	Friendly Hills Hospital (La Habra)
"	Hanalei Hotel (San Diego)
"	Los Angeles City College PD
154.6375	Henley Group (La Jolla)
154.650 R	Basset USD - Police (La Puenta)
"	Long Beach Comm. College PD (F-1)
"	Los Angeles Comm. College PD (F-1)
"	Ontario / Montclair School PD
"	Pasadena Community College PD
"	Saddleback Comm. College PD (F-1)
"	San Bernardino Comm. College
"	Santa Monica Comm. College PD

FREQUENCY	NAME
154.665	State Justice Department (Tactical)
154.680 R	State Justice Department (Primary)
154.695	State Justice Department (Tactical)
154.700	Federal Highway Police (Tijuana & Mexicali)
154.725	Calexico PD (F-1)
"	Covina PD (F-1) Primary
"	Los Angeles City College PD
"	San Diego PD Detectives
154.740	Barstow-Victorville FD
"	Burbank PD (F-4) Tactical
"	Colton PD (F-1) Tactical
"	Los Angeles City College PD
"	Redlands PD (Tactical)
"	Rialto PD (Tactical)
"	San Bernardino PD (Tactical)
154.755	Coachella PD
"	LAPD Burglary Auto Detail BAD
"	Oceanside PD (F-2) Secondary
154.770	LAPD - Tactical 2 (Surveillance)
154.775 R	Directorate of National Security (Mexico)
154.785	Rialto PD (F-2)
"	San Diego PD (Input to 154.055)
154.800	OCSO - Intake Release Center (IRC)
"	OCSO - Honor Camps Security
"	Redland PD (F-1) Primary
154.815 R	Chula Vista PD (F-1) Dispatch
"	Colton PD (F-2) Dispatch
154.815	El Centro PD (F-1) Dispatch
"	Huntington Beach PD (F-1)
"	Imperial PD (F-1) Dispatch
154.830	LAPD - Tactical 1
154.845	Azusa PD (F-1) Dispatch
"	Coronado PD (F-1) Dispatch
"	Glendora PD (F-1) Dispatch
"	Imperial County Sheriff - East Dispatch
154.860	Montclair PD (Inquiry)
154.860	Oceanside PD (Car-to-Car F3)
154.875	Fontana PD (F-2) Tactical
"	San Diego PD (Input to 159.090)
154.890	El Cajon PD (Input to 155.625)
"	Monterey Park PD (F-4)
154.905	C.H.P. Mobile Extenders & Tactical
154.920	Statewide CLEMARS Gold #1
154.935	Statewide CLEMARS Gold #2
154.935	San Diego Police (S.W.A.T.)
154.935	State Justice Department (Surveillance)
154.950	San Diego PD (Input to 158.730)
154.965	LAPD Juvenile Division
"	OCSO - Theo Lacy Jail Security
154.995	Imperial County Sheriff - West Dispatch
"	LAPD - Tactical 3
"	San Diego City Local Government 2
155.010	Brawley / Calipatria PD (F-1)
155.010	San Diego PD (Internal Affairs/Tac)
155.025	Burbank Airport Operations (F-2)
"	Colton PD (F-3) Tactical
"	Station "X" (F-1 EMS Coordination)
155.040	Long Beach City Street Department
"	Montclair/Ontario PD (PRECOM)
"	State Justice Department (Surveillance)
"	Ventura PD
155.055	El Segundo Public Works
"	Hawaiian Gardens Public Works
"	Hemet PD
"	Irwindale PD & Public Works
"	Oceanside PD (F-1) Dispatch
"	Rancho Cucamonga Local Government
"	SD Fire Channel - Red (Command & Tactical)
155.115	Long Beach Community College PD (F-2)
155.130	Holtville PD (F-1)
155.130	Imperial Countywide Mutual Aid
155.145	Chino Local Government
"	La Mesa PD (F-5) Car-to-Car
"	Loma Linda Local Government
155.160	California Radio Assistance Monitors
"	California Search & Rescue
"	De Anza Rescue Unit (El Centro)
"	Hemet Valley Rescue
"	High Desert Mountain Rangers
"	L.A. County Beaches & Harbors
155.160	Long Beach City Lifeguards
"	Los Exploradors Search & Rescue (El Centro)
"	Malibu EMS
"	Rescue 3 Ambulance (Barstow)
"	Saddleback Search & Rescue
"	San Diego County Mountain Rescue Team
"	San Diego Station "X" (F-5) Tactical
"	Sealson Ambulance
"	Sierra Madre Search & Rescue
"	Southwest Desert Rescue (Palmdale)
"	Ventura County Search & Rescue
155.175	Rescue 3 Ambulance (Barstow)
"	Riverside EMS-3
"	San Diego Station "X" (F-3) Hospital-Paramedics
"	Santa Ana Schools (Buses)
"	Simi Valley Ambulance
155.190	Bell / Cudahy PD (F-1) Primary
155.190	National City PD (F-2) Car-to-Car
155.205	Bear Valley Community Medics
"	Beverly Hospital (Montebello)
"	Doctor's Ambulance (Laguna Hills & LA)
"	Hartson's Ambulance (F-1) San Diego
"	Joshua Tree Ambulance
"	Riverside EMS-4
"	Springs Ambulance (Palm Springs)
"	Ventura County (Hospital-Ambulance)
"	Westside Union High School (Lancaster)
155.220	Adam's Ambulance (South Gate)
"	Allen's Professional Ambulance (LA)
"	Competition Ambulance (Riverside)
"	Goodhew Ambulance (Los Alamitos)
"	Emergency Rescue (Monrovia)
"	Hartson's Ambulance (F-4) Supervisors
"	McCormix Ambulance (Inglewood)
"	Mercy Air Ambulance (Fontana)
"	Oxnard Ambulance
"	Southwest Desert Rescue (Palmdale)
155.235	Antelope Valley Schools
"	Courtesy Ambulance (Riverside)
"	Eastside Union Schools (Buses)
"	Imperial County Department of Education
"	Oxnard Ambulance
"	Station "X" (Grossmont Hospital)
"	Station "X" (Scripps La Jolla Hospital)
"	Station "X" (Sharp Memorial Hospital)
"	Saddleback Search & Rescue
"	San Diego Mountain Rescue Team
"	San Dimas USD (Buses)
"	San Marcos USD - Buses
"	Taylor Bus Service (San Diego)
"	Tri-Valley EMS
"	Walnut USD
155.250	Carlsbad PD Tactical
"	Coronado PD (F-2) Car-to-Car
"	El Centro PD (F-2) Car-to-Car
"	Pomona PD (F-5) Detectives
155.265	Centinela Hospital (Inglewood)
"	Chula Vista USD - Buses
"	Eastside Union Schools (Lancaster)
"	Fallbrook USD - Buses
"	Huntington Beach Schools
"	Imperial County Education Department
"	Riverside County EMS-1
"	Southbay USD - Buses
155.280	Borrego USD - Buses
"	Centinela Hospital (Inglewood)
"	Crippen Ambulance (Covina)
"	Courtesy Ambulance (San Bernardino)
"	Desert Ambulance (Barstow)
"	Hartson's Ambulance (F-2) SD South Bay
"	Lemon Grove USD - Buses
"	Lynwood USD
"	Malibu EMS
"	LA & Orange County Hosp. Emergency Network
"	Professional Ambulance (Glendale)
"	Universal Health
155.295	Blythe Ambulance
"	Goodhew Ambulance (Riverside)
"	Hemet Valley Ambulance
"	Holtville USD - Buses

FREQUENCY	NAME
155.295	La Mesa-Spring Valley SD
"	Oceanside USD - Buses
"	Rescue 3 Ambulance (Barstow)
"	Riverside EMS-2
"	Scripps Clinic & Research (La Jolla)
"	Valley Medical (Victorville)
"	Wilsona School District (Lancaster)
155.310	El Cajon PD (F-2) Inquiry / Tactical
"	Oxnard PD (F-3) Tactical
"	Ventura PD (F-1)
155.325	Gold Cross Ambulance (El Centro)
155.325	San Diego Station "X" (F-3) Hospital Paramedics
155.340	Nationwide Hospital H.E.A.R.
155.355	Goodhew Ambulance (LA & Riverside)
155.370	San Diego PD Input/158.895 out
155.385	Hartson's Ambulance - East SD County
155.385	SD Station "X" (F-4) Hospital Paramedic
155.400	Medic Ambulance (LA)
"	San Bernardino County (Hospital-Ambulance)
"	Southwest Desert Rescue (Palmdale)
155.415	Monterey Park (F-2)
"	Oceanside PD (tactical)
155.415	Imperial County Sheriff (Input to F-5)
155.430	Imperial County Sheriff (Tactical)
155.430	San Diego Jack Murphy Stadium
155.445	S.D.S.U. PD (F-3) Secondary
155.460	State Justice Department (Input to 154.680)
155.475	Police NALEMARS Gold #3
155.490	Azusa PD (F-2)
"	Covina PD (F-1) Car-to-Car
"	La Verne PD (F-1) Dispatch
"	National City PD
155.505	U.C. Irvine PD
"	U.C. Riverside PD
"	U.C.S.D. (Campus Police F-1)
155.520	Adelanto PD (F-3) Tactical
"	Imperial County Sheriff (Tactical)
155.535	San Diego Police (Input to 159.045)
155.535	Vemtura County SO
155.550	La Verne PD (F-2) Secondary
"	Pomona PD (F-1) Primary
"	La Verne PD (F-2)
"	Long Beach PD
"	Montclair PD (F-2) Secondary
"	San Diego Police (Input to 158.970)
155.565	Monterey Park PD (F-1) Dispatch
"	Oceanside Police (Input to 154.755)
"	Oxnard PD (F-1) Primary
155.580	Grossmont Community College PD
"	Needles PD (F-1) Primary
155.595	Barstow PD (F-1) Primary
"	Chula Vista PD (F-4) Inquiry
"	Montclair PD (F-1) Dispatch
"	Pomona PD PRECOM
"	Upland PD (F-1) Primary
155.610	Coachella PD
155.625	El Cajon PD
"	Perris PD
"	Ventura PD
155.640	Chula Vista PD (F-5) Tactical
155.640	Downey PD
155.655	Burbank Airport Police (F-1)
"	Golden West College
"	Port Hueneme / Oxnard PD
"	San Bernardino PD (F-1)
155.670	South Gate PD
155.685	Adelanto PD (F-1) Primary
"	Los Angeles Community Colleges PD
"	San Diego PD "Tac"
155.715	Lancaster City Services
155.715 R	S.D. District Attorney (F-1) Station "G"
"	S.D. Marshal's Office (F-1) Station "L"
155.715	West Covina PD
155.730	Barstow PD (F-2) Secondary
155.730	Glendora PD
155.745	Adelanto Local Government
"	Compton PD (F-2) Car-to-Car
"	Imperial Beach Local Government
"	Imperial County Sheriff - Tactical
"	La Palma City Local Government
155.745	Loma Linda City Utilities
"	Oxnard PD (F-2) Inquiry
"	Port Huemene PD
155.760	Burbank PD (F-3) Tactical
"	Imperial County Local Government B
"	Needles Local Government
"	Rialto Local Government
"	Oceanside City Local Government
155.775 R	Big Bear Lake Local Government
155.775 R	Riverside County - Animal Control
155.775	Santa Ana (Building Security)
155.790	Baldwin Park PD (F-1) Tactical
"	Irwindale PD (F-1) Dispatch
"	West Covina PD (B)
155.805	Downey PD (F-3)
"	Redlands PD (F-2)
"	San Juan Capistrano Local Govern.
"	San Marcos Local Government - Portables
"	Seal Beach Local Government
155.820	Inglewood PD Tactical
155.820	San Diego General Services Department (F-1)
155.835	Big Bear Lake MWD Patrol
"	Carlsbad Police Department (F-2)
"	San Bernardino Local Government
"	South Pasadena Local Government
"	UC Irvine (Housing)
155.850	Compton PD (F-1) Dispatch
155.865	Lake Elsinore Local Government
"	San Diego State University (input to 153.740)
"	Ventura FD (tactical)
155.880	Arcadia Local Government
"	Imperial County Local Government A
"	Monrovia PD & Local Government
"	Oceanside City Lifeguards
155.895	Downey PD (F1-R, F2-S) Primary
"	San Diego Unified Port District (F-2)
155.910	Burbank PD (F-2) Inquiry
155.925	Long Beach FD (F-3)
155.940	Big Bear Lake Local Government
155.940 R	San Diego General Services Department (F-2)
155.955	Pasadena Local Government
"	Ventura Local Government
"	Villa Park Public Works
"	Vista Local Government (portables)
155.985	El Cajon Local Government
156.000	Maritime Channel 1 (International Use)
156.015	Oceanside Harbor Patrol
"	U.C. Irvine (Health services)
"	Ventura PD
156.030 R	Carlsbad PD (F-1) Dispatch
"	Imperial County Sheriff - Central Dispatch
"	Maywood PD
"	Rialto PD (F-1) Primary
156.045	Oceanside PD (Input to 155.055)
156.075	Statewide CALCORD (Interagency)
156.090	Los Angeles Community Colleges PD
156.090	Port Hueneme PD
156.105	Bell Gardens City Street Dept.
156.105	Idyllwild Local Government
156.120	San Diego Road Department (F-1)
156.150	San Diego Unified Port District (F-1)
"	Upland PD (F-1) Primary
"	Ventura County SO (East Dispatch)
156.165	San Diego Road Department (F-3)
"	West Covina FD
156.195	Long Beach FD (F-4) Fireground
156.210	Burbank PD (F-1) Primary
"	San Bernardino PD (F-2) Secondary
"	San Diego Co. Court House Security
156.225	Norco Local Government
"	San Diego County Road Department (F-2)
156.250	Maritime Channel 63 (International)
156.275	Maritime Channel 65 (Port Operations)
156.300	Maritime Channel 6 (Intership Safety)
156.325	Maritime Channel 66 (Port Operations)
156.330	San Marino PD (F-1)
156.350	Maritime Channel 7 (Commercial)
156.375	OMBACK Over-The-Line Court Control
156.400	Maritime Channel 8 (Commercial)
156.425	Maritime Channel 68 (Non-Commercial)

FREQUENCY	NAME
156.425	- Marina Del Rey Patrol
156.450	Maritime Channel 9 (All Maritime)
156.475	Maritime Channel 69 (Non-Commercial)
156.500	Maritime Channel 10 (Commercial)
156.525	Maritime Channel 70 (Non-Commercial)
156.550	Maritime Channel 11 (Commercial)
156.575	Maritime Channel 71 (Non-Commercial)
156.575	- Marina Del Rey Patrol
156.600	Maritime Channel 12 (Port Operations)
156.625	Maritime Channel 72 (Non-Commercial)
156.650	Maritime Channel 13 (Navigational)
156.675	Maritime Channel 73 (Port Operations)
156.700	Maritime Channel 14 (Port Operations)
156.725	Maritime Channel 74 (Port Operations)
156.750	Maritime Channel 15 (Environmental)
156.750 R	Tijuana Public Works (East Sector)
156.800	Maritime Channel 16 (Distress, Safety)
156.850	Maritime Channel 17 (State Control)
156.875	Maritime Channel 77 (Commercial)
156.875	Miramar-Point Loma Inspection Teams
156.900	Maritime Channel 18 (Commercial)
156.925	Maritime Channel 78 (Non-Commercial)
156.950	Maritime Channel 19 (Commercial)
156.975	Maritime Channel 79 (Commercial)
157.000	Maritime Channel 20 (Port Operations)
157.025	Maritime Channel 80 (Commercial)
157.050	Maritime Channel 21 (USCG Cutters))
157.075	Maritime Channel 81 (Safety Officer)
157.100	Maritime Channel 22 (USCG Civilian Liason)
157.125	Maritime Channel 82 (USCG)
157.150	Maritime Channel 23 (USCG S.D. Primary)
157.175	Maritime Channel 83 (USCG Auxiliary)
157.425	Maritime Channel 88 (Commercial)
157.450	Canyon General Hospital (Anaheim)
"	Kaiser Foundation (Anaheim)
"	Santa Teresita Hospital (Duarte)
157.470	S. & N. Wards Towing
157.475	Maritime Channel 89 (International Use)
157.500	Al's Towing (Escondido)
157.500	Grossmont Center Shell (La Mesa)
157.530	Calexico Taxi
157.530	R. & L. Taxi (Brawley)
157.530	Yellow Cab Company of San Diego (Mobiles)
157.545	Coast Cab Company (Mobiles)
157.560	Bob Meyer Farm (Brawley)
157.560	Escondido Yellow Cab Company (Mobiles)
157.575	Coast Cab Company (Mobiles)
157.590	Yellow Cab Company of San Diego (Mobiles)
157.600	Social Security Hospital (Mexicali)
157.605	ABC Cab Company (Mobiles)
157.605	U.S.A. Cab Company Inc. (Mobiles)
157.620	Diamond Cab Company (Mobiles)
157.635	ABC Cab Company (Mobiles)
157.635	La Jolla Cab Company (Mobiles)
157.665	Red Top Cab Company (Mobiles)
157.680	Jimsair Aviation Service
157.680	Yellow Cab Company of San Diego (Mobiles)
157.695	American Cab Company (Mobiles)
"	Amigo Cab Company (Mobiles)
"	Fred's Cab Company (leased license)
"	Limousine Service Cab Company
157.710	Oceanside Yellow Cab Co. (Mobiles)
157.710	Yellow Cab Company of San Diego (Mobiles)
157.725	La Cresta Farm Management
157.740	H.H. Restaurant Inc. (San Diego)
"	Little American Westgate Hotel
"	Rodeway Inn (San Diego)
"	Vacation Village Hotel
158.100	American Paging, Inc. (San Marcos)
158.130	Southern Ca. Edison Co.
158.145	Southern Ca. Edison Co.
158.160	Southern Ca. Edison Co.
158.175	Imperial Irrigation District (El Centro)
158.175	Southern Ca. Edison Co.
158.175	Vallecitos Water District
158.190	Imperial Irrigation District
158.190	San Diego City Water District
158.190	Southern Ca. Edison Co.
158.205	Imperial Irrigation District (Imperial)
158.205	Southern Ca. Edison Co.

FREQUENCY	NAME
158.220	Southern Ca. Edison Co.
158.235	Pasadena Water Department
158.235	Southern Ca. Edison Co.
158.250	Imperial Irrigation District
158.250	Southern Ca. Edison
158.265	Southern Ca. Edison
158.295	Lockheed FD (Burbank)
158.295	Rohr Industries (Chula Vista)
158.310	Burroughs Corporation (San Diego)
158.310	Linkabit Corporation (San Diego)
158.310	Rohr Industries (Chula Vista)
158.335	Nelco Refining Company (San Diego)
158.385	Pendelton Farms (San Luis Rey Valley)
158.385	Wimpey-Railco-Kingston (National City)
158.400	Business Itinerate
158.445	Oil Spill Cleanup - San Diego
158.460	Daniel Freeman Hospital (Inglewood)
158.460	Humana West Hospital (Westminster)
158.460	Coronado Crane Co. (San Diego)
158.700	Gencom Incorporated (Pagers)
158.730 R	Indio PD
"	Monrovia PD
"	San Diego PD (F-1) South Dispatch
"	Sierra Madre PD
"	Ventura County SO (Car-to-Car)
158.745	Oceanside PD (input to 154.025)
158.745	Vernon PD (F-1) Primary
158.760	City of Compton Public Works
"	City of Orange Local Government
"	City of Newport Beach Local Government
"	City of San Diego Local Government
"	City of South Gate Local Government
"	City of Vernon Local Government
"	Riverside SO (F-4) Car-to-Car
158.775	Bellflower Local Government
"	Lawndale Local Government
"	Lomita Local Government
"	Riverside SO (F-9) Countywide Tactical
"	Rolling Hills Estates Local Government
"	Theo Lacy Jail Security (OCSO)
"	Vista Local Government
158.790 R	State of California - CLERS
158.805	Hawthorne Public Works
"	Los Alamitos City Local Government
"	San Jacinto Local Government
"	San Marcos City Public Works
"	Simi Valley PD (F-2)
"	Ventura County SO (F-5)
158.820	Pasadena Local Government
158.820	Riverside SO (F-4) Banning, Lake Elsinore
158.835	Carlsbad PD (F-2) Car-to-Car
"	Dana Point City Sanitary Services
"	Santa Paula PD
"	Simi Valley PD (F-2)
"	Ventura County SO (F-4)
"	Westminster Local Government
158.850	Riverside SO (F-1)
158.865 R	LAPD MDT's (West Bureau)
158.865	SD Marshal's Office Tactical (F-2)
158.865	San Diego City - Engineering
158.880	Buena Park City Local Government
"	Newport Beach Local Government
"	Port Hueneme Local Government
"	San Bernardino Local Government
158.895 R	Rosemead Public Works
"	San Diego PD (F-5) West Dispatch
"	San Jacinto PD & FD
"	Westmoreland Local Government
158.910 R	LAPD MDT's (South)
"	San Diego PD (F-6) South Dispatch
158.925	Chula Vista PD (F-3)
"	Chula Vista Local Government
"	Riverside SO (F-6) Countywide Mutual Aid
"	Temple City Local Government
"	Victorville Local Government
158.955	Beaumont / Perris PD's
"	Holtville Local Government
158.970	OCSO - Intake Release Center (IRC)
"	Riverside SO (F-3) Car-to-Car
"	Theo Lacy Jail Security

FREQUENCY	NAME
158.970 R	San Diego PD (F-2) North Dispatch
159.015	San Diego County Road Department
159.030	LAPD MDT's (Citywide)
159.015	S.D. District Attorney Tactical
159.045 R	San Diego PD (F-4) Central Dispatch
159.075 R	State Judicial Police (Baja Calif.)
159.090 R	Riverside SO (F-2) Indio & Palm Desert
159.090 R	San Diego PD (F-3) East Dispatch
159.150 R	LAPD MDT's (Central)
159.165	Corona FD
159.165	Oceanside Fire - Dispatch
159.195	Glendale City Street Department
159.210	Ventura County SO (West Dispatch)
159.225	State C.D.F. (input to 151.190)
159.250	C.D.F. Tactical San Diego
159.270	C.D.F. Tactical Net
159.300	State C.D.F. (input to 151.355)
159.315	C.D.F. Tactical Perris West
159.330	State C.D.F. (input to 151.265)
159.345	C.D.F. Tactical Net
159.375	C.D.F. Tactical Net
159.405	C.D.F. Tactical Net
159.420	State Fish & Game (input to 151.430)
159.435	State Fish & Game (input to 151.415)
159.450 R	Halcones Rescue Team (Tijuana)
159.450	Riverside SO (F-7) & (F-8) Blythe & Moreno V.
159.480	Oil Spill Cleanup - San Diego
159.495	Brinks Armored Car Service
159.525	S.M. Harris Trucking Inc. (San Diego)
159.585	Transcon Lines (San Diego)
159.735	California Messenger Service (San Diego)
159.825	Sky Trucking (Chula Vista)
160.005	Auto Fast Freight Inc. (San Diego)
160.050	Roadrunners Messenger Service (San Diego)
160.095	Expo Stucco Inc. (San Diego)
160.155	Action Courier Inc. (Mt. Otay)
160.185	Emerald Empire Transportation (Mt. Woodson)
160.200	Roadway Express Inc. (Mt. Otay)
160.225 R	State Judicial Police (Mexicali)
160.245 R	A.T.S.F. R.R. - National City PBX
160.260 R	A.T.S.F. R.R. - Sierra Peak PBX
160.260 R	Mexican User
160.275	A.T.S.F. R.R. - Barstow Calling East
160.290 R	U.P. R.R. - PBX Telephones
160.305	A.T.S.F. R.R. - Barstow Mini Hump
160.320	S.P. R.R. - Yuma to Gila Bend, AZ Road
160.335 R	A.T.S.F. R.R. - Crestline / Oceanside PBX's
160.350 R	San Diego Trolley (F-4) Special Units
160.350	S.P. R.R. - Intermodal Container Terminal
160.380 R	San Diego Trolley (F-2) East Line
160.380	S.P. R.R. - Yard Channel
160.395 R	S.P. R.R. - PBX Telephones
160.410	U.P. R.R. - Hump Yard Operations
160.425 R	A.T.S.F. R.R. - Mojave to Four Corners PBX's
160.425 R	MetroRail - Maintenance
160.440 R	A.T.S.F. R.R. - Barstow Carmen
160.455	Pacific Southwest Railway Museum
160.470	S.P. R.R. - Colton Yard Hump
160.515	U.P. R.R. - Road Operations
160.530	A.T.S.F. R.R. - Barstow Calling East
160.530	San Diego & Imperial Valley Railroad (F-7)
160.545	Southern Calif. MetroLink - Road F3
160.560	A.T.S.F. R.R. - Barstow Calling East
160.560	A.T.S.F. R.R. - Los Angeles to San Bernardino
160.605	U.P. R.R. - Maintenance of Way
160.620	S.P. R.R. - Railroad Shops
160.635	MetroRail Red Line
160.650	A.T.S.F. R.R. - Los Angeles to San Diego
160.665 R	San Diego Trolley (F-1) South Line
160.680	U.P. R.R. - Yard Operations
160.695	MetroRail Red Line
160.710 R	San Diego Trolley (F-3) Central Line
160.710	U.P. R.R. - Hump Yard Operations
160.725	MetroRail Blue Line
160.740	U.P. R.R. - Switching & Hump
160.750 R	Federal Highway Patrol (Ensenada)
160.770	A.T.S.F. R.R. - Barstow Calling East
160.785 R	Simi Valley PD (F-1)
160.815	Southern Calif. MetroLink - Yard Ops F2
160.890 R	S.P. R.R. - PBX Telephones
160.935	San Diego Trolley (input to 160.665)
160.950 R	S.P. R.R. - PBX Telephones
160.980	A.T.S.F. R.R. - Barstow Maintenance
160.995	Southern Calif. MetroLink - Maintenance F4
161.010	A.T.S.F. R.R. - Maintenance of Way
161.025	U.P. R.R. - Maintenance
161.055	Amtrack Corporation (Administrative)
161.070	A.T.S.F. R.R. - Hobart Yard
161.100	S.P. R.R. - Yard Channel
161.115	Southern Calif. MetroLink
161.130	L.A. Junction R.R.
161.190	A.T.S.F. R.R. - Lambert to San Bernardino
161.205	Railroads Police (F-1)
161.220	Railroads Police (F-2)
161.280	A.T.S.F. R.R. - Barstow Calling West
161.295	San Diego Trolley (input to 161.565)
161.300 R	State Judicial Police (Tijuana, Ensenada)
161.340	A.T.S.F. R.R. - Barstow Calling West
161.370	A.T.S.F. R.R. - Barstow Calling West
161.370	A.T.S.F. R.R. - Yard Channel
161.400	S.P. R.R. - Intermodal Container Terminal
161.400	S.P. R.R. - Maintenance
161.415	Southern Calif. MetroLink - Road F1
161.430	S.P. R.R. - Switching & Yard
161.505 R	S.D.I.V. Railway (F-1) Primary
161.550	S.P. R.R. - Road Channel
161.565 R	San Diego Trolley (F-5) Security
161.6275	KRTH-AM (Los Angeles)
161.640	KBBY (Ventura)
161.640	KCOP-TV Channel 13 (Secondary)
161.670	KCBQ-AM Radio
161.670	KCOP-TV Channel 13 (Secondary)
161.670	KFSD Radio
161.670	KGIL-AM/FM (Los Angeles)
161.670	KWKW-AM (Los Angeles)
161.6875	KIIS-FM (Los Angeles)
161.700	KG-TV Channel 10 (Primary Dispatch)
"	KHAY-FM (Ventura)
"	KICO-AM (Calexico)
"	KWKW-AM (Los Angeles)
161.730	KG-TV Channel 10 (Car to Car)
161.730	KOWN-FM Radio (Primary)
161.730	KSDO-AM Radio (Primary)
161.760	KAVR-AM/FM
"	KCBQ-AM Radio (Primary)
"	KLZZ-AM/FM Radio
"	KNBC-TV 4 (News Desk Dispatch)
"	KPBS Radio & Television
161.800 R	Marine Operator - San Pedro
161.825 R	Marine Operator - Dana Point
161.850 R	Marine Operator - Santa Barbara
161.875 R	Marine Operator - Redondo Beach
161.900 R	Marine Operator - San Pedro
161.925 R	Marine Operator - San Diego
161.950 R	Marine Operator - Redondo Beach
161.975 R	Marine Operator - Redondo Beach
162.000 R	Marine Operator - San Diego
162.025	U.S. Coast Guard (Anti-Smuggling)
162.050	Mexican User
162.100	National Ocean Service (Nationwide)
162.125	U.S. Coast Guard - Command & Control
162.175	U.S. Coast Guard - Command & Control
162.225	U.S. Coast Guard (Pacific Emergency)
162.400	NOAA Weather Channel 2 (Mt. Woodson)
162.425	NOAA Weather Channel 4 (Nationwide)
162.450	NOAA Weather Channel 5 (Nationwide)
162.475	NOAA Weather Channel 3 (Nationwide)
162.500	NOAA Weather Channel 6 (Nationwide)
162.525	NOAA Weather Channel 7 (Nationwide)
162.550	NOAA Weather Channel 1 (Black Mountain)
162.550	NOAA Weather Channel 1 (Mt. Wilson)
162.625 R	Colorado River Authority (CONAGUA)
162.6375R	F.B.I. - Glendale & Pasadena
162.6875	U.S. Secret Service "Yankee"
162.7125	U.S. Marshal - San Diego Tactical
162.7375R	F.B.I. (Riverside)
162.7625R	F.B.I. (Blythe, Banning)
162.7875	F.B.I. - Tactical (San Diego)
162.7875R	U.S. Marshal F5 - San Diego Dispatch
162.825	I.N.S. (Input to 163.625)

FREQUENCY	NAME
162.875	I.N.S. (Input to 163.725)
162.900	I.N.S. - Brown Field Tactical
162.925	I.N.S. (Input to 163.675)
163.000	Veterans Administration (Hospital Paging)
163.100	U.S. Geologic Survey
163.125 R	B.I.A. Reservation Police (F1-R, F2-S)
163.125	U.S. Bureau of Reclamations (F-4)
163.125 R	U.S. Customs Service (F4-S, F5-R)
163.150	U.S. Geologic Survey (rare use)
163.175	U.S. Coast Guard - Command & Control 1
163.200 R	U.S. Marshal (F1-S, F2-R)
163.250	Naval Hospital Paging
163.275	Mexican Red Cross (Rosarito Beach)
163.325	U.S. Army Corps of Engineers
163.400 R	Mexican Red Cross
163.4125R	U.S. Army Corps of Engineers
163.4375R	U.S. Army Corps of Engineers
163.4625	Los Angeles A.F. Station - Contingency Comm.
163.475 R	Mexican Red Cross
163.4875	Edwards A.F. Base - Test Range Security
163.4875	March A.F. Base - Base Police
163.575	U.S. State Department (Tijuana)
163.5875	Edwards A.F. Base - EOD Teams
163.625 R	I.N.S. - Chula Vista Station
163.625 R	I.N.S. - Yuma Sector
163.650	I.N.S. - Brown Field Office
163.675 R	I.N.S. - North County & Temecula
163.700	I.N.S. - Detentions & Transport Buses
163.725 R	I.N.S. - El Centro Sector
163.725 R	I.N.S. - San Clemente Checkpoint
163.750 R	C.F.E. Light & Power (Tijuana)
163.750	I.N.S. - Alien Detentions (LA & El Centro)
163.775 R	I.N.S. - Calexico / El Centro / Indio Area
163.775 R	I.N.S. - El Cajon Area
163.800 R	C.F.E. Light & Power (Baja Norte)
163.8375	F.B.I. Allocation
163.8625	F.B.I. Allocation
163.875 R	C.F.E. Light & Power (Mexicali)
163.8875	F.B.I. Allocation
163.9125R	F.B.I. - San Diego Repeater
163.9375R	B.L.M. Fire Control (F-5)
163.9375	F.B.I. Allocation
163.9625	F.B.I. Allocation
163.950	I.N.S. - Chula Vista Tactical
163.9875	F.B.I. Allocation
164.025 R	C.F.E. Light & Power (Mexicali)
164.025	National Ocean Service (F-2)
164.050 R	F.B.I. Allocation
164.075	National Ocean Service (F-1)
164.100	U.S. Department of Agriculture (Fire Camp)
164.100	Channel Island National Monument (F-3)
164.150	U.S. Department of Agriculture
164.125	Cleveland National Forest (Fire Camp Net)
164.175	Telemetry - Unknown use
164.200	Department of the Interior (F-1)
164.200	U.S. Customs Service
164.225	Pantec Nuclear Weapons Plants
164.250	F.B.I. Allocation
164.275	U.S. General Services Administration
164.300	U.S. Coast Guard (Long Beach)
164.300	U.S. Health & Human Services (Hospitals)
164.350	F.B.I. Allocation
164.400	U.S. Secret Service "Romeo"
164.425	F.B.I. Allocation
164.450	F.B.I. Allocation
164.450	U.S. E.P.A. F1 - Administration
164.4625	U.S. Customs Tactical
164.475 R	U.S. Bureau of Reclamations (Colorado River)
164.500	Army Corps of Engineers
164.525	F.B.I. Allocation
164.550	U.S. Coast Guard - Law Enforcement
164.5625	F.B.I. Allocation
164.575 R	U.S. Bureau of Reclamation FD (Colorado River)
164.600	F.B.I. Allocation
164.600	U.S. Marshal F4 - Car-to-Car
164.625 R	U.S. Fish & Wildlife Service (Lake Havasu)
164.650	U.S. Secret Service "Tango"
164.675 R	U.S. Bureau of Reclamations (F-7)
164.700	U.S. Housing & Urban Development (HUD)
164.700	U.S. State Department (A-1)
164.725 R	U.S. Bureau of Reclamation (F-8)
164.800	U.S. Agriculture Department (Fire Camps)
164.800	U.S. Geologic Survey
164.825	U.S. Agriculture Department (Fire Camps)
164.8625R	F.B.I. Allocation
164.8875	U.S. Secret Service "Oscar"
164.9125	U.S. Coast Guard - Law Enforcement
164.9125 R	U.S. Agriculture Department (Fire Dispatchers)
165.0125	Los Angeles A.F. Station - Aerospace Corps.
165.2125	U.S. Secret Service "Mike" (Tactical)
165.2375R	U.S. Customs Service (F1-S, F2-R)
165.2625	U.S. Coast Guard (Air Station Security)
165.2625	U.S. Coast Guard (Law Enforcement F-3)
165.2875R	U.S A.T.F. (F2-S, F3-R)
165.3125	U.S. Coast Guard - Oil Spill Cleanup
165.325 R	Mexican Custom's (Tijuana)
165.3375	U.S. Coast Guard (Law Enforcement F1)
165.3375	U.S. Coast Guard (Loran Station)
165.375	U.S. Secret Service "Charlie"
165.400	Rescates Rescue Team (Tijuana)
165.4125	Environmental Protection Agency (S. Calif.)
165.4375	U.S. Customs Service (A7-S, A8-R)
165.4875	U.S. Customs Tactical
165.500	U.S. F.A.A. Airport Security
165.5125	U.S. Customs Tactical (B7-S, B8-R)
165.5625	U.S. Coast Guard - Tactical
165.5875R	F.B.I. Allocation
165.6625	U.S. F.A.A. Airport Security
165.6625	U.S. Nuclear Regulatory Commission
165.6875R	U.S. Secret Service (San Diego Office)
165.7125R	F.B.I. Allocation
165.7125	U.S. Marshal (LA)
165.725	Unknown Mexican User
165.7375	U.S. Customs Service (Air-to-Ground)
165.750 R	National Transportation Safety Board
165.7625	U.S. Custom's Tactical
165.7625	U.S. Secret Service "Golf"
165.7875	U.S. Secret Service "Baker" (Primary)
165.825	I.N.S. (Alien Detentions & Border Crimes)
165.850 R	I.N.S. (Border Crimes)
165.875 R	I.N.S. (Anti-Smuggling Units)
165.900 R	F.B.I. Allocation
165.9125	U.S. A.T.F. (A-5)
165.925 R	F.B.I. Allocation
165.925	U.S. Secret Service "Delta"
165.950	U.S. Internal Revenue Service (F-1) & (F-3)
165.975	I.N.S. (Anti-Smuggling Units)
166.000	U.S. Internal Revenue Service
166.025	NOAA Administrative Operations (San Diego)
166.075 R	National Marine Fisheries Service (F1-R, F2-S)
166.075 R	Mexicali FD (Dispatch)
166.100 R	Tijuana PD (Commercial Division)
166.125 R	Tijuana PD (Downtown & Motorcycles)
166.125	U.S. Dept. of Agriculture (Law Enforcement)
166.150	Federal Maritime Commission (F-1)
166.175	F.A.A. (Airport Security)
166.175	Federal Maritime Commission (F-2)
166.175	Mexicali Local Government
166.175	National Transportation Safety Board
166.200	Mexican State Penitentiary (Tijuana)
166.200	U.S. Bureau of Reclamations (Colorado River)
166.2125	U.S. Secret Service "Hotel"
166.250	KG-TV Back up
166.250	KOWN-FM Radio (San Diego)
166.2625	U.S. Bureau of Land Management (F-6))
166.275	Mexicali PD (F-2)
166.2875	U.S. Customs (A-3)
166.300 R	Tijuana PD (La Mesa)
166.375 R	U.S. Bureau of Land Management (F1-S, F2-R)
166.400	U.S. Secret Service "Golf"
166.425 R	Tijuana PD (San Antonio / South)
166.4625	Treasury Department (Interagency)
166.475	Mexicali Local Government
166.4875R	U.S. Bureau of Land Management (F3-S, F4-R)
166.4875R	U.S. Bureau of Reclamations (F1-R, F2-S)
166.500	Mexicali PD (F-1)
166.5125	U.S. Secret Service "Sierra"
166.525 R	Tijuana PD (Playas / Southwest)
166.5375	U.S. A.T.F. / Customs Tactical
166.575	U.S. Secret Service "Foxtrot"

FREQUENCY	NAME
166.580	Mexican Presidential Protection
166.5875	U.S. A.T.F. / Custom's Tactical
166.605	Tijuana PD (Jail Security)
166.6125	U.S. Bureau of Indian Affairs (F3-Fire Control)
166.6125	U.S. Bureau of Land Management (Fire Control)
166.625 R	Tijuana PD (La Presa / East)
166.650 R	Tijuana PD (Downtown & Patrol Cars)
165.6625	U.S. Customs Tactical
166.675	U.S. Department of Agriculture (Air Tac 1)
166.675	U.S. Bureau of Land Management (F-7)
166.700	Mexicali PD (F-2)
166.700	U.S. Secret Service "Quebec"
166.725 R	Tijuana PD (Rosarito)
166.725	U.S. Bureau of Indian Affairs (F4-Administration)
166.725	U.S. Bureau of Land Management (Fire Tac 1)
166.750 R	Ensenada PD (Rural Police)
166.750	U.S. Bureau of Land Management (Administ.)
166.775	U.S. Bureau of Land Management (Fire Tac 2)
166.800	Mexicali Police Department (F-4)
166.875	Ensenada PD (F-1)
166.9375	U.S. Bureau of Land Management (F-7)
167.000	U.S. Bureau of Land Management (F-8)
167.000	U.S. Internal Revenue Service (F-2)
167.025	U.S. Secret Service "November"
167.050 R	U.S. F.C.C. (F1-R, F2-S)
167.075	U.S. F.C.C. (F3-S)
167.075	U.S. Bureau of Reclamations
167.100 R	Sonora-Baja Railroad (Mexicali)
167.100	U.S. Bureau of Land Management (Fire Tac 6)
167.125	U.S. Bureau of Reclamations
167.175	U.S. Bureau of Land Management (Yuma, AZ)
167.1875R	U.S. E.P.A. (Los Angeles Operations)
167.2125	F.B.I. - Surveillance San Diego
167.2375R	F.B.I. - Repeaters
167.250	U.S. Bureau of Land Management (Fire Tac 1)
167.2625R	F.B.I. - Repeaters (San Diego)
167.2875R	F.B.I. - Repeaters (San Diego)
167.300 R	Federal Highway Maintenance (Tijuana)
167.325	F.B.I. - Surveillance
167.3375R	F.B.I. - Repeaters
167.360 R	Federal Highway Patrol (Mexico-Remote Links)
167.3875R	F.B.I. - Repeaters (San Diego)
167.400 R	Federal Highway Patrol (Mexico-Remote Links)
167.4125R	F.B.I. - Repeaters (Los Angeles)
167.4375	F.B.I. - Surveillance
167.4625R	F.B.I. - Repeaters (San Diego)
167.475	F.B.I. - Surveillance
167.500	F.B.I. - Surveillance
167.5125R	F.B.I. - Repeaters (San Diego)
167.5375R	F.B.I. - Repeaters (San Diego)
167.5625	F.B.I. - Nationwide Car-to-Car
167.5875R	F.B.I. - Repeaters
167.600	F.B.I. - Surveillance
167.6125R	F.B.I. - Repeaters
167.625	Sonora-Baja California (Mexicali)
167.650	F.B.I. - Surveillance
167.6625	F.B.I. - Surveillance
167.6875R	F.B.I. - Repeaters
167.7375	F.B.I. - Surveillance
167.7875R	F.B.I. - Repeaters
167.825	F.B.I. - Surveillance
167.850	U.S. Department of Energy (Military Bases)
167.875	Federal Emergency Warning Radio System
167.900	U.S. Coast Guard - Tactical Primary
167.900	U.S. E.P.A. (Los Angeles Operations)
167.925 R	Federal Emergency Warning Radio System
167.950	U.S. Bureau of Land Management
167.975 R	Federal Emergency Warning Radio System
167.975	U.S. Department of Energy (NEST)
168.000	U.S. Alcohol Tobacco & Firearms (A-7)
168.050	U.S. Department of Agriculture (Tactical 1)
168.075	U.S. Department of Agriculture (Command 3)
168.100	U.S. Department of Agriculture (Command 2)
168.125	U.S. Department of Agriculture (Fire Tac 5)
168.125	Cleveland National Forest (F-7)
168.150	Cleveland National Forest (F-3)
168.200	Angeles National Forest (F-4)
168.200	Cleveland National Forest (F-5)
168.200	U.S. Department of Agriculture (Tactical 2)
168.250	U.S. Bureau of Land Management (Fire Tac 3)
168.325	U.S. Bureau of Indian Affairs (Parker, AZ)
168.350	U.S. Department of Agriculture (Interagency)
168.400	U.S. Bureau of Land Management (Fire Tac 4)
168.450	C.F.E. Light & Power (Mexicali)
168.450	U.S. Department of Energy (Military Bases)
168.475	U.S. Bureau of Land Management (F-10)
168.525	U.S. Veterans Administration (F1-Security)
168.550	U.S. Bureau of Land Management
168.575	U.S. Veterans Administration (F2-Security)
168.600	U.S. Department of Agriculture (Tactical 3)
168.625	U.S. Department of Agriculture (Supply Net)
168.650	U.S. Department of Agriculture (Region Calling)
168.675	U.S. Department of Agriculture (Inspectors)
168.700 R	Tijuana Water Department
168.700	U.S. Department of Agriculture (Command 1)
168.750 R	Cleveland National Forest (F-1, F-2)
168.775	U.S. Department of Agriculture (Cache Primary)
168.875 R	I.N.S. - Yuma Area
168.950	Federal Light & Power (Tijuana)
169.000	North Island NAS (telemetry)
169.050	Housing & Urban Development (HUD)
169.075	U.S. Federal Maritime Commission (F-2)
169.100	U.S. Housing & Urban Development (HUD)
169.100	U.S. Department of Energy (Input to 170.075)
169.100	U.S. State Department (B)
169.125	U.S. Department of Agriculture (Interagency)
169.150	U.S. Department of Agriculture (Air Tac 2)
169.175	Riverside Forest Fire Lab (Operations Net)
"	San Dimas Forest Project (Experimental)
"	U.S. Department of Agriculture (Command 5)
169.200	U.S. Department of Agriculture (Air Tac 3)
169.250	Federal Executive Emergency Radio System
169.300	U.S. F.A.A. (Tactical)
169.300	U.S. Veteran's Administration
169.350	U.S. Bureau of Land Management (Fire Tac 8)
169.375	U.S. Postal Service "Yellow"
169.425	U.S. Army - Corps of Engineers
169.450 R	U.S. Customs Tactical (C1-S, C2-R)
169.550	U.S. Customs Tactical
169.575	U.S. Department of Agriculture
169.6125	U.S. State Department (San Diego)
169.625 R	U.S. State Department (Orange County)
169.650	U.S. Postal Inspectors "Blue"
169.700	U.S. Housing & Urban Development (HUD)
169.725	Cleveland National Forest (Input to F-4)
169.725	Ensenada PD (F-2) Traffic
169.750	U.S. Bureau of Land Management (F-6)
169.750 R	XEWT-TV Channel 12 (news & remotes)
169.850	U.S. Postal Inspectors "Red"
169.875	U.S. Department of Agriculture (Command 4)
169.900	U.S. Forestry Administrative (F1-R, F2-S)
169.925 R	U.S. Department of Agriculture (Tactical 4)
169.925	U.S. Secret Service "Able"
169.925	U.S. State Department (C)
169.950	Angeles National Forest (F-3) Administration
169.975	U.S. Forestry Administrative (F3-Car-to-Car)
170.000	U.S. Department of Agriculture (Air-to-Ground)
170.075	LoJack Stolen Vehicle Tracking
170.075	U.S. Department of Energy (Blythe)
170.100 R	Death Valley National Monument (F1-R, F2-S)
170.125	F.A.A. (Tactical)
170.150	KFI-AM (Los Angeles)
170.200 R	Federal Emergency Warning Radio System
170.220	Guardian Angels (Security Ops)
170.375	U.S. Department of Agriculture
170.450	U.S. Department of Agriculture (Tactical)
170.500	Cleveland National Forest (Input to F-2)
170.650	U.S. Bureau of Prisons (F-3)
170.675	I.N.S. - El Centro Sector (Tactical)
170.700	I.N.S. - Chula Vista Sector (Alarms)
170.775	Delta 7 Rescue Team (Tijuana)
170.775	I.N.S. - Chula Vista Sector (Alarms)
170.825	U.S. Bureau of Prisons (F-4) Security
170.850	U.S. Marshal (Court Service)
170.875	U.S. Bureau of Prisons (F-1) Security
170.900 R	F.B.I. Repeaters
170.900	U.S. Bureau of Prisons (F-5) Security
170.925	U.S. Bureau of Prisons (F-2) Security
170.925	U.S. Department of Agriculture (Tactical)
171.100 R	Tijuana Water Department

FREQUENCY	NAME
171.150 R	Tijuana Water Department
171.200	U.S. Department of Energy (NEST)
171.2375	U.S. Coast Guard (Stations)
171.2875	U.S. Secret Service "Zulu"
171.3375	U.S. Coast Guard - Law Enforcement (F-4)
171.3625	U.S. Coast Guard - Law Enforcement (F-5)
171.3875	U.S. Department of Agriculture
171.3875	U.S. Veterans Administration (Doctor Paging)
171.450	D.E.A. - Nationwide Use
171.475 R	San Bernardino National Forest (F1-S, F2-R)
171.500	Cabrillo National Monument (F-2)
171.500	San Dimas Forest Project (Experimental)
171.500	U.S. Department of Agriculture (Fire Camp)
171.525	State of California - Plant Industry Division
171.525	U.S. Dept. of Agriculture (Inspectors)
171.550	U.S. Dept. of Agriculture (Fire Camp)
171.5625	U.S. Dept. of Agriculture (Inspectors)
171.575 R	Angeles National Forest (F1-S, F2-R)
171.650	D.E.A. - Nationwide Use
171.675	Channel Islands National Park (F3-Tactical)
171.675	Joshua Tree National Park (F1-S, F2-R)
171.700	Channel Islands National Park (F1-R, F2-S)
171.700	U.S. Department of Agriculture (Fire Camp)
171.750	Cabrillo National Monument (F-1)
171.750	U.S. Bureau of Indian Affairs (Imperial Reserv.)
171.845	OCSO - Wireless Microphones
171.900 R	Tecate Police Department (F-1, F-2)
171.950	Tecate Fire Department (Dispatch)
172.000 R	Federal Judicial Police (Tijuana)
172.100	D.E.A. - Nationwide Use
172.125	F.A.A. - Flight Standards Control
172.150	F.A.A. - Airport Security Nationwide
172.175	F.A.A. - Airport Maintenance
172.200	D.E.A. - Nationwide Use
172.225 R	San Bernardino National Forest (F-3, F-4)
172.250	San Dimas Forest Project (Experimental)
172.275	U.S. Mint-Assay Office (F-1)
172.300 R	U.S. Department of Labor (Los Angeles)
172.300 R	U.S. Department of Labor (San Bernardino)
172.325	U.S. Dept. of Agriculture (Fire Camp)
172.400	U.S. Dept. of Agriculture (Fire Camp)
172.525 R	Santa Monica Mountains National Park
172.550	Tecate Police Department
172.575	San Bernardino National Forest (Tactical)
172.600	U.S. Department of Agriculture
172.600	U.S. Bureau of Land Management (Air Safety)
172.750	North Island Naval Air Station
172.775	U.S. Bureau of Reclamation (F-5)
172.800	Federal Communications Commission
172.825 R	F.A.A. (Operations Orange County)
172.850	F.A.A. (Operations Nationwide)
172.850 R	Federal Judical Police (Input to 172.000)
172.875 R	F.A.A. (Operations San Diego)
172.900 R	F.A.A. (Law Enforcement Intersystem)
172.925 R	F.A.A. (Operations Riverside)
172.950 R	F.A.A. (Operations Lindbergh Field
172.950 R	F.A.A. (Operations Los Angeles))
172.975 R	F.A.A. (Operations Los Angeles)
173.050 R	F.B.I - Repeaters
173.075	Lo-Jack Anti-Theft Transponders
173.125 R	F.B.I. - Repeaters
173.150 R	F.B.I. - Repeaters
173.175 R	F.B.I. - Repeaters
173.210	Anaheim Water & Power Dept.
173.225	CBS Studio City
173.225	Dexter Film Service
173.225	Kenny Loggins, Inc.
173.2375	Carlsbad Race Track - Security
173.275	Los Angeles Times
173.300	Claremont Municipal Water District
173.3125	Laguna Beach City Water
173.325	Dailey News Los Angeles
173.3375	Burbank Airport Maintenance
173.3375	Huntington Beach Local Government
173.3625	Carlsbad Race Track - Security
173.3625	Long Beach City Water Department
173.375	Santa Monica Evening Outlook
173.4875	Edwards A.F. Base - Medical Services
173.5625	U.S. Department of Agriculture (Inspectors)
173.5875	Edwards A.F. Base - Fire & Crash
173.7625	U.S. B.I.A. - Palm Springs
173.800	U.S. Federal Communications Commission
173.8875R	U.S. A.T.F. (A6-S) & (A9-R)
173.9125	U.S. Department of Agriculture
173.9375	U.S. Department of Agriculture
173.9625	U.S. Department of Agriculture
173.9875	U.S. Department of Agriculture
174.000	Wireless Headsets (KG-TV 10)
186.000	Wireless Headsets (KFMB-TV 8)
202-205 MHz	Wireless Headsets (All news media)
209.750	Television Channel 12 (Audio)
216-218 MHz	AMTS (Coastal Stations)
217.000	Naval Space Surveillance System
218-220 MHz	AMTS (Ship/Offshore oil rigs)
220.0-222.0	Land Mobile Band (ACSSB / FM)
222.0-225.0	1.25 Meter Amateur Band - See Page 259
225.500	U.S. Navy Aircraft (tactical)
228.900	TAC ROCC "Sierra Pete" (CH-1)
230.400	El Centro NAF (air-to-ground)
230.800	Los Alamitos NAS (approach Control)
230.900	U.S. Army (Air-to-Air)
231.300	U.S. Navy "Bravo Zulu"
233.900	U.S. Navy FACSF (Air-to-Air)
234.100	Los Alamitos (approach Control)
234.400	U.S. Navy FACSF (Air-to-Air)
234.600	TAC ROCC "Sierra Pete" CH-2
235.400	Miramar NAS (air-to-air advisory)
236.300	Camp Pendleton (ground utility)
236.600	Edwards A.F. Base - Tower UHF
237.200	California Air National Guard "Operations"
237.800	San Diego Naval Division HQ
237.900	Camp Pendleton (ground utility)
237.900	U.S. Coast Guard-Search & Rescue
238.000	North Island NAS (undetermined)
238.050	U.S. Navy FACSF (Air-to-Air)
238.300	VAW117 Wallbangers Air Squadron
238.400	TAC ROCC "Sierra Pete" (CH-11)
239.000	Van Nuys Tower UHF
239.300	Los Angeles Intl. Tower UHF
239.400	El Centro NAF (air-to-ground)
239.700	Miramar NAS (air-to-ground)
239.800	Military Aircraft - Pilot Weather 1
240.200	El Centro NAF (air-to-ground)
240.400	El Centro NAF (air-to-ground)
240.600	El Centro NAF (air-to-ground)
241.000	California Army National Guard (Aircraft)
241.000	Yuma Proving Ground - Control Tower
241.400	El Centro NAF (air-to-ground, Blue Angels)
241.600	El Centro NAF (air-to-ground)
241.800	El Centro NAF (air-to-ground)
243.000	Military Aircraft Distress (MAD)
244.100	North Island NAS "Beaver" (air-to-air)
245.000	Salton Sea Naval Test Base
246.800	U.S. Navy (Air Intercepts)
248.600	El Centro NAF (air-to-ground)
249.300	Camp Pendleton (air-to-ground)
249.800	VAW113 Black Hawks Air Squadron
249.900	Camp Pendleton Aircraft / War Games
250.000	El Centro NAF (air-to-ground)
250.100	North Island NAS (Air-to-Air)
250.100	Twentynine Palms USMC
250.150	U.S. Navy BEAVER (Air-to-Air)
250.200	VF124 Gunfighters Air Squadron
250.400	North Island NAS (air-to-air)
250.500	VAW114 Hormel Hawks Air Squadron
250.750	U.S. Navy BEAVER (Air-to-Air)
250.800	VAW112 Golden Hawks Air Squadron
250.800	U.S. Navy Blue Angels (Tactical)
250.900	Long Beach Naval Shipyard (undetermined)
251.400	El Centro NAF (air-to-ground)
251.400	Long Beach Naval Shipyard (undetermined)
251.600	U.S. Navy Blue Angels (Tactical)
251.700	Camp Pendleton (air-to-ground)
251.900	Air National Guard - UHF Common
252.800	U.S. Air Force (Tactical)
253.000	North Island NAS (air-to-air)
253.050	U.S. Navy FACSF (Air-to-Air)
253.100	VF124 Hormel Hawks Air Squadron
253.300	VF2 Bountyhunters
253.500	March AFB - Control Tower UHF

FREQUENCY	NAME
253.800	VAW117 Wallbangers Air Squadron
253.900	North Island NAS (air-to-air)
253.900	U.S. Coast Guard - Air Operations
254.200	TAC ROCC "Sierra Pete" (Ch-5)
254.300	Burbank Tower UHF
254.900	North Island NAS (air-to-air tactical)
255.000	Miramar NAS (visiting squadrons)
255.100	Coast Approach
255.200	U.S. Navy FACSF (Air-to-Air)
255.300	San Diego Naval OSC "Happy Hunter"
255.400	FAA Flight Service Stations
255.600	U.S. Navy "Happy Hunter" Intercept Control
256.300	Imperial Beach Naval Radio Receiving
256.400	North Island NAS (air-to-air)
256.800	Imperial Beach Naval Radio Receiving
257.600	Long Beach Tower UHF
257.800	Civilian Airports - Control Tower UHF
257.800	Military Aircraft - Control Towers Common
259.400	Space Shuttle - Astronauts
259.100	Ontario - Approach Control
259.700	Space Shuttle - Astronauts
260.800	Custom's Service (Aircraft Tracking)
260.800	TAC ROCC "Sierra Pete" (Ch-12)
261.100	U.S. Navy (Air Fueling)
261.500	North Island NAS "Beaver" (discrete)
262.700	Miramar NAS (Top Gun School)
263.100	San Diego Approach / Departure Control
263.300	PLEAD Fighter Discrete
263.400	El Centro Naval Air Facility (Ground)
263.400	North Island Naval Air Station
263.500	VF51 Screaming Eagles Air Squadron
263.700	VAW110 Firebirds Air Squadron
263.800	PLEAD Fighter Discrete
264.000	U.S. Navy "Starburst" ASW
264.300	U.S. Navy (ground utility)
264.500	El Centro Naval Air Facility - ATIS
265.000	U.S. Navy FACSF (Air-to-Air)
265.100	San Clemente NALF - Approach Control
265.250	VF213 Black Lions Air Squadron
265.300	U.S. Navy Top Gun "Vandy"
265.800	Camp Pendleton (air-to-ground)
265.800	Miramar NAS (approach control)
265.900	San Diego Naval Division HQ
266.500	Air Force - Refueling
266.700	North Island NAS "Sideflare"
266.900	North Island "Beaver" (Flight Check-in)
267.000	TAC ROCC "Sierra Pete" (Ch-20)
267.100	U.S. Army (air-to-air)
267.300	U.S. Army (air-to-air)
267.400	San Diego Naval OSC "Happy Hunter"
267.450	U.S. Navy "Starburst" ASW
267.600	Camp Pendleton Marine Corps Base - ATIS
267.700	VF-1 Wolfback Air Squadron
267.800	Air Force - Refueling
268.400	Long Beach Naval Station (air-to-ground)
268.500	North Island NAS "Beaver" (air-to-air primary)
268.600	San Clemente NALF - ATIS
268.600	U.S. Navy (aircraft carriers common)
268.800	PLEAD Fighter Discrete
269.000	Los Angeles Intl. - Approach Control East
269.100	San Diego Departure Control
269.150	Imperial Beach OLF (approach control)
269.200	Edwards A.F. Base - Flight Service Station
269.300	Ontario - Approach Control
269.400	Montgomery Field - Control Tower UHF
269.600	Coast Approach
269.900	Edwards A.F. Base - ATIS
270.100	March A.F.B. - ATIS
270.200	TAC ROCC "Sierra Pete" (Ch-16)
270.500	U.S. Navy FACSF (Air-to-Air)
270.600	North Island NAS "Beaver" (discrete)
270.800	TAC ROCC "Sierra Pete" (discrete)
270.900	Lindbergh Field - Control Tower UHF
271.400	Imperial Beach OLF - Control Tower UHF
271.700	El Toro Air Station - Control Tower UHF
272.300	Camp Pendleton (air-to-ground)
272.400	Camp Pendleton (ground control)
272.600	San Diego Approach/Departure discrete
272.700	U.S. F.A.A. - Flight Service
272.900	Chocolate Mountain Range Control (El Centro)
272.900	Long Beach Naval OSC (undetermined)
273.000	Camp Pendleton (air-to-ground)
273.100	Miramar NAS (Top Gun School)
273.100	North Island NAS "Beaver" (air-to-air)
273.200	Yuma Air Station - ATIS
273.300	PLEAD Fighter Discrete
273.500	Air Force Thunderbirds Demonstrations (F-2)
273.700	North Island NAS (experimental)
274.000	Yuma MCAS - Range Control
274.400	TAC ROCC "Sierra Pete" (Ch-6)
274.600	U.S. Navy "Starburst" ASW
275.100	U.S. Navy / Coast Guard Rescue Beacons
275.350	U.S. Navy Blue Angels (Tactical)
275.400	U.S. Army / Navy "Tropical discrete"
275.500	VF124 Bandits Air Squadron
275.600	North Island NAS (air-to-air)
275.700	U.S. Navy - Fleet Common Secure
276.100	Air Force - Refueling
276.200	Imperial Beach Naval OLF - ATIS
276.400	El Centro NAF (air-to-ground)
276.500	Air Force Refueling
276.500	North Island NAS "Beaver" (air refueling)
276.800	North Island NAS (air-to-ground)
277.000	Camp Pendleton (approach control)
277.100	Camp Pendleton (air-to-ground)
277.100	San Diego Naval OSC (air-to-air data)
277.200	El Centro NAF (air-to-ground)
277.250	U.S. Navy Blue Angels (Training)
277.400	Oceanside Sector (enroute)
277.500	U.S. Army (air-to-air)
277.800	U.S Navy - Fleet Warning & Tactical
277.900	Camp Pendleton (weapon supplies)
278.100	U.S. Navy BEAVER (Air to Air)
278.200	U.S. Navy "Warwag"
278.300	Ontario - Approach Control
278.400	TAC ROCC "Sierra Pete" (Discrete)
278.800	San Clemente NALF - Control Tower UHF
278.900	Camp Pendleton (undetermined)
279.000	Space Shuttle - Astronauts to Orbitor
279.100	U.S. Navy FACASF (Air-to-Air)
279.400	TAC ROCC "Sierra Pete" (Discrete)
279.800	Air Force - Refueling
280.050	U.S. Navy BEAVER (Air-to-Air)
280.200	U.S. Navy - Fleet Common
280.400	Miramar NAS - ATIS (low power)
280.600	Camp Pendleton (air-to-ground)
280.900	U.S. Army (air-to-air)
281.000	Air Force (Refueling, Tankers)
281.100	U.S. Navy (Fleet Operations)
281.300	North Island NAS (air-to-air)
281.400	Coast Approach
281.600	TAC ROCC "Sierra Pete" (Ch-3)
281.700	Camp Pendleton (Approach Control)
281.800	San Diego Approach / Departure Control
281.900	U.S. Navy (air-to-ground)
282.000	North Island / San Clemente Island (ATIS)
282.100	San Diego Approach Control
282.425	U.S. Custom's Service Aircraft
282.800	U.S. Coast Guard-Search & Rescue
282.800	U.S. Navy
283.000	North Island Naval Air Station (ATIS)
283.200	El Centro NAF (air-to-ground)
283.300	VF111 Sundowners Air Squadron
283.400	Camp Pendleton (weapon supplies)
283.500	Air Force Thunderbirds Demonstrations (F-3)
283.900	Air Force - Refueling
284.200	El Toro Air Station - ATIS
284.400	Camp Pendleton / North Island NAS
284.500	VF-31 Tomcatters Air Squadron
284.600	San Diego Approach / Departure Control)
284.700	Military Aircraft (Enroute)
284.900	San Diego Naval OSC "Happy Hunter"
285.100	North Island NAS (air-to-air)
285.200	San Diego Approach / Departure Control
285.700	North Island NAS "Beaver" (Flight Check-in)
285.800	U.S. Navy Intercept Control "Happy Hunter"
285.900	Imperial Beach OLF - Ground Control
286.200	Air Force - Refueling
286.800	Space Shuttle - Astronauts
286.900	Air Force - Refueling

FREQUENCY	NAME
287.400	Air Force - Refueling
287.700	TAC ROCC "Sierra Pete" (Discrete)
288.100	Brown Field - Control Tower
289.800	Miramar NAS (approach control)
289.900	North Island NAS "Beaver" (Flight Checkin))
290.100	Yuma MCAS - Range Control Moving Sand
290.300	Edwards A.F. Base - Departure Control
290.400	San Diego Approach / Departure Control
291.200	U.S. Navy "Starburst" ASW
291.300	VF31 Tomcaters Air Squadron
291.600	Edwards AFB - Approach Control
291.700	Los Angeles Center (Julian)
291.900	Air Force - Refueling
293.800	TAC ROCC "Sierra Pete" (Discrete)
294.700	Camp Pendleton (air-to-air, war games)
295.400	Air Force - Refueling
295.700	Air Force Thunderbirds Demonstrations (F-1)
295.700	Ontario - Approach Control
295.800	Air Force - Refueling
296.200	Air Force - Refueling
296.800	Space Shuttle Downlink - Primary
297.300	Air Force - Refueling
297.600	Air Force - Refueling
299.000	TAC ROCC "Sierra Pete" (Discrete)
299.300	Camp Pendleton (air-to-ground)
299.550	Camp Pendleton (air-to-ground)
299.600	Coast Approach
299.700	VF24 Renegedes Air Squadron
300.000	Camp Pendleton (air-to-ground)
300.200	Miramar NAS (air-to-ground)
300.400	San Diego Approach / Departure Control
300.500	El Centro NAF (air-to-ground)
300.500	Long Beach Naval Station (undetermined)
301.000	U.S. Navy (ground utility)
301.100	North Island NAS "Beaver" (air-to-air)
301.300	Miramar NAS - Clearance Delivery
301.900	Camp Pendleton (air-to-ground)
302.100	U.S. Navy San Diego (undetermined)
302.600	U.S. Navy FACSF (Air-to-Air)
303.625	U.S. Navy FACSF (Air-to-Air)
303.700	Camp Pendleton (air-to-ground)
304.000	Camp Pendleton (air-to-ground)
304.100	Miramar NAS (air-to-air)
304.200	U.S. Navy Intercept Control "Happy Hunter"
304.500	U.S. Navy FACSF (Air-to-Air)
304.900	North Island NAS "Beaver" (air-to-air)
305.000	U.S. Navy FACSF (Air-to-Air)
305.100	Miramar NAS (air-to-air)
305.200	Camp Pendleton (air-to-ground)
305.250	U.S. Navy FACSF (Air-to-Air)
305.300	San Clemente NALF (Approach Control)
305.700	Air Force - Refueling
305.800	North Island NAS (data link)
305.900	Camp Pendleton (air-to-ground)
306.000	Miramar NAS (air-to-air)
306.100	Lindbergh Field - Departure Control
306.700	San Diego Approach Control
306.800	Camp Pendleton (air-to-ground)
307.400	Camp Pendleton (weapons supplies)
307.500	Camp Pendleton (air-to-ground)
307.600	North Island NAS (air-to-air)
308.100	North Island NAS "Beaver" (air-to-air)
308.300	Twentynine Palms Air Station - Weather
308.500	U.S. Navy (Fleet Operations)
309.300	Miramar NAS (Top Gun School)
309.400	TAC ROCC "Sierra Pete" (Ch-9)
310.000	North Island NAS (air-to-ground)
310.100	Camp Pendleton (air to ground)
310.200	Camp Pendleton (air to ground)
310.300	North Island NAS (air-to-ground)
310.600	Camp Pendleton (air-to-ground)
310.700	El Centro NAF (air-to-ground)
310.800	Miramar NAS (approach control)
311.000	SAC Primary
311.150	U.S. Navy FACSF (Air-to-Air)
311.400	U.S. Navy FACSF (Air-to-Air)
311.700	North Island NAS (air-to-air)
311.800	Miramar NAS (air-to-ground)
312.700	Camp Pendleton (Weapon Supplies)
313.000	VF211 Flying Checkmates Air Squadron
313.200	Miramar NAS (approach control)
313.300	Camp Pendleton (air-to-ground)
313.700	U.S. Navy (Fleet Operations)
313.800	Camp Pendleton (air-to-ground)
314.300	Air Force - Refueling
314.600	Flight Tests (all military)
314.700	North Island NAS "Beaver" (UHF Checkin)
314.900	Miramar NAS (air-to-ground)
315.300	North Island NAS "Beaver" (air-to-ground)
315.400	San Diego Naval Division HQ
315.500	North Island NAS (air-to-air)
315.600	Miramar NAS - Control Tower
315.700	U.S. Navy San Diego (undetermined)
316.300	TAC ROCC "Sierra Pete" (Ch-17)
316.900	North Island NAS "Beaver" (air-to-air)
317.000	U.S. Navy San Diego (undetermined)
317.500	Burbank - Approach Control
317.600	Palmdale - Approach Control
318.500	U.S. Navy (Fleet Operations)
318.100	Edwards A.F. Base - Tower UHF
318.200	Ontario - Approach Control
318.600	San Diego Naval Division HQ
318.700	U.S. Navy "Happy Hunter" Discrete
318.800	North Island NAS - Approach Control
318.800	San Clemente NALF - Operations
318.900	Camp Pendleton (air-to-ground)
319.500	North Island NAS "Beaver" (Air Refueling)
320.000	Tactical Air Command (TAC)
320.200	North Island NAS "Black Knights"
320.300	Camp Pendleton (air-to-ground)
320.600	TAC ROCC "Sierra Pete" (Ch-4)
321.000	SAC Secondary
321.800	Long Beach Naval Shipyard (undetermined)
321.900	Camp Pendleton (air-to-ground)
322.000	Miramar NAS (approach / departure control)
322.100	VF11 Red Rippers Air Squadron
322.300	Air Force Thunderbirds Demonstrations (F-4)
322.600	Air Force Thunderbirds Demonstrations (F-5)
323.000	San Diego Approach / Departure Control
323.100	Coast Approach
323.500	U.S. Navy (Fleet Operations)
324.000	TAC ROCC "Sierra Pete" (Ch-18)
324.400	Air Force - Refueling
324.500	El Centro NAF (air-to-ground)
324.600	North Island NAS "Gunbearers"
325.100	North Island NAS (air-to-air)
325.200	Miramar NAS (approach control)
325.400	Camp Pendleton (air-to-ground)
326.100	U.S. Navy Intercept Control "Happy Hunter"
326.500	North Island NAS "Beaver" (air-to-air)
326.600	U.S. Navy San Diego (undetermined)
326.800	PLEAD Fighter Discrete
326.900	North Island NAS (air-to-air)
327.900	TAC ROCC "Sierra Pete" (Ch-19)
328.100	El Toro Air Station - Operations
328.200	U.S. Navy - Command Early Warning Net
328.300	VF213 Black Lions Air Squadron
328.400	Miramar NAS (approach control)
328.500	San Diego Approach / Departure Control
332.350	Miramar NAS (Top Gun School)
336.100	Air Force - Refueling
336.200	Camp Pendleton (air-to-ground)
336.400	North Island NAS - Control Tower UHF
336.500	Camp Pendleton (air-to-ground)
336.600	TAC ROCC "Sierra Pete" (Ch-7)
336.900	Camp Pendleton (air-to-ground)
337.000	PLEAD Check-in/Return to Base
337.200	Coast Approach
337.900	Yuma Air Station - Operations
338.000	PLEAD Check-in/Return to Base
338.100	Camp Pendleton (approach control)
338.200	Burbank - Approach Control
338.300	VAW88 Cottonpickers Air Squadron
338.400	North Island NAS - Control Tower UHF
338.600	U.S. Navy(undetermined)
339.500	VF302 Stallions Air Squadron (reserve)
339.600	Camp Pendleton (air-to-ground)
339.700	U.S. Navy (Fleet Operations)
340.200	Camp Pendleton - Control Tower UHF
"	El Centro NAF - Control Tower UHF

FREQUENCY	NAME
340.200	Miramar NAS - Control Tower UHF
"	North Island NAS - Control Tower UHF
"	Twentynine Palms MC - Control Tower UHF
340.300	San Clemente NALF - Control Tower UHF
340.300	U.S. Navy Towers (Common frequency)
340.300	Twentynine Palms Air Station - Control Tower
340.300	U.S. Navy San Diego (undetermined)
340.400	Camp Pendleton (air-to-ground)
340.500	U.S. Navy San Diego (Fleet Operations)
341.000	U.S. Navy San Diego (undetermined)
341.100	San Diego Naval Division HQ
341.200	Camp Pendleton (air-to-ground)
341.300	PLEAD Fighter Discrete
341.400	Air Force - Refueling
341.800	TAC ROCC "Sierra Pete" (Ch-13)
342.100	U.S. Navy (training)
342.500	Military Aircraft - Pilot Weather 2
342.600	San Diego Naval Division HQ
342.700	North Island NAS (air-to-ground)
342.900	North Island NAS "Beaver" (air-to-air)
342.900	San Clemente NALF (air-to-air)
343.800	Imperial Beach OLF (approach control)
343.900	Coast Approach
344.000	North Island NAS (air squadrons)
344.100	North Island NAS (air-to-ground)
344.100	San Clemente NALF "Beaver" (discrete)
344.300	VAW116 Sun Kings Air Squadron
344.400	VF301 Devils Disciples Air Squadron
344.600	Pilot to Metro Service - Weather
344.600	El Toro Air Station - Weather
344.600	North Island NAS - Weather
344.600	Military Aircraft - Pilot Weather 3
344.600	Tustin Air Station - Weather
345.000	U.S. Navy FACSF (Air-to-Air)
345.100	U.S. Navy (Fleet Operations)
345.300	Camp Pendleton (air-to-ground)
345.800	U.S. Navy San Diego (Fleet Operations)
346.000	Miramar NAS (approach control)
346.200	TAC ROCC "Sierra Pete" (Ch-8)
346.400	F.A.A. - Los Angeles Center
346.500	U.S. Navy San Diego (undetermined)
346.600	U.S. Navy San Diego (Fleet Operations)
346.700	North Island NAS "Beaver" (air-to-ground)
346.900	U.S. Navy San Diego (Fleet Operations)
347.900	North Island NAS (air-to-ground)
348.100	U.S. Navy "Starburst" ASW
348.700	Edwards A.F. Base - Approach/Departure
348.800	U.S. Navy San Diego (undetermined)
349.800	U.S. Navy "Starburst" ASW
349.900	North Island NAS (data link)
349.900	Yuma Air Station - Weather
350.000	San Diego Naval Division HQ
350.100	Tustin Air Station - Control Tower UHF
350.500	Camp Pendleton (approach Control)
350.800	San Diego Approach / Departure Control
350.900	Miramar NAS (Top Gun School)
351.100	Ontario - Approach Control
352.200	San Diego Naval OSC - Range Operations
352.300	Air Squadrons Common Channel
352.400	North Island NAS (ground control)
352.400	San Clemente NALF - Ground Control
352.500	North Island NAS (air-to-air)
352.900	Air Force - Refueling
353.000	U.S. Navy San Diego (Fleet Operations)
353.100	U.S. Navy San Diego (Fleet Operations)
353.200	Camp Pendleton (air-to-ground)
353.400	U.S. Navy San Diego (Fleet Operations)
353.500	North Island Air Station (App/Dep. Control)
353.900	U.S. Coast Guard (Narcotics Interdiction)
354.100	Coast Approach
354.200	North Island NAS "Beaver" (air-to-air)
354.600	U.S. Navy San Diego (Fleet Operations)
354.400	Edwards A.F. Base - Air Refueling
354.700	VFC13 Saints Air Squadron (adversary)
354.900	North Island NAS "Beaver" (air-to-air)
355.000	U.S. Navy - Fleet Operations (RTTY)
355.100	U.S. Navy Intercept Control "Happy Hunter"
355.400	San Diego Naval Division HQ
355.500	North Island Air Station - Field Operations
355.600	Miramar NAS (Radar Control)
355.600	San Clemente NALF
355.700	VFC13 Saints Air Squadron (adversary)
355.800	Camp Pendleton (air-to-ground)
355.900	San Diego Naval OSC - Range Operations
356.100	Camp Pendleton (air-to-ground)
356.200	San Diego Naval Support Facility
356.800	Imperial Beach OLF - Ground Utility
356.800	North Island NAS - Clearance Del.
357.800	U.S. Navy San Diego (Fleet Operations)
357.900	U.S. Navy "Starburst" ASW
358.000	U.S. Navy "Starburst" ASW
358.100	U.S. Navy San Diego (Fleet Operations)
358.200	Air Force - Refueling
358.600	Yuma MCAS - Range Control Cactus West
358.700	U.S. Navy San Diego (Fleet Operations)
358.900	U.S. Navy (air-to-air)
359.100	Air Force - Refueling
359.400	U.S. Navy Fleet Mult-Channel
359.500	North Island NAS "Beaver" (air-to-air)
359.600	U.S. Navy San Diego (undetermined)
359.700	U.S. Navy San Diego (Fleet Operations)
360.200	Camp Pendleton Air Field - Ground Control
360.200	El Centro Naval Air Facility - Control Tower
360.000	Burbank - Approach Control
360.200	North Island NAS - Control Tower UHF
360.300	Camp Pendleton - North/South Range Control
360.400	U.S. Navy Blue Angels (Tactical)
360.700	Los Angeles Intl. - Approach Control
360.800	FAA - Radar Control
360.800	Imperial Beach Naval Radio Receiving
360.800	Miramar NAS - Approach / Departure Control
361.000	U.S. Navy San Diego (Fleet Operations)
361.100	Camp Pendleton - South Range Control
361.300	Camp Pendleton (air-to-ground)
361.700	Air Force - Refueling
361.800	El Centro Naval Air Facility "Nickel"
361.800	Long Beach Naval Shipyard (undetermined)
361.800	U.S. Coast Guard (Narcotics Interdiction)
361.900	Miramar NAS - altimeter information
362.000	San Diego Naval OSC (undetermined)
362.100	Miramar NAS - Weather
362.200	Camp Pendleton (approach control)
362.200	Twentynine Palms Air Station - Ground Control
362.500	Miramar NAS (air-to-air)
362.600	Miramar NAS (approach control) "Paddles"
362.900	Camp Pendleton (air-to-ground)
363.100	San Diego Approach / Departure Control
363.300	Camp Pendleton (air-to-ground)
363.400	Camp Pendleton (air-to-ground)
363.500	U.S. Navy San Diego (undetermined)
363.600	Miramar NAS (approach control)
364.200	TAC ROCC "Sierra Pete" (CH-21)
364.800	U.S. Navy FACSF (Air-to-Air)
366.300	Air Force - Refueling
368.600	Air Force - Refueling
369.200	Air Force - Refueling
369.700	Camp Pendleton (air-to-ground)
369.900	FAA - High Altitude Approach (Julian)
369.900	FAA - High Altitude Approach (Mt. Laguna)
370.400	Air Force - Refueling
372.300	Air Force - Refueling
372.700	Camp Pendleton (air-to-ground)
373.800	North Island NAS "Beaver" (air-to-air)
374.000	TAC ROCC "Sierra Pete" (Discrete)
374.900	Miramar NAS (approach control)
375.000	El Centro NAF / North Island NAS
375.200	Military Aircraft - Pilot Weather 4
375.500	Air Force - Refueling
375.700	Air Force - Refueling
376.000	Air Force - Refueling
376.800	Miramar NAS (air-to-ground)
376.800	North Island NAS
376.800	San Clemente Island NALF
376.900	U.S. Navy San Diego (undetermined)
378.500	U.S. Air Force - Point Loma
379.100	Los Angeles Intl. - Control Tower UHF
379.900	John Wayne - Control Tower UHF
380.200	Camp Pendleton (air-to-ground)
380.300	Miramar NAS (approach control)
380.500	San Diego Navy FACSF (Air-to-Air)

FREQUENCY	NAME
380.550	U.S. Navy (air-to-air)
380.700	U.S. Navy (Fleet Operations)
380.800	Miramar NAS - Ground Control
380.800	Tustin Air Station - Ground Control
380.900	U.S. Navy San Diego (Fleet Operations)
381.300	Tactical Air Command (TAC)
381.400	U.S. Coast Guard-Search & Rescue Ops
381.500	San Diego Approach / Departure Control
381.700	U.S. Coast Guard-Air Support
381.800	U.S. Coast Guard-Air Support Primary
382.000	San Diego Approach / Departure Control
382.100	Camp Pendleton (air-to-air)
382.200	Camp Pendleton Air Field - Control Tower
382.500	U.S. Navy San Diego (Fleet Operations)
382.700	U.S. Navy San Diego (undetermined)
382.800	Yuma Air Station - Control Tower UHF
382.900	Air Force Thunderbirds Demonstrations (F-6)
382.900	U.S. Navy (Fleet Operations)
383.400	Long Beach Naval Station (undetermined)
383.500	U.S. Navy (Fleet Operations)
383.700	Miramar NAS (air-to-ground)
383.800	El Toro Air Station - Ground Control
383.900	U.S. Coast Guard-Air Support Secondary
384.000	El Centro NAF (air-to-ground)
384.100	Long Beach Naval OSC (undetermined)
384.200	El Centro NAF (air-to-ground)
384.300	Tustin Air Station - ATIS
384.400	North Island NAS (air-to-air)
384.400	U.S. Navy Blue Angels
384.500	San Diego Approach / Departure Control
384.600	Air Force - Refueling
384.800	El Centro NAF (air-to-ground)
385.000	Camp Pendleton (weapons supplies)
385.000	San Diego Approach / Departure Control
385.100	San Clemente Island - Radar Control
385.200	U.S. Navy San Diego (Fleet Operations)
385.250	U.S. Navy (air-to-air)
385.300	San Diego OSC - Range Operations
385.400	Los Angeles Intl. - Departure Control
385.500	Miramar NAS - Approach / Departure Control
385.600	Ontario - Control Tower UHF
385.800	Air Force - Refueling
386.000	TAC ROCC "Sierra Pete" (Discrete)
387.400	Camp Pendleton (Experimental)
387.800	U.S. Coast Guard (Narcotics Interdiction)
387.800	U.S. Custom's Service (Aircraft Tracking)
387.900	Camp Pendleton (weapons supplies)
388.400	Air Force - Refueling
390.200	TAC ROCC "Sierra Pete" (Discrete)
390.300	San Diego Naval OSC - Air/Ships Wideband
391.800	Air Force - Refueling
391.900	U.S. Navy Blue Angels (Tactical)
392.000	McClellan-Palomar - Control Tower UHF
393.800	U.S. Navy FACSF (Air-to-Air)
394.000	Air Force Thunderbirds Demonstrations (F-7)
394.200	TAC ROCC "Sierra Pete" (Discrete)
394.600	Air Force - Refueling
394.900	Air Force - Refueling
395.900	Burbank - Approach Control
396.200	Air Force - Refueling
397.800	TAC ROCC "Sierra Pete" (Discrete)
399.600	State LEAF "Red"
399.650	State LEAF "White"
399.700	State LEAF "White"
399.750	State LEAF "Yellow"
399.900	State LEAF "Tactical 1"
399.950	State LEAF "Tactical 2"
406.100	U.S. E.P.A. - Nationwide Portables
406.200 R	U.S. Department of Labor (Long Beach)
406.200 R	U.S. Department of Labor (Los Angeles)
406.325 R	U.S. Veterans Administration (F1-R, F2-S)
406.350 T	Military Trunking System - San Diego
406.375	U.S. Navy San Diego (Undetermined)
406.475 R	U.S. Bureau of Indian Affairs (Riverside)
406.950 T	Military Trunking System - San Diego
407.125 R	U.S. Department of Labor (F3-R)
407.150 T	Military Trunking System - San Diego
407.175 R	U.S. Department of Labor (F2-R, F4-S)
407.200 R	U.S. State Department NET 3 (San Diego)
407.225 R	U.S. Army - Corps of Engineers (Los Angeles)
407.250	Camp Pendleton USMC
407.425	Coronado Amphibious Base -SEAL Teams
407.425	Salton Sea Naval Test Base - SEAL Teams
407.525	Salton Sea Naval Test Base - SEAL Teams
407.550 T	Military Trunking System - San Diego
407.575	North Island NAS (Public Works)
407.600 R	U.S. State Department NET 2 (Los Angeles)
407.675	U.S. Secret Service "White"
407.700	U.S. Department of Health
407.750	U.S. Secret Service "Gray"
407.825	U.S. Alcohol Tobacco & Firearms
407.850	U.S. Alcohol Tobacco & Firearms
407.875	U.S. Alcohol Tobacco & Firearms
407.925	U.S. Secret Service "India"
407.950 T	Military Trunking System - San Diego
408.350 T	Military Trunking System - San Diego
408.575	U.S. BLM (Link to 166.4875)
408.600 R	U.S. State Department NET 1
408.675 R	U.S. Fish & Wildlife Service (F1-R, F2-S)
408.950 T	Military Trunking System - San Diego
409.100	North Island NAS (airfield)
409.125	San Diego Naval Station (Weapons Testing)
409.125	San Diego Navy (Shipboard Refueling)
409.150 T	Military Trunking System - San Diego
409.200	U.S. Department of Energy (F-1)
409.200	U.S. Interstate Commerce Commission
409.300	U.S. Department of Energy (F-2) Los Angeles
409.500	U.S. Department of Energy (F-3) San Diego
409.525	Federal Reserve Bank (Los Angeles)
409.950 T	Military Trunking System - San Diego
409.625 R	U.S. State Department NET-4
409.800	U.S. Department of Energy (F-4)
410.150	U.S. International Trade Commission
410.725 R	U.S. International Trade Commission
410.750	U.S. International Trade Commission
410.850	U.S. Attorney (F-3)
410.850	U.S. International Trade Commission (F-3)
410.850	U.S. International Trade Commission
410.875	U.S. Attorney (F-1) & (F-2)
410.875	U.S. International Trade Commission
410.900	U.S. International Trade Commission
410.975	U.S. International Trade Commission
411.075	Nuclear Regulatory Commission (F-1)
411.150	U.S. State Department (F-5)
411.200	Nuclear Regulatory Commission (F-2)
411.750	U.S. Bureau of Land Management (Command 1)
411.925	U.S. Bureau of Land Management (Command 2)
412.150	U.S. Bureau of Land Management (Command 3)
412.200	U.S. Bureau of Land Management (Command 4)
412.375	U.S. Bureau of Land Management
412.400	U.S. General Services Administration (F-6)
412.400	U.S. Agriculture SYSTEM 1 (F-3)
412.425 R	F.B.I. - Remote Link
412.525 R	F.B.I. - Remote Link
412.750	N.R.C. - San Diego Operations
412.800	U.S. Agriculture SYSTEM 2 (Car-to-Car)
412.800	U.S. Army (nationwide use)
412.900	U.S. Army (CID Tactical)
412.925	U.S. Army (Portables)
412.975	U.S. Army (Portables)
413.000	Coronado Amphibious Base - SEAL Teams
413.025	Air Force Thunderbirds (Ground Support)
413.075	Air Force Thunderbirds (Ground Support)
413.100	Air Force Thunderbirds (Ground Support)
413.150	Coronado Naval Amphibious Base - SEAL
413.300	Air Force - Civil Engineers
413.425	U.S. Department of Health (Nationwide)
413.475	March A.F. Base - Ramp Control
413.500	California Army National Guard (ML-5)
413.525	California Army National Guard (ML-5)
413.550	North Island NAS (undetermined)
413.600	U.S. Postal Service Inspectors (F-5, F-6)
413.675	I.N.S. - Border Sensors
413.750	Immigration & Naturalization
413.825	U.S. Veterans Administration
414.425 R	F.B.I. - Link to 167.6875
414.625	U.S. International Trade Commission (F-1)
414.650	U.S. Department of Agriculture (Logistics 1)
414.6625	Treasury Department UHF Tactical
414.675	U.S. Secret Service "Yellow"

FREQUENCY	NAME
414.750 R	U.S. Postal Service SYSTEM 1 (F1-R, F2-S)
414.775	U.S. Secret Service (Foreign Dignitaries)
414.800	U.S. Secret Service (Foreign Missions)
414.850 R	U.S. Secret Service (Foreign Dignitaries)
414.875	U.S. Secret Service (Training Division)
414.950	U.S. Secret Service "Orange"
415.025 R	U.S. Postal Service Inspectors (F-3, F-4)
415.050 R	U.S. Postal Service (F3-R, F4-S)
415.075	U.S. B.L.M. - Link to 167.000
415.100	Federal Reserve Banks (F-1, F-3)
415.150	U.S. B.L.M. - Link to 168.550
415.200 R	Federal Postal Service SYSTEM 2 (F-1,F-2)
415.250 R	Cleveland National Forest (Link)
415.275 R	Cleveland National Forest (Link)
415.300 R	U.S. Dept. of Agriculture (Grain Inspectors)
415.375 R	U.S. Forestry - Air Dispatch Link
415.400	U.S. Department of Agriculture (Logistics 2)
415.425 R	Cleveland National Forest (Link)
415.450 R	Cleveland National Forest (Link)
415.475 R	San Bernardino National Forest - Remote Link
415.500	U.S. Department of Agriculture (Logistics 3)
415.525 R	U.S. Department of Agriculture (Link)
415.550 R	U.S. Forestry (Air Dispatch Link, Ops)
415.575 R	San Bernardino National Forest (Link)
415.650 R	U.S. Secret Service (Communications)
415.675	U.S. Secret Service "Black"
415.925	U.S. Coast Guard - Special Operations (F-1)
416.175 R	U.S. Attorney (F1-R, F2-S)
416.375 R	N.O.A.A. Weather (San Diego Control Link)
416.500	U.S. Air Force - Mt. Laguna
416.550	F.A.A. Maintenance (Lindbergh Field)
416.850	U.S. Coast Guard - Special Operations F-2
417.125	U.S. Coast Guard - Special Operations F-3
417.200 R	General Services Administration (F-1, F-2)
417.200	Central Intelligence Agency (Security)
417.275	D.E.A. Surveillance
417.300 R	Interior Fire Cache - Command
417.350 R	Interior Fire Cache - Command
417.425	General Services Administration (F-5)
417.500 R	Interior Fire Cache - Command
417.800 R	Interior Fire Cache - Command
418.050	D.E.A. Wide Area Itinerate
418.075	D.E.A. Interagency Common
418.100	U.S. Postal Service (Vehicle Control)
418.175	ATF/IRS UHF (F-3)
418.175	D.E.A. Surveillance FLINT, Body Mikes
418.175	U.S. Postal Service (Car-to-Car)
418.200	D.E.A. Surveillance FLINT reported
418.225 R	ATF/IRS UHF (F1-R, F2-S)
418.250	IRS UHF (F3-S)
418.300 R	U.S. Postal Service (Building Security)
418.325	Narcotics Task Force (East County)
418.500	D.E.A. Surveillance
418.575	D.E.A. Wide Area Itinerate
418.625 R	D.E.A. Repeaters (Channel 1)
418.675	D.E.A. Surveillance (Channel 4)
418.700	D.E.A. Surveillance
418.750	D.E.A. Surveillance (Channel 3)
418.775	D.E.A. Surveillance
418.800	D.E.A. Surveillance
418.825 R	D.E.A. Repeaters (Channel 5)
418.850 R	I.N.S. (Link to 163.675)
418.875	D.E.A. Surveillance
418.900 R	D.E.A. Repeaters (Channel 2)
418.950 R	D.E.A. Repeaters (Channel 6)
418.975 R	D.E.A. Repeaters (Channel 7)
419.000	D.E.A. Surveillance
419.225 R	D.E.A. Surveillance (San Marcos Peak)
419.250 R	D.E.A. (Descanso / LA / Tecate)
419.350 R	F.B.I. Link to SD Office (Mt. Soledad)
419.975 R	A.T.F. Repeaters
420.0-450.0	75 CM Amateur Band - See Page 263
450.010	KBON / KBBV / KCKC AM/FM
450.020	KCMJ-FM (Palm Springs)
"	KPFK-FM (Los Angeles)
"	KTTY-TV Channel 69 (F-1)
450.0375R	KDAY-TV Channel 63
"	KUSI-TV Channel 51 (F-1) & (F-2)
450.050 R	KABC-AM 790 (Traffic Watch)
450.050	KBBV-AM/FM
450.0625R	KUSI-TV Channel 51 (F-3) & (F-4)
450.0875R	KCBS-TV Channel 2 (F-6)
"	KNX-AM (News & Traffic)
450.1125R	KCRW-FM (Remotes)
"	KDOC-TV Channel 56 (Remotes)
"	KMEX-TV Channel 34 (News Desk)
"	KPBS-TV Channel 15 (F-1) & (F-2)
"	KSCI-TV Channel 18 (News)
450.1375R	KABC-TV Channel 7 (LA News Desk)
450.150	CNN American
"	KABC-TV Channel 7 (Simplex Ops)
"	Orange Coast College
450.1625R	KABC-TV Channel 7 (ABC News Desk)
450.1875R	NBC-TV (Network News)
450.2125R	KCAL-TV Channel 9 (F-1)
"	KLZZ-FM Radio
450.225	KABC-TV Channel 7
450.2375R	KG-TV Channel 10 (Engineering)
450.250R	KTLA-TV Channel 5 (F-2)
450.250	Time Magazine (San Diego)
450.2625R	KFMB-AM/FM (B-100)
450.2875R	KCBS-TV Channel 2 (News Desk)
450.3125R	KABC-TV Channel 7 (News Crews)
450.3375R	KFMB-TV Channel 8 (F-1)
450.350 R	KTLA-TV Channel 5 (F-1)
450.3625R	KG-TV Channel 10 (Secondary Dispatch)
450.3875R	KNBC-TV Channel 4 (News Base)
450.4125R	CBS-TV (Network News)
"	KFMB-TV Channel 8 (F-3)
450.4375R	KDAY-TV Channel 56
"	KNSD-TV Channel 39 (F-1, F-2)
450.450	KFMB-TV Channel 8 (CBS Link)
"	KRLA-AM / KBBV -AM (News & Remotes)
450.4625R	KNSD-TV Channel 39 (F-3, F-4)
450.4875R	KPBS-TV Channel 15 (F-2)
"	Orange Coast College
450.5125R	KCBS-TV Channel 2 (F-2)
450.5375R	KNBC-TV Channel 4 (Remotes)
450.550 R	CNN America
"	Diario 29 Periodico (Mexico)
"	KABC-TV Channel 7
450.5625R	KNBC-TV Channel 4 (Remotes)
450.5875R	KTTV-TV Channel 11 (News Desk)
450.600	KG-TV Channel 10 (Live Audio)
450.6125R	CBS-TV (Network News)
450.650 R	KFWB-AM (News & Traffic)
450.6875R	KSON Radio (F-1)
450.700 R	KNX-AM 1070 (News & Traffic)
450.700 R	State Governor's Office (Baja Calif.)
450.7125	Airwatch Communications
450.725 R	KFI-AM/FM (delayed audio)
450.750 R	KBBV-AM/FM
450.750 R	KLIT-FM (Glendale)
450.7625R	KFMB - AM/FM (B-100)
450.775 R	CNN America (LA Repeater)
450.800 R	KCBS-TV Channel 2 (Remotes)
450.800 R	KNX-AM 1070 (Remotes)
450.825 R	KABC-AM 790 (Traffic Watch)
450.825 R	KLOS-FM (Los Angeles)
450.850 R	KFWB-AM (News & Traffic)
450.850 R	KGMG-AM/FM Radio
450.8875	KKOS-FM Radio (Traffic Reports)
450.900 R	KCPB / KPWR-FM (Remotes)
450.925 R	KGMG-AM/FM (Oceanside)
450.925	Monica Zeck Traffic (F-2)
450.950 R	KTTV-TV Channel 11 (Secondary)
450.980	KCMJ-FM (Palm Springs)
"	KGMG-FM (Los Angeles)
"	KIIS-FM (Los Angeles)
450.990	KGGI-FM (Riverside)
450.990	KMEN-AM 1290 (San Bernardino)
451.025 R	Southern Ca. Edison (San Onofre Power Plant)
451.0375	Olivenhain Municipal Water District (Data)
451.050 R	Burbank City Water & Utilities
451.050 R	Southern Ca. Edison (San Onofre Power Plant)
451.0625	San Diego County Water Authority
451.075 R	Imperial Irrigation District
451.100 R	Azusa City Water & Utilities
"	Calabasas Water District
"	Las Virgenes Municipal Water

FREQUENCY	NAME
451.100 R	Palmdale Water District
451.1125	Imperial Irrigation District
451.125 R	Imperial Irrigation District
451.1375	Imperial Irrigation District
451.150 R	Long Beach City Water Department
"	Sweetwater Authority Water District
451.1625	Texaco Mineral Company (San Diego)
451.175	Industrial / Utility Radio
451.200 R	City of Industry Water Department
"	Santa Ana Water District
"	Southern Ca. Edison (San Onofre Power Plant)
451.225	Southwest Marine Inc.
451.250	Pomona Water District
"	Southern Ca. Edison (San Onofre Power Plant)
451.2625	Olivenhain Municipal Water District
451.275	Fontana Steel Corporation
451.2875	Fallbrook Public Utility District
451.300 R	American Telephone & Telegraph (AT&T)
451.300 R	American Telephone & Telegraph (Mt. Palomar)
451.325 R	American Telephone & Telegraph (Imperial)
451.3375	Sports Car Club of America
451.3625	Imperial Irrigation District
451.425 R	Industrial / Utility Radio.
451.475 R	Anaheim Water & Power Dept.
451.525 R	Encino Power Plant
451.525 R	Pomona Water District
451.5375	Southwest Marine Inc.
451.5625	Long Beach City Water Department
451.600 R	San Diego Pipe Lines Inc.
451.625 R	Encino Power Plant
451.6375	Southwest Marine Inc.
451.675 R	Union Oil Company of California
451.7125	Ellis Farms Inc. (Borrego Springs)
451.725 R	Industrial Radio
451.750 R	Industrial Radio
451.775 R	H. & H. Avocado (Escondido)
"	Jess Hutchings Nursery & Grove Service
451.7875	Durling Nursery (Fallbrook)
451.8125	Seawest Energy Group Inc.
451.825 R	Don and Dave Brock Farming (Glamis)
"	Econo Rooter (San Diego)
"	Pre-Mixed Concrete Co. (Escondido)
451.8375	San Marcos County Water District
451.850 R	La Brucheria Ranch (Jacumba)
"	Sully Miller Contracting Co.
"	V.R. Dennis Construction Co.
451.875 R	D. Arrigo Brothers of California (Glamis)
"	Del Rio Farms Inc. (Glamis)
"	Hesters Granite Co. (San Diego)
"	Neill Paving Company (Glamis)
451.900 R	Signs & Pinnick Inc. (EL Cajon)
451.900 R	Southcoast Asphalt Products Inc. (Carlsbad)
451.9125	Seawest Energy Group Inc.
451.925 R	H.G. Fenton Material Co. (Dulzura)
451.950 R	Special Industrial Radio
451.975 R	Special Industrial Radio
451.9875	Seawest Energy Group (San Diego)
452.000 R	B.K.K. Corporation (San Diego)
452.000 R	Chevron Telecommunications Co. (Heber)
452.0125	Andre Nursery Company (Fallbrook)
452.025 R	Calmat Co. (El Cajon/San Diego/Santee)
452.0375	Independent Cab Owners (San Diego)
452.050 R	Coachella Valley Water District
452.050 R	Limousine Service Cab Co. (San Diego)
452.075 R	Marco Crane & Rigging Inc. (San Diego)
452.100 R	Ray's Taxi (Chula Vista)
452.125 R	Bock Co. (Escondido)
452.125 R	Special Industrial Radio
452.150	Amigo Cab Co. (San Diego)
452.175 R	Special Industrial Radio
452.200	Red Top Cab Co. (Mt. Woodson)
452.225	Cab Companies (various)
452.250	Parviz Cab Co. (San Diego)
452.275	Independent Cab Owners (San Diego)
452.325	Sports Car Club of America (San Diego)
452.350	Port Sunny Cab (San Diego)
452.400	El Cajon Cab Co.
452.425	Sports Car Club of America (San Diego)
452.450	United Cab Co. (San Diego)
452.475	Gray Lines Tours Co. (San Diego)

FREQUENCY	NAME
452.5125	Independent Cab Owners (San Diego)
452.525	Automobile Club of Southern Ca.
452.5375	Automobile Club of Southern Ca.
452.550	Automobile Club of Southern Ca.
452.5625	Automobile Club of Southern Ca.
452.575	Automobile Club of Southern Ca.
452.5875	Automobile Club of Southern Ca.
452.600	Automobile Club of Southern Ca.
452.625	Omni-Buses (Montclair)
452.650	Omni-Buses (Montclair)
452.675	Purolator Courier Corp. (San Diego)
452.700	Transportation
452.725	Sports Car Club of America (San Diego)
452.7373	North County Transit District (Maintenance)
452.750	Priority 1 Express (San Diego)
452.775 R	Armored Transpot Inc. (San Diego)
452.800	Sports Car Club of America (San Diego)
452.8125	North County Transit District (Maintenance)
452.825	Sports Car Club of America (San Diego)
452.850	American Messenger Service Inc. (San Diego)
452.875	Smiser Freight Service (Escondido)
452.900	Sports Car Club of America (San Diego)
452.925	Sports Car Club of America (San Diego)
452.950	Sports Car Club of America (San Diego)
452.975 R	Times Advocate Newspaper (Deliveries)
452.975 R	Times Mirror Co. (LA Times)
453.000	San Diego Union Newspaper (Security)
453.025	Hospital Paging (Numerous)
453.0375R	Rancho Fire Dispatch (Remote link)
453.050 R	Los Angeles Local Government 1
"	Palm Desert Local Government
453.0625R	Rancho Fire Dispatch (Remote link)
453.075	Hospital Paging (Numerous)
453.0875R	Escondido Local Government (Telemetry)
453.100 R	LAPD - Narcotics Task Force - TBD
"	Long Beach PD (F-5) Special Operations
"	SDSO - Secondary Dispatch (F-8)
453.125	Hospital Paging (Numerous)
453.125	SDSO - S.E.D. (F-2)
453.150	L.A. County Beach & Harbors BROWN
453.175	Hospital Paging (Numerous Hospitals)
453.1875	San Diego Air Pollution Control District
453.200 R	City of Industry Local Government
453.200 R	San Diego Jack Murphy Stadium (Maint.)
453.200 R	San Diego Wild Animal Park (F-1)
453.200 R	Redondo Beach Local Government
453.200	San Diego Zoo (F-1)
453.225 R	Beverly Hills Tactical (F5-S, F6-R)
453.225 R	Camarillo Local Government
"	Long Beach City Beaches & Parks
453.250 R	Los Angeles Local Government 1
"	SDSO - S.E.D. Operations (F-1)
453.275	SDSO - South Bay Detention Facility (F-1)
453.275 R	So. California RTD - Maintenance
453.2875	La Canada-Flintridge Public Works
453.2875	SD Air Polution Control District
453.300 R	San Diego School District PD (F-1)
453.300 R	Whittier Local Government
453.325 R	San Diego Convention Center (Shows)
453.325 R	So. California RTD - Supervisors
453.325	SDSO - Barrett Honor Camp
453.325	SDSO - Vista Detention Facility
453.3375	Ontario PD (Surveillance)
453.3375	Rancho Fire Dispatch (Remote link)
453.350 R	LAPD - Vice / Special Units - TBD
453.350 R	San Diego School District PD (F-2)
453.350 R	San Diego Wild Animal Park (F-3)
453.375 R	Los Angeles Airport Operations
453.3875	San Diego Lifeguard Portables (F-3)
453.400 R	SDSO - Administration/Paging (F-6)
453.4125	San Diego Lifeguard Portables (F-4)
453.4125	Santa Ana Public Works
453.425 R	Gardena Transit District
453.425	SDSO - Tactical North (F-9 & (F-11)
453.425	Riverside (Local Government)
453.4375	San Diego Lifeguard Portables (F-5)
453.450 R	Alhambra Local Government
"	Carson Local Government
"	San Diego Community Colleges PD
453.475 R	So. California RTD - Supervisors

FREQUENCY	NAME
453.500 R	Chino School District PD
"	Long Beach City Airports
"	SDSO - East Dispatch (F-4)
453.525 R	Beverly Hills PD (F-1) Primary
"	San Diego Lifeguards (Mission Bay)
453.5375	Saddleback Comm. College (Phone)
"	SD Air Pollution Control District
453.550 R	Long Beach Marine Boat Patrol
"	Ventura County (Building Security)
"	Victorville Local Government
453.575 R	Los Angeles District Attorney
"	SDSO - Inquiry/MDT's (F-7)
453.600 R	Cerritos Local Government
453.600	Downey City Street Department
"	Ontario Local Government
"	SDSO - Imperial Beach Detention Center
453.6125	Rancho Fire Dispatch (Remote Link)
453.625	SDSO - Central Detention Facility
"	So. California RTD - Transit Police
453.6375	Rancho Fire Dispatch (Remote Link)
453.650 R	Beverly Hills PD (F3-R, F4-S)
"	Lancaster Local Government
"	San Diego City Lifeguards (F1-R, F2-S)
453.675 R	CLERS - Statewide system
453.700	Long Beach City Animal Control
"	SDSO - El Cajon Detention Facility (F-1)
"	Santa Ana Local Government
453.725	SDSO - Car to Car (F-1)
453.750 R	Commerce Public Works
"	Moreno Valley Local Government
"	SDSO - Coastal Dispatch (F-2)
453.775 R	Lifeguards Point Loma
453.800	S.D. Jack Murphy Stadium (F-1)
453.800 R	S.D. Wild Animal Park / SD Zoo (F-2)
453.800 R	Los Angeles School District PD
453.800	San Diego Zoo (F-2)
453.825 R	Moreno Valley Local Government
"	SDSO - Tactical South (F-10)
"	SDSO - Surveillance (F-12)
"	SDSO - El Cajon Detention Facility F2
453.850 R	Long Beach City Beaches & Parks
"	Long Beach Local Government
"	Palmdale Mosquito Abatement Control
"	SDSO - South Bay Detention Facility F2
"	San Bernardino City Street Department
"	Santa Monica Local Government
453.8625R	Rancho Fire (Remote link)
453.875	LAPD - Special Units
"	SDSO - Surveillance Operations
453.900 R	Riverside (Local Government)
"	State of California-Caltrans (Coronado Bridge)
453.900 R	San Diego Convention Center (Security)
"	Victorville Local Government
453.9125R	Santa Ana Public Works
453.925 R	Bell Gardens PD (F-1) Primary
"	SDSO - North Inland Dispatch (F-3)
453.950 R	Anaheim Convention Center - Security
"	Norwalk USD
"	Pasadena Street Department
"	SDSO - South Bay Dispatch (F-5)
453.9625	San Clemente Public Works (Data)
453.9875	San Clemente Public Works (Data)
453.975	SDSO - Surveillance Operations
454.000 R	Clean Coastal Waters Inc. (Coastline)
454.025-650	PAGING SERVICES EVERY 25 kHz
454.550	San Diego Armored Transport Service
454.675	Pacific Telephone (Aircraft Signaling)
454.825	Pacific Telephone (Aircraft Telephone)
455.0875R	KCAL-TV Channel 9 (F-2)
455.0875	KSON Radio (F-2)
455.010	KCBS Channel 2 (simplex)
455.050	Metro Traffic Control
455.230	KABC-TV Channel 7
455.2375	KSDO-AM/FM Radio
455.3125R	KCOP-TV Channel 13 (F-1)
455.5125	KEZY-FM (Remotes)
455.5125	Metro Traffic Control
455.5875	Fox Television (LA)
455.6375	Metro Traffic Control
455.650	KCBQ-FM Radio

FREQUENCY	NAME
455.925	Monica Zeck Traffic (F-1)
455.980	KECY-TV Channel 9 (El Centro)
456.750 R	"Z" Periodico (Mexico)
458.2625	Orange Co. Environmental Management
458.4625	Orange Co. Environmental Management
458.600	Glendale FD (Nozzle Control)
458.725	SDSO Tactical
458.800	S.D. Jack Murphy Stadium (F-2)
459.825	Pacific Telephone (Aircraft Telephone)
460.025	Statewide CLEMARS Gold #4 (s) Gold #5 (r)
460.050 R	Anaheim PD (26)
"	Inglewood PD (F-3) Tactical
"	Pasadena Community College
"	Riverside PD (Special Enforcement)
460.075 R	Costa Mesa PD (44)
"	Cypress PD (36)
"	San Gabriel PD (F-2)
"	Stanton PD
"	Redondo Beach PD (F-1)
"	San Bernardino Schools PD
460.0875	Riverside FD (Fireground)
"	San Diego PD (Bugs, Surveillance)
460.100 R	Brea PD (22)
"	Huntington Beach PD (42)
"	Los Angeles Airport Maintenance
"	Yorba Linda PD
460.1125	Riverside FD (Fireground)
"	SDSO Narcotics Task Force Cache
460.125	SDSO Las Colinas Detention Facility
460.125 R	Long Beach PD (F-1) Westside
460.125	SDSO Narcotics Task Force Cache
460.125	San Fernando PD (F-3) Tactical
460.1375	SDSO Narcotics Task Force Cache
460.150 R	Inglewood PD (F-1) Primary
460.150	OCSO - Car-to-Car WHITE
460.150	San Bernardino Community College
460.1625	SDSO Narcotics Task Force Cache
460.175 R	Fullerton PD (24)
"	Inglewood PD (F-4) Tactical
"	OCSO - South Area Dispatch
"	Riverside PD (F-1) North Dispatch
"	San Gabriel PD (F-1)
"	Seal Beach PD (41)
460.1875	SDSO Narcotics Task Force Cache
460.200 R	Orange Co. PD's South ORANGE
"	So. California RTD - Police 1
460.225 R	Long Beach PD (F-2) Eastside
"	Riverside Co. Jail
"	San Fernando PD (F-2) Tactical
460.250 R	Buena Park PD (23)
"	Inglewood PD (F-5) Detectives
"	Orange PD (33)
"	Riverside PD (Special Enforcement)
460.250 R	Westminster PD (47)
460.2625	San Diego PD (Bugs, Vehicle Tracking)
460.275 R	Garden Grove PD (32)
"	Industry Convention Center
"	Inglewood PD (F-2) Tactical
"	Newport Beach PD (43)
460.300 R	OCSO Inquiry Channel PURPLE
"	Riverside PD (F-2) Inquiry
460.325 R	Fountain Valley PD (40)
"	Inglewood PD (F-6) Narcotic Units
"	La Palma PD (23)
"	Los Alamitos PD (48)
"	Placentia PD (25)
"	Riverside PD (F-3) South
"	Tustin PD (34)
460.350 R	Long Beach PD (F-3) Ca-to-Car
"	Riverside Co. Jail
"	San Fernando PD (F-1) Primary
460.375 R	California State Police (F1-R, F2-S)
460.3875	Verdugo Fire - Burbank FD Firegound
460.400 R	Orange Co. PD's North ORANGE
460.425 R	OCSO - Dispatch YELLOW
460.425 R	So. California RTD - Police 2
460.450 R	Long Beach PD (F-4) Tactical
460.475 R	Laguna Beach PD
"	La Habra PD
"	Redondo Beach PD (F-2)

FREQUENCY	NAME
460.475 R	Riverside PD (F-4)
"	Santa Ana PD (31)
460.500 R	OCSO - Task Force BLUE
460.525 R	Los Angeles Airport Police
"	OCSO - Surveillance Operations
460.550 R	Anaheim PD (Tactical)
"	Baldwin Park USD (Police)
"	Los Angeles Airport Police & Convention Ctr.
"	OCSO Tactical GRAY
"	Wilson Ambulance (Palmdale)
460.5625	Verdugo Fire - Glendale FD Fireground
460.575 R	Beverly Hills FD Secondary
"	Riverside FD (F-1)
460.5875	Verdugo Fire - Pasadena FD Fireground
460.600 R	Irvine PD
"	Riverside FD (F-2)
460.625 R	Beverly Hills FD Primary
"	OCSO - Tactical
"	Riverside FD (Fireground)
"	San Juan Capistrano SO
460.650 R	Frontier Airlines Inc. (San Diego)
460.650 R	Jims Air Aviation (San Diego)
460.650 R	Torrence Farm Implements (Brawley)
460.6625R	San Diego Convention Center
460.675 R	Sanbon Inc. (Jacumba)
460.675 R	Trans World Airlines (San Diego)
460.700 R	Air California (San Diego)
460.700 R	Ross Flying Service (Imperial)
460.725 R	United Airlines (San Diego)
460.750 R	Southwest Hay Exchange (El Centro)
460.750 R	Western Airlines Inc. (San Diego)
460.775 R	American Airlines (San Diego)
460.775 R	Mar Aviation Inc. (Brawley)
460.800	Jims Air Aviation (F-1) (San Diego)
460.800 R	Sky West Airlines Inc. (San Diego)
460.825 R	Pacific Southwest Airlines (Delta)
460.850 R	Jims Air Aviation (F-2) (San Diego)
460.875 R	Brogdex Company (Glamis)
"	El Centro Air & Electric (El Centro)
"	James Brock Produce (El Centro)
"	Northeastern International Airways
460.900 R	Alarm Companies
460.9125	San Diego Alarm Company Inc.
460.925 R	Honeywell Inc. (Escondido)
460.950 R	A.D.T. Security (Alarms)
460.975 R	San Diego Alarm Company Inc
461.000	Motion Picture Coordination Channel
461.0125	Century City Mall
"	Congregational Tower (Chula Vista)
"	San Diego Convention Center
"	San Diego Alarm Company Inc.
461.025 R	Brawley Stockmans Supply (Brawley)
461.025 R	Centinela Cattle Co. (Jacumba)
"	Hollywood Presbyterian Hopital (LA)
461.0375	Grace Tower (San Diego)
461.050 R	Beverly Security (Alarms)
"	Farm Air Service (Calipatria)
"	Huntington Memorial Hospital
"	Wells Fargo Armored Car Service (San Diego)
"	Whittier Union High School
461.0625	Corona Community Hospital
"	Green Manor (San Diego)
461.075 R	Alhambra USD
"	Cam-Mar Growers (Encinitas)
"	Coronado Landmark Inc. (Coronado)
"	H.R. Medical Center (Santa Paula)
"	Julian Chevron Service (Towing)
"	Star Towing (San Diego)
461.0875	Sea World Inc.
461.100 R	Ann's Trucking (Oceanside)
"	Beverly Medical Center (Beverly Hills)
"	Cannon Farm Equipment (Brawley)
"	College Grove Shopping Center
461.1125	Glendale Memorial Hospital
461.125 R	A.B.M.I. Security Service (Vista)
"	Alamo Chemical Company (Jacumba)
"	Gardena Memorial Hospital
"	Russel Brothers Ranches (Jacumba)
"	Universal Refuse Removal (El Cajon)
461.1375	El Conquistador (San Diego)

FREQUENCY	NAME
461.150	Motion Picture Coordination Channel
461.150 R	San Diego Hospitality Group (San Diego)
461.1625	North County Fair Shopping Center
461.175 R	A.D.T. Security (Operations)
"	Downey USD
"	Ripco Security (Oceanside)
"	San Diego Convention Center
461.1875	Boardwalk Condos (La Jolla)
"	National Steel & Shipbuilding (NASCO)
461.200 R	Camacho Aerial Service (Holtville)
"	Miola Farms (El Centro)
"	Rockwood Chemical Co. (El Centro)
"	West Coast Veterinary (Imperial)
461.225 R	Bay Harbor Hospital
"	Circle T Security Service (National City)
"	Del Mar Fair (Security F2)
"	Downey USD
"	Hawthorne Community Hospital
"	Quality Towing (Escondido)
"	Seabreeze Farms (Ramona
461.250 R	Desert Gardens Towing (Escondido)
461.2625	Century Club (Jack Murphy Stadium)
461.275 R	Anaheim Convention Center
"	Stephen J. Cannel Productions
"	Ramona Taxi
461.2875	Gifford Engineering (San Diego)
"	Servomation Corp. (Jack Murphy Stadium)
461.300 R	Armstrong Farms Inc. (Valley Center)
"	Grinnell Fire Protection Systems
"	Holiday Inn (Mission Valley)
"	Sky Surfer Balloon Co.
461.3125	Sheraton Inn (San Diego)
461.325 R	D. & D. Farms (El Centro)
461.325 R	Equity Investigative Service (San Diego)
461.350 R	A.D.C. Cable TV (Lawrence Welk Village)
"	Clancy's Towing (Lemon Grove)
"	Granfield Security Systems (Bonita)
"	Inglewood USD
461.3625	B.P.S. Guard Services Inc.
461.3625	Servomation Corp. (Jack Murphy Stadium)
461.375 R	Bridwell Patrol Service (San Diego)
"	California Convention Events (San Diego)
"	Colonial Gardens Inc. (Chula Vista)
"	Duck Tours Inc. (San Diego)
"	Garden Convention Services (San Diego)
"	Hawthorne Plaza
"	La Jolla Village Inn
"	Park & Fly Inc. (San Diego)
"	Sundance Trucks Inc. (Spring Valley)
"	Vacation Village (San Diego)
461.3875	Red Lobster Inn (La Mesa)
461.400 R	Bill Pound Automotive (San Diego)
"	Glenview Towing (El Cajon)
461.425 R	Baker Electric (Escondido)
"	Escondido Veterinary Hospital (old)
"	Rodriguez Ranch (Escondido)
461.4375	Gifford Engineering (San Diego)
"	Servomation Corp. (Jack Murphy Stadium)
461.450 R	Beverly Security (Operations)
"	Pleasant Valley Hospital (Santa Paula)
461.4625R	Citizens Agency for Joint Intelligence
461.475 R	Cam-Mar Growers (Encinitas)
"	Hortencias Fine Mexican Food (Carlsbad)
"	Lion Messenger Service (San Diego)
"	Kaiser Foundation (LA)
"	Pedus Security (San Diego)
461.500 R	Business Radio Service (IB)
461.5125	Anaheim Convention Center
"	Gifford Engineering (San Diego)
"	Servomation Corp. (Jack Murphy Stadium)
461.525 R	Brown Field Auto Wrecking
"	Chula Vista Shell (Towing)
"	L.E. Morrison Sand Plant (Spring Valley)
461.550 R	A.B.A. Recovery Inc. (San Diego)
"	Claremont Square Shopping Center
"	Jack Kalman & Associates (El Cajon)
"	Martinez Farms (Chula Vista)
"	Pacific Security Service (San Diego)
"	Servomation Corporation (San Diego)
"	Shield Signal Corporation (La Jolla)

FREQUENCY	NAME
461.550 R	Valley Presbyterian Hospital (Van Nuys)
"	Whispering Trees Apartments (Chula Vista)
"	Whittier Union High School
461.5625	Gifford Engineering Inc. (San Diego)
"	Servomation Corp. (Jack Murphy Stadium)
461.575 R	Assurance Patrol (San Diego)
"	Holiday Inn (San Diego)
"	Mount Miguel Covenant Village
"	Timbers Apartment Management (El Cajon)
461.5875	Carlsbad Security Inc.
"	Sea World
461.600 R	Gabriel's Towing (Pasadena)
"	H. & F. Sheep Corporation (Holtville)
"	Holland Trucking Co. (National City)
"	San Diego Sports Arena (Parking/Security)
461.6125	Hydro Products (San Diego)
461.6125	Delta Airlines (San Diego)
461.625 R	Britschgi Farms (Holtville)
"	Clarklift of El Centro
"	John C. Veysey Farms (Jacumba)
461.625 R	Wilson AG Consulting Service (Jacumba)
461.6375	Carousel Isle Condominiums (San Diego)
"	East County Large Animal Practice
"	G.M.I. Security Service (Encinitas)
"	Half Moon Inn (San Diego)
"	Miller Patrol Service (San Diego)
"	Mission Ridge Condominiums (San Diego)
"	North Coast Village (Oceanside)
"	Point Loma Tennis Club (Security)
"	San Diego Stadium - Parking
"	San Diego Women's Medical Clinic
"	Seapoint Townhomes (Del Mar)
"	Shelter Cove Marina (San Diego)
"	Southern California Security Service
"	Summer House Inn (San Diego)
"	Sunburst Homes (San Diego)
"	Teleseat Ticket Sales (San Diego)
"	Villa La Jolla Condominiums
"	Warner Springs Ranch
461.650 R	A.B.A. Recovery (San Diego)
461.6625	Sea World Inc.
461.675 R	G.M.I. Security (San Diego)
"	Greenwood Memorial Park (San Diego)
"	Grossmont Hospital (La Mesa)
"	Kaiser Foundation (Panorama City)
"	Quaranty Chevrolet Towing (El Cajon)
"	Roger's Police Patrol (San Diego)
"	Valley Presbyterian Hospital (Van Nuys)
"	White Memorial Med Center (LA)
461.6875	Seven Seas Lodge (San Diego)
461.700 R	A.A. Towing (Spring Valley)
461.700 R	Golden State Farms (Jacumba)
461.7125	Borrego Communications
461.725 R	Border Turf Lines (San Ysidro)
"	Bradley's Towing (Chula Vista)
"	California Edison Utilities (San Diego)
"	Columbia Security Patrol (San Diego)
"	Crop Protection Service (Holtville)
"	Friendship Manor (National City)
"	N.C.I. Protective (San Diego)
"	New Side Farms (Imperial)
"	Pacific Media Corporation (El Centro)
"	Tucker & Sons Towing (Glamis)
461.7375	Century Club (Jack Murphy Stadium)
461.750 R	Gordon Sentry Dogs (San Diego)
"	J.R. Benson Farms (Glamis)
"	Northridge Hospital
"	P.T.L. Towing (Pala)
461.775 R	Bostic Harvesting Co. (Holtville)
"	G. Randy Health Entomology (Jacumba)
"	La Mirada USD
"	Norwalk USD
"	Osterkamp Ranches (Jacumba)
461.7875	Old Globe Theater (San Diego)
"	Sea World
461.800 R	Burns & Sons Trucking (Spring Valley)
461.825 R	California Plant Protection Security
"	Fashion Island Shopping Ctr. (Newport Beach)
"	Pinkerton Security
461.8375	University of San Diego

FREQUENCY	NAME
461.850 R	Alert Security Service (Escondido)
"	Calguard Patrol Service (San Diego)
"	Central Towing (San Diego)
"	Executive Air Service (San Diego)
"	Fargo Detective Agency (San Diego)
"	Globe Security Systems (Escondido)
"	Fallbrook Irrigation District
"	Marine Services Commercial Diving
"	Ochoa Packing Co. (Imperial)
"	Select Seed Co. (Ogilby)
"	T.J. Fallis Investigations (San Diego)
461.850 R	U.S. Grant Hotel (San Diego)
461.8625	Borrego Communications
461.875 R	Calguard Patrol Service (San Diego)
"	El Monte UHSD
"	Jamul Towing (Spring Valley)
"	Point Loma Security (San Diego)
"	Tipton Patrol Service (San Diego)
461.8875	San Diego Stadium
461.900 R	Bernstein Security Service (Chula Vista)
"	Broadbent Brothers Sheep Co. (Calexico)
"	C. & C. Transportation (Lakeside)
"	Centurion Guard Dog (San Diego)
"	Daniel's Tire Service (San Diego)
"	Fashion Island Shopping Ctr (Newport Beach)
"	Imed Corporation (San Diego)
"	Mike Moore's 24 Hour Towing (San Diego)
"	Nakasawa Farms (Holtville)
"	Seven Sea Lodges (San Diego)
461.925 R	Grower & Shipper (Holtville)
"	Hayden's Towing (San Diego)
"	Lake Park Condominiums (La Mesa)
"	Omni Mega Security Patrol (Spring Valley)
"	Pacific Radio Engineers (Carlsbad)
"	Pasadena USD
"	Quigley Communications (Oceanside)
"	Ramada Inn (San Diego)
461.950 R	Danenburg Farms (El Centro)
"	J.R. Callens & Sons Ranch (Monument Peak)
"	Marlin Medearis Ranches (Monument Peak)
"	Scheafer's Ambulance (La Crescenta)
"	York Animal Care (El Centro)
461.975 R	Coronado Medical Ctr. (Montrose)
"	Mojave Farms (El Centro)
"	S.L.R. Farm (Bonsall)
461.9875	Rohr Industries (Chula Vista)
462.000 R	Action Security Service (San Diego)
"	Belflower USD
"	Bennett Trucking (El Cajon)
"	Culver City USD (Buses)
"	Honor Marine Communications (San Diego)
"	J.C. Towing (Chula Vista)
"	Rich Man Poor Man Limousine Service
"	Temple City USD
462.0125	Calstate Long Beach
"	Kaiser Foundation (LA)
"	San Diego Convention Center
462.025 R	San Gabriel School District
462.0375	Kaiser Foundation (LA)
"	Welk Park North (San Diego)
462.050 R	Checkmate Protective (San Diego)
"	El Camino Real Farms (Oceanside)
"	Hugh's Auto Recovery (Oceanside)
"	San Diego Stadium - Paramedics UHF
462.0625	Sea World (Park Operations)
462.075 R	Champagne Chevron Service (Escondido)
"	Emmanuel Faith Community Church (Escondido)
"	Laborers Intl. Union (San Diego)
"	R.S. Garewal Farms (Holtville)
"	Torrance USD
462.0875	Sea World
462.100 R	Allegiance Security Inc. (La Mesa)
"	Brawley Animal Clinic
"	De Anza Investigations (El Centro)
"	Imperial Hardware Co. (El Centro)
"	Loduca Trucking (Jacumba)
"	Poway Valley Senior Citizens
"	York's Animal Clinic (Jacumba)
462.1125	Hydro Products (San Diego)
462.125 R	L. & G. Towing (Chula Vista)

FREQUENCY	NAME
462.125 R	Laborers Intl. Union (San Diego)
"	Paul's Line Buses (El Cajon)
"	Protectors Security Services (San Diego)
"	Tijuana Oil Company (San Ysidro)
462.1375	Century Club (Jack Murphy Stadium)
462.150 R	Beach Towing (Imperial Beach)
462.150 R	Shield Security (San Diego)
462.1625	Rohr Industries (Chula Vista)
462.175-525	INDUSTRIAL RADIO SERVICE (IX)
462.150	Electrocomm Radio Service (San Diego)
462.200 R	Solar Turbines Incorporated (San Diego)
462.225 R	Loral Instrumentation (San Diego)
462.250 R	General Dynamics Corp. (San Diego)
462.250 R	National Steel & Shipbuilding (San Diego)
462.300 R	Solar Turbines Incorporated (San Diego)
462.325 R	Chrysler Pacifica (Carlsbad)
462.325 R	General Dynamics (San Diego)
462.3375	United States Robots (Carlsbad)
462.350 R	National Steel & Shipbuilding (San Diego)
462.375 R	Rohr Industries (Chula Vista)
462.400 R	General Dynamics (San Diego)
462.400 R	Frazee Industries Inc. (San Diego)
462.425 R	Triple A Towing South (San Diego)
462.450 R	National Steel & Shipbuilding (San Diego)
462.475 R	Rincon Del Diablo Municipal Water District
"	U.S. Sprint Communications (Oceanside)
"	U.S. Sprint Communications (San Diego)
462.500 R	Hughes Aircraft Company (Carlsbad)
462.525 R	M.C.I. Telecommunications (El Centro)
462.550 R	G.M.R.S. Channel 1
462.5125	G.M.R.S. Simplex
462.575 R	G.M.R.S. Channel 2
462.5875	G.M.R.S. Simplex
462.600 R	G.M.R.S. Channel 3
462.6125	G.M.R.S. Simplex
462.625 R	G.M.R.S. Channel 4
462.6375	G.M.R.S. Simplex
462.650 R	G.M.R.S. Channel 5
462.6625	G.M.R.S. Channel 5A
462.675 R	G.M.R.S. Channel 6 (R.E.A.C.T.)
462.6875	G.M.R.S. Channel Simplex
462.700 R	G.M.R.S. Channel 7
462.7125	G.M.R.S. Simplex
462.725 R	G.M.R.S. Channel 8
462.750 R	Torrance USD
462.800 R	Centinela Hospital (Inglewood/LA)
"	Hoag Memorial Hospital (Newport Beach)
"	I.B.M. Research & Development (San Diego)
462.825	Hotel Del Coronado (Security)
"	Methodist Hospital (Arcadia)
"	St. Joseph Medical Center (Burbank)
462.8375	Honor Mobile Communications (San Diego)
462.850 R	Bay Harbor Hospital (LA)
"	Sea World Inc. (Field Operations)
"	Whittier Hospital
462.8875	Gas Lamp Quarter (San Diego)
"	East County Large Animal Practice
462.8875	Honor Mobile Communications (San Diego)
"	Pacific Towboat Co. (San Diego)
"	Point Loma Tennis Club
"	San Diego Stadium
"	San Diego Womens Medical Clinic
"	Seapoint Townhomes (Del Mar)
"	Teleseat Ticket Sales
"	Villa La Jolla Condominiums
"	Warner Springs Ranch
462.900	Paging Services
462.9125	Boardwalk Condos (La Jolla)
"	Mission Ridge Condos (San Diego)
"	National Steel & Shipbuilding (San Diego)
"	Welk Park North (San Diego)
462.925 R	U.C.S.D. Paging Services
462.950-463.2	MEDICAL RADIO SERVICES
462.950 R	MED-10 (Hospital Calling - Orange County)
462.950 R	MED-10 (San Diego County-San Diego)
462.975 R	MED-09 (Hospital Calling - LA County)
462.975 R	MED-09 (U.C.S.D. Life Flight Helicopters)
462.9875	Sea World (Production Department)
463.000	MED-1
463.0125	San Diego Convention Center

FREQUENCY	NAME
463.025	MED-2
463.050	MED-3
463.075	MED-4
463.100	MED-5 (Station "X" Coordination)
463.125	MED-6
463.150	MED-7
463.175	MED-8
463.200-465.0	BUSINESS RADIO SERVICE
463.200	Hansen & Son Security (Ramona)
"	Mission Bay Toyota (Towing)
"	Odona Central Security (Vista)
463.2125	Calabasas USD
"	Honor Mobile Communications (San Diego)
"	Salk Institute (San Diego)
463.225 R	International Union of Engineers (San Diego)
"	Pala Mesa Cable Co. (Fallbrook)
"	San Diego Motor Patrol (Spring Valley)
"	SPRINT Telephone Service
463.250 R	Alliance Police Patrol (La Mesa)
"	Arnold's Garage (Chula Vista)
"	Associated Towing (San Diego)
"	Dehesa School District (El Cajon)
"	Duber International Security (San Diego, Old)
"	Jamul Dulzura School District (Jamul)
"	Seaside Buick (San Diego)
463.2625	Honor Mobile Communications (San Diego)
463.275 R	American Home Protection (San Diego)
"	Arcamex Avionics Inc. (Glamis)
"	Community Transit Services (Chula Vista)
"	Stardust Country Club (San Diego)
"	Wheel Chair Transport of San Diego Inc.
463.2875	Honor Mobile Communications (San Diego)
463.300 R	Loma Farms (Brawley)
"	Rose Toyota Towing (San Diego)
"	Times Mirror Communications (Escondido)
463.3125	Honor Mobile Communications (San Diego)
463.325 R	Del Rey Avocado Co. (Fallbrook)
"	Desert Agricultural Chemical (El Centro)
"	Fashion Island Shopping Ctr. (Newport Beach)
"	Imperial Irrigation District (Imperial)
"	Knight Security (Escondido)
"	Pacific Southwest Seed (Winterhaven)
"	Saint Maio Beach Association (Carlsbad)
"	Universal Investigations (San Diego)
463.3375	Honor Mobile Communications (San Diego)
463.350 R	Border Turf Lines (San Ysidro)
463.3625	Chula Vista Community Hosptital
"	Honor Mobile Communications (San Diego)
463.375 R	Evergreen Nursery (Escondido)
"	Knight Security & Fire Systems (San Marcos)
"	St. Francis Hospital (Lynwood)
"	U.C.L.A. Medical Center (Indio)
463.400 R	American Burglary & Fire Alarm (Escondido)
"	Bestway Security Towing (Escondido)
463.400 R	Drewery Chemical Co. (Imperial)
"	Long Beach USD
463.4125	Honor Mobile Communications (San Diego)
463.425 R	American Towing Inc. (Chula Vista)
"	Coronado Hospital (Operations)
"	East County Transit (El Cajon)
"	Pala Mesa Cable (Fallbrook)
"	South Bay Towing (Chula Vista)
463.4375	Honor Mobile Communications (San Diego)
"	Nordstrom's (Fashion Valley)
463.450 R	Advanced Communications Systems
"	California Video Cable (Alpine)
"	Carlsbad Security Patrol
"	Fairbanks Landscape Co. (Rancho Santa Fe)
"	I.T. Corp. Hazardous Waste (San Diego)
"	Jim Newlowill Hay (Brawley)
"	Montebello USD
"	Sorrento Sand Company (San Diego)
463.475 R	American Red Cross (LA Repeaters)
463.4875	Mission Valley Shopping Center (San Diego)
463.500 R	American Red Cross (LA Repeaters)
"	Apple City Transport (Glamis)
"	Double Eagle Patrol (Solana Beach)
"	Joe L. Lucio Dairy (Pala)
"	La Costa Hotel & Spa (Carlsbad)
463.5125	Chula Vista Alarm Systems

FREQUENCY	NAME
463.525 R	Aqua Dynamics (El Centro)
"	El Paseo Tours (Chula Vista)
"	Hertz Rent A Car (San Diego)
"	Holtville Alfalfa Mills Inc.
"	Industrial Waste Systems (San Diego)
"	Kelly's Towing (El Cajon)
"	National Ordnance Inc. (National City)
"	Pala Mesa Cable Co. (Fallbrook)
"	Southern California Security (San Diego)
"	Whittemores Auction (San Diego)
463.5375	Honor Mobile Communications (San Diego)
463.550 R	Allied Barracade Co. (Lakeside)
"	Kobey Swap Meet (may have changed)
"	Mesa Vista Hospital - Operations
"	R.M.I. Inc. Shipyard (National City)
"	San Diego Convention Center (Audio/Visual)
"	Sharp Memorial Hospital - Operations
463.5625	Los Angeles Doctors Hospital
"	Nordstrom's (Fashion Valley)
"	Paradise Valley Hospital (National City)
463.575 R	Abe's / Kiley's Towing (San Marcos)
"	Greybor Ambulance (LA)
"	Mountain Gas (San Ysabel)
"	Pala Mesa Cable Co. (Fallbrook)
"	Sentinel Fire Protection (San Diego)
463.5875	San Diego Trolley (Mt. Otay Link)
463.600 R	Escondido National Bank (ENB)
"	Hendrix Electronics (Imperial)
"	Paramount Citrus Association (Escondido)
463.6125	Honor Mobile Communications (San Diego)
"	Knox Attorney Service (San Diego)
"	San Dieg Barge Service
463.625 R	Ampco Marine Emergency (San Diego)
"	Bartlett Farm Management (Fallbrook)
"	Fullerton High School
"	L. & R. Towing Service (Chula Vista)
"	Plaza International Hotel (San Diego)
"	Washington Patrol Service (San Diego)
"	Woodward Sand Company (Lakeside)
463.6375	Hotel Del Coronado (Security)
463.650 R	Cedars Sinai Medical Center (LA)
"	Nickerson Nursery (Fallbrook)
463.6625	Hotel Inter-Continental (San Diego)
"	Motorola Communications (San Diego)
463.675 R	California Recoveries Co. (Oceanside)
"	Hofer Trucking Inc. (San Diego)
"	Marini Farms (Holtville)
"	Midcal Harvesting Company (Escondido)
"	Parrot's Trucking (Vista)
"	Shepard Brothers Trucking (Escondido)
463.6875	Border's Auction Center (San Diego)
"	Summer House Inn (San Diego)
463.700 R	Aero Specialties (San Marcos)
"	El Cajon Towing (changed)
463.7125	Honor Mobile Communications (San Diego)
463.725 R	Action Fire Systems (San Diego)
"	Fox Transport Systems (San Diego)
"	Gene's Towing (Imperial Beach)
"	Holt Dental Laboratory (San Diego)
"	Hughes Aircraft Co. (National City)
"	I.T. Corporation (San Diego)
"	Jack Kalman & Associates (El Cajon)
463.7375	Scripps Clinic Foundation (La Jolla)
463.750 R	Countywide Communications (San Diego)
"	Saint Vincent De Paul Center (San Diego)
"	San Diego Institute of Pathology
"	San Pasqual USD (Winterhaven)
463.7625	Scripps Clinic Foundation (La Jolla)
463.775 R	A.B.M.I. Security Service (San Diego)
"	Anaheim Plaza
"	Area Wide Services (San Diego)
"	Las Encinas Hospital (Pasadena)
"	Nave's Towing (San Diego)
"	Tony Abatti Farms (El Centro)
463.7875	Chula Vista Alarm Systems
463.800 R	Border Turf Lines (San Ysidro)
"	San Dieguito Towing (Solana Beach)
"	Trusthouse Forte Viscount Hotel
463.8125	Half Moon Inn (San Diego)
"	Shelter Cove Marina (San Diego)

FREQUENCY	NAME
463.825 R	Baker Protective Services (Escondido)
"	Beach & Sons Towing (El Centro)
"	Rapid Harvest Co. (Brawley)
463.8325	Nordstrom's (Fashion Valley)
463.850 R	California Grove Management (Fallbrook)
"	Jack Powell Dodge (Escondido)
"	Strahm Farms Inc. (Holtville)
463.875 R	Borders Auction Center (San Diego)
463.875 R	Reliance Steel & Aluminum (National City)
463.900 R	Pacific Communications (National City)
"	Production Farm Management (Jacumba)
"	Rockwood Chemical (Brawley)
463.9125	Honor Mobile Communications (San Diego)
463.925 R	Tarzana Medical Center (Tarzana)
463.9375	Air Resorts Airlines (Carlsbad)
"	Expo Stucco Inc. (San Diego)
463.950 R	Automobile Club of Southern Ca. (Ramona)
"	C. Martinez Farm Labor (Calexico)
"	Johnson-McWane Entomological (Imperial)
463.9625	Air Resorts Airlines (Carlsbad)
463.975 R	Alpine Equine Service
"	Automation & Protection Service (El Cajon)
"	Central Brave Agriculture (Brawley)
"	Nitelite Inns (San Diego)
"	Oceanside Disposal
"	Rollies Service Center (National City)
"	Sands Hotel (San Diego)
463.9875	Air Resorts Airlines (Carlsbad)
"	Honor Mobile Communications (San Diego)
"	May Company Centers
464.000 R	Agricultural Grove Management (Fallbrook)
"	Brentwood Park Mobile Homes (Chula Vista)
"	Green Thumb Nurseries (El Cajon)
"	Jewish Community Center (San Diego)
"	Lilac Farm Management (Bonsall)
"	McCune Audio Visual (San Diego)
"	Performance Datsun (San Diego)
464.0125	Air Resorts Airlines (Carlsbad)
"	California Crane Co.
"	Honor Mobile Communications (San Diego)
"	San Diego Convention Center
464.025 R	Coronado Shores Apartments (Security)
"	Rainbow Pools
"	Sentinel Security Systems (Escondido)
464.0375	Honor Mobile Communications (San Diego)
464.050 R	Beverly Security Service (San Diego)
"	California Coast Nursery (Vista)
"	Castle Farm Service (Brawley)
"	Sun Security Services (Lakeside)
464.0625	Honor Mobile Communications (San Diego)
464.0625	Hotel Del Coronado
464.075 R	AA Parking and Trucking (El Centro)
"	Cavalli Security (San Diego)
"	Communications Center (Jacumba)
"	Laidlaw Transit (La Jolla)
"	Locater Services Inc. (San Diego)
"	Smith Private Security? (Chula Vista)
464.0875	Honor Mobile Communications (San Diego)
"	Nordstroms (Fashion Valley)
"	Welk Park North (San Diego)
464.100 R	California Exotic Towing (San Diego)
"	Columbia Pest Control (National City)
"	Coronado Shores Apartments
"	Gas Lamp Quarter Detectives (San Diego)
"	Holmquist Investigations (San Diego)
"	Pacific Medical Center (Chula Vista)
"	Sharp Cabrillo Hospital (San Diego)
464.1125	Gifford Engineering (San Diego)
"	Honor Mobile Communications (San Diego)
464.125 R	Betcourt Property Management (San Diego
"	Motorola Communications (San Diego)
464.1375	Glendale USD (Security)
"	Honor Mobile Communications (San Diego)
"	Hotel Inter Continental (San Diego)
464.150 R	Claremont City Schools
"	Nim's Security Patrol (San Diego)
"	Reynold's Farms (Calipatria)
"	San Diego County Processing Service
"	Ted Drysdale Hay (Brawley)
464.1625	Calstate Pomona (Telephone)

FREQUENCY	NAME
464.1625	Gifford Engineering (San Diego)
"	Lake Park Condominiums (La Mesa)
"	Woodhaven (El Cajon)
464.175 R	Castro Trucking (Winterhaven)
"	De Anza Investigations (El Centro)
"	General Electric Radio Services (San Diego)
"	Gifford Engineering (San Diego)
"	Neuman Seed Co. (El Centro)
"	Seaport Village Shopping Ctr. (San Diego)
"	West Covina USD
464.1875	Gifford Engineering (San Diego)
"	Lake Park Condominiums (La Mesa)
"	Sea World
464.200 R	Antenna Engineering (San Diego)
"	Bank of San Diego
"	California Plant Protection (San Diego)
"	Coronado Cays Landscape & Maintenance
"	Pro Tect Security (National City)
"	Quality Hay Co. (El Centro)
"	Royal Packing (Glamis)
"	Telophase Society (Medical Service)
464.2125	Honor Mobile Communications (San Diego)
"	Lake Park Condominiums (La Mesa)
"	Sea World (Security)
464.225 R	Sidhu Ranches (El Centro)
"	Southeast Signal Service (El Cajon)
"	West Valley Packing (Winterhaven)
464.2375	Campbell Industries (San Diego Plant)
"	Consumer Protection Alliance (La Mesa)
"	East County Large Animal Practice (El Cajon)
"	G.M.I. Security Service (San Diego)
"	Gaslamp Quarter (San Diego)
"	Glendale USD (Security)
"	Half Moon Inn (San Diego)
"	Harbor Vista Condos (San Diego)
"	Lake Park Condominiums (La Mesa)
"	Miller Patrol Service (San Diego)
"	Mission Ridge Condos (San Diego)
"	North Coast Village (Oceanside)
"	Shelter Cove Marina (San Diego)
"	Southern California Security Services
"	Summer House Inn (San Diego)
"	Sunburst Homes (San Diego)
"	Teleseat Ticket Sales (San Diego)
464.250 R	Cow Doc, Inc. (El Centro)
"	Imperial Communications (El Centro)
"	Knightwatch Security (Escondido)
"	Los Angeles School District (Security)
"	Paine Aero Service Inc. (El Centro)
"	Pauma Valley Pump Service
"	Westmoreland Cattle Co. (Imperial)
464.2625	G..M.I. Security Service (San Diego)
"	Nordstrom's (Fashion Valley)
"	North Coast Village (Oceanside)
464.275 R	Alcoa Defense Systems (San Diego)
464.275 R	Bernardo Heights Country Club
"	Claremont City Schools
"	Oak Communications Inc. (San Diego)
"	Productions for Entertainment & Learning
"	Redondo Beach School District
"	Sea Breeze Airport Service (Carlsbad)
"	Sky Surfer Balloon Co. (San Diego)
"	Yuma Regional Medical Center & Ambulance
464.300 R	Alex Abatti Ranches (El Centro)
"	Blanket Security (San Diego)
"	Exotic Limes & Fruits Inc. (Bonsall)
"	Fleetwood Airporter Service (San Diego)
"	La Jolla Coast Apartment Motel Security
"	Quality Royale (San Diego)
"	T. & M. Ranches (Oceanside)
464.3125	Bay View Farms (Chula Vista)
"	Lake Park Condominiums (La Mesa)
"	Long Beach Convention Center
"	Novento FM 90 (Chula Vista)
"	Welk Park North (San Diego)
464.325 R	Bogosian's Security Service (Campo)
"	California Hospital (LA)
"	Country Estates Security (Diamond Bar)
"	Coast Management Corp (La Jolla)
"	Donald N. Sharp Memorial Hospital
464.325 R	E.F. Hutton Life Insurance (La Jolla)
"	Glendale Adventist Hospital
"	Holiday Inn Embarcadero (San Diego)
"	Hotel San Diego (security)
"	Marriot Corporation (La Jolla)
"	Pomona Humane Society
"	R.M.I. Inc. Shipyard (National City)
"	Saint Joseph Hospital (Orange)
"	Walt Disney Imagination (Glendale)
464.3375	Honor Mobile Communications (San Diego)
464.350 R	Circle T Security Service (National City)
"	Rowland USD (La Puente)
"	San Diego Motor Patrol (El Cajon)
"	United California Dental Lab (Valley Center)
464.3625	Novento 90 FM (Chula Vista)
"	Old Nester Smokehouse (San Diego)
"	Welk Park North (San Diego)
464.375 R	Antelope Valley USD (Lancaster)
"	Bernardo Heights Country Club (San Diego)
"	Brea Mall (Westminster)
"	Buena Park Mall
"	Compton USD (Security)
"	Fullerton High School
"	Grossmont Hospital (La Mesa)
"	Horton Plaza Security (F-2)
"	Park & Ride Co. (San Diego)
"	Pomerado Hospital (Operations)
"	Queen Mary (Security)
"	Universal Studios Security "Control"
464.3875	Honor Mobile Communications (San Diego)
464.400 R	Buckeye Investigations (Escondido)
"	Cimmaron Truck Lines (San Diego)
"	Dune Company (Brawley)
"	Reliable Waste Inc. (Jamul)
"	Superior Crane & Rigging (San Diego)
464.4125	Honor Mobile Communications (San Diego)
"	Hotel Inter Continental (San Diego)
"	Paradise Valley Hospital (National City)
464.425 R	Baldwin Park USD
"	Beverly Mall (Beverly Hills)
"	Buena Park Schools (Buses)
"	Cerritos USD
"	Chevron Industries (San Diego)
"	Glendale Mall
"	Paradise Valley Hospital (National City)
"	Point Loma Nazarene College Security
464.425 R	Rowland USD (La Puente)
"	Saddleback Comm. College
"	San Diego Zoo (Operations)
"	Scripps Memorial Hospital (La Jolla)
464.4375	Honor Mobile Communications (San Diego)
"	Long Beach USD
"	Sea World (Night Services)
464.450 R	Tonka Tow (Oceanside)
464.4625	Honor Mobile Communications (San Diego)
"	Sea Port Village Security (San Diego)
464.4625	Sea World (Food Services)
"	Southwestern Community College (Security)
464.475 R	Glendale Mall
"	Grossmont Shopping Center (La Mesa-Security)
"	La Jolla Village Square (Security)
"	Plaza Camino Real Shopping Center (Carlsbad)
"	San Diego Bulk Terminal
"	Westlake High School (West Covina)
464.4875	Honor Mobile Communications (San Diego)
"	Tustin Schools (Security)
"	Warner Springs Ranch
464.500 R	Atmospheric Space Corp. (San Diego)
"	Valhalla High School (El Cajon)
464.5125	Bay View Farms (Chula Vista)
"	Carousel Isle Condominiums (San Diego)
"	East County Large Animal Practice (El Cajon)
"	G.M.I. Security Service (Encinitas)
"	Gaslamp Quarter Detectives (San Diego)
"	Half Moon Inn (San Diego)
"	Harbor Vista Condominiums (San Diego)
"	Honor Mobile Communications (San Diego)
"	Humphrey Mortuary (Chula Vista)
"	Lockheed Air Terminal (San Diego)
"	North Coast Village (Oceanside)

FREQUENCY	NAME
464.5125	Palm Desert Town Center (Rancho Santa Fe)
"	Point Loma Tennis Club (Security)
"	San Diego Jack Murphy Stadium (Lot D)
"	San Diego Womens' Medical Clinic
"	Sea Point Townhomes Community (San Diego)
"	Shelter Cove Marina (San Diego)
"	Southern California Security Services
"	Summer House Inn (San Diego)
"	Teleseat Ticket Sales (San Diego)
"	Villa La Jolla Condominiums
464.525 R	Atlas Hotels (Security)
"	Buena Park Mall
"	Calstate Long Beach (Student Union)
"	Chevron Industries (San Diego)
"	El Cajon Valley Hospital (Operations)
"	Mercy Hospital (Operations)
"	Scripps Memorial - La Jolla (Operations)
"	Western Medical Ctr. (Santa Ana)
464.5375	G.M.I. Security (San Diego)
"	Miller Patrol Service (San Diego)
"	Mission Ridge Condos (San Diego)
"	Safety School of Driving (Chula Vista)
"	Southern California Security Services
"	Teleseat Ticket Sales (San Diego)
"	Walt Disney Productions (Burbank)
464.550 R	BUSINESS ITINERATE
"	Ampco Marine Emergency (San Diego)
"	Calstate Long Beach (Student Union)
"	Marriott Hotel (San Diego)
"	San Diego Padre Productions (F-1)
464.5625	Border's Auction Center (San Diego)
"	Carousel Isle Condominiums (San Diego)
"	East County Large Animal Practice (El Cajon)
"	G.M.I. Security Service (San Diego)
"	Gaslamp Quarter (San Diego)
"	Half Moon Inn (San Diego)
"	Harbor Vista Condominiums (San Diego)
"	Honor Mobile Communications (San Diego)
464.5625	Miller Patrol Service (San Diego)
"	North Coast Village (Oceanside)
"	Point Loma Tennis Club Security
"	San Diego Jack Murphy Stadium
"	San Diego Womens' Medical Clinic
"	Sea Point Townhomes Community (San Diego)
"	Shelter Cove Marina (San Diego)
"	Summer House Inn (San Diego)
"	Sunburst Homes (San Diego)
"	Teleseat Ticket Sales (San Diego)
"	Villa La Jolla Condominiums
"	Warner Springs Ranch
464.575 R	Bechtel Power Corporation (San Onofre)
"	Buena Park Mall
"	Choc Hospital (Orange)
"	Coronado Cays Security
"	Cypress Community College
"	El Segundo USD
"	High Desert Mountain Rangers
"	Lockheed Air Terminal (San Diego)
"	Mesa Vista Hospital (Operations)
"	Mira Costa College Security (Oceanside)
"	Neiman Marcus Department Store (San Diego)
"	Sharp Hospital (Operations)
"	Travelodge Sports Arena (San Diego)
"	Universal Studios (Security Secondary)
464.5875	Boardwalk Condos (La Jolla)
"	Mission Ridge Condominiums (San Diego)
"	National Steel & Shipbuilding (San Diego)
"	U.D.C. Homes (San Diego)
464.600 R	American West Airlines (Lindbergh Field)
"	Azusa USD
"	Largo Security Services (Escondido)
"	Station "A" Unknown Security
"	"Yankee" - Unknown Security
464.6125	Honor Mobile Communications (San Diego)
"	Nordstrom's (Fashion Valley)
464.625 R	Desert Cab Co. (El Centro)
"	Ed C. Wiest Ranch (Brawley)
"	Golds Field Mining (Winterhaven)
"	Huntington Beach, City of
"	Scheafer's Ambulance (La Crescenta)

FREQUENCY	NAME
464.6375	Border's Auction Center (San Diego)
464.650 R	Jack Bros & McBurney Inc. (Brawley)
"	Pinky's Transportation (Brawley)
"	Santec Security (San Diego)
"	Slater Air Inc. (Brawley)
464.6625	Century Club (San Diego Stadium)
"	Honor Mobile Communications (San Diego)
464.675 R	Bay General Hospital (Chula Vista)
"	Camino Real Shopping Center
"	Carson Mall (Carson)
"	Compton USD (Security)
"	Coronado Shores Coucil
"	Hewlett Packard Security (San Diego)
"	Mercy Hospital (Operations)
"	North County Fair Shopping Center
"	Robert F. Kennedy Hospital (Hawthorne)
"	Salton Sea Air Service
"	U.C.S.D. (Medical Center)
464.6875	Bay View Farms (Chula Vista)
"	Border's Auction Center (San Diego)
"	Harbor Vista Condominiums (San Diego)
"	Summer House Inn (San Diego)
464.700 R	Del Mar Security (JP)
"	Grower's Aerial Service (Jacumba)
464.7125	Honor Mobile Communications (San Diego)
"	Safety School of Driving (Chula Vista)
464.725 R	Farmers Land Leveling (Brawley)
"	Hyster Co. (Imperial)
"	Offshore Marine Enterprises (National City)
"	River Bottom Farms (Imperial)
"	Woodhaven Estates (El Cajon)
464.7375	Bechtel Power Corporation (San Onofre)
"	Honor Mobile Communications (San Diego)
"	National City USD (Security)
"	Woodhaven Homes (El Cajon)
"	B.P.S. Guard Services (San Diego)
"	Baker Protective Service (San Diego)
"	Broadbent Livestock Co. (El Centro)
464.750 R	Del Mar Fair (Security F1)
464.7625	Chula Vista Alarm Systems
"	Honor Mobile Communications (San Diego)
"	Safety School of Driving (Chula Vista)
464.775 R	Calstate Long Beach (Student Union)
"	Coast Waste Management (Carlsbad)
"	Desert Solar (Brawley)
"	Queen Mary (Long Beach)
"	Saint Joseph Hospital (Orange)
"	Walt Disney Productions (Anaheim)
464.7625	Bay View Farms (Chula Vista)
"	Border's Auction Center (San Diego)
"	Warner Springs Ranch (portables)
464.800 R	Boy Scouts of America (San Diego)
"	Carlson Security (San Diego)
"	Coronado Shores Las Palmas
"	Master Security Services (San Diego)
"	Scaroni Ranches Inc. (Heber)
"	Shell Beach Motel (La Jolla)
464.8125	Honor Mobile Communications (San Diego)
464.8125	Long Beach Convention Center
464.825 R	Fashion Valley Shopping Center (San Diego)
"	Frederick's Manor (Chula Vista)
"	Horton Plaza Security (San Diego)
"	Inglewood USD
"	Lynwood USD
"	Marriot Corporation (La Jolla)
"	Plaza Bonita Center (National City)
"	Saint Joseph Hospital (Orange)
"	Universal Studios (Security Tactical)
"	West Covina USD
"	Western Medical Ctr. (Santa Ana)
464.8375	Welk Park North (San Diego)
464.850 R	American Security (San Diego)
464.8625R	Honor Mobile Communications (San Diego)
464.8625R	Pomerado Hospital (Poway)
464.875 R	Beverly Hills USD
"	Kaiser Foundation (Anaheim & San Diego)
"	Mission Community Hospital
"	Parkway Plaza Mall (El Cajon)
"	Saint Joseph Hospital (Orange)
"	University Town Center (La Jolla)

FREQUENCY	NAME
464.8875	American Cablevision (San Diego)
464.900 R	Calipatria USD
"	Gifford Engineering (San Diego)
"	Holmquist Investigations (San Diego)
"	Northend Helicopters (Calipatria)
"	Salvation Army (San Diego)
"	Spartan Security (La Mesa)
464.9125	General Dynamics Corp. (San Diego)
464.925 R	Birger Ranches (Brawley)
"	Buena Park Mall
"	De Anza Harbor Inc. (San Diego)
"	Saint Joseph Hospital (Orange)
"	Universal Studios (Parking Control)
464.9375	Bellflower USD
"	General Dynamics Corp. (San Diego)
"	National City USD (Security)
464.950 R	Desert Agriculture (Calipatria)
"	Limousine Chauffer & Guide (San Diego)
464.9625	Lusardi Construction (handie-talkies)
"	Try Security (Oceanside)
464.975 R	Centinela Hospital (Inglewood)
"	De Anza Harbor (San Diego)
"	Glendale USD
"	Manhattan Beach, City of
"	Mission Valley Shopping Center (Security)
"	St. Joseph Medical Center (Burbank)
"	U.S. Grant Hotel (San Diego)
464.9875	General Dynamics Corp. (San Diego)
"	Rancho Santa Fe Inn
465.000	Bechtel Power Corp. (San Onofre Paging)
"	Nazareth House (San Diego)
"	Scottish Rite Center (San Diego)
"	Town & Country Hotel (San Diego)
"	U.S. Grant Hotel (San Diego)
465.1125	Riverside FD (Fireground)
465.150	OCSO - Surveillance GOLD
465.550	OCSO - Surveillance
465.9125	Goodyear Blimp (Low Power)
465.9375	Goodyear Blimp (Low Power)
465.9625	Goodyear Blimp (Low Power)
467.0125	San Diego Convention Center
467.800	BUSINESS ITINERATE
467.825	BUSINESS ITINERATE
468.000	MED-1 (Paramedics-Mobiles)
468.025	MED-2 (Paramedics-Mobiles)
468.050	MED-3 (Paramedics-Mobiles)
468.075	MED-4 (Paramedics-Mobiles)
468.100	MED-5 (Paramedics-Mobiles)
468.125	MED-6 (Paramedics-Mobiles)
468.150	MED-7 (Paramedics-Mobiles)
468.175	MED-8 (Paramedics-Mobiles)
468.4125	Federal Reserve Bank (Security)
469.325	Barstow Mall
469.325	Hotel San Diego (Security)
469.550	San Diego Padre Productions (F-2)
469.8375	Lorimar Productions (Culver City)
475.000	Tijuana Grading & Paving Service
470.0125R	Burbank, City of
470.0625R	Burbank, City of
470.0875R	Verdugo Fire - Glendale FD Incident Command
470.1125R	Burbank Transit Buses
470.1375R	Verdugo Fire - Burbank FD Fireground
470.1625R	Burbank, City of
470.1625R	Whittier PD
470.175 R	Los Angeles County
470.1875R	Hermosa Beach PD (Dispatch)
470.1875	Verdugo Fire - Burbank FD Incident Command
470.200 R	Los Angeles County
470.2125R	Verdugo Fire - Primary Dispatch
470.225 R	Los Angeles County
470.2375R	Burbank, City of
470.250 R	Los Angeles County
470.2625R	Verdugo Fire - Glendale FD Fireground
470.2875R	Burbank, City of
470.3125R	Fullerton City Public Works
470.3375R	Palmdale Local Government 2
470.3375R	Whittier PD (F-1)
470.3625R	Glendale Public Works
"	LACoFD BLUE10 - Antelope Fireground
470.3875R	South Bay Reg. Pub. Comm. (Police 2)

FREQUENCY	NAME
470.4125R	LACoFD BLUE 6 - Valley Dispatch
470.4375R	LACoFD BLUE 2 - LA Basin Fireground
470.4625	LACoFD BLUE 7 - Valley Tactical 2
470.475 R	Los Angeles County
470.4875R	LACoFD BLUE 12 - Antelope Battalions 6,11
470.5125R	LACoFD BLUE 4 - Administrative
470.5375R	LACoFD BLUE 8 - County Dispatch
470.5625R	LACoFD BLUE 1 - Battalions 7,8,9,14
470.5875	LACoFD BLUE 11 - Antelope Fireground
470.6125R	LACoFD BLUE 3 - Battalions 1,3,5,13
470.6375	LACoFD BLUE 9 - Antelope Tactical 1
470.6375	South Bay Reg. Pub. Comm. (Police 1)
470.6625	LACoFD BLUE 5 - Valley Tactical 1
470.6875R	LASO - Data West 1
470.7125R	El Monte PD (F-1)
470.7375R	LASO - Data West 2
470.7625R	LASO - Data South 2
470.7875R	LASO - Data Paging
470.8125R	LASO - Data North
470.8125R	South Bay Reg. Pub. Comm. (Police 3)
470.8375R	LASO - Data East 2
470.8625R	LASO - Data East 1
470.8875R	LASO - Data South 1
470.9125R	LASO - SRC Access
470.9375R	LASO - Data 1
470.9625R	El Monte PD (F-2)
470.975 R	Los Angeles County
470.9875R	South Pasadena PD
471.000 R	Santa Clarita, City of
471.0125R	Alhambra PD (F-2)
"	Santa Monica PD (F-3) Tactical
471.0375R	La Mirada Public Works
"	Palmdale Local Government 1
471.0375R	Santa Monica Airport Operations
"	South El Monte Public Works
471.0625R	Santa Monica PD (F-1) Primary
471.075 R	Burbank City Public Works
471.0875R	Santa Monica PD (F-2) Secondary
471.100 R	Santa Clarita Valley, City of
471.1125R	Whittier PD (F-2)
471.1375R	Alhambra PD (F-1)
471.1875R	Pepperdine University - Security
471.3375	So. California RTD - Central
471.3625	So. California RTD - Southeast
471.8375	Lorimar Productions (LA)
471.9875	So. California RTD - MetroRail Blue Line
472.5375	So. California RTD - Valley
472.5625	So. California RTD - South
472.5875	So. California RTD - Central/Downtown
472.6125	So. California RTD - San Gabriel Valley
472.6375	So. California RTD - East
472.6625	So. California RTD - South
472.6875	So. California RTD - Central/Downtown
472.7125	So. California RTD - East
481.750	Television Station Broadcast-Audio (15)
482.0125R	Santa Monica FD (TAC-1)
482.025 R	Burbank City Animal Control (F-1)
482.0375R	Montebello Local Government
482.050 R	Burbank City Animal Control (F-2)
482.050 R	Galleria Mall (Glendale)
482.0625R	Culver City Public Works
482.0875R	Glendale, City of
482.100 R	Los Angeles County
482.1125R	Montebello, City of
482.125 R	Los Angeles County
482.1375R	Glendale, City of
482.150	LA Interagency Task Force
482.1625R	Culver City Transit District
482.175 R	Los Angeles County
482.1875R	Glendale PD
482.200 R	Los Angeles County
482.2125R	Montebello PD Dispatch
482.2375	Verdugo Fire - Pasadena FD Fireground
482.2625R	Culver City PD (F-1) Primary
482.2875R	Glendale PD
482.300 R	Los Angeles County
482.3125R	Montebello PD Tactical
482.325 R	Santa Monica PD
482.3375R	Pasadena PD (F-1) Primary
482.3625R	Culver City FD (F-1) Primary

FREQUENCY	NAME
482.3875R	Verdugo Fire - Command
482.4125R	Montebello FD Dispatch
482.4375R	Pasadena PD (F-2) Tactical
482.4625R	Culver City PD (F-2) Secondary
482.475 R	Los Angeles County
482.4875R	Glendale PD
482.500 R	Torrance PD
482.5125R	Montebello Local Government
482.525 R	Santa Monica FD
482.5375R	Verdugo Fire - Pasadena FD Incident Command
482.5625R	Culver City FD (F-2) Secondary
482.575 R	La Verne FD
482.5875R	Glendale PD
482.600 R	Torrance PD
482.6375R	Glendale PD (F-4)
482.6625R	LA County Fire MDT's
482.6875R	LA County Fire MDT's
482.7125R	LA County Fire MDT's
482.7375R	LA County Fire MDT's
482.7625R	LA County Fire MDT's
482.7875R	LA County Fire MDT's
482.8125R	LASO - C-TAC 1 (Countywide Tactical 1)
482.8375R	LASO - C-TAC 2 (Countywide Tactical 2)
482.8625	LASO - DISPATCH 5 (Antelope Vlly/Santa Clarita)
482.8875R	LASO - L-TAC 2 (Avalon/Lennox/Lomita)
482.9125	LASO - DISPATCH 10 (Malibu/Marina Del Rey)
482.9375	LASO - DISPATCH 6 (San Dimas/Walnut)
482.9625R	LASO - SV-2 (Special Operations 2)
482.9875	LASO - DISPATCH 11 (Temple)
483.0125R	LASO - A-TAC 3 (Area Tactical East)
483.0375	LASO - SIMPLEX B (Operations Direct)
483.0625R	LASO - C-TAC 3 (Countywide Tactical 3)
483.0875R	LASO - A-TAC 1 (Area Tactical North)
483.1125R	LASO - L-TAC 8 (Altadena/Antelope Valley)
483.1375	LASO - DISPATCH 14 (Lakewood)
483.1625R	LASO - A-TAC 2 (Area Tactical West)
483.1875	LASO - SIMPLEX D (Operations Direct)
483.2125	LASO - DISPATCH 3 (East Los Angeles)
483.2375R	LASO - L-TAC 7 (San Dimas/Temple/Walnut)
483.2625	LASO - DISPATCH 4 (Firestone/Lynwood)
483.2875R	LASO - EMERG (Emergencies)
483.3125R	LASO - INV 1 (Investigators Countywide)
483.3375R	LASO - INV 2 (Investigators Central/North)
483.3625	LASO - DISPATCH 2 (West Hollywood)
483.3875R	LASO - L-TAC 3 (Lakewood)
483.4125	LASO - SIMPLEX F (Operations Direct)
483.4375	LASO - DISPATCH 12 (Lennox)
483.4625R	LASO - L-TAC 5 (Carson)
483.4875R	LASO - L-TAC 6 (East Los Angeles)
483.5125R	LASO - INV 7 (Investigators East)
483.5375R	LASO - SPC UNIT (Special Unit Dispatch)
483.5625R	LASO - SRC ACC (Sheriff Comm Center)
483.5875R	LASO - M-AID 1 (Mutual Aid North)
483.6125R	LASO - INV 4 (Investigators South)
483.6375R	LASO - Command Posts 1
483.6625R	LASO - INV 6 (Investigators Central)
483.6875R	LASO - DISPATCH 8 (City of Industry)
483.7125R	LASO - CUST (Custody Procedures)
483.7375R	LASO - INV 5 (Investigators Southeast)
483.7625R	LASO - DISPATCH 13 (Norwalk/Pico Rivera)
483.7875R	LASO - M-AID 3 (Mutual Aid East/Metro)
483.8125R	LASO - INV 3 (Investigators West)
483.8375R	LASO - Command Posts 2
483.8625R	LASO - L-TAC 9 (Firestone/Lynwood)
483.8875R	LASO - SV-1 (Special Operations 1)
483.9125R	LASO - L-TAC 1 (Malibu/Marina/W. Hollywood)
483.9375R	LASO - CRT SVC (Court Service)
483.9625R	LASO - L-TAC 4 (City of Industry)
483.9875R	LASO - DISPATCH 1 (Altadena/Crescenta)
484.0125R	LASO - L-TAC 10 (Norwalk/Pico Rivera)
484.0375R	LASO - DISPATCH 7 (Lomita/Avalon)
484.0625R	LASO - M-AID 5 (Mutual Aid Central/Metro)
484.0875R	LASO - M-AID 2 (Mutual Aid Metro/West)
484.1125R	LASO - A-TAC 4 (Area Tactical South)
484.1375R	LASO - M-AID 4 (Mutual Aid San Gabriel)
484.1625R	LASO - DISPATCH 9 (Carson & Metro Blue Line)
484.1875R	Glendale PD
484.2125	Small Cities PD Mutual Aid Channel
484.2375	CLEMARS GOLD
484.2625R	Glendale PD

FREQUENCY	NAME
484.2875-484.9875	LAPD - New digital-voice - See P.87
486.0375	LASO - SIMPLEX A (Operations Direct)
486.1875	LASO - SIMPLEX C (Operations Direct)
486.4125	LASO - SIMPLEX E (Operations Direct)
487.2375	Statewide FIREMARS
506.1125R	LAPD - New digital system
506.1375R	Torrance, City of
506.150 R	Claremont PD (F-1) Primary
506.175 R	Baldwin Park PD (F-1) Dispatch
506.200 R	Pasadena Local Government
506.2125R	Torrance PD (F-2)
506.225 R	Pasadena Local Government
506.2375R	Torrance, City of
506.250 R	Baldwin Park PD (F-2) Secondary
506.2625R	Torrance FD (F-1)
506.275 R	Baldwin Park PD (F-2) Secondary
506.2875R	Torrance FD (F-2)
506.3125R	Fire MDT's
506.3375R	West Covina PD (F-1)
506.3625R	Fire MDT's
"	Orange Co. Marshal's Office (F-2)
506.3875R	Claremont Local Government
"	Los Angeles Housing Dept.
"	Santa Ana Street Department
"	Yorba Linda Public Works
506.4125R	City of Industry Street Department
"	City of Long Beach Public Transit
506.4375R	Fire MDT's
"	Glendale PD (MDT's)
"	Irvine Local Government
506.4625R	Fire MDT's
506.4875R	West Covina PD & Local Government
506.500 R	Torrance, City of
506.5125R	West Covina PD (F-2)
506.5375R	Fire MDT's
506.5625R	Arcadia PD (F-2)
506.5875R	LAPD - ROVER 4 (Central Tactical)
506.6125R	LAPD - New digital system
506.6125	Orange Co. Marshal's Office (F-3)
506.6375R	LAPD - New digital system
506.6625R	Santa Monica Transit District
506.6875R	LAPD - New digital system
506.7125R	LAPD - Van Nuys/West Valley Div. (F-4)
506.7875R	LAPD - Harbor/Southeast Division (F-8)
506.7375R	LAPD - Central Division (F-1)
506.7625R	LAPD - West Los Angeles Division (F-10)
506.8125R	LAPD - ROVER 6 (West Tactical)
506.8375R	LAPD - Metro Division/Special Ops (F-18_
506.8625R	LAPD - New digital system
506.8625	Orange Co. Marshal's Office (F-1)
506.8875R	LAPD - Devonshire Division (F-5)
506.9125R	LAPD - New digital system
506.9375R	LAPD - Northeast/Rampart Division (F-2)
506.9625R	LAPD - Pacific/Wilshire Division (F-11)
506.9875R	LAPD - Southwest Division (F-9)
507.0125	LAPD - New digital system
507.0375	LAPD - ROVER 7 (South Tactical)
507.0625R	Arcadia PD (F-1)
507.0875	LAPD - Citywide Emergency Trigger
507.1125	LAPD - New digital system
"	Orange Co. Marshal's Office (F-4)
507.1375R	LAPD - New digital system
507.1625R	LAPD - North Hollywood/Foothill (F-6)
507.1875R	LAPD - Hollenbeck/Newton Division (F-3)
507.2125R	LAPD - Hollywood Division (F-12)
507.2375R	LAPD - 77th Street Division (F-7)
507.2625R	LAPD - ROVER 5 (Valley Tactical)
507.3125R	Pico Rivera Water District
507.7375R	Fountain Valley Medical Ctr. (Tustin)
508.0875R	Huntington Beach - USD
508.1875R	California Institute of Technology
508.3875R	Warner Bros. Studios (Burbank)
508.4125R	KCOP-TV (News & Remotes)
508.4625R	Orange Co. Dial A Ride
508.5125R	Orange Co. Dial A Ride
508.6125R	Orange Co. Dial A Ride
508.6625R	Orange Co. Dial A Ride
508.7125R	Orange Co. Dial A Ride
625.750	Television Station Broadcast-Audio (51)
697.750	Television Station Broadcast-Audio (69)

FREQUENCY	NAME
805.770	Television Station Broadcast-Audio
824.0-847.0	Cellular Telephones (Mobiles)
851.0625	Antelope Valley Union HSD
851.0625	Orange County (L)
851.3625	Newport Mesa USD
851.4125	Westminster SD
851.4625	North County Community College District
851.4875	Yorba Linda SD
851.500	San Bernardino County (Valley System)
851.5875	Risher Ambulance
851.6375	Orange City Schools - Buses
851.7875	LAPD Communications Bureau
851.9375	Cypress USD
851.9625	Compton USD
852.0375	UC-Irvine
852.1125	Fullerton Elementary SD
"	Los Angeles Harbor Department
"	Dr. Micheal Steppe
852.200	San Bernardino County SO (Valley System)
852.3125	Los Angeles County (L)
852.4375R	Riverside County Schools
852.4625R	Fresno USD
852.8875R	Risher Ambulance
853.3125R	Los Angeles USD
853.3375R	Los Angeles USD
853.3625R	Los Angeles USD
853.3875R	Los Angeles USD
853.4125R	Los Angeles USD
853.450	San Bernardino County (Valley System)
853.4875R	California State
"	City of Downey
"	City of Norwalk
"	City of Palm Springs
"	Event Medical Services
"	Sharmax Freightlines (Santee)
853.5125R	Fontana USD
853.5125R	Torrance Transit District
853.5375R	Bob Hubbard Horse Transport (Riverside)
853.8625R	Colton USD
"	Lake Elsinore USD
854.1125R	Montebello, City of
854.3375R	Glendale School District (Buses)
854.6375R	Redlands USD
"	UC-Irvine
854.7375R	Pacific Telephone (San Diego)
854.9125R	Compton USD (Buses)
854.950 R	San Bernardino County (Valley System)
854.9625R	Los Angeles Harbor Department
"	Riverside County
"	Soledad Agua Dulce Union USD
854.975 R	San Bernardino County SO (Valley System)
854.9875R	Antelope Valley Schools Transportation
"	Montebello, City of
"	Moreno Valley, City of
"	Orange County FD (Laguna Beach)
855.125 R	State Parks & Recreation (Frontera District)
855.2125R	Orange Co. (MDT's)
"	Palmdale School District
855.225 R	San Bernardino County (Needles)
855.2375R	South Bay Regional PCA (MDT's)
855.375 R	San Bernardino County (Valley System)
855.4625R	Orange County Fire (MDT's)
"	Murrietta FPD
855.4875R	State Parks & Recreation
855.5875R	Adam's Ambulance (South Gate)
855.7125R	California State - Parks & Recreation
"	Orange Co. (MDT's)
"	Palmdale School District
855.7375R	State Parks & Recreation
"	Redondo Beach FD
855.9625R	California State Parks & Recreation
"	Redondo Beach FD
855.9875R	Costa Mesa City (Mobile Printers & Voice)
"	Riverside County (L)
856.025 T	San Diego PD / FD (trunking system)
856.050 T	San Diego PD / FD (trunking system)
856.075 T	San Diego PD / FD (trunking system)
856.100 T	Caltrans (Maintenance)
856.125 R	California State - Parks & Recreation
856.150 T	S.D.G.& E. (trunking system)

FREQUENCY	NAME
856.175 T	Heartland Fire (trunking system)
856.1875 T	Advanced Mobilecomm SMR (Elsinore Peak)
856.200 T	Heartland Fire (trunking system)
856.2125R	Orange County Fire - Countywide Net
856.225	UC-San Diego
856.2375R	LAFD - Rescue Units
856.250 T	SMR San Diego
856.2625	California Multi-Agency Radio System (CMARS)
856.2625R	Huntington Beach PD - MDT's
856.275 R	Barnacle Oil (El Cajon)
"	Southern Ca. Edison Co.
856.2875T	Signal Hill Municipal System
856.300 R	S.D.G. & E. (trunking system)
856.325 T	San Diego Transit Corporation
856.350 R	SMR San Diego
856.375 R	SMR San Diego
856.400 R	SMR San Diego
"	- Metropolitan Towing (San Diego)
856.425 R	U.C.S.D. - CPPS Operations
856.4375R	Los Angeles FD Fireground (F-14)
856.450 R	S.D.G & E. (trunking system)
856.450 R	San Bernardino County SO (Valley System)
856.4625R	Hall Ambulance (Isabella)
856.4625R	Orange County Fire - Central Net
856.475 R	SMR Trunked - San Diego
"	- Basic Vegetable Products (Glamis)
"	- Lamb Chevrolet (National City)
"	San Bernardino County SO (Desert System)
856.4875	Orange County Transit District
856.500 T	SMR Trunked - San Diego
"	- Anytime Towing (San Diego)
"	- Cat Corporation (San Diego)
"	- National UFO Watch
"	- Jones Towing (Vista)
"	- Lusardi Construction Inc. (San Marcos)
"	- National Multiple Sclerosis Society
"	- Pepsi Cola of San Diego
"	- Rancho Santa Fe Security (Encinitas)
"	- Terra Trucking (Escondido)
"	- Twins Carpet & Flood Service (San Marcos)
"	- Wright Oil Company (Santee)
856.525 R	Arizona Southern Coach Lines (San Diego)
"	Matheny Communications (San Diego)
"	Oceania Tours (San Diego)
"	Radiotelephone Comm. of Southern Ca.
"	San Bernardino County SO (Valley System)
856.550 R	Federal Express Corp. (San Diego)
856.575 R	SMR Trunked - San Diego
856.625 R	Federal Express Corporation (San Diego)
856.650 R	IBM Research & Development (San Diego)
856.675 T	Boylan & Cannon SMR (Mt. Palomar)
"	- Desert Cab Co.
"	- Economy Tow
"	- Lusardi Construction
"	- Rick's Concrete Pumping
"	- Southern Calif. Med. Transport "SC"
"	- Union Tribune Newspapers
856.700 T	Loomis Radio Comm. SMR (San Miguel)
"	- Aztec Paving
"	- Doubletree Hotel
"	- Quality Inspection Service
"	- San Diego Mini Tours
856.7125	Orange County Fire - Countywide Net
856.725 T	Barbara & Shade Ross SMR (Mt. Palomar)
"	- Barrett Fence Co.
"	- Davey Tree Surgery
"	- Janis Trucking
"	- Motorola (Sales & Technicians)
"	- Mallet & Sons Trucking
"	- Pacific Disposal
"	- Pauma Vista Ranch
"	- Super Shuttle Intl. (very busy)
"	- Southern California Med. Transport
"	- Vadnais Corp. "Vadco"
"	Black Mountain Comm. SMR (Glamis)
"	- American Farms
"	- Bonanza Farms
"	- Desert Security Farm
"	- Hayhurst Farming
"	- Hughes Cattle Co.

FREQUENCY	NAME
856.725 T	- KECY TV News
856.7375	Caltrans - Maintenance
856.750 T	Communications Sites SMR (Glamis)
"	- Western Farm Service
"	Jerry & Kay M. Gastil SMR (Mt Otay)
"	- Allied Gardens Towing
"	- Edco Disposal
"	- Irish Construction
"	- Knox Attorney Service
"	- San Diego Airporter Express
"	- Space Age Delivery
"	- U.S. Recovery Service
856.7625	Los Angeles FD
856.775 T	B. & L. Communications SMR (Mt. Woodson)
"	- California Stage Lines
"	- Cal Pacific Transport
"	- Golden State Concrete
"	- Union Tribune Newspapers
856.800 T	Mountain Relay Company SMR (Mt Otay)
"	- M.E.D.S. San Diego
856.825 T	San Bernardino County SO (Valley System)
"	SMR Trunking (San Miguel)
"	- Able Transportation
"	- Bob Stall Chevrolet Towing
"	- Duber Security
"	- Goodwill Industries
"	- Hughes Attorney Service
"	- Industrial Waste Systems
"	- Jones Towing
"	- La Salle Trucking
"	- Loomis Armored Transport
"	- National Ordnance Co.
"	- Western Waste
"	- Rhodeway Tow Service
856.850 T	Woodward Communications SMR (San Miguel)
"	- Aztec Paving
"	- Doubletree Hotel
"	- North County Transit District
856.875 T	Kahle, Frederick SMR (Mt. Palomar)
"	- Cable Co.
"	- Metro Cab Co.
"	- Rick's Concrete Pumping
"	- San Diego Mini Tours
856.900 T	Lagorio Comm. SMR (Mt. Palomar)
"	- J.R.S. Tree Service
"	- Jones Towing
856.925 T	Mushroom Enterprises SMR (Mt. Woodson)
"	- Allied Gardens Towing
"	- Cranney Veterinary
"	- Edco Disposal
"	- Escondido Veterinary
"	- Fallbrook Refuse District
"	- Irish Construction
"	- Knox Attorney Service
"	- Palomar Transit Mix
"	- Robert Lynch Trucking
"	- Space Age Delivery
"	- U.S. Recovery Service
856.9375	California State - Parks & Recreation
856.9375	LAFD - Fireground & Inspectors
856.950 T	Motorola Inc. SMR (San Miguel)
"	- Embassy Hotels
"	- Executive Rent A Car
"	- Fuller Ford Towing
"	- Home Merchants Delivery
"	- Jon Hunt Trucking
"	- Junes Attorney Service
"	- Pony Express Courier
"	- Premier Recovery Service
"	- San Diego Housing Co.
"	- San Diego Trucking
"	- Sweetwater Union High School
"	- Swiss Mountain View Egg Farms
"	- Two Bit Tow
"	- W.T.C. Air Freight
856.9625T	Orange County Fire - North Net
856.9625	Orange County Fire
856.975 T	Palomar Comm. SMR (Mt. Palomar)
"	- Allied Gardens Towing
"	- B.W. Alarms
856.975 T	- Cranney Veterinary
"	- Edco Disposal
"	- Escondido Veterinary
"	- Fallbrook Refuse District
"	- Ferris Communications
"	- Irish Construction
"	- Knox Attorney Service
"	- Rancho Del Oro Towing
"	- Robert Lynch Trucking
"	- Space Age Delivery
"	- U.S. Recovery Service
"	- Valley Center Union School District
856.9875	State of California - Caltrans
857.000 T	San Diego PD / FD (trunking system)
857.025 T	San Diego PD / FD (trunking system)
857.050 T	San Diego PD / FD (trunking system)
857.075	San Diego PD / FD (MDT's)
857.100 T	Caltrans (Maintenance)
857.125 R	State Parks & Recreation (Montane District)
857.150 T	S.D.G.& E. (Trunking System)
857.175	S.D.G.& E. (Trunking System)
857.1875	Advanced Mobilecomm SMR (Elsinore Peak)
857.200	Heartland Fire (T)
857.2125	Orange County - North Net
857.225	UC-San Diego
857.2375	LAFD - Brush Fire Dispatch
857.2625	California State
"	Costa Mesa City (Mobile Printers & Voice)
"	Torrance PD (MDT's)
857.250	North SD County Region PSD (future system)
857.275 R	Southern Ca. Edison Co.
857.2875T	Signal Hill Municipal System
857.300	S.D.G. & E. (trunking system)
857.325 T	San Diego Transit Corporation
857.375 R	Conventional Systems
857.400 R	C.E. Wilson Corporation (Spring Valley)
857.425 R	Sweetwater Union High School District
857.4375R	Los Angeles FD Fireground (F-13)
857.4625	Orange County Fire - North Net
857.475 R	San Bernardino County SO (Desert System)
"	Texas Street Towing (San Diego)
857.4875	Orange County Transit District
"	Palm Springs PD
857.500 R	Motorola Communications & Electronics
857.525 R	Clover B Ranch (Valley Center)
"	San Bernardino County SO (Valley System)
857.550 R	Baltruanas Company (San Diego)
"	Hicks Chemical (San Marcos)
"	Jims Auto Repair Towing (Oceanside)
"	Pauma Farm Management (Pauma Valley)
857.550 R	Thermodyne Corporation (San Marcos)
857.575 R	California Best Builders (San Diego)
857.600 T	(Part of trunking system; same as 857.500)
857.625 R	C.Q. Communications Inc. (San Diego)
857.625 R	Rescue Industries (San Diego)
857.650 R	Palomar Transit Mix (Ramona)
857.650 R	Wells Trucking (Escondido)
857.675 T	Part of trunking system - See 856.675
857.700 T	Part of trunking system - See 856.700
857.7125	Orange County - Countywide Net
857.725 T	Part of trunking system - See 856.700
857.7375	Caltrans (Orange / Riverside)
857.750 T	Part of trunking system - See 856.750
857.7625R	LAFD - Fireground (F-17)
857.775 T	Part of trunking system - See 856.775
857.800 T	Part of trunking system - See 856.800
857.825 T	Part of trunking system - See 856.825
857.850 T	Part of trunking system - See 856.850
857.875 T	Part of trunking system - See 856.875
857.900 T	Part of trunking system - See 856.900
857.925 T	Part of trunking system - See 856.925
857.9375	California State - Parks & Recreation
857.9375R	LAFD - Metro Ambulances
857.950 T	Part of trunking system - See 856.950
857.9625	Orange County Fire - Central Net
857.975 T	Part of trunking system - See 856.975
"	San Bernardino County (Needles & Transp.)
857.9875	State of California - Caltrans
858.000 T	San Diego PD / FD (trunking system)
858.025 T	San Diego PD / FD (trunking system)

FREQUENCY	NAME
858.0375	La Habra Schools
858.050 T	San Diego PD / FD (trunking system)
858.075	San Diego PD / FD (MDT's)
858.100	State of California - Caltrans
858.125 R	State of California-Parks (Frontera District)
858.1375	Los Angeles FD
858.150 T	S.D.G.& E. (trunking system)
858.175 T	Heartland Fire (trunking system)
858.1875T	Advanced Mobilecomm SMR (Elsinore Peak)
858.200 T	Heartland Fire (trunking system)
858.2125	Orange County Fire -Talkaround
858.225	San Bernardino County SO (Desert System)
"	UC-San Diego
858.2375	Los Angeles FD - Command 2 (F-12)
858.250 T	North SD County Regional PSD
858.2625T	California Multi-Agency Radio System (CMARS)
"	Los Angeles County (L)
858.275 R	Chilcote Inc. (El Cajon)
858.2875T	Signal Hill Municipal System
858.300 R	S.D.G.& E. (trunking system)
858.325 T	San Diego Transit Corporation (trunked)
858.3375	Garden Grove PD (MDT's)
858.350 R	Conventional Systems
858.375 R	Conventional Systems
858.400 R	Action Courier Inc. (Spring Valley)
858.4125	Orange County Transportation Authority
858.425 R	Southern Ca. Edison Co.
858.4375R	LAFD - San Fernando Valley Dispatch (F-8)
858.450 R	Western Hvy. Equipment Rentals (Escondido)
858.4625	Orange County Fire - North Net
858.475 R	SMR Trunked - San Diego
858.4875	Orange County Transit District
858.500 R	Carlsbad Unified School District
"	Coastal Communications (Mt. Otay)
"	Lomack Service (Vista)
858.525 R	McBrides Mission Valley Towing
"	North County Pool & Spa Repair (Oceanside)
"	Radio Baja Inc. (San Diego)
"	San Bernardino County SO (Valley System)
858.550 R	A.R.A. Transportation Services (San Diego)
858.575 R	Sanitainer Inc. (San Diego)
858.600 T	Part of trunking system; See 856.500
858.625 R	Larrys Truck & Auto Repair (San Marcos)
858.625 R	Thompson Type (San Diego)
858.650 R	A Equipment Rental (Chula Vista)
858.650 R	Bonneville Satellite Corp. (San Diego)
858.650 R	Butlers Bestway Towing (Chula Vista)
"	San Diego Mini Tours (San Diego)
"	Southwest Paving (La Mesa)
"	University Ford Chrysler Towing (San Diego)
"	Wilson Equipment (Spring Valley)
858.675 T	Part of trunking system, See 856.675
858.700 T	Part of trunking system, See 856.700
858.7125	Orange County Fire - Central Net
858.725 T	Part of trunking system, See 856.725
858.7375	State of California - Caltrans
858.750 T	Part of trunking system, See 856.750
858.7625	LAFD - Fireground (F-18)
858.775 T	Part of trunking system, See 856.775
"	San Bernardino County SO (Valley System)
858.800 T	Part of trunking system, See 856.800
858.8125	Ontario Montclair SD
858.825 T	Part of trunking system, See 856.825
858.850 T	Part of trunking system, See 856.850
858.875 T	Part of trunking system, See 856.875
858.900 T	Part of trunking system, See 856.900
858.925 T	Part of trunking system, See 856.925
858.9375	California State - Caltrans
858.9375R	LAFD - Division 3 (F-3)
858.950 T	Part of trunking system, See 856.950
"	San Bernardino County (Needles & Transp.)
858.9625	Orange County Fire - Central Net
"	Palm Springs PD
858.975 T	Part of trunking system, See 856.975
858.9875	State of California - Caltrans
859.000 T	San Diego PD / FD (trunking system)
859.025 T	San Diego PD / FD (trunking system)
859.050 T	San Diego PD / FD (trunking system)
859.075	San Diego PD / FD (MDT's)
859.100	State of California - Caltrans
859.125 R	State Parks & Recreation (Frontera District)
859.150 T	S.D.G.& E. (trunking system)
859.175 T	Heartland Fire (trunking system)
859.1875T	Advanced Mobilecomm SMR (Elsinore Peak)
859.200 T	Heartland Fire (trunking system)
859.2125	Orange County Fire - Countywide Net
859.225 R	San Diego PD (conventional system)
859.2375	Antelope Valley Schools Transportation
"	Fontana & Glendale PD (MDT's)
859.250	North SD County Regional PSD (lost)
859.2625	California Multi-Agency Radio System (CMARS)
"	Montebello FD (MDT's)
"	Yorba Linda Water District
859.275	Conventional Systems
859.300	Conventional Systems
859.325 T	San Diego Transit Corporation
859.400 R	Jet Delivery Service (Lemon Grove)
859.425 R	Southern Ca. Edison
859.4375R	LAFD - Metro Fire Dispatch (F-7)
859.450 R	Pacific Bell Telephone (San Diego)
859.4625	Orange County Fire - North Net
"	Palm Springs PD
859.475 R	Premier Products (San Diego)
859.475 R	Westec Security Services (Mt. Soledad)
859.4875R	Orange County Transit District
859.4875R	Palm Springs PD
859.500 R	SMR San Diego
859.525 R	West Pacific Communications (San Diego)
859.550 T	Part of trunking system; See 856.500
859.575 R	San Diego Ready Mix (Mt. Otay)
859.600 R	Advanced Mobilecomm (Mt. Woodson)
859.625 R	SMR Trunked - San Diego
"	- Jim Brown Trucking (Bonsall)
"	- Laines People Mover (San Diego)
"	- Radiology Medical Group (San Diego)
"	- Strong Arm Protection Agency (Escondido)
859.675 T	Part of trunking system; See 856.675
859.700 T	Part of trunking system, See 856.700
859.7125	Orange County Fire - Central Net
859.725 T	Part of trunking system; See 856.725
859.7375	State of California - Caltrans
859.750 T	Part of trunking system; See 856.750
859.7625R	LAFD - Fireground (F-15)
859.775 T	Part of trunking system, See 856.775
859.7875T	Bonneville Satellite Corporation (San Diego)
859.800 T	Part of trunking system, See 856.800
859.825 T	Part of trunking system, See 856.825
856.825 T	San Bernardino County (Valley System)
859.850 T	Part of trunking system, See 856.850
"	San Bernardino County (Valley System)
859.875 T	Part of trunking system, See 856.875
859.8875T	Aeronautical Radio, Inc. (Los Angeles)
859.900 T	Part of trunking system, See 856.900
859.9125	Los Angeles, City of
859.925 T	Part of trunking system, See 856.925
"	San Bernardino County (Valley System)
859.9375	California State - Parks & Recreation
859.9375R	LAFD - Division 2 (F-2)
859.950 T	Part of trunking system, See 856.950
859.9625	Orange County Fire - North Net
859.975 T	Part of trunking system; See 856.975
"	San Bernardino County (Needles & Transp.)
859.9875	State of California - Caltrans
860.000 T	San Diego PD / FD (trunking system)
860.025 T	San Diego PD / FD (trunking system)
860.050 T	San Diego PD / FD (trunking system)
860.075	San Diego PD / FD (MDT's)
860.100 T	California Multi-Agency Radio System (CMARS)
860.125 T	California Multi-Agency Radio System (CMARS)
860.150 T	S.D.G.& E. (trunking system)
860.175 T	S.D.G.& E. (trunking system)
860.1875T	Advanced Mobilecomm SMR (Elsinore Peak)
860.200 T	Heartland Fire (trunking system)
860.2125	Orange County Fire - Countywide Net
860.225 T	Heartland Fire "Conventional 1"
"	San Bernardino County (Desert System)
860.2375	Long Beach PD (MDT's)
"	Murrietta PD
860.250 T	SMR San Diego - See 856.250
"	San Bernardino County (Desert System)

FREQUENCY	NAME
860.2625T	California Multi-Agency Radio System (CMARS)
"	Los Angeles County (L)
860.275 T	Southern Ca. Edison Co.
860.300 R	Conventional systems
860.325 T	San Bernardino County (Valley System)
"	San Diego Transit Corporation (trunked)
860.375 R	C.E. Wilson Corporation (Escondido)
860.375 R	Mar Con Concrete (San Marcos)
860.3875	Garden Grove USD
860.400 R	Mexicoach (San Diego)
860.425 R	S.D.G. & E. (trunking system)
860.4375R	LAFD - Fireground & Administrative
860.450 R	Penhall International (San Diego)
860.4625T	Orange County Fire - North Net
860.475 R	California Meter Service (San Diego)
"	Davey Tree Surgery (San Marcos)
"	Oliver H. Atkins Ranches (Fallbrook)
"	Vista Towing
860.4875R	Orange County Transit District
860.500 R	American Television & Communications
860.525 R	Conely, Dwayne (Escondido)
860.550 R	Cases Lock & Key (San Diego)
"	Courtesy Chevrolet (San Diego)
"	Professional Mortuary Services (San Diego)
860.575 R	San Diego Ready Mix
860.600 T	SMR Trunked - San Diego
"	Colonial Ford Towing (National City)
"	Lyons Construction (Coronado)
860.600 R	Ralphs Towing (San Diego)
860.625 R	McNamara Pump & Electric
860.650 T	Part of trunking system, See 856.500
860.675 T	Part of trunking system, See 856.675
860.700 T	Part of trunking system, See 856.700
860.7125	Orange County Fire - Central Net
860.7125R	Palm Springs PD
860.725 T	Part of trunking system, See 856.725
860.7375	State of California - Caltrans
860.750 T	Part of trunking system, See 856.750
860.7625R	LAFD - Command 1 (F-11)
860.775 T	Part of trunking system, See 856.775
860.800 T	Part of trunking system, See 856.800
860.825 T	Part of trunking system, See 856.825
860.825 T	San Bernardino County (Valley System)
860.850 T	Part of trunking system, See 856.850
860.875 T	Part of trunking system, See 856.875
860.8875T	Aeronautical Radio, Inc. (Los Angeles)
860.900 T	Part of trunking system, See 856.900
860.925 T	Part of trunking system, See 856.925
860.9375R	LAFD - Division 1 (F-1)
860.950 T	SMR San Diego - See 856.950
860.950 T	San Bernardino County (Desert System)
860.9625R	Orange County Fire - Laguna Beach
860.975 T	SMR San Diego - See 856.975
860.9875R	Caltrans (East LA)
861.000 T	Escondido Municipal System
861.050	San Diego PD / FD (trunking system)
861.100	San Diego PD / FD (trunking system)
861.150 T	Caltrans (El Centro/San Diego)
861.200	Chuckawalla State Prison (Blythe)
"	San Bernardino County (Valley System)
"	San Diego County Regional Radio System
861.250 T	Escondido Municipal System (trunked)
861.250 T	National City Municipal System (trunked)
861.300 R	Evergreen Nursery (Escondido)
861.325	San Bernardino County (Valley System)
861.350 R	Nelson & Sloan Concrete (Chula Vista)
861.3625R	Fidelco Communications Corp. (San Ysidro)
861.400 R	Yellow Cab of San Diego
861.450 R	Mark IV Charter Lines (San Diego)
861.500 R	Grossmont Union High School District
861.550 R	Federal Express Corporation
861.575 T	San Bernardino County (Valley System)
861.600 T	Aeronautical Radio, Inc. (San Diego)
861.650 T	SMR Trunking System (Mt. Palomar)
861.700 T	Clement, AJF SMR (Mt. Palomar)
861.750 T	Stream, Kenneth B. SMR (Mt. Woodson)
"	- Pacific Cablevision
861.800 T	Stream, Kenneth B. SMR
861.825	San Bernardino County (Valley System)
861.850	San Bernardino County (Cities System)

FREQUENCY	NAME
861.950 T	Tropico Development SMR (Mt. Palomar)
"	- Peerless Shuttle (very busy)
862.000	Vista USD (Buses)
862.050 T	San Diego PD / FD (trunking system)
862.100 T	San Diego PD / FD (trunking system)
862.150 T	Caltrans (El Centro/San Diego)
862.200	Chuckwalla State Prison (Blythe)
"	San Bernardino County (Cities System)
"	San Diego County Regional Radio System
862.250 T	Escondido Municipal System (trunked)
"	National City Municipal System (trunked)
862.300 R	Airborne Freight Corporation (San Diego)
862.325	San Bernardino County (Valley System)
862.350 T	Southern Ca. Edison Co.
862.600 T	Aeronautical Radio, Inc.
862.650 T	SMR Trunked San Diego - See 861.650
862.800 T	SMR Trunked San Diego - See 861.800
862.825 T	San Bernardino County (Valley System)
862.850 T	San Bernardino County (Cities System)
862.950 T	SMR Trunked San Diego - See 861.950
863.000 R	Cajon Valley USD (Buses)
863.050 T	San Diego PD / FD (trunking system)
863.100 T	San Diego PD / FD (Conventional)
863.150 T	Caltrans (El Centro/San Diego)
"	San Diego County Government (MDT's)
863.200	Calipatria & Otay Mesa State Prisons
863.200 T	Chuckawalla State Prison (Blythe)
863.250 T	Escondido Municipal System (trunked)
"	National City Municipal System (trunked)
863.325 T	San Bernardino County (Valley System)
863.350 T	Southern Ca. Edison Co.
863.575 T	San Bernardino County (Valley System)
863.600 T	Aeronautical Radio, Inc.
863.650 T	SMR San Diego - See 861.650
863.800 T	SMR San Diego - See 861.800
863.825 T	San Bernardino County (Valley System)
863.850 T	San Bernardino County (Cities System)
863.950 T	SMR San Diego - See 861.950
864.000 R	La Mesa Spring Valley School District
864.050 T	San Diego PD / FD (trunking system)
864.100 T	San Diego PD / FD (trunking system)
864.150 T	Caltrans (El Centro/San Diego)
864.150 R	Escondido & Oceanside PD (MDT's)
864.200	Calipatria & Otay Mesa State Prisons
"	Chuckawalla State Prison (Bythe)
864.250 T	Escondido Municipal System (trunked)
"	National City Municipal System (trunked)
864.500 R	Premier Products Inc. (San Diego)
864.550 R	Intl. Labor Union 86 (San Marcos)
864.600 T	Aeronautical Radio, Inc. (trunked)
864.650 T	SMR San Diego - See 861.650
864.800 T	SMR San Diego - See 861.800
864.850 T	San Bernardino County (Cities System)
865.000 T	Escondido Municipal System (trunked)
865.050 T	San Diego PD / FD (trunking system)
865.100 T	San Diego PD / FD (trunking system)
865.150 T	Caltrans Future System (El Centro/San Diego)
"	Poway USD (Buses)
865.200	Calipatria & Otay Mesa State Prisons
"	Chuckawalla State Prison (Blythe)
865.250 T	Escondido Municipal System (trunked)
865.250 R	National City Police / Mall Channel
865.500 R	Baker Protection Systems (San Diego)
865.800 T	SMR San Diego - See 861.800
865.825 T	San Bernardino County (Valley System)
865.850 T	San Bernardino County (Cities System)
865.950 T	SMR San Diego - See 861.950
866.0125	SMARS - National Public Safety Channel 1
866.0375	State of California - CalTrans
866.0625	Los Angeles County (LA Basin System)
"	San Diego County Regional Radio System
866.075	Irvine FD - MDT's
866.0875	Los Angeles County (Countywide System)
866.100	Imperial County (L)
"	Orange County (L)
866.1125	Los Angeles (L)
"	San Bernardino County (Valley System)
866.125	Orange County (L)
866.1375	Los Angeles (L)
"	San Bernardino County (Valley System)

FREQUENCY	NAME
866.1375	San Diego County Regional Radio System
866.150	Orange County (L)
866.1625	Los Angeles (L)
"	San Bernardino County (Jails System)
866.175	Orange County (L)
"	Ventura County (L)
866.1875	Los Angeles (L)
"	San Bernardino County (Valley System)
866.200	SMARS - National Common 8
"	Statewide CLEMARS GOLD #20/#21
866.2125	Riverside County SO (Indio & West Areas)
866.225	Los Angeles County (LA Basin System)
866.2375	Riverside County SO (Chuckawalla, Whitehall)
866.250	Los Angeles County (Countywide System)
866.2625	Riverside County SO (Box Springs, Corona)
866.275	Los Angeles County (Countywide System)
866.2875	Riverside County SO (Mountain Areas)
866.300	Los Angeles County (Countywide System)
866.3125	Riverside County SO (Banning & West)
866.325	Orange County (L)
866.3375	Los Angeles County (L)
"	San Bernardino County (Valley System)
866.350	Orange County (L)
866.3625	Los Angeles County (L)
"	San Bernardino County (Valley System)
866.375	Orange County (L)
866.3875	Calstate San Diego & San Marcos
"	Imperial County (L)
"	Los Angeles County
"	San Bernardino County (Valley System)
866.400	Orange County (L)
866.4125	Los Angeles (L)
"	San Bernardino County (Valley System)
"	San Diego County Sheriff (future)
866.425	Orange County (L)
866.4375	Los Angeles County (LA Basin System)
"	San Diego County Sheriff (future)
866.450	Orange County
866.4625	State Prisons (Chino, East LA, Lancaster)
"	State Parks & Recreation (Santa Monica, SD)
866.4875	UC - Irvine, LA, Riverside, San Diego
866.5125	SMARS - National Public Safety Channel 2
866.5375T	CMARS MultiAgency Radio System
866.5625	Pasadena FD MDT's
"	State Parks & Recreation (Orange Coast)
"	State Rehabilitation Center (Norco)
866.5875	Los Angeles County (Countywide System)
866.600	Imperial County (L)
"	Orange County (L)
866.6125	Los Angeles (L)
"	San Bernardino County (Valley System)
866.625	Orange County (L)
866.6375	Los Angeles (L)
"	San Bernardino County (Cities System)
"	San Diego County Sheriff (future)
866.650	Orange County (L)
866.6625	Los Angeles (L)
"	San Bernardino County (Jails System)
866.675	Orange County (L)
866.6875	Orange County (L)
"	San Bernardino County (Mountain System)
866.700	Orange County (L)
"	Ventura County (L)
866.7125	Riverside County (L)
866.725	Los Angeles County (LA Basin System)
866.7375	Riverside County (Palm Springs)
866.750	Los Angeles County (LA Basin System)
866.7625	Riverside County SO (Indio & West)
866.775	Los Angeles County (Countywide System)
866.7875	Riverside County (MDT's)
866.800	Los Angeles County (Countywide System)
866.8125	Riverside County SO (Banning & West)
866.825	Orange County (L)
866.8375	Los Angeles (L)
"	San Bernardino County (Valley System)
866.850	Orange County (L)
866.8625	Los Angeles (L)
"	San Bernardino County (Cities System)
866.875	Orange County (L)
866.8875	Imperial County (L)

FREQUENCY	NAME
866.8875	Los Angeles (L)
"	San Bernardino County (Valley System)
866.900	Orange County (L)
866.9125	Los Angeles (L)
"	San Bernardino County (Cities System)
"	San Diego County Regional Radio System
"	SMARS - National Public Safety Channel 9
866.925	Orange County (L)
866.9375	Los Angeles County (L)
866.950	San Diego County Regional Radio System
866.9625	Los Angeles County (LA Basin System)
866.9875	State Rehabilitation Center (Norco)
"	UC-Irvine, LA, Riverside, San Diego
867.0125	SMARS - National Public Safety Channel 3
867.0375	CalTrans
867.0625	Los Angeles County (LA Basin System)
867.075	San Diego County Regional Radio System
867.0875	Los Angeles County (LA Basin System)
867.100	Imperial County
"	Orange County (L)
867.1125	Los Angeles (L)
"	San Bernardino County (Mountain System)
867.125	Orange County (L)
867.1375	Los Angeles County (L)
"	San Bernardino County (Cities System)
867.1375	San Diego County SO
867.150	Orange County (L)
867.1625	Los Angeles (L)
"	San Bernardino County (Valley System)
867.175	Orange County (L)
867.1875	Los Angeles (L)
"	San Bernardino County (Cities System)
867.200	Orange County (L)
"	Ventura County (L)
867.2125	Riverside County (L)
867.225	Los Angeles County (Basin System)
867.2375	Riverside County (L)
867.250	Los Angeles County (Countywide System)
867.2625	Riverside County (MDT's)
867.275	Los Angeles County (Countywide System)
867.2875	Riverside County SO (Banning & West)
867.300	Los Angeles County (LA Basin System)
867.3125	Riverside County SO (Banning & West)
867.325	Orange County (L)
867.3375	Los Angeles (L)
"	San Bernardino County (Mountain System)
867.350	Orange County (L)
867.3625	Los Angeles (L)
"	San Bernardino County (Valley System)
867.375	Orange County
867.3875	Imperial County
"	Los Angeles (L)
"	San Bernardino County (Cities System)
867.400	Orange County (L)
867.4125	Los Angeles (L)
"	San Bernardino County (Valley System)
"	San Diego County Sheriff (future)
867.425	Orange County (L)
867.4375	Los Angeles County (LA Basin System)
"	San Diego County Regional Radio System
867.450	Orange PD (MDT's)
867.4625	State Prisons (Chino, East LA, Lancaster)
867.4625	State Parks & Recreation (Orange / Pendleton)
867.4875	UC-Irvine, LA, Riverside, San Diego, St. Barbara
867.5125	SMARS - National Public Safety Channel 4
867.5375T	CMARS Multi-Agency Radio System
867.5625	State Parks & Recreation (Orange Coast)
"	State Rehabilitation Center (Norco)
867.5875	Los Angeles County (LA Basin System)
"	San Diego County Sheriff (future)
867.600	Orange County (L)
867.6125	Imperial County (L)
"	Los Angeles (L)
"	San Bernardino County (Cities System)
867.625	Orange County (L)
867.6375	Los Angeles (L)
"	San Bernardino County (Cities System)
"	San Diego County Regional Radio System
867.650	Orange County (L)
867.6625	Los Angeles (L)

FREQUENCY	NAME
867.6625	San Bernardino County (L)
867.675	Orange County (L)
867.6875	Los Angeles (L)
"	San Bernardino County (L)
867.700	Orange County (L)
"	Ventura County (L)
867.7125	Riverside County SO (Indio & West)
867.725	Los Angeles County (Countywide System)
867.7375	Riverside County (L)
867.750	Los Angeles County (LA Basin System)
867.7625	Riverside County SO (Banning & West)
867.775	Los Angeles County (Countywide System)
867.7875	Riverside County (L)
867.800	Los Angeles County (Countywide System)
867.8125	Riverside County SO (Banning & West)
867.825	Orange County (L)
867.8375	Los Angeles (L)
"	San Bernardino County (Mountain System)
867.850	Orange County (L)
867.8625	Los Angeles (L)
"	San Bernardino County (L)
867.875	Imperial County
"	Orange County (L)
867.8875	Los Angeles (L)
867.8875	San Bernardino County (Valley System)
867.900	Orange County (L)
867.9125	Los Angeles (L)
"	San Bernardino County (L)
"	San Diego County Regional System
867.925	Orange County (L)
867.9375	Los Angeles County (LA Basin System)
867.950	San Diego County Regional Radio System
867.9625	Los Angeles County (LA Basin System)
867.9875	State Corrections Dept. (Corona)
"	UC-Irvine, LA, San Diego, Riverside
868.0125	SMARS - National Public Safety Channel 5
868.0375	CalTrans
868.0625	Los Angeles County (LA Basin System)
868.075	San Diego County Regional Radio System
868.0875	Los Angeles County (Countywide System)
868.100	Imperial County (L)
"	Orange County (L)
868.1125	Los Angeles County (L)
"	San Bernardino County (L)
868.125	Orange County (L)
868.1375	Los Angeles (L)
"	San Bernardino County (L)
"	San Diego County Regional Radio System
868.150	Orange County (L)
868.1625	Los Angeles (L)
"	San Bernardino County (L)
868.175	Orange County (L)
868.1875	Los Angeles (L)
"	San Bernardino County (L)
868.200	Orange County (L)
"	Ventura County (L)
868.2125	Riverside County SO (Indio & West)
868.225	Los Angeles County (LA Basin System)
868.2375	Riverside County (Palm Desert)
868.250	Los Angeles County (Countywide System)
868.2625	Riverside County SO (Banning & West)
868.275	Los Angeles County (Countywide System)
868.2875	Riverside County SO (Mountain Areas)
868.300	Los Angeles County (LA Basin System)
868.3125	Riverside County SO (Banning & West)
868.325	Orange County (L)
868.3375	Los Angeles (L)
"	San Bernardino County (Valley System)
868.350	Orange County (L)
868.3625	Los Angeles (L)
"	San Bernardino County (Valley System)
868.375	Imperial County (L)
"	Orange County (L)
868.3875	Los Angeles (L)
"	San Bernardino County (Valley System)
868.400	Orange County (L)
868.4125	Los Angeles (L)
"	San Bernardino County (L)
"	San Diego County Regional Radio System
868.425	Orange County (L)
868.4375	Los Angeles County (LA Basin System)
"	San Diego County Regional Radio System
868.450	Santa Ana, City of (MDT's)
868.4625	State Parks & Recreation (Frontera)
"	State Prisons (Chino, East LA, Lancaster)
868.4875	UC-Irvine, LA, Riverside, San Diego
868.5125	CLEMARS Gold #8 (s) & Gold #9 (r)
"	SMARS - National Public Safety Channel 6
868.5375	State Parks & Recreation
"	(High Desert, Mt. San Jacinto, Orange County)
868.550	Garden Grove PD (MDT's)
868.5625	State Parks & Recreation (Frontera, Santa Mon.)
"	State Rehabilitation Center (Norco)
868.575	Calstate Fullerton
"	Calstate San Diego & San Marcos
868.5875	Los Angeles County (LA Basin System)
868.600	Imperial County (L)
"	Orange County (L)
868.6125	San Bernardino County (Valley System)
"	Signal Hill PD
868.625	Orange County (L)
868.6375	Arcadia (L)
"	San Bernardino County (MDT's)
868.650	Orange County (L)
868.6625	Long Beach PD (MDT's)
"	San Bernardino County (MDT's)
868.675	Orange County (L)
868.6875	Los Angeles (L)
"	San Bernardino County (Mutual Aid System)
868.700	Orange County (L)
"	Ventura County (L)
868.7125	Riverside County SO (Indio & West)
868.725	Los Angeles County (Countywide System)
868.7375	Riverside County (Palm Springs)
868.750	Los Angeles County (LA Basin System)
868.7625	Riverside County (L)
868.775	Los Angeles County (Countywide System)
868.7875	SMARS - National Public Safety Channel 10
868.7875	Riverside County (Mountain Areas)
868.800	Los Angeles County (Countywide System)
868.8125	Riverside County SO (Banning & West)
868.825	Orange County (L)
868.8375	Beverly Hills PD (MDT's)
"	San Bernardino County (Cities System)
868.850	Orange County (L)
868.8625	San Bernardino County (Air-to-Ground)
868.875	Orange County (L)
868.8875	San Bernardino County (MDT's)
"	Santa Monica Public Works (MDT's)
868.900	Orange County Transportation Authority
868.9125	City of Downey (MDT's)
"	San Bernardino County (MDT's)
868.925	Imperial County (L)
"	Orange County (L)
868.9375	San Bernardino County (Mutual Aid System)
"	Signal Hill PD
"	Ventura County (L)
868.950	Orange County (L)
"	San Diego County Regional Radio System
868.9625	State Correction Department (Corona)
"	Culver City (MDT's)
"	Ventura County (L)
868.9875	California State - FIREMARS 800 MHz
"	CLEMARS GOLD #20 (s) & #21 (r)
869.0-892.0	Cellular Telephones
902.0-928.0	33 CM Amateur Band - See Page 264
902.0-928.0	Cordless Telephones
929.0-930.0	Digital Technologies
929.0-930.0	Paging Services
935.0-940.0	Land Mobile - Trunking Systems
935.225 R	Long Beach USD (Buses)
935.3878-935.500	Los Angeles Department of Water & Power (trunked)
936.6375	Placentia City Schools (Buses)
938.2125	Huntington Beach UHSD
938.775-9375	So. California Edison
939.9375	Fullerton High School
944.0-952.0	Studio-Transmitter Links (News Media)
1240.0-1300.0	23 CM Amateur Band - See Page 264
1296.1	CW, SSB Calling Frequency (rare use)

ScanFan Publications
P.O.Box 544
San Marcos, Ca. 92079

To obtain additional copies of this directory, please send $18.95 + $2.00 for tax and handling to ScanFan Publications. We also recommend our "Scanner Mates" for a handy quick reference guide. Priced at under $5.00 they compliment the directory and cover all the major San Diego public safety frequencies. If you have purchased a prior ScanFan edition and would like the current one - just send us a check for $15.00 and tell us where you bought it - we will gladly send you our new 5th edition!

Copyright © 1988, 1990, 1992, 1993, 1994

Remember these fine dealers who carry "ScanFan" products:

HAM RADIO OUTLET
933 North Euclid Street
Anaheim, CA 92801
(714-533-7373)

HAM RADIO OUTLET
6265 Sepulveda Blvd.
Van Nuys, CA 91411
(818-988-2212)

AVAILABLE AT
MANY
INDEPENDENTLY
OWNED
RADIO SHACK STORES!

HAM RADIO OUTLET
5375 Kearny Villa Road
San Diego, CA 92123
(619-560-4900)

MUD SHACK
6512 El Cajon Blvd. #D
San Diego, CA 92115
(619-464-6855)

CORONET
NEWS STAND
111 So. Hill St.
Oceanside, CA 92054
(619-722-3233)

JIM DICKENS RADIO SALES
244 Main Street
El Centro, CA 92243
(619-353-6084)

O' NEALS C.B.
208 No. Freeman St.
Oceanside, CA 92054
(619-722-1931)

EDC ELECTRONICS
307 So. Hill St.
Oceanside, CA 92054
(619-722-5855)
(619-722-8711)

LION ELECTRONICS
457 Broadway
Chula Vista, CA 92010
(619-427-6131)

WILLY'S ELECTRONICS
1636 D Avenue
National City, CA 92050
(619-477-2119)

RADIO SHACK
1853 Main St. #A
Ramona, CA 92065
(619-789-0066)

RADIO SHACK
12035 Woodside Ave.
Lakeside, CA 92040
(619-443-0074)

RADIO SHACK
2960 Navajo Road
El Cajon, CA 92020
(619-465-6006)

Available at most
B. DALTON – WALDEN
BOOKSTAR
BARNES & NOBLE
BOOK STORES

RADIO SHACK
1662 No. Magnolia #3
El Cajon, CA 92020
(619-447-8106)

RS ELECTRONICS
12845 Poway Road
Poway, CA 92064
(619-748-4641)

HENRY RADIO
2050 So. Bundy Drive
Los Angeles, CA 90025
(310-820-1234)